NOMENCLATURE OF ORGANIC CHEMISTRY

Sections A, B, C, D, E, F and H

1979 Edition

INTERNATIONAL UNION OF PURE AND APPLIED CHEMISTRY

IUPAC Secretariat: Bank Court Chambers, 2-3 Pound Way,
Cowley Centre, Oxford OX4 3YF, UK

IUPAC COMMISSION ON NOMENCLATURE
OF ORGANIC CHEMISTRY

Members for varying periods from 1947 to 1977

M. BETTI*	D. HELLWINKEL	W. H. POWELL
K. BLÁHA	K. HIRAYAMA	V. PRELOG
R. S. CAHN	G. KERSAINT	J. C. RICHER
L. T. CAPELL	S. P. KLESNEY	F. RICHTER*
L. C. CROSS	W. KLYNE*	J. RIGAUDY
G. DUPONT*	K. L. LOENING	S. VEIBEL*
G. M. DYSON*	N. LOZAC'H	P. E. VERKADE
D. R. ECKROTH	R. MARQUIS*	F. VÖGTLE
J. H. FLETCHER	A. D. MITCHELL*	E. VOTOČEK*
C. S. GIBSON*	H. S. NUTTING	H. J. ZIEGLER
H. GRÜNEWALD	A. M. PATTERSON*	

* Deceased

INTERNATIONAL UNION OF PURE AND APPLIED CHEMISTRY
ORGANIC CHEMISTRY DIVISION
COMMISSION ON NOMENCLATURE OF ORGANIC CHEMISTRY

NOMENCLATURE OF ORGANIC CHEMISTRY

Sections A, B, C, D, E, F and H

1979 Edition

Prepared for publication by

J. RIGAUDY
Université Pierre et Marie Curie, Paris, France

and

S. P. KLESNEY
The Dow Chemical Company, Midland, Michigan, USA

PERGAMON PRESS

OXFORD · NEW YORK · TORONTO · SYDNEY · PARIS · FRANKFURT

U.K.	Pergamon Press Ltd., Headington Hill Hall, Oxford OX3 0BW, England
U.S.A.	Pergamon Press Inc., Maxwell House, Fairview Park, Elmsford, New York 10523, U.S.A.
CANADA	Pergamon Press Canada Ltd., Suite 104, 150 Consumers Rd., Willowdale, Ontario M2J 1P9, Canada
AUSTRALIA	Pergamon Press (Aust.) Pty. Ltd., P.O. Box 544, Potts Point, N.S.W. 2011, Australia
FRANCE	Pergamon Press SARL, 24 rue des Ecoles, 75240 Paris, Cedex 05, France
FEDERAL REPUBLIC OF GERMANY	Pergamon Press GmbH, 6242 Kronberg-Taunus, Hammerweg 6, Federal Republic of Germany

SECTIONS A AND B
First edition 1958, Second edition 1966
Third edition 1971 (combined with Section C)
Fourth edition 1979 (combined with Sections C, D, E, F & H)

SECTION C
First edition 1965
Second edition 1971 (combined with Sections A & B)
Third edition 1979 (combined with Sections A, B, D, E, F & H)

SECTIONS D, E, F, H
First edition 1979 (combined with Sections A, B & C)

Sections A-F, H Reprinted 1982

British Library Cataloguing in Publication Data
International Union of Pure and Applied Chemistry.
Commission on the Nomenclature of Organic Chemistry
Nomenclature of organic chemistry. - 4th ed.
1. Chemistry, Organic - Nomenclature
I. Title II. Rigaudy, J III. Klesney, S P
547'.001'4   QD291   79-40358
ISBN 0-08-022369-9

The contents of the first edition of Section C appeared in *Pure and Applied Chemistry*, Vol. 11, Nos. 1-2 (1965). The contents of the first edition of Section E appeared in *Pure and Applied Chemistry*, Vol. 45, No. 1 (1976) and of the first edition of Section H in Vol. 51, No. 2 (1979)

Printed in Great Britain by A. Wheaton & Co. Ltd., Exeter

CONTENTS

SECTION A

Hydrocarbons

SECTION D

Organic Compounds Containing Elements that are not Exclusively Carbon, Hydrogen, Oxygen, Nitrogen, Halogen, Sulfur, Selenium and Tellurium

SECTION E

Stereochemistry

SECTION F

General Principles for the Naming of
Natural Products and Related Compounds

SECTION H

Isotopically Modified compounds

INTRODUCTION TO THE IUPAC REVISED AND COLLECTED RECOMMENDATIONS FOR THE NOMENCLATURE OF ORGANIC CHEMISTRY, 1978

The progress of the IUPAC Commission on the Nomenclature of Organic Chemistry during the period 1969 to 1977 has been reported in the Comptes Rendus of the Conferences (Recommended Rules) or in the Information Bulletin of the Union, Appendices on Provisional Nomenclature, Symbols, Units and Standards (Provisional Rules). While work for improving and completing nomenclature rules continues, we have now reached the point where it is possible and convenient to collect in a single volume a compendium of existing recommendations providing improved guidance to internationally agreed nomenclature.

The Sections included in this compendium are :

Section A. Hydrocarbons

Section B. Fundamental Heterocyclic Systems

Section C. Characteristic Groups containing Carbon, Hydrogen, Oxygen, Nitrogen, Halogen, Sulfur, Selenium and/or Tellurium

These Rules are the IUPAC Organic Nomenclature Rules Sections A, B and C, 1969, reprinted from the combined edition, Butterworths, London (1971), corrected for material errors.

Section D. Organic Compounds containing elements that are not exclusively Carbon, Hydrogen, Oxygen, Nitrogen, Halogen, Sulfur, Selenium and Tellurium.

These Rules are issued jointly by the Organic and Inorganic Nomenclature Commissions, and were published originally in IUPAC Information Bulletin, Appendix No.31, August 1973. Corrections for material errors and a few revisions have been introduced into the provisional text after evaluation of comments.

INTRODUCTION

Section E. Stereochemistry

These Rules are IUPAC Recommendations 1974 and are reprinted from *Pure and Applied Chemistry*, Vol. 45, pp. 11-30 (Pergamon, 1976) with a few corrections for material errors.

Section F. General Principles for the Naming of Natural Products and Related Compounds

These Rules have not been finally approved, but their inclusion here seems appropriate (some minor errors have been corrected).

Section H. Isotopically Modified Compounds

These Rules, now approved, were originally published in the IUPAC Information Bulletin, Appendix No. 62, July 1977. Corrections for material errors have been introduced after evaluation of comments.

Although the titles of these Sections are to a large extent self-explanatory, it is thought useful here to make some comments on the arrangement of material. Glossaries of terms used are to be found in the Introduction to the 1969 Rules, p. xviii and in Section C, p. 81.

The attention of a reader having a nomenclature problem to solve is drawn first to Section C, which deals initially with general principles for the construction of a name. The operations described in this Section are usually to be applied to the names for hydrocarbons and heterocyclic systems provided by application of the Rules in Sections A and B.

Section D contains various types of material (see contents pp. ix-xii), some of which would be more appropriately included in earlier sections ; for example, much of Rules D-1 (Nomenclature Systems, p. 326), D-4 (Chains and Rings with Regular Patterns of Heteroatoms, p. 373), D-6 and D-7 (Organosilicon and Organoboron Compounds, pp. 409-458), and the Appendix Tables I, II and IV (pp. 459-471) falls into this category. We hope to reorganise the subject matter in a later edition.

INTRODUCTION TO 1969 EDITION OF SECTIONS
A, B AND C

The first international proposals on the nomenclature of organic chemistry, made at Geneva in 1892, were revised and extended by the Definitive Report of the Commission for the Reform of Nomenclature in Organic Chemistry of the International Union of Chemistry (I.U.C.) which appeared after the meeting at Liége in 1930 (Liége Rules), and was supplemented by less extensive reports from the meetings at Lucerne in 1936 and at Rome in 1938. Although these proposals rendered great service, it was apparent at the meeting of the International Union of Pure and Applied Chemistry (I.U.P.A.C.) at London in 1947 that extension and revision of the nomenclature rules for organic chemistry were required.

Those who have served on the Commission on the Nomenclature of Organic Chemistry for varying periods from 1947 to 1969 are M. Betti*, R. S. Cahn, L. T. Capell, L. C. Cross, G. Dupont*, G. M. Dyson, C. S. Gibson*, H. Grünewald, G. Kersaint, S. P. Klesney, K. L. Loening, N. Lozac'h, R. Marquis*, A. D. Mitchell*, H. S. Nutting, A. M. Patterson*, V. Prelog, F. Richter*, J. Rigaudy, S. Veibel, P. E. Verkade, and E. Votoček*, and, as observers, K. A. Jensen (chairman, I.U.P.A.C. Commission on the Nomenclature of Inorganic Chemistry), W. Klyne (member of the IUPAC/IUB Commission for Biochemical Nomenclature).

The Commission's progress in the period 1947 to 1969, inclusive, has been reported in successive issues of the Comptes Rendus of the Conferences of the Union. Relevant parts of those reports are included, with a few revisions, in the rules which form the body of this publication.

Comments on these rules should be sent to the Secretary, S. P. Klesney, 3609 Boston, Midland, Michigan 48640, U.S.A. or to any other member of the Commission.

GENERAL PRINCIPLES

The Commission believes that differences in nomenclature frequently hinder the accurate and intelligible conveyance of information from one chemist to another, so tending to hamper understanding and progress. The Commission urges conformity with internationally agreed nomenclature even when this nomenclature may not seem the best possible from the point of view of the chemists of a particular nation or group.

The rules now presented are intended to be suitable for textbooks, journals and patents, for lexicons and similar compilations, and for indexes, even if not always wholly so for conversation or lectures. The rules will be issued in parts, as they become approved by the Union. They constitute recommendations for the naming of types of compounds and of individual compounds. They are not exhaustive, except in specified cases. Where, for various reasons, limitation to a single method of nomenclature appears undesirable or impossible, alternatives are given; but the Commission hopes that elimination of alternatives may become acceptable as the merits of one method become more generally recognized. The Commission hopes also that each nation will try to reduce the variations in nomenclature with regard to spelling, the position of numbers, punctuation, italicizing, abbreviations, elision of vowels, certain terminations, and so forth; the

* Deceased.

present rules are not to be held as making recommendations in these matters.

Owing to the very extensive nomenclature which has come into being since the last revision, the Commission has, in the main, confined its efforts to codifying sound practices which already existed, rather than to originating new nomenclature—the latter may form a later stage of the Commission's activities.

In so doing, the Commission had in mind the following main principles: (*a*) as few changes as possible should be made in existing nomenclature, though utility is more important than priority; (*b*) rules and names should be unequivocal and unique, but simple and concise; (*c*) records in journals, abstracts, compendia, and industry should be used to assess the relative extent of past use of various alternatives; (*d*) rules should be consistent with one another, yet aid expression in the particular field of chemistry involved and be capable of extension with the progress of science; (*e*) trivial names, and names having only a very small systematic component, cannot be eliminated when in very common use, but those of less value should be replaced by systematic (or at least more systematic) ones, and the creation of new trivial names should be discouraged by provision of extensible systematic nomenclature; (*f*) names should be adaptable to different languages. The Commission is aware that acceptance of its recommendations depends in large measure on the success which has attended its attempts to assess, for each particular case, the relative merits of these often conflicting claims.

GLOSSARY

The Commission considered it unnecessary to define chemical terms in common use. However, certain terms which have special meaning in nomenclature merit brief description; namely:

Parent compound: the principal chain or ring system from which a name is derived by substitution of hydrogen with other atoms or groups; e.g. methylcyclohexane has cyclohexane as its parent.

Systematic name: a name composed wholly of specially coined or selected syllables, with or without numerical prefixes; *e.g.*, pentane, oxazole.

Trivial name: a name no part of which is used in a systematic sense; *e.g.*, xanthophyll.

Semi-systematic name or semi-trivial name: a name of which only a part is used in a systematic sense; *e.g.*, methane (-ane), butene (-ene), calciferol (-ol). (Most names in organic chemistry belong to this class.)

Substitutive name: a name involving replacement of hydrogen by a group or by another element; *e.g.*, 1-methylnaphthalene, 1-pentanol.

Replacement name: an " a " name, where C, CH, or CH_2 is replaced by a hetero atom; *e.g.*, 2,7,9-triazaphenanthrene. Also, certain names involving thio- (also seleno- or telluro-) to indicate replacement of oxygen by sulfur (or selenium or tellurium, respectively); *e.g.*, thiopyran.

Subtractive name: a name involving removal of specified atoms; *e.g.*, in the aliphatic series names ending in -ene or -yne. Also names involving anhydro-, dehydro-, deoxy-, *etc.*, or nor-.

Radicofunctional name: a name formed from the name of a radical and the name of a functional class; *e.g.*, acetyl chloride, ethyl alcohol.

Additive name: a name signifying addition between molecules and/or atoms; *e.g.*, styrene oxide.

Conjunctive name: a name formed by placing together the names of two molecules, it being understood that the two molecules are linked by loss of one hydrogen atom from each; *e.g.*, naphthaleneacetic acid.

Fusion name: a name for a cyclic system formed by use of a linking " o " between the names of two ring systems, denoting that the two systems are fused by two or more common atoms; *e.g.*, benzofuran.

Hantzsch–Widman name: a name for a heterocyclic system, derived from the original proposals of Hantzsch and Widman, and formed from a prefix or prefixes (to denote one or more hetero atoms) and a suffix -ole or -ine (to denote a five- or a six-membered ring, respectively); *e.g.*, triazole, thiazole.

Bibliography

International Union of Pure and Applied Chemistry
    Comptes rendus of the 9th Conference   1928, pp. 63–71
International Union of Chemistry
    Comptes rendus of the 10th Conference   1930, pp. 57–64
    Comptes rendus of the 12th Conference   1936, pp. 39–42
    Comptes rendus of the 13th Conference   1938, pp. 36–37
    Comptes rendus of the 14th Conference   1947, pp. 129–137
International Union of Pure and Applied Chemistry
    Comptes rendus of the 15th Conference   1949, pp. 127–186
    Comptes rendus of the 16th Conference   1951, pp. 100–104
    Comptes rendus of the 18th Conference   1955, pp. 120–184
Patterson, A. M., Capell, L. T., and Walker, D. F. *The Ring Index*, 2nd ed., American Chemical Society, Washington, D.C., 1960
Hantzsch, A., and Weber, J. H., *Ber. Dtsch. Chem. Ges.* **20**, 3119 (1887)
Widman, O., *J. Prakt. Chem.*, [2], **38**, 185 (1888)
Baeyer, A., *Ber. Dtsch. Chem. Ges.* **33**, 3771 (1900)

Definitive Rules* for Nomenclature of Organic Chemistry
Section A. Hydrocarbons
Section B. Fundamental Heterocyclic Systems

issued by the Commission on
the Nomenclature of Organic Chemistry
of the
International Union of Pure and Applied Chemistry

Fourth Edition
1979

*These Rules shall be known as the I.U.P.A.C. Organic Nomenclature Rules A and B, 1979.

PREFACE TO THE THIRD EDITION
OF SECTIONS A AND B

This Third Edition of Sections A and B contains considerable changes from the Second Edition, but these have been confined, in the main, to correction of errors, to clarifications, in a few cases to expansion of existing Rules to cover special cases, and to provision of better or further examples.

However, major changes are the deletion of the Rules for order of complexity of side chains [Rule A–2.3 (a) and parts of Rules A–2.4 and A–2.5] and of the Stelzner method of naming heterocyclic systems (Rule B–4) by replacement nomenclature ("a" nomenclature), because it is planned to abandon these procedures in Beilstein's *Handbuch der organischen Chemie* and they have been little used recently elsewhere.

Attention is also drawn to new Rules, in accord with principles of the other Rules, for naming heterocyclic ring assemblies (Rule B–13), for naming radicals derived from bridged compounds (A–34.5 and B–15), from spiro compounds (A–43 and B–12), from ring assemblies (A–55 and B–13), and from compounds named by the von Baeyer system (B–14); there is also a new Rule, embodying *Ring Index* practice, for naming heterocyclic systems that contain one benzene ring and one hetero ring (B–3.5).

Other changes and additions that might have been made are reserved for inclusion in a major revision of nomenclature systems that is under consideration by the Commission.

The following constituted the Commission responsible for this (the Third) Edition: P. E. Verkade (Chairman), S. P. Klesney (Secretary), L. C. Cross, G. M. Dyson, K. L. Loening, N. Lozac'h, J. Rigaudy, S. Veibel, with, as Associate Members, R. S. Cahn, H. Grünewald, and, as Observers, K. A. Jensen (Chairman, I.U.P.A.C. Commission on the Nomenclature of Inorganic Chemistry) and W. Klyne (Member of the I.U.P.A.C./I.U.B. Commission for Biochemical Nomenclature).

A. HYDROCARBONS

ACYCLIC HYDROCARBONS

Rule A–1. Saturated Unbranched-chain Compounds and Univalent Radicals

1.1—The first four saturated unbranched acyclic hydrocarbons are called methane, ethane, propane and butane. Names of the higher members of this series consist of a numerical term, followed by "-ane" with elision of terminal "a" from the numerical term. Examples of these names are shown in the table below. The generic name of saturated acyclic hydrocarbons (branched or unbranched) is " alkane ".

Examples of names:

(n = total number of carbon atoms)

n		n		n	
1	Methane	15	Pentadecane	29	Nonacosane
2	Ethane	16	Hexadecane	30	Triacontane
3	Propane	17	Heptadecane	31	Hentriacontane
4	Butane	18	Octadecane	32	Dotriacontane
5	Pentane	19	Nonadecane	33	Tritriacontane
6	Hexane	20	Icosane*	40	Tetracontane
7	Heptane	21	Henicosane	50	Pentacontane
8	Octane	22	Docosane	60	Hexacontane
9	Nonane	23	Tricosane	70	Heptacontane
10	Decane	24	Tetracosane	80	Octacontane
11	Undecane	25	Pentacosane	90	Nonacontane
12	Dodecane	26	Hexacosane	100	Hectane
13	Tridecane	27	Heptacosane	132	Dotriacontahectane
14	Tetradecane	28	Octacosane		

1.2—Univalent radicals derived from saturated unbranched acyclic hydrocarbons by removal of hydrogen from a terminal carbon atom are named by replacing the ending " -ane " of the name of the hydrocarbon by "-yl". The carbon atom with the free valence is numbered as 1. As a class, these radicals are called normal, or unbranched chain, alkyls.

Examples:

Pentyl
$$\overset{5}{CH_3}-\overset{4}{CH_2}-\overset{3}{CH_2}-\overset{2}{CH_2}-\overset{1}{CH_2}-$$

Undecyl
$$\overset{11}{CH_3}-\overset{10-2}{[CH_2]_9}-\overset{1}{CH_2}-$$

Rule A–2. Saturated Branched-chain Compounds and Univalent Radicals

2.1—A saturated branched acyclic hydrocarbon is named by prefixing the designations of the side chains to the name of the longest chain present in the formula.

*The prefix for twenty is spelled "icosa" not "eicosa" as in earlier editions (cf. Rule D-7.11).

Example:

$$CH_3—CH_2—CH—CH_2—CH_3$$
$$|$$
$$CH_3$$

Methylpentane

The following names are retained for unsubstituted hydrocarbons only:

Isobutane	$(CH_3)_2CH—CH_3$
Isopentane	$(CH_3)_2CH—CH_2—CH_3$
Neopentane	$(CH_3)_4C$
Isohexane	$(CH_3)_2CH—CH_2—CH_2—CH_3$

2.2—The longest chain is numbered from one end to the other by Arabic numerals, the direction being so chosen as to give the lowest numbers possible to the side chains. When series of locants containing the same number of terms are compared term by term, that series is " lowest " which contains the lowest number on the occasion of the first difference. This principle is applied irrespective of the nature of the substituents.

Examples:

$$\overset{5}{CH_3}—\overset{4}{CH_2}—\overset{3}{CH}—\overset{2}{CH_2}—\overset{1}{CH_3}$$
$$|$$
$$CH_3$$

3-Methylpentane

$$\overset{6}{CH_3}—\overset{5}{CH}—\overset{4}{CH_2}—\overset{3}{CH}—\overset{2}{CH}—\overset{1}{CH_3}$$
$$\qquad | \qquad\qquad |\quad\;\; |$$
$$\qquad CH_3 \qquad\;\; CH_3\;\; CH_3$$

2,3,5-Trimethylhexane (not 2,4,5-Trimethylhexane)

$$\overset{10}{CH_3}—\overset{9}{CH_2}—\overset{8}{CH}—\overset{7}{CH}—\overset{6}{CH_2}—\overset{5}{CH_2}—\overset{4}{CH_2}—\overset{3}{CH_2}—\overset{2}{CH}—\overset{1}{CH_3}$$
$$\qquad\qquad\;\; |\quad\;\; | \qquad\qquad\qquad\qquad\qquad |$$
$$\qquad\qquad CH_3\;\; CH_3 \qquad\qquad\qquad\qquad\;\; CH_3$$

2,7,8-Trimethyldecane (not 3,4,9-Trimethyldecane)

$$\overset{9}{CH_3}—\overset{8}{CH_2}—\overset{7}{CH_2}—\overset{6}{CH_2}—\overset{5}{CH}—\overset{4}{CH}—\overset{3}{CH_2}—\overset{2}{CH_2}—\overset{1}{CH_3}$$
$$\qquad\qquad\qquad\qquad\quad |\quad\;\; |$$
$$\qquad\qquad\qquad\quad CH_3\;\; CH_2—CH_2—CH_3$$

5-Methyl-4-propylnonane
(not 5-Methyl-6-propylnonane because 4,5 is lower than 5,6)

2.25—Univalent branched radicals derived from alkanes are named by prefixing the designation of the side chains to the name of the unbranched alkyl radical possessing the longest possible chain starting from the carbon atom with the free valence, the said atom being numbered as 1.

6

Examples:

	5 4 3 2 1
1-Methylpentyl	$CH_3CH_2CH_2CH_2CH(CH_3)$—
2-Methylpentyl	$CH_3CH_2CH_2CH(CH_3)CH_2$—
5-Methylhexyl	$(CH_3)_2CHCH_2CH_2CH_2CH_2$—

The following names may be used for the unsubstituted radicals only:

Isopropyl	$(CH_3)_2CH$—
Isobutyl	$(CH_3)_2CH—CH_2$—
sec-Butyl	$CH_3—CH_2—CH—$
	$\qquad\qquad\quad \mid$
	$\qquad\qquad\quad CH_3$
tert-Butyl	$(CH_3)_3C$—
Isopentyl	$(CH_3)_2CH—CH_2—CH_2$—
Neopentyl	$(CH_3)_3C—CH_2$—

$$\qquad\qquad\qquad\qquad CH_3$$
$$\qquad\qquad\qquad\qquad \mid$$
$$tert\text{-Pentyl} \qquad CH_3—CH_2—C—$$
$$\qquad\qquad\qquad\qquad \mid$$
$$\qquad\qquad\qquad\qquad CH_3$$

Isohexyl	$(CH_3)_2CH—CH_2—CH_2—CH_2$—

2.3—If two or more side chains of different nature are present, they are cited in alphabetical order.*

The alphabetical order is decided as follows:

(*i*) The names of simple radicals are first alphabetized and the multiplying prefixes are then inserted.

Example:

$$\qquad\qquad\qquad\qquad CH_3—CH_2\ \ CH_3$$
$$\quad 7 \quad\ 6 \quad\ 5 \quad 4\mid \quad 3\mid \ 2 \quad\ 1$$
$$CH_3—CH_2—CH_2—CH—C—CH_2—CH_3$$
$$\qquad\qquad\qquad\qquad\qquad \mid$$
$$\qquad\qquad\qquad\qquad\quad CH_3$$

ethyl is cited before methyl, thus   4-Ethyl-3,3-dimethylheptane

*Use of an order of complexity given as alternative in the First and Second Editions is abandoned.

(*ii*) The name of a complex radical is considered to begin with the first letter of its complete name.

Example:

$$CH_3$$
$$\overset{1}{C}H_3—\overset{2}{C}H—\overset{3}{C}H—\overset{4}{C}H_2—\overset{5}{C}H_2—CH_3$$

$$\overset{13}{C}H_3—\overset{12-8}{[CH_2]_5}—\overset{7}{C}H—\overset{6}{C}H_2—\overset{5}{C}H—\overset{4}{C}H_2—\overset{3}{C}H_2—\overset{2}{C}H_2—\overset{1}{C}H_3$$
$$CH_2—CH_3$$

dimethylpentyl (as a complete single substituent) is alphabetized under " d ", thus 7-(1,2-Dimethylpentyl)-5-ethyltridecane

(*iii*) In cases where names of complex radicals are composed of identical words, priority for citation is given to that radical which contains the lowest locant at the first cited point of difference in the radical.

Example:

$$CH_3 \qquad\qquad CH_3$$
$$CH_3—CH_2—CH—CH_2 \qquad CH—CH_2—CH_2—CH_3$$
$$\overset{13}{C}H_3—\overset{12-9}{[CH_2]_4}—\overset{8}{C}H—\overset{7}{C}H_2—\overset{6}{C}H—\overset{5}{C}H_2—\overset{4}{C}H_2—\overset{3}{C}H_2—\overset{2}{C}H_2—\overset{1}{C}H_3$$

6-(1-Methylbutyl)-8-(2-methylbutyl)tridecane

2.4—If two or more side chains are in equivalent positions, the one to be assigned the lower number is that cited first in the name.

Examples:

$$\overset{8}{C}H_3—\overset{7}{C}H_2—\overset{6}{C}H_2—\overset{5}{C}H—\overset{4}{C}H—\overset{3}{C}H_2—\overset{2}{C}H_2—\overset{1}{C}H_3$$
$$CH_3 \quad CH_2—CH_3$$

4-Ethyl-5-methyloctane

$$\overset{8}{C}H_3—\overset{7}{C}H_2—\overset{6}{C}H_2—\overset{5}{C}H—\overset{4}{C}H—\overset{3}{C}H_2—\overset{2}{C}H_2—\overset{1}{C}H_3$$
$$CH_2 \quad CH—CH_3$$
$$CH_3—CH_2 \quad CH_3$$

4-Isopropyl-5-propyloctane

2.5—The presence of identical unsubstituted radicals is indicated by the appropriate multiplying prefix di-, tri-, tetra-, penta-, hexa-, hepta-, octa-, nona-, deca-, undeca-, *etc.*

Example:

$$\begin{array}{cccccc} & & & \overset{}{CH_3} & & \\ \overset{5}{CH_3}-\overset{4}{CH_2}-\overset{3|}{C}-\overset{2}{CH_2}-\overset{1}{CH_3} \\ & & & \underset{|}{} & & \\ & & & CH_3 & & \end{array}$$

3,3-Dimethylpentane

The presence of identical radicals each substituted in the same way may be indicated by the appropriate multiplying prefix bis-, tris-, tetrakis-, pentakis-, *etc.* The complete expression denoting such a side chain may be enclosed in parentheses or the carbon atoms in side chains may be indicated by primed numbers.

Examples:

$$\begin{array}{c} \overset{}{CH_3} \\ \overset{3}{CH_3}-\overset{2}{CH_2}-\overset{1|}{C}-CH_3 \\ \overset{10}{CH_3}-\overset{9}{CH_2}-\overset{8}{CH_2}-\overset{7}{CH_2}-\overset{6}{CH_2}-\overset{5|}{C}-\overset{4}{CH_2}-\overset{3}{CH_2}-\overset{2}{CH}-\overset{1}{CH_3} \\ CH_3-CH_2-\overset{|}{C}-CH_3 \qquad \overset{|}{CH_3} \\ \overset{|}{CH_3} \end{array}$$

(*a*) Use of parentheses and unprimed numbers:
 5,5-Bis(1,1-dimethylpropyl)-2-methyldecane

(*b*) Use of primes:
 5,5-Bis-1′,1′-dimethylpropyl-2-methyldecane

$$\begin{array}{c} \overset{}{CH_3} \\ \overset{4}{CH_3}-\overset{3}{CH_2}-\overset{2}{CH_2}-\overset{1|}{C}-CH_3 \\ \overset{13}{CH_3}-\overset{12-10}{[CH_2]_3}-\overset{9}{CH_2}-\overset{8}{CH_2}-\overset{7|}{C}-\overset{6}{CH_2}-\overset{5}{CH_2}-\overset{4}{CH_2}-\overset{3}{CH_2}-\overset{2}{CH_2}-\overset{1}{CH_3} \\ \overset{5}{CH_3}-\overset{4}{CH_2}-\overset{3}{CH_2}-\overset{2}{CH_2}-\overset{1|}{C}-CH_3 \\ \overset{|}{CH_3} \end{array}$$

(a) Use of parentheses and unprimed numbers:

7-(1,1-Dimethylbutyl)-7-(1,1-dimethylpentyl)tridecane

(b) Use of primes:

7-1′,1′-Dimethylbutyl-7-1″,1″-dimethylpentyltridecane

2.6—If chains of equal length are competing for selection as main chain in a saturated branched acyclic hydrocarbon, then the choice goes in series to:

(a) The chain which has the greatest number of side chains.

Example:

$$
\begin{array}{ccccccc}
7 & 6 & 5 & 4 & 3 & 2 & 1 \\
CH_3 & \!\!-CH_2- & CH- & CH- & CH- & CH- & CH_3 \\
& & | & | & | & | & \\
& & CH_3 & CH_2 & CH_3 & CH_3 & \\
& & & | & & & \\
& & & CH_2\!-\!CH_3 & & &
\end{array}
$$

2,3,5-Trimethyl-4-propylheptane

(b) The chain whose side chains have the lowest-numbered locants.

Example:

$$
\begin{array}{ccccccc}
7 & 6 & 5 & 4 & 3 & 2 & 1 \\
CH_3 & \!\!-CH_2- & CH- & CH- & CH_2- & CH- & CH_3 \\
& & | & | & & | & \\
& & CH_3 & CH_2 & & CH_3 & \\
& & & | & & & \\
& & & CH\!-\!CH_3 & & & \\
& & & | & & & \\
& & & CH_3 & & &
\end{array}
$$

4-Isobutyl-2,5-dimethylheptane

(c) The chain having the greatest number of carbon atoms in the smaller side chains.

Example*:

$$
\begin{array}{l}
\qquad CH_3 \qquad\qquad CH_3 \qquad\qquad\qquad\qquad CH_3 \qquad\qquad CH_2\!-\!CH_3 \\
\qquad | \qquad\qquad\quad | \qquad\qquad\qquad\qquad\quad | \qquad\qquad\quad | \\
CH_3CH_2CH\!-\!CH_2\!-\!CH\!-\!CH_2 \qquad CH_2\!-\!CH\!-\!CH_2\!-\!CHCH_2CH_3 \\
\;13 \quad 12 \quad 11 \qquad 10 \qquad 9 \qquad 8 \;\;\backslash\, 7\;/\,6 \qquad 5 \qquad 4 \qquad 3 \quad 2 \quad 1 \\
\qquad\qquad\qquad\qquad\qquad\qquad\qquad\quad C \\
\qquad\qquad\qquad\qquad\qquad\qquad\;/\qquad\backslash \\
CH_3CH_2CH\!-\!CH_2\!-\!CH\!-\!CH_2 \qquad CH_2\!-\!CH\!-\!CH_2\!-\!CHCH_2CH_3 \\
\qquad\quad | \qquad\qquad\quad | \qquad\qquad\qquad\qquad | \qquad\qquad\quad | \\
\qquad\quad CH_3 \qquad\qquad CH_3 \qquad\qquad\qquad CH_3 \qquad\qquad CH_3
\end{array}
$$

7,7-Bis(2,4-dimethylhexyl)-3-ethyl-5,9,11-trimethyltridecane

*See footnote on facing page

(d) The chain having the least branched side chains.

$$
\begin{array}{c}
\overset{\text{CH}_2-\text{CH}_2-\text{CH}_3}{\underset{5|}{}} \\
\overset{1}{\text{CH}_3}-\overset{2\text{-}4}{(\text{CH}_2)_3}-\overset{5|}{\text{CH}}-\overset{6}{\text{CH}}-\overset{7\text{-}11}{(\text{CH}_2)_5}-\overset{12}{\text{CH}_3} \\
| \\
\text{CH}_3-(\text{CH}_2)_3-\text{CH}-\text{CH}-\text{CH}_3 \\
| \\
\text{CH}_3
\end{array}
$$

6-(1-Isopropylpentyl)-5-propyldodecane

Rule A–3. Unsaturated Compounds and Univalent Radicals

3.1—Unsaturated unbranched acyclic hydrocarbons having one double bond are named by replacing the ending " -ane " of the name of the corresponding saturated hydrocarbon with the ending " -ene ". If there are two or more double bonds, the ending will be " -adiene ", " -atriene ", *etc.* The generic names of these hydrocarbons (branched or unbranched) are "alkene", "alkadiene", "alkatriene", *etc.* The chain is so numbered** as to give the lowest possible numbers to the double bonds. When, in cyclic compounds or their substitution products, the locants of a double bond differ by unity, only the lower locant is cited in the name; when they differ by more than unity, one locant is placed in parentheses after the other (see Rules A–31.3 and A–31.4).

Examples:

2-Hexene
$$\overset{6}{\text{CH}_3}-\overset{5}{\text{CH}_2}-\overset{4}{\text{CH}_2}-\overset{3}{\text{CH}}=\overset{2}{\text{CH}}-\overset{1}{\text{CH}_3}$$

1,4-Hexadiene
$$\overset{6}{\text{CH}_3}-\overset{5}{\text{CH}}=\overset{4}{\text{CH}}-\overset{3}{\text{CH}_2}-\overset{2}{\text{CH}}=\overset{1}{\text{CH}_2}$$

The following non-systematic names are retained:

Ethylene $\quad \text{CH}_2=\text{CH}_2 \qquad$ Allene $\quad \text{CH}_2=\text{C}=\text{CH}_2$

---

* Here the choice lies between two possible main chains of equal length, each containing six side chains in the same positions. Listing in increasing order, the number of carbon atoms in the several side chains of the first choice as shown and of the alternate second choice results as follows:

first choice	1, 1, 1, 2, 8, 8
second choice	1, 1, 1, 1, 8, 9

The expression, " the greatest number of carbon atoms in the smaller side chains ", is taken to mean the largest side chain at the first point of difference when the size of the side chains is examined step by step. Thus, the selection in this case is made at the fourth step where 2 is greater than 1.

**Only the lower locant for a double bond is cited in the name of an acyclic compound.

3.2—Unsaturated unbranched acyclic hydrocarbons having one triple bond are named by replacing the ending " -ane " of the name of the corresponding saturated hydrocarbon with the ending " -yne ". If there are two or more triple bonds, the ending will be " -adiyne ", " -atriyne ", *etc.* The generic names of these hydrocarbons (branched or unbranched) are " alkyne ", " alkadiyne ", " alkatriyne ", *etc.* The chain is so numbered as to give the lowest possible numbers to the triple bonds. Only the lower locant for a triple bond is cited in the name of a compound.

The name " acetylene " for $HC{\equiv}CH$ is retained.

3.3—Unsaturated unbranched acyclic hydrocarbons having both double and triple bonds are named by replacing the ending " -ane " of the name of the corresponding saturated hydrocarbon with the ending " -enyne ", " -adienyne ", " -atrienyne ", " -enediyne ", *etc.* Numbers as low as possible are given to double and triple bonds even though this may at times give " -yne " a lower number than " -ene ". When there is a choice in numbering, the double bonds are given the lowest numbers.

Examples:

1,3-Hexadien-5-yne

$$\overset{6}{HC}{\equiv}\overset{5}{C}{-}\overset{4}{CH}{=}\overset{3}{CH}{-}\overset{2}{CH}{=}\overset{1}{CH_2}$$

3-Penten-1-yne

$$\overset{5}{CH_3}{-}\overset{4}{CH}{=}\overset{3}{CH}{-}\overset{2}{C}{\equiv}\overset{1}{CH}$$

1-Penten-4-yne

$$\overset{5}{HC}{\equiv}\overset{4}{C}{-}\overset{3}{CH_2}{-}\overset{2}{CH}{=}\overset{1}{CH_2}$$

3.4—Unsaturated branched acyclic hydrocarbons are named as derivatives of the unbranched hydrocarbons which contain the maximum number of double and triple bonds. If there are two or more chains competing for selection as the chain with the maximum number of unsaturated bonds, then the choice goes to (*1*) that one with the greatest number of carbon atoms; (*2*) the number of carbon atoms being equal, that one containing the maximum number of double bonds. In other respects, the same principles apply as for naming saturated branched acyclic hydrocarbons. The chain is so numbered as to give the lowest possible numbers to double and triple bonds in accordance with Rule **A–3.3**.

Examples:

3,4-Dipropyl-1,3-hexadien-5-yne

$$\begin{array}{c} \overset{3}{CH_2}{-}\overset{2}{CH_2}{-}\overset{1}{CH_3} \\ | \\ \overset{6}{CH}{\equiv}\overset{5}{C}{-}\overset{4}{C}{=}C{-}CH{=}CH_2 \\ | \\ CH_2{-}CH_2{-}CH_3 \end{array}$$

5-Ethynyl-1,3,6-heptatriene

$$\begin{array}{c} \overset{7}{CH_2}{=}\overset{6}{CH}{-}\overset{5}{CH}{-}\overset{4}{CH}{=}\overset{3}{CH}{-}\overset{2}{CH}{=}\overset{1}{CH_2} \\ | \\ C{\equiv}CH \end{array}$$

5,5-Dimethyl-1-hexene

$$\begin{array}{c} CH_3 \\ | \\ \overset{6}{CH_3}{-}\overset{5}{C}{-}\overset{4}{CH_2}{-}\overset{3}{CH_2}{-}\overset{2}{CH}{=}\overset{1}{CH_2} \\ | \\ CH_3 \end{array}$$

$$\overset{7}{CH_3}-\overset{6}{C}\equiv\overset{5}{C}-\overset{4}{CH}-\overset{3}{CH_2}-\overset{2}{CH}=\overset{1}{CH_2}$$

4-Vinyl-1-hepten-5-yne

$$CH=CH_2$$

The following name is retained for the unsubstituted compound only:

$$CH_3$$
$$|$$
Isoprene $CH_2=CH-C=CH_2$

3.5—The names of univalent radicals derived from unsaturated acyclic hydrocarbons have the endings " -enyl ", " -ynyl ", " -dienyl ", *etc.*, the positions of the double and triple bonds being indicated where necessary. The carbon atom with the free valence is numbered as 1.

Examples:

Ethynyl	$CH\equiv C-$
2-Propynyl	$CH\equiv C-CH_2-$
1-Propenyl	$CH_3-CH=CH-$
2-Butenyl	$CH_3-CH=CH-CH_2-$
1,3-Butadienyl	$CH_2=CH-CH=CH-$
2-Pentenyl	$CH_3-CH_2-CH=CH-CH_2-$
2-Penten-4-ynyl	$CH\equiv C-CH=CH-CH_2-$

Exceptions:

The following names are retained:

Vinyl (for ethenyl)	$CH_2=CH-$
Allyl (for 2-propenyl)	$CH_2=CH-CH_2-$
Isopropenyl (for 1-methylvinyl)	$CH_2=C-$ (for unsubstituted radical only)

$$CH_3$$

3.6—When there is a choice for the fundamental chain of a radical, that chain is selected which contains (*1*) the maximum number of double and triple bonds; (*2*) the largest number of carbon atoms; and (*3*) the largest number of double bonds.

Examples:

$$\overset{10}{CH_3}-\overset{9}{CH}=\overset{8}{CH}-\overset{7}{CH}=\overset{6}{CH}-\overset{5}{CH}-\overset{4}{CH}=\overset{3}{CH}-\overset{2}{C}\equiv\overset{1}{C}-$$

$$CH_2-CH_2-CH=CH-CH_3$$

5-(3-Pentenyl)-3,6,8-decatrien-1-ynyl

$$\overset{12}{CH_3}-\overset{11}{CH_2}-\overset{10}{C}\equiv\overset{9}{C}-\overset{8}{CH}=\overset{7}{CH}-\overset{6}{CH}-\overset{5}{CH}=\overset{4}{CH}-\overset{3}{CH}=\overset{2}{CH}-\overset{1}{CH_2}-$$

$$CH=CH-CH=CH-CH_3$$

6-(1,3-Pentadienyl)-2,4,7-dodecatrien-9-ynyl

$$\overset{11}{CH_3}-\overset{10}{CH}=\overset{9}{CH}-\overset{8}{CH}=\overset{7}{CH}-\overset{6}{CH}-\overset{5}{CH}=\overset{4}{CH}-\overset{3}{CH}=\overset{2}{CH}-\overset{1}{CH_2}-$$

$$CH=CH-C\equiv C-CH_3$$

6-(1-Penten-3-ynyl)-2,4,7,9-undecatetraenyl

$$\overset{4}{CH_3}-\overset{3}{CH}=\overset{2}{C}-\overset{1}{CH_2}-$$

$$CH_2-CH_2-CH_2-CH_2-CH_2-CH_2-CH_2-CH_2-CH_3$$

2-Nonyl-2-butenyl

Rule A–4. Bivalent and Multivalent Radicals **

4.1—Bivalent and trivalent radicals derived from univalent acyclic hydrocarbon radicals whose authorized names end in " -yl " by removal of one or two hydrogen atoms from the carbon atom with the free valences are named by adding " -idene " or " -idyne ", respectively, to the name of the corresponding univalent radical. The carbon atom with the free valence is numbered as 1.

The name " methylene " is retained for the radical $CH_2=$.

Examples:

Methylidyne*	$CH\equiv$
Ethylidene	$CH_3-CH=$
Ethylidyne	$CH_3-C\equiv$
Vinylidene	$CH_2=C=$
Isopropylidene†	$(CH_3)_2C=$

4.2—The names of bivalent radicals derived from normal alkanes by removal of a hydrogen atom from each of the two terminal carbon atoms of the chain are ethylene, trimethylene, tetramethylene, *etc.*

Examples:

Pentamethylene	$-CH_2-CH_2-CH_2-CH_2-CH_2-$
Hexamethylene	$-CH_2-CH_2-CH_2-CH_2-CH_2-CH_2-$

Names of the substituted bivalent radicals are derived in accordance with Rules A–2.2 and A–2.25.

Example:

Ethylethylene $\qquad -\overset{2}{CH_2}-\overset{1}{CH}-$
$$CH_2-CH_3$$

The following name is retained:

Propylene $\qquad CH_3-CH-CH_2-$

* The group $=CH-$ may be referred to as the " methine " group.

† For unsubstituted radical only.

**Rule D-4.14 introduces an alternate method of naming radicals derived from any position of unbranched chains or ring systems by adding "-yl", "-diyl", "-triyl", *etc.* to the name of the chain or ring system with elision of "e" before "-yl". Examples: 2-pentanyl CH_3-CH_2-CH_2-CH-CH_3; 1,6-hexanediyl -CH_2-$(CH_2)_4$-CH_2-.

4.3—Bivalent radicals similarly derived from unbranched alkenes, alkadienes, alkynes, *etc.*, by removing a hydrogen atom from each of the terminal carbon atoms are named by replacing the endings " -ene ", " -diene ", " -yne ", *etc.*, of the hydrocarbon name by " -enylene ", " -dienylene ", " -ynylene ", *etc.*, the positions of the double and triple bonds being indicated where necessary.

Example:

$$\overset{3}{-CH_2}-\overset{2}{CH}=\overset{1}{CH}-$$

Propenylene

The following name is retained:

Vinylene (for ethenylene) —CH=CH—

Names of the substituted bivalent radicals are derived in accordance with Rule **A–3.4**.

Example:

4-Propyl-2-pentenylene $\overset{5}{-CH_2}-\overset{4}{CH}-\overset{3}{CH}=\overset{2}{CH}-\overset{1}{CH_2}-$

$$| \\ CH_2-CH_2-CH_3$$

4.4—Trivalent, quadrivalent and higher-valent acyclic hydrocarbon radicals of two or more carbon atoms with the free valences at each end of a chain are named by adding to the hydrocarbon name the terminations " -yl " for a single free valence, " -ylidene " for a double, and " -ylidyne " for a triple free valence on the same atom (the final " e " in the name of the hydrocarbon is elided when followed by a suffix beginning with " -yl "). If different types are present in the same radical, they are cited and numbered in the order " -yl ", " -ylidene ", " -ylidyne ".

Examples:

Butanediylidene $\overset{4}{=CH}-\overset{3}{CH_2}-\overset{2}{CH_2}-\overset{1}{CH}=$

Butanediylidyne $\overset{4}{\equiv C}-\overset{3}{CH_2}-\overset{2}{CH_2}-\overset{1}{C}\equiv$

1-Propanyl-3-ylidene $\overset{3}{=CH}-\overset{2}{CH_2}-\overset{1}{CH_2}-$

Propadienediylidene $\overset{3}{=C}=\overset{2}{C}=\overset{1}{C}=$

2-Pentenediylidyne $\overset{5}{\equiv C}-\overset{4}{CH_2}-\overset{3}{CH}=\overset{2}{CH}-\overset{1}{C}\equiv$

1-Butanyliden-4-ylidyne $\overset{4}{\equiv C}-\overset{3}{CH_2}-\overset{2}{CH_2}-\overset{1}{CH}=$

4.5—Multivalent radicals containing three or more carbon atoms with free valences at each end of a chain and additional free valences at intermediate carbon atoms are named by adding the endings " -triyl ", " -tetrayl ", " -diylidene ", " diyl-ylidene ", *etc.*, to the hydrocarbon name.

Examples:

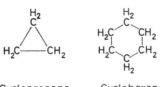

    1,2,3-Propanetriyl   1,3-Propanediyl-2-ylidene.

MONOCYCLIC HYDROCARBONS

Rule A–11. Unsubstituted Compounds and Radicals *

11.1—The names of saturated monocyclic hydrocarbons (with no side chains) are formed by attaching the prefix " cyclo " to the name of the acyclic saturated unbranched hydrocarbon with the same number of carbon atoms. The generic name of saturated monocyclic hydrocarbons (with or without side chains) is " cycloalkane ".

Examples:

Cyclopropane   Cyclohexane

11.2—Univalent radicals derived from cycloalkanes (with no side chains) are named by replacing the ending " -ane " of the hydrocarbon name by " -yl ", the carbon atom with the free valence being numbered as 1. The generic name of these radicals is "cycloalkyl".

Examples:

Cyclopropyl   Cyclohexyl

11.3—The names of unsaturated monocyclic hydrocarbons (with no side chains) are formed by substituting " -ene ", " -adiene ", " -atriene ", " -yne ", " -adiyne ", *etc.*, for " -ane " in the name of the corresponding cycloalkane. The double and triple bonds are given numbers as low as possible as in Rule A–3.3.

Examples:

Cyclohexene   1,3-Cyclohexadiene   1-Cyclodecen-4-yne

*See footnote to Rule A-4.

The name " benzene " is retained.

11.4—The names of univalent radicals derived from unsaturated monocyclic hydrocarbons have the endings " -enyl ", " -ynyl ", " -dienyl ", *etc.*, the positions of the double and triple bonds being indicated according to the principles of Rule A-3.3. The carbon atom with the free valence is numbered as 1, except as stated in the rules for terpenes (see Rules A-72 to A-75).

Examples:

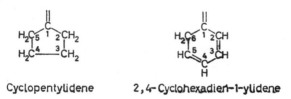

2-Cyclopenten-1-yl   2,4-Cyclopentadien-1-yl

The radical name " phenyl " is retained.

11.5—Names of bivalent radicals derived from saturated or unsaturated monocyclic hydrocarbons by removal of two atoms of hydrogen from the same carbon atom of the ring are obtained by replacing the endings " -ane ", " -ene ", " -yne ", by " -ylidene ", " -enylidene " and " -ynylidene ", respectively. The carbon atom with the free valences is numbered as 1, except as stated in the rules for terpenes.

Examples:

Cyclopentylidene   2,4-Cyclohexadien-1-ylidene

11.6—Bivalent radicals derived from saturated or unsaturated monocyclic hydrocarbons by removing a hydrogen atom from each of two different carbon atoms of the ring are named by replacing the endings " -ane ", " -ene ", " -diene ", " -yne ", *etc.*, of the hydrocarbon name by " -ylene ", " -enylene ", " -dienylene ", " -ynylene ", *etc.*, the positions of the double and triple bonds and of the points of attachment being indicated. Preference in lowest numbers is given to the carbon atoms having the free valences.

Examples:

1,3-Cyclopentylene   3-Cyclohexen-   2,5-Cyclohexadien-
                       1,2-ylene   1,4-ylene

The following name is retained:

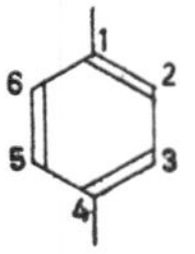

Phenylene (*p*- shown)

Rule A–12. Substituted Aromatic Compounds

12.1—The following names for monocyclic substituted aromatic hydrocarbons are retained:

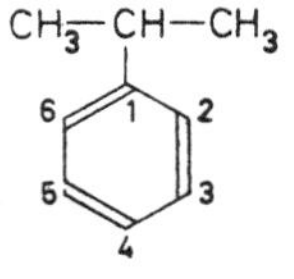

Cumene 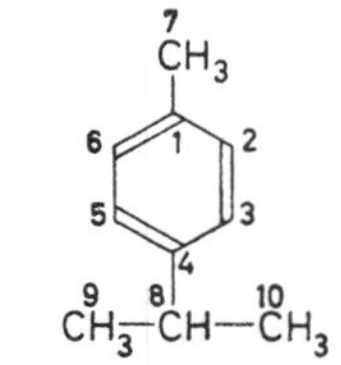   Cymene (*p*- shown) 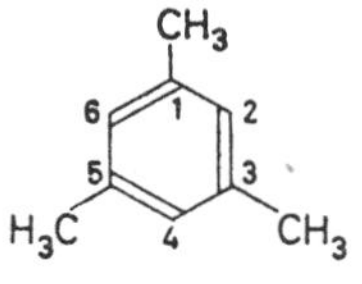   Mesitylene

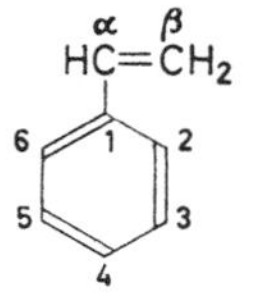

Styrene 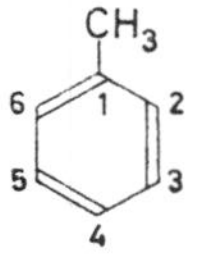   Toluene 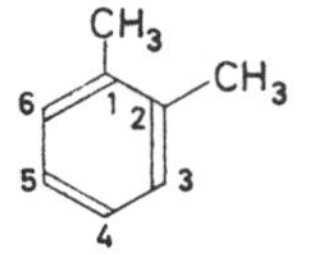   Xylene (*o*- shown)

12.2—Other monocyclic substituted aromatic hydrocarbons are named as derivatives of benzene or of one of the compounds listed in Part .1 of this rule. However, if the substituent introduced into such a compound is identical with one already present in that compound, then the substituted compound is named as a derivative of benzene (see Rule **61.4**).

12.3—The position of substituents is indicated by numbers except that *o*- (*ortho*), *m*- (*meta*) and *p*- (*para*) may be used in place of 1,2-, 1,3-, and 1,4-, respectively, when only two substituents are present. The lowest numbers possible are given to substituents, choice between alternatives being governed by Rule **A–2** so far as applicable, except that when names are based on those of compounds listed in Part .1 of this rule the first priority for lowest numbers is given to the substituent(s) already present in those compounds.

Examples:

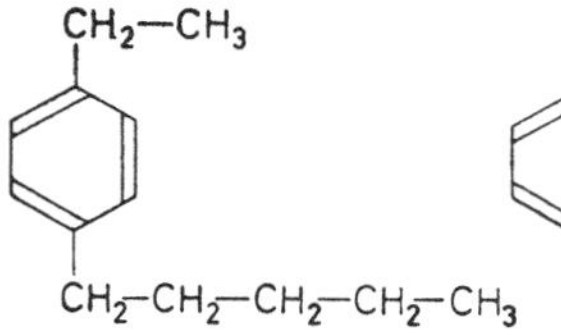

1- Ethyl- 4- pentylbenzene   1,4-Diethylbenzene 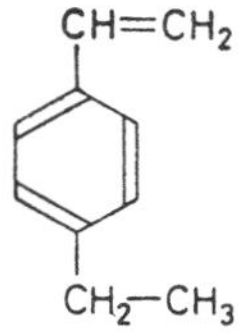   4-Ethylstyrene
or *p*-Ethylpentylbenzene   or *p*- Diethylbenzene   or *p*-Ethylstyrene

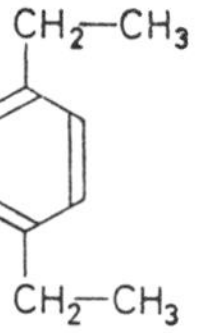

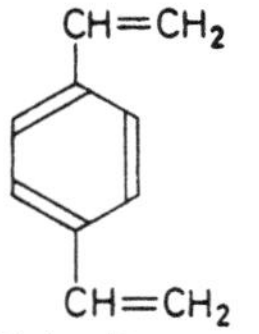

1,4-Divinylbenzene
or *p*-Divinylbenzene
not *p*-Vinylstyrene

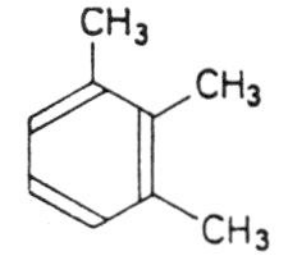

1,2,3-Trimethylbenzene
not Methylxylene
not Dimethyltoluene

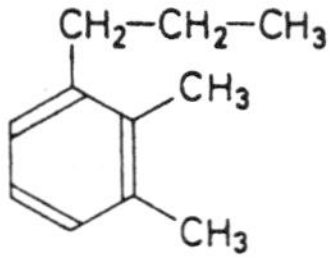

1,2-Dimethyl-
3-propylbenzene
or 3-Propyl-*o*-xylene

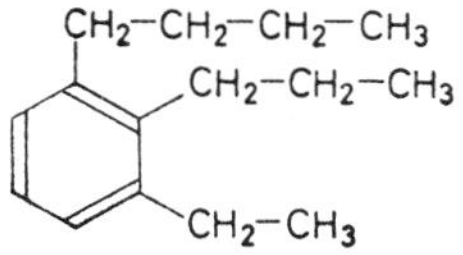

1-Butyl-3-ethyl-2-propylbenzene

12.4—The generic name of monocyclic and polycyclic aromatic hydrocarbons is " arene ".

Rule A–13. Substituted Aromatic Radicals

13.1—Univalent radicals derived from monocyclic substituted aromatic hydrocarbons and having the free valence at a ring atom are given the names listed below. Such radicals not listed below are named as substituted phenyl radicals. The carbon atom having the free valence is numbered as 1.

Phenyl C_6H_5-

Cumenyl (*m*- shown)

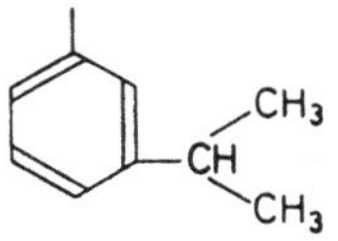

Mesityl

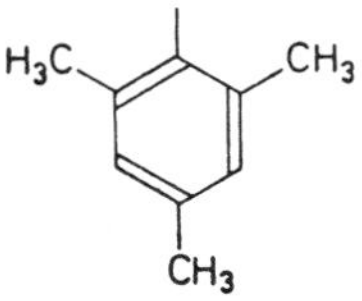

Tolyl (*o*- shown)

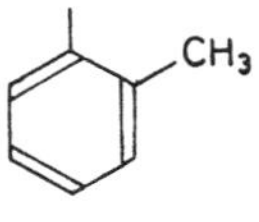

Xylyl (2,3- shown)

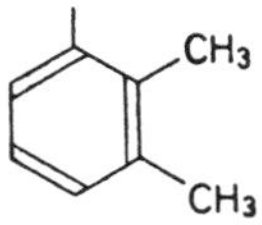

13.2—Since the name phenylene (*o*-, *m*- or *p*-) is retained for the radical —C_6H_4— (exception to Rule A–11.6), bivalent radicals formed from substituted benzene derivatives and having the free valences at ring atoms are named as substituted phenylene radicals. The carbon atoms having the free valences are numbered 1,2-, 1,3- or 1,4- as appropriate.

13.3—The following trivial names for radicals having a single free valence in the side chain are retained:

Benzyl	$C_6H_5\overset{\alpha}{-CH_2}-$
Benzhydryl (alternative to Diphenylmethyl)	$(C_6H_5)_2\overset{\alpha}{CH}-$
Cinnamyl	$C_6H_5\overset{\gamma}{-CH}=\overset{\beta}{CH}-\overset{\alpha}{CH_2}-$
Phenethyl	$C_6H_5\overset{\beta}{-CH_2}-\overset{\alpha}{CH_2}-$
Styryl	$C_6H_5\overset{\beta}{-CH}=\overset{\alpha}{CH}-$
Trityl	$(C_6H_5)_3C-$

13.4—Multivalent radicals of aromatic hydrocarbons with the free valences in the side chain are named in accordance with Rule **A–4**.

Examples:

Benzylidyne	$C_6H_5-C\equiv$
Cinnamylidene	$C_6H_5\overset{\gamma}{-CH}=\overset{\beta}{CH}-\overset{\alpha}{CH}=$

13.5—The generic names of univalent and bivalent aromatic hydrocarbon radicals are " aryl " and " arylene ", respectively.

FUSED POLYCYCLIC HYDROCARBONS

Rule A–21. Trivial and Semi-trivial names

21.1—The names of polycyclic hydrocarbons with maximum number of non-cumulative* double bonds end in " -ene ". The names listed on pp. 21 and 22 are retained.

21.2—The names of hydrocarbons containing five or more fused benzene rings in a straight linear arrangement are formed from a numerical prefix as specified in Rule A–1.1 followed by " -acene ". [Examples on p. 22]

---

* Cumulative double bonds are those present in a chain in which at least three contiguous carbon atoms are joined by double bonds; non-cumulative double bonds comprise every other arrangement of two or more double bonds in a single structure. The generic name " cumulene " is given to compounds containing three or more cumulative double bonds.

Examples:

$$CH_2=C=C=C=CH_2 \qquad\qquad CH_3-CH=CH-CH=CH-CH=CH_2$$

Cumulative   or

Non-cumulative

Examples (to Rule **A–21.2**):

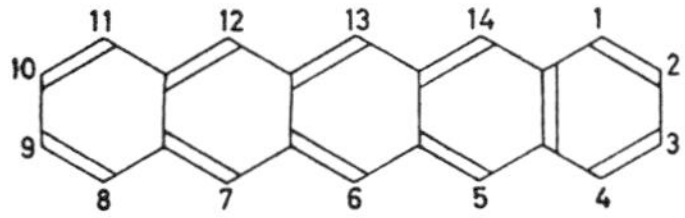

Pentacene

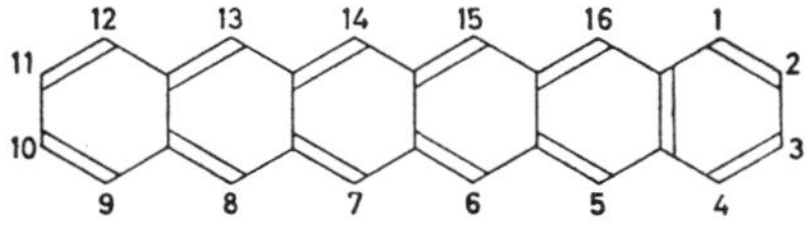

Hexacene

The following list contains the names of polycyclic hydrocarbons which are retained (see Rule **A–21.1**). This list is not limiting.

(1) Pentalene

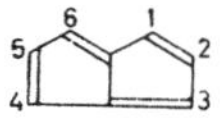

(2) Indene

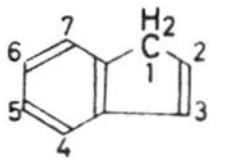

(3) Naphthalene

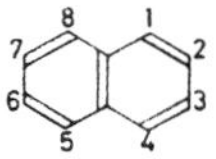

(4) Azulene

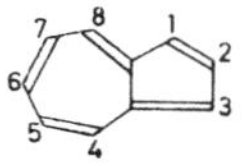

(5) Heptalene

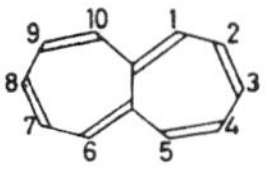

(6) Biphenylene

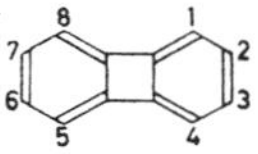

(7) *as* - Indacene

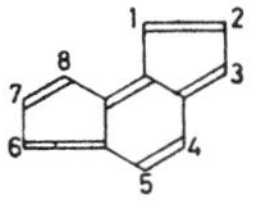

(8) *s* - Indacene

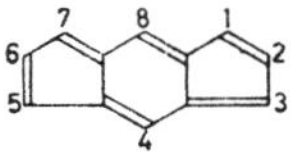

(9) Acenaphthylene

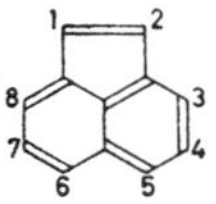

(10) Fluorene

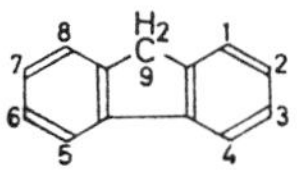

(11) Phenalene

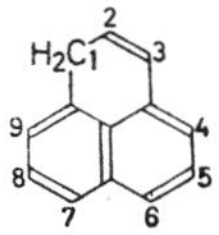

(12) Phenanthrene*

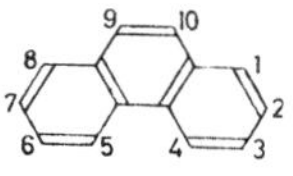

(13) Anthracene*

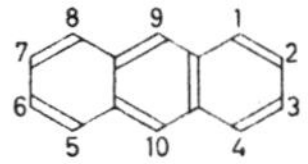

(14) Fluoranthene

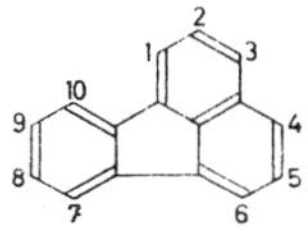

(15) Acephenanthrylene

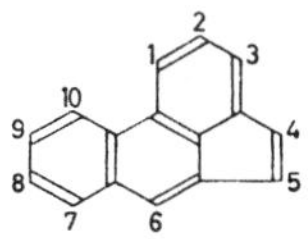

(16) Aceanthrylene

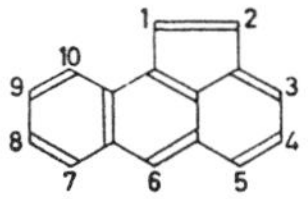

(17) Triphenylene

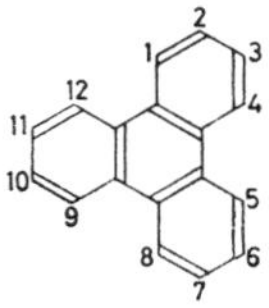

(18) Pyrene

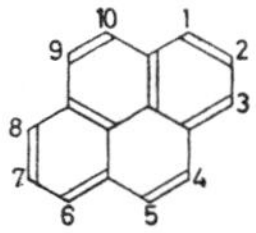

(19) Chrysene

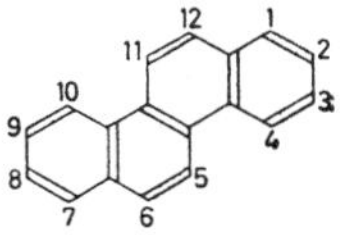

(20) Naphthacene

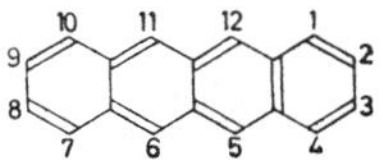

* Denotes exception to systematic numbering.   [*Continued overleaf*

continued]

(21) Pleiadene

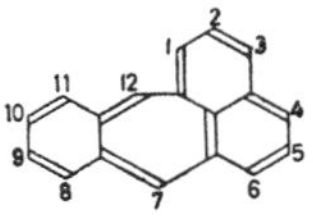

(22) Picene

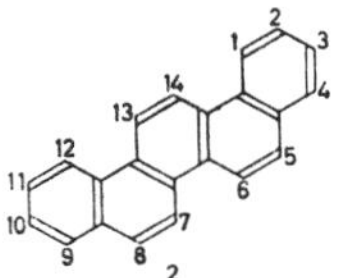

(23) Perylene

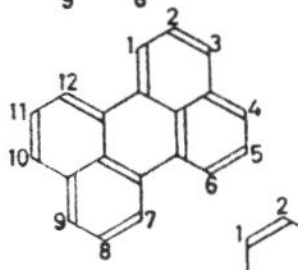

(24) Pentaphene

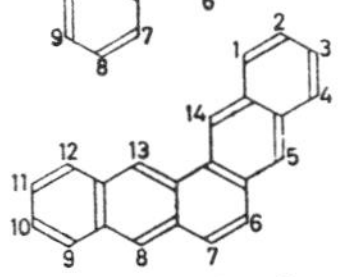

(25) Pentacene**

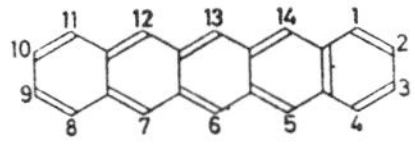

(26) Tetraphenylene***

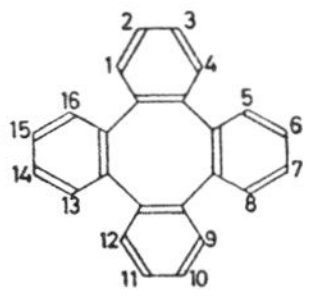

(27) Hexaphene

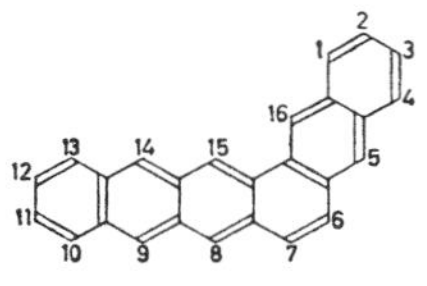

(28) Hexacene**

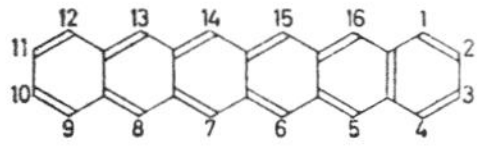

(29) Rubicene

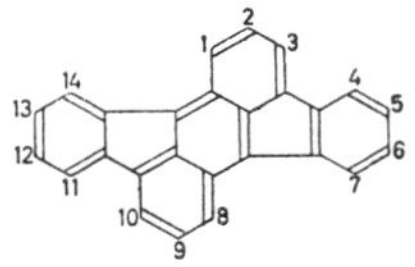

(30) Coronene

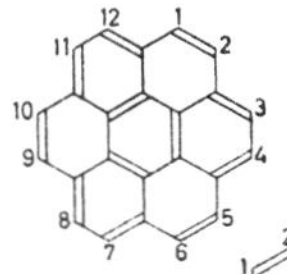

(31) Trinaphthylene***

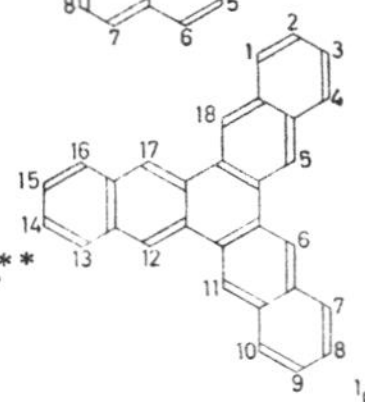

(32) Heptaphene

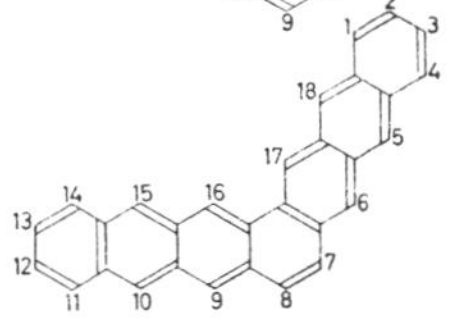

(33) Heptacene**

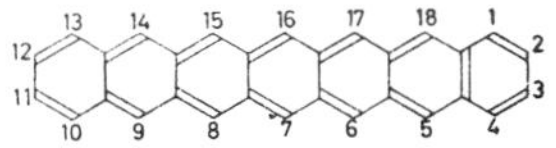

(34) Pyranthrene

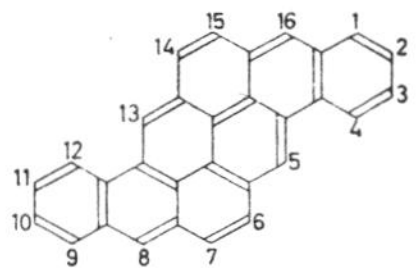

(35) Ovalene

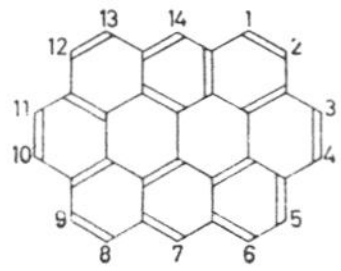

** See Rule A-21.2.

*** For isomer shown only.

21.3—"*Ortho*-fused"* or "*ortho*- and *peri*-fused"† polycyclic hydro-carbons with maximum number of non-cumulative double bonds which contain at least two rings of five or more members and which have no accepted trivial name such as those of Part **.1** of this rule, are named by prefixing to the name of a component ring or ring system (the base component) designations of the other components. The base component should contain as many rings as possible (provided it has a trivial name), and should occur as far as possible from the beginning of the list of Rule **A-21.1**. The attached components should be as simple as possible.

*† See footnotes on facing page.

Example:

(not Naphthophenanthrene; benzo is "simpler" than naphtho, even though there are two benzo rings and only one naphtho)

Dibenzophenanthrene

21.4—The prefixes designating attached components are formed by changing the ending " -ene " of the name of the component hydrocarbon into " -eno "; *e.g.,* " pyreno " (from pyrene). When more than one prefix is present, they are arranged in alphabetical order. The following common abbreviated prefixes are recognized (see list in Part .**1** of this rule):

Acenaphtho	from	Acenaphthylene	Naphtho	from	Naphthalene
Anthra	from	Anthracene	Perylo	from	Perylene
Benzo	from	Benzene	Phenanthro	from	Phenanthrene

For monocyclic prefixes other than " benzo ", the following names are recognized, each to represent the form with the maximum number of non-cumulative double bonds: cyclopenta, cyclohepta, cycloocta, cyclonona, *etc.* When the base component is a monocyclic system, the ending " -ene " signifies the maximum number of non-cumulative double bonds, and thus does not denote one double bond only. **

Examples:

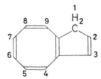
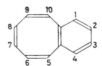

·1*H*–Cyclopentacyclooctene   Benzocyclooctene

21.5—Isomers are distinguished by lettering the peripheral sides of the base component *a, b, c, etc.,* beginning with " *a* " for the side " 1,2 ", " *b* "

* Polycyclic compounds in which two rings have two, and only two, atoms in common are said to be " *ortho*-fused ". Such compounds have *n* common faces and 2*n* common atoms (Example I).

† Polycyclic compounds in which one ring contains two, and only two, atoms in common with each of two or more rings of a contiguous series of rings are said to be " *ortho*- and *peri*-fused ". Such compounds have *n* common faces and less than 2*n* common atoms (Examples II and III).

Examples:

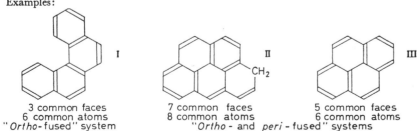

3 common faces   7 common faces   5 common faces
6 common atoms   8 common atoms   6 common atoms
" *Ortho*-fused " system   " *Ortho* - and *peri* - fused " systems

**The final "o" of acenaphtho, benzo, naphtho and perylo and the "a" of the monocyclic prefixes cyclopropa, cyclopenta, cyclohepta, *etc.* are elided before another vowel, as benz(o) [*a*] anthracene. In all other cases the final "o" or "a" is retained.

23

for " 2,3 " (or in certain cases " 2,2a ") and lettering every side around the periphery.   To the letter as early in the alphabet as possible, denoting the side where fusion occurs, are prefixed, if necessary, the numbers of the positions of attachment of the other component.   These numbers are chosen to be as low as is consistent with the numbering of the component, and their order conforms to the direction of lettering of the base component (see Examples II and IV).   When two or more prefixes refer to equivalent positions so that there is a choice of letters, the prefixes are cited in alphabetical order according to Rule A–21.4 and the location of the first cited prefix is indicated by a letter as early as possible in the alphabet (see Example V).   The numbers and letters are enclosed in square brackets and placed immediately after the designation of the attached component.   This expression merely defines the manner of fusion of the components.

Examples:

Benz[a]anthracene   I

Anthra[2,1-a]naphthacene   II

Dibenz[a,j]anthracene   III
(not Naphtho[2,1-b]phenanthrene)

Indeno[1,2-a]indene   IV

1H-Benzo[a]cyclopent[j]anthracene   V

The completed system consisting of the base component and the other components is then renumbered according to Rule A–22, the enumeration of the component parts being ignored.

Example:

Benzene
Pentaphene

Benzene

9H-Dibenzo[de,rst]pentaphene

24

21.6—When a name applies equally to two or more isomeric condensed parent ring systems with the maximum number of non-cumulative double bonds and when the name can be made specific by indicating the position of one or more hydrogen atoms in the structure, this is accomplished by modifying the name with a locant, followed by italic capital *H* for each of these hydrogen atoms. Such symbols ordinarily precede the name. The said atom or atoms are called "indicated hydrogen". The same principle is applied to radicals and compounds derived from these systems.*

Examples:

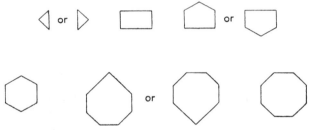

3*H*–Fluorene   2*H*–Indene

Rule A–22. Numbering

22.1—For the purposes of numbering, the individual rings of a polycyclic " *ortho*-fused " or " *ortho*- and *peri*-fused " hydrocarbon system are normally drawn as follows:

and the polycyclic system is oriented so that (*a*) the greatest number of rings are in a horizontal row and (*b*) a maximum number of rings are above and to the right of the horizontal row (upper right quadrant). If two or more orientations meet these requirements, the one is chosen which has as few rings as possible in the lower left quadrant.

Example:

Correct   Incorrect   Incorrect
orientation   orientation   orientation

The system thus oriented is numbered in a clockwise direction commencing with the carbon atom not engaged in ring-fusion in the most

*See Rule B-5.12 for examples of "indicated hydrogen" in monocyclic rings.

counter-clockwise position of the uppermost ring, or if there is a choice, of the uppermost ring farthest to the right, and omitting atoms common to two or more rings.

Example:

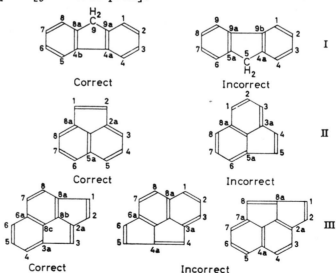

22.2—Atoms common to two or more rings are designated by adding roman letters " a ", " b ", " c ", *etc.*, to the number of the position immediately preceding. Interior atoms follow the highest number, taking a clockwise sequence wherever there is a choice.

Example:

22.3—When there is a choice,* carbon atoms common to two or more rings follow the lowest possible numbers.

Examples [*cf.* notes on p. 27]:

I

II

III

*If after the requirements of the rules on orientation (cf Rule A-22.1) are met a choice remains, Rule A-22.3 is applied.

Notes: I. 4, 4, 8, 9 is lower than 4, 5, 9, 9.
         II. 2, 5, 8 is lower than 3, 5, 8.
         III. 2, 3, 6, 8 is lower than 3, 4, 6, 8 or 2, 4, 7, 8.

22.4—When there is a choice, the carbon atoms which carry an indicated hydrogen atom are numbered as low as possible.

Example:

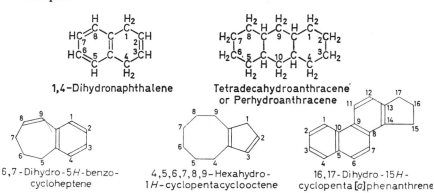

Correct   Incorrect

22.5—The following are recommended exceptions to the above rules on numbering:

Anthracene   Phenanthrene   Cyclopenta [a] phenanthrene
                                                      (15H– shown)
                                          See also rules on steroids *

Rule A–23.   Hydrogenated Compounds

23.1—The names of " *ortho*-fused " or " *ortho*- and *peri*-fused " polycyclic hydrocarbons with less than maximum number of non-cumulative double bonds are formed from a prefix " dihydro- ", " tetrahydro- ", *etc.,* followed by the name of the corresponding unreduced hydrocarbon. The prefix " perhydro- " signifies full hydrogenation. When there is a choice for *H* used for indicated hydrogen it is assigned the lowest available number.

Examples:

1,4–Dihydronaphthalene   Tetradecahydroanthracene
                            or Perhydroanthracene

6,7 -Dihydro- 5*H*-benzo-   4,5,6,7,8,9–Hexahydro-   16,17–Dihydro -15*H*-
   cycloheptene   1*H*–cyclopentacyclooctene   cyclopenta [a] phenanthrene

*Definitive Rules for Nomenclature of Steroids,
Pure and Applied Chemistry,* Vol. 31, Nos. 1-2 (1972), pp. 285-322

Exceptions:

The following names are retained:

Indan　　　　Acenaphthene　　　　Cholanthrene

Aceanthrene　　　　Acephenanthrene

Violanthrene　　　　Isoviolanthrene

23.2—When there is a choice, the carbon atoms to which hydrogen atoms are added are numbered as low as possible.

Example:

Correct　　　　　　　Incorrect

23.3—Substituted polycyclic hydrocarbons are named according to the same principles as substituted monocyclic hydrocarbons (see Rules **A–12** and **A–61**).

23.5 (Alternate to part of Rule **A–23.1**)—The names of " *ortho*-fused " polycyclic hydrocarbons which have (a) less than the maximum number of non-cumulative double bonds, (b) at least one terminal unit which is most conveniently named as an unsaturated cycloalkane derivative, and (c) a double bond at the positions where rings are fused together, may be derived by joining the name of the terminal unit to that of the other component by means of a letter " o " with elision of a terminal " e ". The abbreviations for fused aromatic systems laid down in Rule A–21.4 are used, and the exceptions of Rule A–23.1 apply.

Examples:

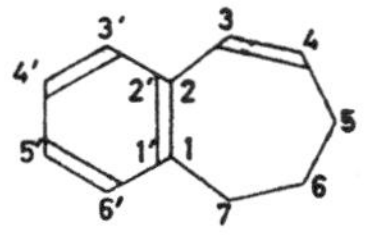

1,2-Benzo-1,3-cycloheptadiene

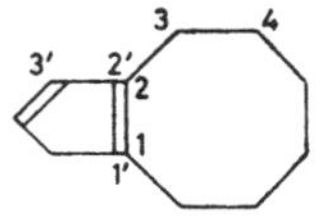

1,2-Cyclopenta-1',3'-dienocyclooctene

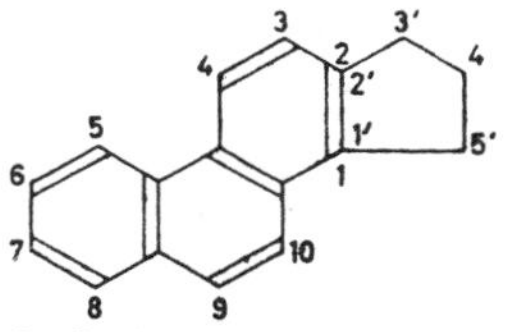

1,2-Cyclopentenophenanthrene

Rule A–24. Radical Names* from Trivial and Semi-trivial Names

24.1—For radicals derived from polycyclic hydrocarbons, the numbering of the hydrocarbon is retained. The point or points of attachment are given numbers as low as is consistent with the fixed numbering of the hydrocarbon.

24.2—Univalent radicals derived from " *ortho*-fused " or " *ortho*- and *peri*-fused " polycyclic hydrocarbons with names ending in " -ene " by removal of a hydrogen atom from an aromatic or alicyclic ring are named in principle by changing the ending " -ene " of the names of the hydrocarbons to " -enyl ".

Examples:

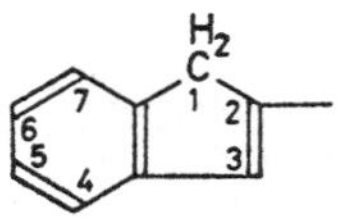

2-Indenyl

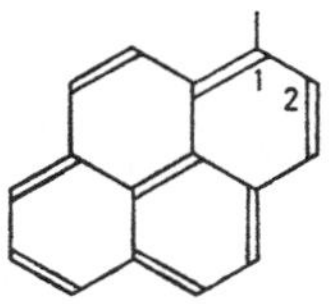

1-Pyrenyl

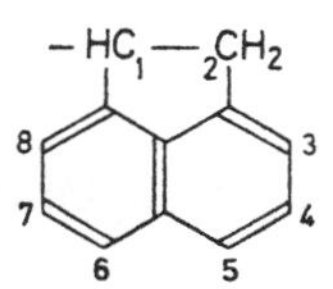

1-Acenaphthenyl

*See footnote to Rule A-4

Exceptions:

Naphthyl
(2- shown)

Anthryl
(2- shown)

Phenanthryl
(2- shown)

5,6,7,8-Tetrahydro-2-naphthyl

24.3—Bivalent radicals derived from univalent polycyclic hydrocarbon radicals whose names end in " -yl " by removal of one hydrogen atom from the carbon atom with the free valence are named by adding " -idene " to the name of the corresponding univalent radical.

Examples:

1 - Acenaphthenylidene

1 (4H)-Naphthylidene
(for 4H see Rule A-21.6)
or 1,4-Dihydro-1-naphthylidene

24.4—Bivalent radicals derived from " *ortho*-fused " or " *ortho*- and *peri*-fused " polycyclic hydrocarbons by removal of a hydrogen atom from each of two different carbon atoms of the ring are named by changing the ending "yl" of the univalent radical name to "-ylene" or by adding "-diyl" to the name of the ring system Multivalent radicals, similarly derived, are named by adding "-triyl", "-tetrayl", *etc.*, to the name of the ring system.

Examples:

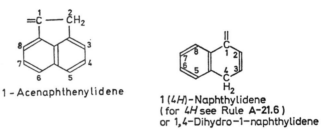

2,7- Phenanthrylene
or 2,7-Phenanthrenediyl

1,4,5,8 -Anthracenetetrayl

30

Rule A–28. Radical Names for Fused Cyclic Systems with Side Chains

28.1—Radicals formed from hydrocarbons consisting of polycyclic systems and side chains are named according to the principles of the preceding rules.

BRIDGED HYDROCARBONS

EXTENSION OF THE VON BAEYER SYSTEM

Rule A–31. Bicyclic Systems

31.1—Saturated alicyclic hydrocarbon systems consisting of two rings only, having two or more atoms in common, take the name of an open chain hydrocarbon containing the same total number of carbon atoms preceded by the prefix " bicyclo- ". The number of carbon atoms in each of the three bridges* connecting the two tertiary carbon atoms is indicated in brackets in descending order.

Examples:

Bicyclo [1.1.0] butane   Bicyclo[3.2.1]octane   Bicyclo [5.2.0]nonane

31.2—The system is numbered commencing with one of the bridge-heads, numbering proceeding by the longest possible path to the second bridgehead; numbering is then continued from this atom by the longer unnumbered path back to the first bridgehead and is completed by the shortest path from the atom next to the first bridgehead.

Examples:

Bicyclo[3.2.1]octane   Bicyclo[4.3.2]undecane

Note: Longest path 1, 2, 3, 4, 5
Next longest path 5, 6, 7, 1
Shortest path 1, 8, 5

---

*A bridge is a valence bond or an atom or an unbranched chain of atoms connecting two different parts of a molecule. The two tertiary carbon atoms connected through the bridge are termed " bridgeheads ".

31.3—Unsaturated hydrocarbons are named in accordance with the principles set forth in Rule **A-11.3.** When after applying Rule A-31.2 a choice in numbering remains unsaturation is given the lowest numbers.

Examples:

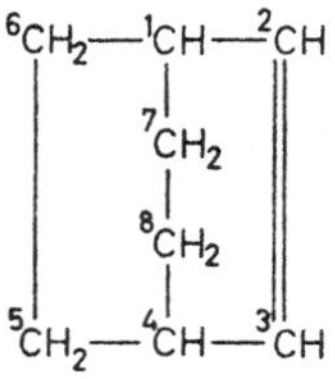

Bicyclo[2.2.2]oct-2-ene

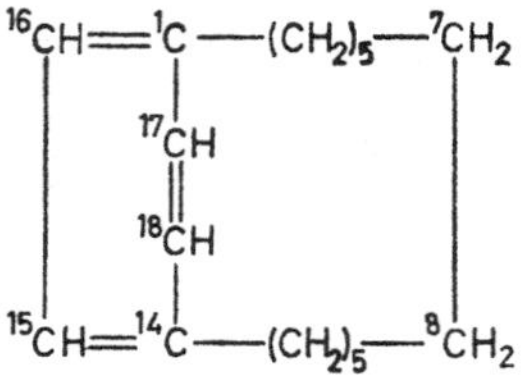

Bicyclo[12.2.2]octadeca – 1(16),14, 17 – triene
or Bicyclo[12.2.2]octadeca – 14,16(1), 17 – triene
(See Rule A-3.1 for double locants)

31.4—Radicals derived from bridged hydrocarbons are named in accordance with the principles set forth in Rule **A–11.** The numbering of the hydrocarbon is retained and the point or points of attachment are given numbers as low as is consistent with the fixed numbering of the saturated hydrocarbon.

Examples:

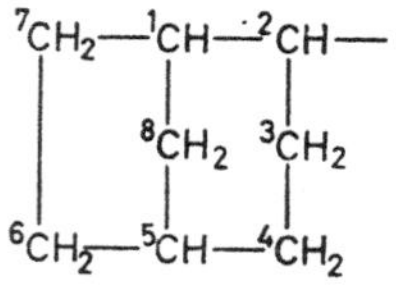

Bicyclo[3.2.1]oct-2-yl

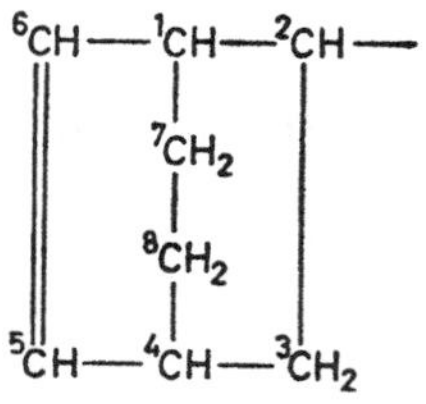

Bicyclo[2.2.2]oct-5-en-2-yl

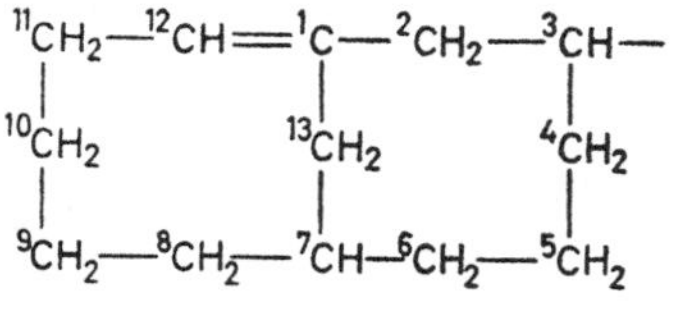

Bicyclo[5.5.1]tridec-1(12)-en-3-yl
or Bicyclo[5.5.1]tridec-12(1)-en-3-yl
(See Rule A-3.1 for double locants)

Rule A–32.　Polycyclic Systems

32.11—Cyclic hydrocarbon systems consisting of three or **more rings** may be named in accordance with the principles stated in Rule **A–31.**

The appropriate prefix " tricyclo- ", " tetracyclo- ", *etc.*, is substituted for " bicyclo- " before the name of the open-chain hydrocarbon containing the same total number of carbon atoms. Radicals derived from these hydrocarbons are named according to the principles set forth in Rule **A–31.4**.

32.12—A polycyclic system is regarded as containing a number of rings equal to the number of scissions required to convert the system into an open-chain compound.

32.13—The word " cyclo " is followed by brackets containing, in decreasing order, numbers indicating the number of carbon atoms in:

the two branches of the main ring,
the main bridge,
the secondary bridges.

Examples:

Tricyclo[2.2.1.0*]heptane   Tricyclo[5.3.1.1*]dodecane

* For location and numbering of the secondary bridge see Rules **A–32.22**, **A–32.23**, **A–32.31**.

32.21—The main ring and the main bridge form a bicyclic system whose numbering is made in compliance with Rule **A–31**.

32.22—The location of the other or so-called secondary bridges is shown by superscripts following the number indicating the number of carbon atoms in the said bridges.

32.23—For the purpose of numbering, the secondary bridges are considered in decreasing order. The numbering of any bridge follows from the part already numbered, proceeding from the highest-numbered bridgehead. If equal bridges are present, the numbering begins at the highest-numbered bridgehead.

32.31—When there is a choice, the following criteria are considered in turn until a decision is made:

(*a*) The main ring shall contain as many carbon atoms as possible, two of which must serve as bridgeheads for the main bridge.

Tricyclo[5.4.0.0^{2,9}]undecane
Correct numbering

Tricyclo[4.2.1.2^{7,9}]undecane
Incorrect numbering

33

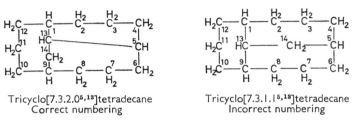

Tricyclo [5.3.2.0^{4,9}]dodecane
Correct numbering

Tricyclo [5.2.3.0^{4,11}]dodecane
Incorrect numbering

(b) **The main bridge shall be as large as possible.**

Tricyclo[7.3.2.0^{5,13}]tetradecane
Correct numbering

Tricyclo[7.3.1.1^{5,13}]tetradecane
Incorrect numbering

(c) **The main ring shall be divided as symmetrically as possible by the main bridge.**

Tricyclo [4.4.1.1^{1,5}]dodecane
Correct numbering

Tricyclo [5.3.1.1^{1,6}]dodecane
Incorrect numbering

(d) **The superscripts locating the other bridges shall be as small as possible (in the sense indicated in Rule A–2.2).**

Tricyclo [5.5.1.0^{3,11}] tridecane
Correct numbering

Tricyclo [5.5.1.0^{5,9}] tridecane
Incorrect numbering

Rule A–34. Hydrocarbon Bridges

34.1—Polycyclic hydrocarbon systems which can be regarded as " *ortho*-fused " or " *ortho*- and *peri*-fused " systems according to Rule **A–21** and which, at the same time, have other bridges*, are first named as " *ortho*-fused " or " *ortho*- and *peri*-fused " systems. The other bridges are then indicated by prefixes derived from the name of the corresponding hydrocarbon by replacing the final " -ane ", " -ene ", *etc.*, by " -ano ", " -eno ", *etc.*, and their positions are indicated by the points of attachment in the parent compound. If bridges of different types are present, they are cited in alphabetical order.

Examples of bridge names:

Butano	$-CH_2-CH_2-CH_2-CH_2--$
Benzeno (*o-*, *m-*, *p-*)	$-C_6H_4-$
Ethano	$-CH_2-CH_2-$
Etheno	$-CH=CH-$
Methano	$-CH_2-$
Propano	$-CH_2-CH_2-CH_2-$

Examples:

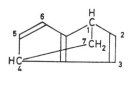

1,4-Dihydro-1,4-methanopentalene

9,10-Dihydro-9,10-[2]butenoanthracene

7,14-Dihydro-7,14-ethano-dibenz[*a*,*h*]anthracene

* The term " bridge ", when used in connection with an " *ortho*-fused " or " *ortho*- and *peri*-fused " polycyclic system as defined in the note to Rule **A–31.1** also includes " bivalent cyclic systems ".

34.2—The parent " *ortho*-fused " or " *ortho*- and *peri*-fused " system is numbered as prescribed in Rule **A–22**. Where there is a choice, the position numbers of the bridgeheads should be as low as possible. The remaining bridges are then numbered in turn starting each time with the bridge atom next to the bridgehead possessing the highest number.

Example:

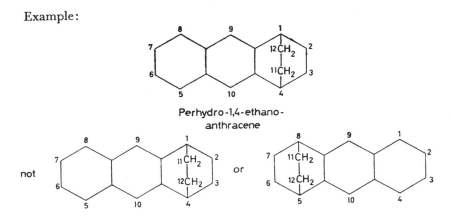

Perhydro-1,4-ethano-
anthracene

34.3—When there is a choice of position numbers for the points of attachment for several individual bridges, the lowest numbers are assigned to the bridgeheads in the order of citation of the bridges and the bridge atoms are numbered according to the preceding rule.

Example:

Perhydro-1,4-ethano-5,8-methanoanthracene

34.4—When the bridge is formed from a bivalent cyclic hydrocarbon radical, low numbers are given to the carbon atoms constituting the shorter bridge and numbering proceeds around the ring.

Example:

10,11-Dihydro-5,10-o-benzeno-5H-benzo[b]fluorene

34.5—Names for radicals derived from the bridged hydrocarbons considered in Rule A–34.1 are constructed in accordance with the principles set forth in Rule A–24. The abbreviated radical names naphthyl, anthryl, phenanthryl, naphthylene, etc., permitted as exceptions to Rules A–24.2 and A–24.4, are replaced in such cases by the regularly formed names naphthalenyl, anthracenyl, phenanthrenyl, naphthalenediyl, etc.

Examples:

9,10-Dihydro-9,10-[2]butenoanthracen-2-yl

1.4-Dihydro-1,4-[2]butenoanthracen-6-yl

SPIRO HYDROCARBONS

A " spiro union " is one formed by a single atom which is the only common member of two rings. A " free spiro union " is one constituting the only union direct or indirect between two rings*. The common atom is designated as the " spiro atom ". According to the number of spiro atoms present, the compounds are distinguished as monospiro-, dispiro-, trispiro-compounds, etc. The following rules apply to the naming of compounds containing free spiro unions.

* An example of a compound where the spiro union is *not* free is:

This compound is named by previous rules as dodecahydrobenz[*c*]indene.

Rule A–41. Compounds: Method 1

41.1—Monospiro compounds consisting of only two alicyclic rings as components are named by placing "spiro" before the name of the normal acyclic hydrocarbon of the same total number of carbon atoms. The number of carbon atoms linked to the spiro atom in each ring is indicated in ascending order in brackets placed between the spiro prefix and the hydrocarbon name.

Examples:

Spiro[3.4]octane

Spiro[3.3]heptane

41.2—The carbon atoms in monospiro hydrocarbons are numbered consecutively starting with a ring atom next to the spiro atom, first through the smaller ring (if such be present) and then through the spiro atom and around the second ring.

Example:

Spiro[4.5]decane

41.3—When unsaturation is present, the same enumeration pattern is maintained, but in such a direction around the rings that the double and triple bonds receive numbers as low as possible in accordance with Rule A–11.

Example:

Spiro[4.5]deca-1,6-diene

41.4—If one or both components of the monospiro compound are fused polycyclic systems, "spiro" is placed before the names of the components arranged in alphabetical order and enclosed in brackets. Established numbering of the individual components is retained. The lowest possible number is given to the spiro atom, and the numbers of the second component are marked with primes. The position of the spiro atom is indicated by placing the appropriate numbers between the names of the two components.

Example:

Spiro[cyclopentane-1,1′-indene]

41.5—Monospiro compounds containing two similar polycyclic components are named by placing the prefix " spirobi " before the name of the component ring system. Established enumeration of the polycyclic system is maintained and the numbers of one component are distinguished by primes. The position of the spiro atom is indicated in the name of the spiro compound by placing the appropriate locants before the name.

Example:

1,1′-Spirobiindene

41.6—Polyspiro compounds consisting of a linear assembly of three or more alicyclic systems are named by placing " dispiro- ", " trispiro- ", " tetraspiro- ", *etc.*, before the name of the unbranched-chain acyclic hydrocarbon of the same total number of carbon atoms. The numbers of carbon atoms linked to the spiro atoms in each ring are indicated in brackets in the same order as the numbering proceeds about the ring. Numbering starts with a ring atom next to a terminal spiro atom and proceeds in such a way as to give the spiro atoms as low numbers as possible after numbering all the carbon atoms of the first ring linked to the terminal spiro atom.

Example:

Dispiro[5.1.7.2]heptadecane

41.7—Polycyclic compounds containing more than one spiro atom and at least one fused polycyclic component are named in accordance with Part .4 of this rule by replacing "spiro" with "dispiro", "trispiro", *etc.*, and choosing the end components by alphabetical order.

Example:

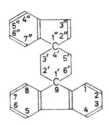

Dispiro[fluorene-9,1′- cyclohexane-4′,1″-indene]

Rule A–42. Compounds: Method 2

42.1 (Alternate to Rules **A–41.1** and **A–41.2**)—When two dissimilar cyclic components are united by a spiro union, the name of the larger component is followed by the affix " spiro " which, in turn, is followed by the name of the smaller component. Between the affix " spiro " and the name of each component system is inserted the number denoting the spiro position in the appropriate ring system, these numbers being as low as permitted by any fixed enumeration of the component. The components retain their respective enumerations but numerals for the component mentioned second are primed. Numerals 1 may be omitted when a free choice is available for a component.

Examples:

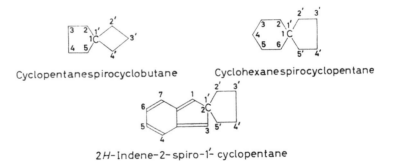

Cyclopentanespirocyclobutane   Cyclohexanespirocyclopentane

2H-Indene-2-spiro-1′- cyclopentane

42.2 (Alternate to **A–41.3**)—Rule **A–41.3** applies also with appropriate different enumeration, where nomenclature is according to Rule **A–42.1**, but the spiro junction has priority for lowest numbers over unsaturation.

Example:

2-Cyclohexenespiro-(2′-cyclopentene)

42.3 (Alternate to **A–41.5**)—The nomenclature of Rule **A–41.5** is applied also to monocyclic components with identical saturation, the spiro union being numbered 1.

Example:

Spirobicyclohexane   but   2-Cyclohexenespiro-(3'-cyclohexene)

42.4 (Alternate to **A–41.6** and **A–41.7**)—Polycyclic compounds containing more than one spiro atom are named in accordance with Rule **A–42.1** starting from the senior* end-component irrespective of whether the components are simple or fused rings.

Examples:

Cyclooctanespirocyclopentane-3'-
spirocyclohexane

Fluorene-9-spiro-1'-cyclohexane-4'-
spiro-1″-indene

Rule A–43.   Radicals

43.1—Radicals derived from spiro hydrocarbons are named according to the principles set forth in Rules **A–11** and **A–24**.

Examples:

Spiro [4.5] deca-1,6-dien-2-yl
(cf. Rules **A–41·3** and **A–11**)
or 2-Cyclohexenespiro-2'-cyclopenten-3'-yl
(cf. Rule **A–42·2**)

* " Seniority " in respect to spiro compounds is based on the principles:
   (*i*) an aggregate is senior to a monocycle;
   (*ii*) of aggregates, the senior is that containing the largest number of individual rings;
   (*iii*) of aggregates containing the same number of individual rings, the senior is that containing the largest ring;
   (*iv*) if aggregates consist of equal numbers of equal rings the senior is the first occurring in the alphabetical list of names.

Spiro [cyclopentane -1,1'- inden]- 2'-yl
(cf. Rules A-41.4 and A-24)

HYDROCARBON RING ASSEMBLIES

Rule A–51. Definition

51.1—Two or more cyclic systems (single rings or fused systems) which are directly joined to each other by double or single bonds are named " ring assemblies " when the number of such direct ring junctions is one less than the number of cyclic systems involved.

Examples:

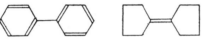

Ring   assemblies

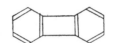

Fused polycyclic system

Rule A–52. Two Identical Ring Systems

52.1—Assemblies of two identical cyclic hydrocarbon systems are named in either of two ways: (*a*) by placing the prefix " bi- " before the name of the corresponding radical, or (*b*) for systems joined by a single bond by placing the prefix " bi- " before the name of the corresponding hydrocarbon.   In each case, the numbering of the assembly is that of the corresponding radical or hydrocarbon, one system being assigned unprimed numbers and the other primed numbers.   The points of attachment are indicated by placing the appropriate locants before the name.

Examples:

1,1'-Bicyclopropyl
or 1,1'-Bicyclopropane

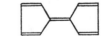

1,1'-Bicyclopentadienylidene
or $\Delta^{1,1'}$ - Bicyclopentadienylidene*
(cf. Footnote to Rule B - 1.2)

52.2—If there is a choice in numbering, unprimed numbers are assigned to the system which has the lower-numbered point of attachment.

*A Greek capital delta (Δ) followed by superscript locants is used to denote the double bond.

Example:

1,2'– Binaphthyl
or 1,2'– Binaphthalene

52.3—If two identical hydrocarbon systems have the same point of attachment and contain substituents at different positions, the locants of these substituents are assigned according to Rule A–2.2; for this purpose an unprimed number is considered lower than the same number when primed. Assemblies of primed and unprimed numbers are arranged in ascending numerical order.

Examples:

2,3,3',4',5'–Pentamethylbiphenyl
(not 2',3,3',4,5 – Pentamethylbiphenyl)

2 – Ethyl – 2'–propylbiphenyl

52.4—The name " biphenyl " is used for the assembly consisting of two benzene rings.

Biphenyl

Rule A–53. Non-identical Ring Systems

53.1—Other hydrocarbon ring assemblies are named by selecting one ring system as the base component and considering the other systems as substituents of the base component. Such substituents are arranged in alphabetical order. The base component is assigned unprimed numbers and the substituents are assigned numbers with primes.*

53.2—The base component is chosen by considering the following characteristics in turn until a decision is reached: **

(*a*) The system containing the larger number of rings.

Examples:

2 – Phenylnaphthalene

4–Cyclooctyl –4'–cyclopentylbiphenyl

*As in Rule A-2.5 the substituents may be numbered using unprimed numbers and enclosed in parentheses if substituted (See examples in Rules C-16.36 and C-203.1).
**For a more complete order of seniority see Subsection C-0.14.

(*b*) The system containing the larger ring.

Examples:

2-(2'-Naphthyl)azulene

1,4-Dicyclopropylbenzene
or *p*-Dicyclopropylbenzene

(*c*) The system in the lowest state of hydrogenation (see also Part .3 of this rule).

Example:

Cyclohexylbenzene

(*d*) The order of ring systems as set forth in the list of Rule A–21.1.

53.3—Compounds covered by Part .2(*c*) of this rule may also be named as hydrogenation products according to Rule A–23.

Example:

1,2,3,3',4,4'-Hexahydro-1,1'- binaphthyl
or 1, 2, 3, 3', 4, 4'-Hexahydro—1, 1'-binaphthalene

Rule A–54.   Three or More Identical Ring Systems

54.1—Unbranched assemblies consisting of three or more identical hydrocarbon ring systems are named by placing an appropriate numerical prefix before the name of the hydrocarbon corresponding to the repetitive unit.   The following numerical prefixes are used:

3. ter-	7. septi-
4. quater-	8. octi-
5. quinque-	9. novi-
6. sexi-	10. deci-

Example:

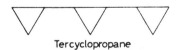

Tercyclopropane

54.2—Unprimed numbers are assigned to one of the terminal systems, the other systems being primed serially.   Points of attachment are assigned the lowest numbers possible.

Examples:

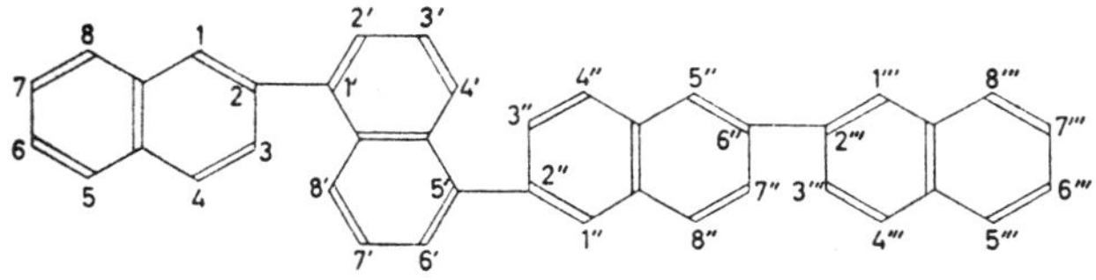

2,1′:5′,2″:6″,2‴-Quaternaphthalene

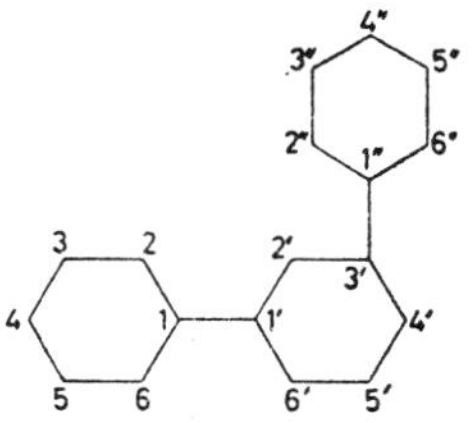

1,1′:3′,1″-Tercyclohexane

54.3—As exceptions, unbranched assemblies consisting of benzene rings are named by using the appropriate prefix with the radical name " phenyl ".

Examples:

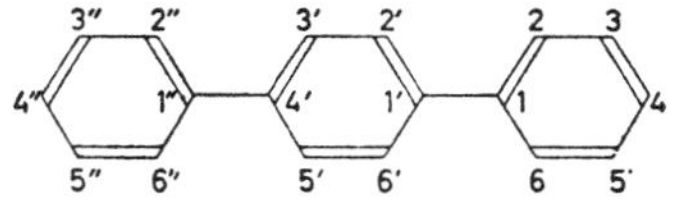

p-Terphenyl
or 1,1′:4′,1″-Terphenyl

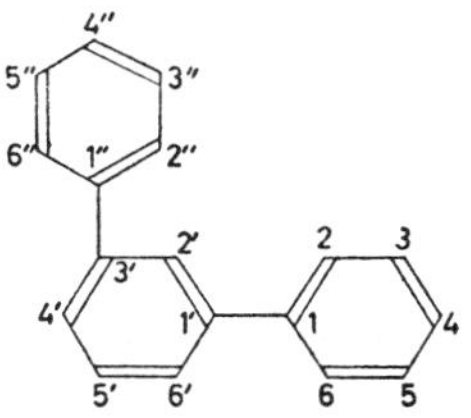

m-Terphenyl
or 1,1′:3′,1″-Terphenyl

Rule A–55. Radicals for Identical Ring Systems (Alternative in part to Rule A–56.1)

55.1—Univalent and multivalent radicals derived from assemblies of identical hydrocarbon ring systems are named by adding " -yl ", " -ylene " or " -diyl ", " -triyl ", *etc.*, to the name of the ring assembly.

Examples:

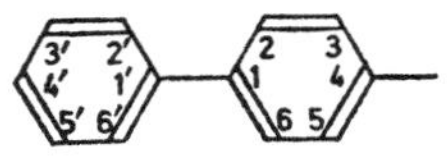

4 – Biphenylyl

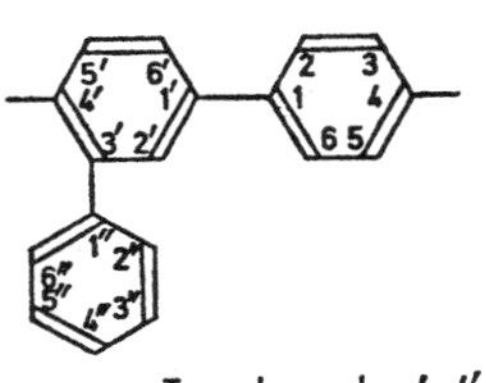

m – Terphenyl—4,4′-ylene
or m – Terphenyl—4,4′-diyl

45

[1,2'-Binaphthalene]-4,5,5'-triyl

Rule A–56. Radicals for Non-benzenoid Ring Systems (Alternative in part to Rule A–55.1)

56.1—Radicals derived from hydrocarbon ring assemblies other than benzene ring assemblies by removal of one or more hydrogen atoms from only one ring are named with that ring as the parent radical, the remaining rings being named as substituents.

Examples:

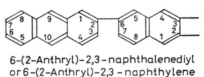

6-(2-Anthryl)-2,3-naphthalenediyl
or 6-(2-Anthryl)-2,3-naphthylene

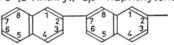

7-(2-Naphthyl)-2-naphthyl

Note: This method is used for assemblies of non-identical systems; also it is sometimes preferable to that of Rule A–55 for assemblies of identical systems when a group to be specified as a suffix or as a separate word is present in a chain attached to the ring assembly.

CYCLIC HYDROCARBONS WITH SIDE CHAINS
(Note: *cf.* Rules A–12 and A–13)

Rule A–61. General Principles

61.1—Hydrocarbons more complex than those envisioned in Rule **A–12**, composed of cyclic nuclei and aliphatic chains, are named according to one of the methods given below. Choice is made so as to provide the name which is the simplest permissible or the most appropriate for the chemical intent.

61.2—When there is no generally recognized trivial name for the hydrocarbon, then (*1*) the radical name denoting the aliphatic chain is prefixed to the name of the cyclic hydrocarbon, or (*2*) the radical name for the cyclic hydrocarbon is prefixed to the name of the aliphatic compound. Choice between these methods is made according to the more appropriate of the following principles: (*a*) the maximum number of substitutions into a single unit of structure; (*b*) treatment of a smaller unit of structure as a substituent into a larger. Numbering of double and triple bonds in chains or non-aromatic rings is assigned according to the principles of Rule A–3; numbering and citation of substituents are effected as described in Rule A–2.

61.3—In accordance with the principle (*a*) of Part .2 of this rule, hydrocarbons containing several chains attached to one cyclic nucleus are generally named as derivatives of the cyclic compound; and compounds containing several side chains and/or cyclic radicals attached to one chain are named as derivatives of the acyclic compound.

Examples:

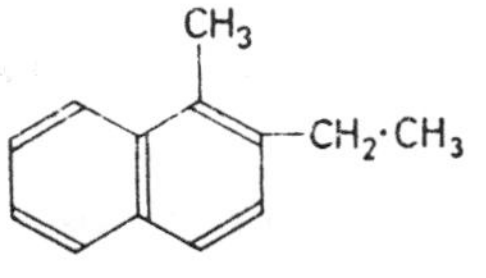

2-Ethyl-1-methylnaphthalene

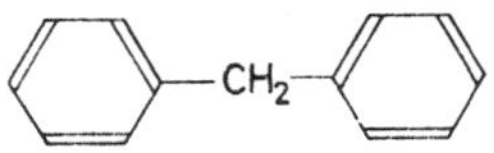

Diphenylmethane

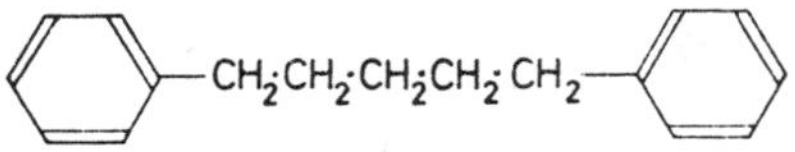

1,5-Diphenylpentane

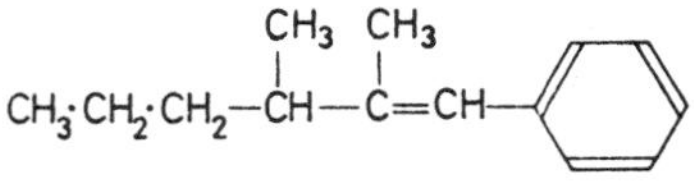

2,3-Dimethyl-1-phenyl-1-hexene

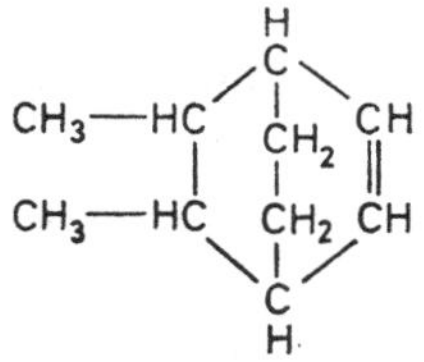

5,6-Dimethylbicyclo[2.2.2]oct-2-ene

61.4—In accordance with principle (*b*) of Part .2 of this rule, a hydrocarbon containing a small cyclic nucleus attached to a long chain is generally named as a derivative of the acyclic hydrocarbon; and a hydrocarbon containing a small group attached to a large cyclic nucleus is generally named as a derivative of the cyclic hydrocarbon.

Examples:

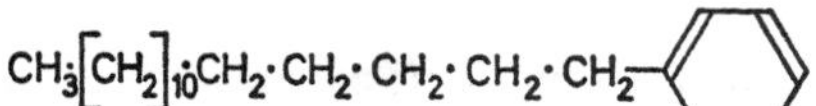

1-Phenylhexadecane

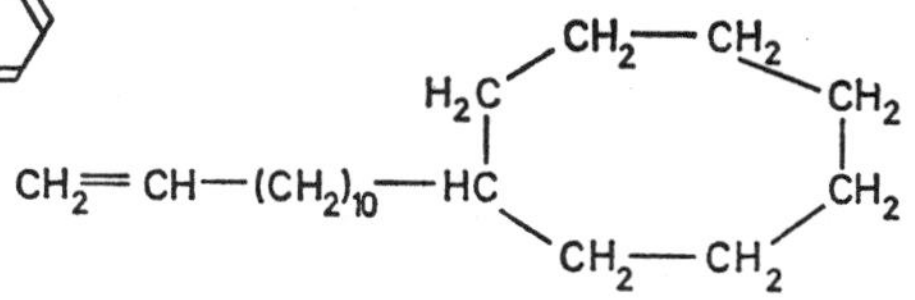

12-Cyclooctyl-1-dodecene

47

$$CH_3$$
$$|$$
$$CH_3 \cdot CH \cdot CH \cdot CH_2 \cdot CH_2 \cdot CH_3$$
$$1 2 3 4 5$$

9-(1,2-Dimethylpentyl)anthracene

$$CH_2 \cdot CH_2 \cdot CH_2$$
$$1 2 3$$

7-(3-Phenylpropyl)benz[*a*]anthracene

61.5—Recognized trivial names for composite radicals are used if they lead to simplifications in naming.

Examples:

$$CH_2$$

1-Benzylnaphthalene

$$CH_2 - \!\!\!-\!\!\!- CH_3$$
$$CH_2$$
$$CH_2$$
$$-CH_2 \cdot CH_2 \cdot CH_2 - \!\!\!-\!\!\!- CH_3$$
$$CH_2 \cdot CH_2 \cdot CH_2 - \!\!\!-\!\!\!- CH_3$$

1,2,4-Tris(3-*p*-tolylpropyl)benzene

61.6—The following are among trivial names retained for cyclic hydrocarbons with side chains: "fulvene" (for methylenecyclopentadiene) and "stilbene" (for 1,2-diphenylethylene) (see also Rule **A–12.1**).

$$^6CH_2$$
$$\|$$
$$C$$

Fulvene

$$-CH = CH-$$
$$\alpha \beta$$

Stilbene

TERPENE HYDROCARBONS

Owing to long-established custom, terpenes are given exceptional treatment in these rules.

Rule A–71. Acyclic Terpenes

71.1—The acyclic terpene hydrocarbons are named in a manner similar to that used for other unsaturated acyclic hydrocarbons when compounds with known structures are involved.

Example:

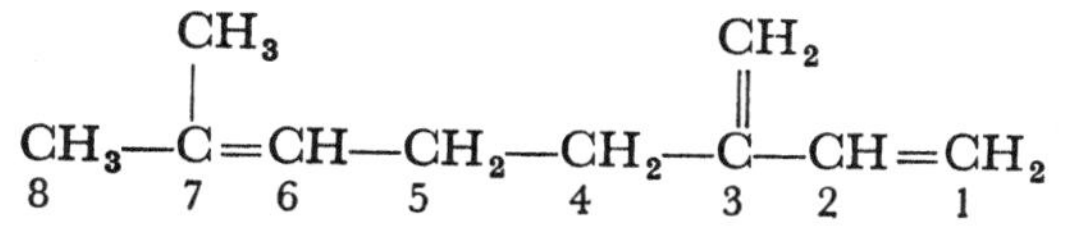

$$\underset{8}{CH_3}-\underset{7}{\overset{\overset{\displaystyle CH_3}{|}}{C}}=\underset{6}{CH}-\underset{5}{CH_2}-\underset{4}{CH_2}-\underset{3}{\overset{\overset{\displaystyle CH_2}{||}}{C}}-\underset{2}{CH}=\underset{1}{CH_2}$$

7-Methyl-3-methylene-1,6-octadiene

Rule A–72. Cyclic Terpenes

72.1—The following structural types with their special names and special systems of numbering are used as the basis for the specialized nomenclature of monocyclic and bicyclic terpene hydrocarbons. The name " bornane " replaces camphane and bornylane; " norbornane " replaces norcamphane and norbornylane.*

Fundamental terpene types:

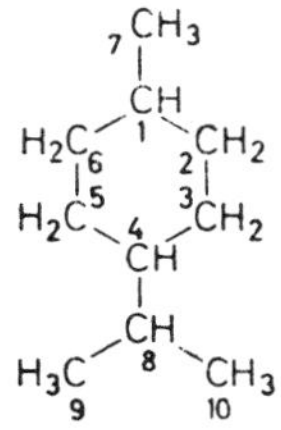

I
Menthane (p - form)

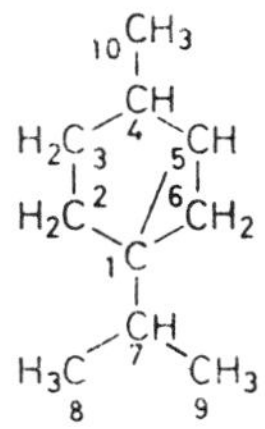

II
Thujane

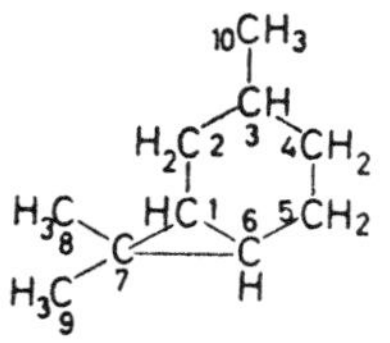

III
Carane

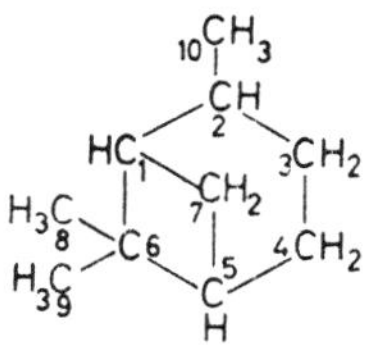

IV
Pinane

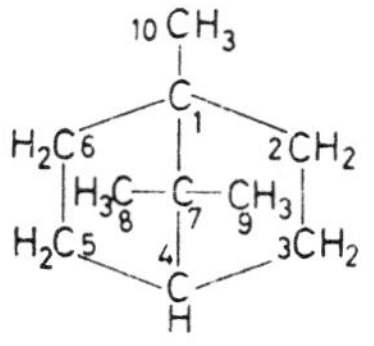

V
Bornane

Nor-structures:

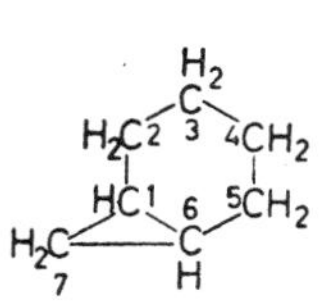

VI
Norcarane *

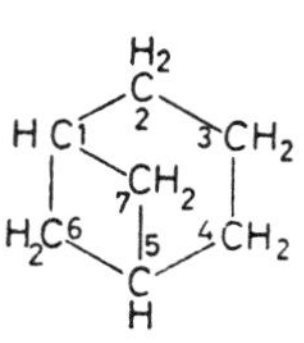

VII
Norpinane *

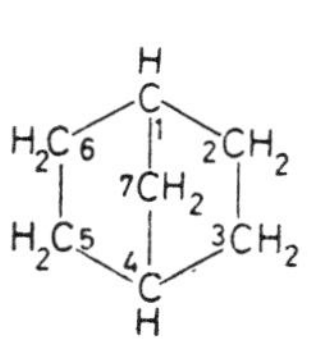

VIII
Norbornane *

*These names have been superseded (cf. Rule F-4.2).

Rule A-73. Monocyclic Terpenes

73.1—Menthane Type: Monocyclic terpene hydrocarbons of this type (*ortho-*, *meta-*, and *para-*isomers) are named menthane, menthene, menthadiene, *etc.*, and are given the fixed numbering of menthane (Formula I). Such compounds substituted by additional alkyl groups are named in accordance with Rules **A-11** and **A-61**.

Examples:

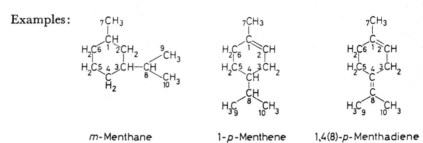

m-Menthane   1-*p*-Menthene   1,4(8)-*p*-Menthadiene

73.2—Tetramethylcyclohexane Type: Monocyclic terpene hydrocarbons of this type are named systematically as derivatives of cyclohexane, cyclohexene, and cyclohexadiene (see Rule **A-11**).

Examples:

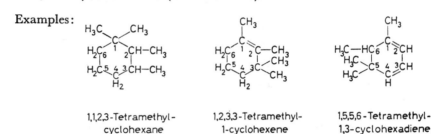

1,1,2,3-Tetramethyl-cyclohexane   1,2,3,3-Tetramethyl-1-cyclohexene   1,5,5,6-Tetramethyl-1,3-cyclohexadiene

Rule A-74. Bicyclic Terpenes

74.1—Bicyclic terpene hydrocarbons having the skeleton of Formula II or this skeleton and additional side chains except methyl or isopropyl (or methylene if one methylene group is already present) are named as thujane, thujene, thujadiene, *etc.*, and are given the fixed numbering shown for thujane (Formula II). Other hydrocarbons containing the thujane ring-skeleton are named from bicyclo[3.1.0]hexane and are given systematic bicyclo numbering (*cf.* Rule **A-31**).

Examples:

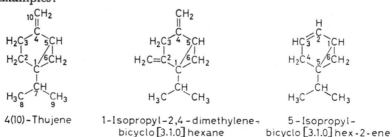

4(10)-Thujene   1-Isopropyl-2,4-dimethylene-bicyclo[3.1.0]hexane   5-Isopropyl-bicyclo[3.1.0]hex-2-ene

50

74.2—Bicyclic terpene hydrocarbons having the skeleton of Formula III, IV, or V and additional side chains except methyl (or methylene if one methylene group is already present) are named, respectively, as carane, carene, caradiene, *etc.*; pinane, pinene, pinadiene, *etc.*; bornane, bornene, bornadiene, *etc.* They are given, respectively, the fixed numbering shown for carane (Formula III), pinane (Formula IV), and bornane (Formula V). Other hydrocarbons containing the ring-skeleton of carane, pinane, or bornane are named, respectively, from norcarane (Formula VI), norpinane (Formula VII), or norbornane (Formula VIII). These names are preferred to those from bicyclo[4.1.0]heptane, bicyclo[3.1.1]heptane, or bicyclo-[2.2.1]heptane. The nor-names[*]are given systematic bicyclo numbering (*cf.* Rule A–31).

Examples:

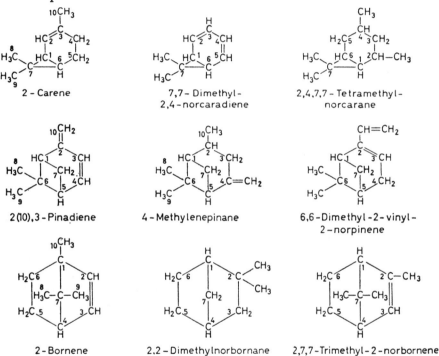

2 - Carene	7,7– Dimethyl– 2,4–norcaradiene	2,4,7,7 – Tetramethyl– norcarane

2 (10),3 – Pinadiene	4 – Methylenepinane	6,6 –Dimethyl –2– vinyl– 2 –norpinene

2 - Bornene	2,2 – Dimethylnorbornane	2,7,7 –Trimethyl–2 –norbornene

74.3—The name "camphene" is retained for the unsubstituted compound 2,2-dimethyl-3-methylenenorbornane.

Camphene

*These names have been superseded (cf. Rule F-4.2).

Rule A–75. Terpene Radicals

75.1—Simple acyclic hydrocarbon terpene radicals are named and numbered according to Rule **A–3.5**. The trivial names geranyl, neryl, linalyl and phytyl [for (E)-$(7R, 11R)$-3,7,11,15-tetramethyl-2-hexadecenyl] are retained for the unsubstituted radicals.

75.2—Radicals derived from menthane, pinane, thujane, carane, bornane, norcarane, norpinane, and norbornane are named in accordance with the principles set forth in Rules **A–1.2** and **A–11.4** except that the saturated radicals of pinane are named pinanyl, pinanylene, and pinanylidene. The numbering of the hydrocarbon is retained and the point or points of attachment, whether in the ring or side chain, are given numbers as low as is consistent with the fixed numbering of the hydrocarbon.

Examples:

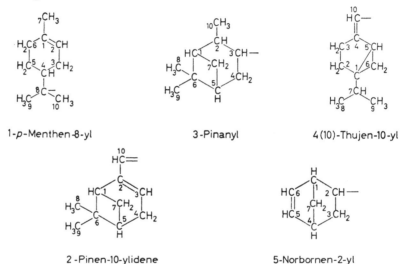

1-*p*-Menthen-8-yl   3-Pinanyl   4(10)-Thujen-10-yl

2-Pinen-10-ylidene   5-Norbornen-2-yl

75.3—Radicals not named in Rules **A–75.1** and **A–75.2** are named as described in Rules **A–11** and **A–31.4**.

B. FUNDAMENTAL HETEROCYCLIC SYSTEMS

SPECIALIST HETEROCYCLIC NOMENCLATURE

Rule B–1. Extension of Hantzsch–Widman System

1.1—Monocyclic compounds containing one or more hetero atoms in a three- to ten-membered ring are named by combining the appropriate prefix or prefixes from Table I (eliding " a " where necessary) with a stem from Table II. The state of hydrogenation is indicated either in the stem, as shown in Table II, or by the prefixes " dihydro- ", " tetrahydro- ", *etc.*, according to Rule **B–1.2**.

TABLE I. In decreasing order of priority

Element	Valence	Prefix	Element	Valence	Prefix
Oxygen	II	Oxa	Antimony	III	Stiba*
Sulfur	II	Thia	Bismuth	III	Bisma
Selenium	II	Selena	Silicon	IV	Sila
Tellurium	II	Tellura	Germanium	IV	Germa
Nitrogen	III	Aza	Tin	IV	Stanna
Phosphorus	III	Phospha*	Lead	IV	Plumba
Arsenic	III	Arsa*	Boron	III	Bora
			Mercury	II	Mercura

* When immediately followed by " -in " or " -ine ", " phospha- " should be replaced by " phosphor- ", " arsa- " should be replaced by " arsen- " and " stiba- " should be replaced by " antimon- ". In addition, the saturated six-membered rings corresponding to phosphorin and arsenin are named phosphorinane and arsenane. Further exceptions: borin is replaced by borinane.

TABLE II

No. of members in the ring	Rings containing nitrogen Unsaturation (a)	Saturation	Rings containing no nitrogen Unsaturation (a)	Saturation
3	-irine	-iridine	-irene	-irane (e)
4	-ete	-etidine	-ete	-etane
5	-ole	-olidine	-ole	-olane
6	-ine (b)	(c)	-in (b)	-ane (d)
7	-epine	(c)	-epin	-epane
8	-ocine	(c)	-ocin	-ocane
9	-onine	(c)	-onin	-onane
10(f)	-ecine	(c)	-ecin	-ecane

(a) Corresponding to the maximum number of non-cumulative double bonds, the hetero elements having the normal valences shown in Table I.

(b) For phosphorus, arsenic, antimony and boron, see the special provisions of Table I.

(c) Expressed by prefixing "perhydro" to the name of the corresponding unsaturated compound.

(d) Not applicable to silicon, germanium, tin and lead. In this case, " perhydro- " is prefixed to the name of the corresponding unsaturated compound.

(e) The syllables denoting the size of rings containing 3, 4 or 7-10 members are derived as follows: " ir " from *tri*, " et " from *tetra*, " ep " from *hepta*, " oc " from *octa*, " on " from *nona*, and " ec " from *deca*.

(f) Rings with more than ten members are named by replacement nomenclature (cf. Rule B-4).

+ It is necessary to elide the final "a" of the prefix when followed immediately by a vowel, e.g. ox(a)azole.

Examples:

Oxirane   Aziridine   2H-Azepine

1.2—Heterocyclic systems whose unsaturation is less than the one corresponding to the maximum number of non-cumulative double bonds are named by using the prefixes " dihydro- ", " tetrahydro- ", *etc.*

In the case of 4- and 5-membered rings, a special termination is used for the structures containing one double bond, when there can be more than one non-cumulative double bond.

No. of members of the partly saturated rings	*Rings containing nitrogen*	*Rings containing no nitrogen*
4	-etine	-etene
5	-oline	-olene

Examples:

Δ^3-1,2-Azarsetine*   3-Silolene

1.3—Multiplicity of the same hetero atom is indicated by a prefix " di- ", ' tri- ", *etc.*, placed before the appropriate " a " term (Table I).

Example:

1,3,5-Triazine

1.4—If two or more kinds of " a " terms occur in the same name, their order of citation is in order of their appearance in Table I of Rule B-1.1.**

Examples:

1,2-Oxathiolane   1,3-Thiazole

*As exceptions, Greek capital delta ($\triangle$), followed by superscript locant(s), is used to denote a double bond in a compound named according to Rule B–1.2 if its name is preceded by locants for hetero atoms; and also to denote a double bond uniting components in an assembly of rings (cf. Examples to Rules A–52.1 and C–71.1) or in conjunctive names (cf. Rule C–55.1).

**This Rule is extended by Rule D-1.60.

1.51—The position of a single hetero atom determines the numbering in a monocyclic compound.

Example:

Azocine

1.52—When the same hetero atom occurs more than once in a ring, the numbering is chosen to give the lowest locants to the hetero atoms.

Example:

1,2,4 - Triazine

1.53—When hetero atoms of different kinds are present, the locant 1 is given to a hetero atom which is as high as possible in Table I. The numbering is then chosen to give the lowest locants to the hetero atoms.

Examples:

6H-1,2,5-Thiadiazine
(not: 2,1,4-Thiadiazine)
(not: 1,3,6-Thiadiazine)

2H,6H-1,5,2 - Dithiazine
(not:1,3,4-Dithiazine)
(not:1,3,6-Dithiazine)
(not:1,5,4-Dithiazine)

The numbering must begin with the sulfur atom. This condition eliminates 2,1,4-thiadiazine. Then the nitrogen atoms receive the lowest possible locant, which eliminates 1,3,6-thiadiazine.

The numbering has to begin with a sulfur atom. The choice of this atom is determined by the set of locants which can be attributed to the remaining hetero atoms of any kind.

As the set 1,2,5 is lower than 1,3,4 or 1,3,6 or 1,5,4 in the usual sense, the name is 1,5,2-dithiazine.

Rule B-2.　Trivial and Semi-trivial Names*

2.11—The following trivial and semi-trivial names constitute a partial list of such names which are retained for the compound and as a basis of

*For a more complete list see Table IV of Section D.

fusion names. The names of the radicals shown are formed according to Rule **B–5**.

Parent Compound		Radical Name

(1)

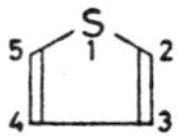

Thiophene

Thienyl
(2- shown)

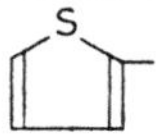

(2)

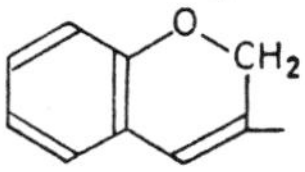

(2) and (3) in earlier editions have been deleted.

(3)

(4)

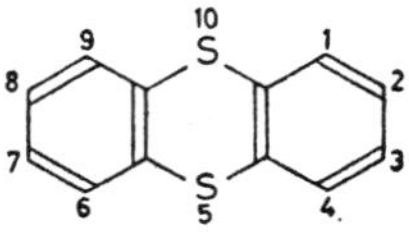

Thianthrene

Thianthrenyl
(2- shown)

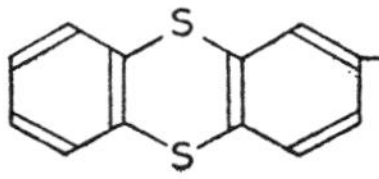

(5)

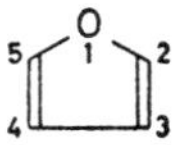

Furan

Furyl
(3- shown)

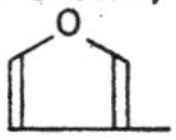

(6)

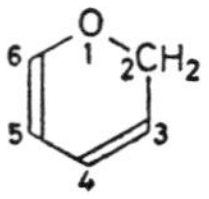

Pyran
(2*H*- shown)

Pyranyl
(2*H*-Pyran-3-yl shown)

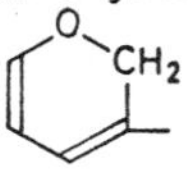

(7)

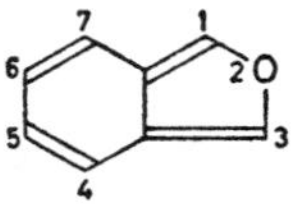

Isobenzofuran

Isobenzofuranyl
(1- shown)

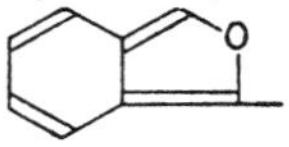

(8)

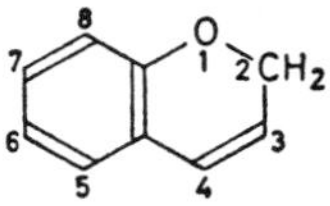

Chromene
(2*H*- shown)

Chromenyl
(2*H*-Chromen-3-yl shown)

Parent Compound   Radical Name

(9) Xanthene*

Xanthenyl *
(2- shown)

*Denotes exceptions to systematic numbering.

(10) Phenoxathiin

Phenoxathiinyl
(2- shown)

(11) 2H-Pyrrole

2H-Pyrrolyl
(2H-Pyrrol-3-yl shown)

.(12) Pyrrole

Pyrrolyl
(3- shown)

(13) Imidazole

Imidazolyl
(2- shown)

(14) Pyrazole

Pyrazolyl
(1- shown)

(14a) Isothiazole

Isothiazolyl
(3- shown)

(14b) Isoxazole

Isoxazolyl
(3- shown)

(15) Pyridine

Pyridyl
(3- shown)

(16) Pyrazine

Pyrazinyl

Parent Compound | Radical Name

(17) Pyrimidine — Pyrimidinyl (2- shown)

(18) Pyridazine — Pyridazinyl (3- shown)

(19) Indolizine — Indolizinyl (2- shown)

(20) **Isoindole** — Isoindolyl (2- shown)

(21) **3H–Indole** — 3H–Indolyl (3H-Indol-2-yl shown)

(22) Indole — Indolyl (1- shown)

(23) 1H–Indazole — Indazolyl (1H-Indazol-3-yl shown)

(24) Purine* — Purinyl* (8- shown)

* Denotes exceptions to systematic numbering.

Parent Compound	Radical Name

(25)

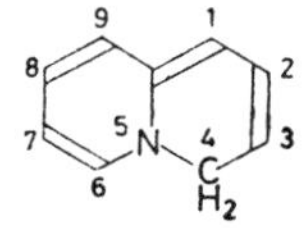

4H-Quinolizine

4H-Quinolizinyl
(4*H*-Quinolizin-2-yl shown)

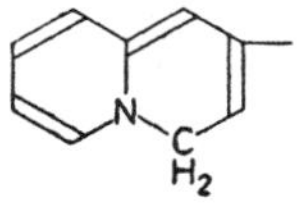

(26)

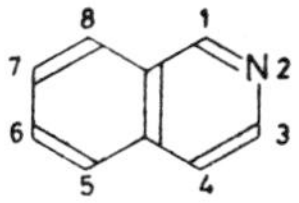

Isoquinoline

Isoquinolyl
(3- shown)

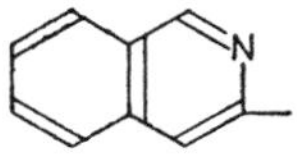

(27)

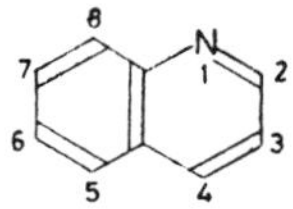

Quinoline

Quinolyl
(2- shown)

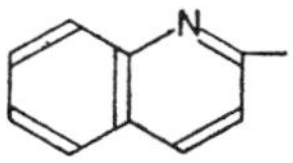

(28)

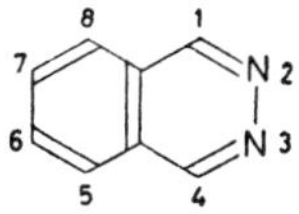

Phthalazine

Phthalazinyl
(1- shown)

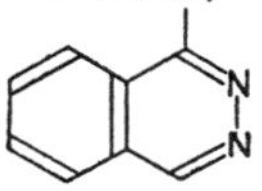

(29)

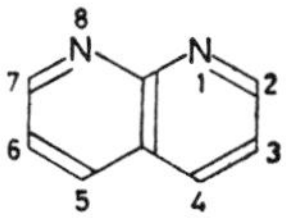

Naphthyridine
(1,8- shown)

Naphthyridinyl
(1,8-Naphthyridin-2-yl shown)

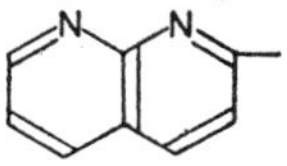

(30)

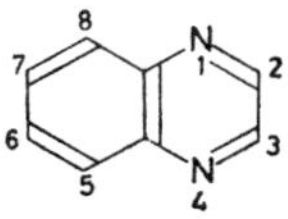

Quinoxaline

Quinoxalinyl
(2- shown)

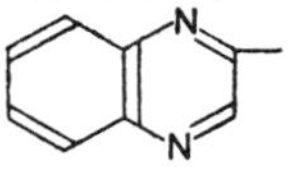

(31)

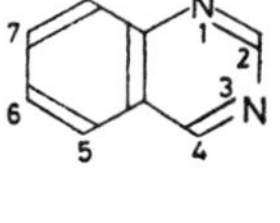

Quinazoline

Quinazolinyl
(2- shown)

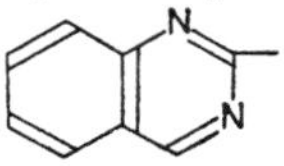

(32)

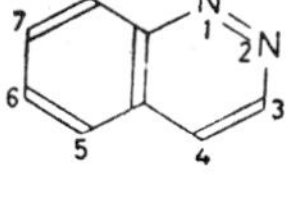

Cinnoline

Cinnolinyl
(3- shown)

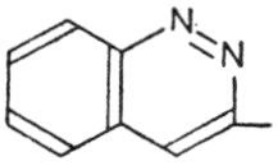

59

Parent Compound   Radical Name

(33)   Pteridine   Pteridinyl
(2- shown)

(34)   4aH-Carbazole*   4aH-Carbazolyl*
(4aH-Carbazol-2-yl shown)

(35)   Carbazole*   Carbazolyl*
(2- shown)

(36)   β-Carboline
(See page 466, paragraph 2)   β-Carbolinyl
(β-Carbolin-3-yl shown)

(37)   Phenanthridine   Phenanthridinyl
(3- shown)

(38)   Acridine*   Acridinyl*
(2- shown)

(39)   Perimidine   Perimidinyl
(2- shown)

* Denotes exceptions to systematic numbering.

Parent Compound   Radical Name

(40)

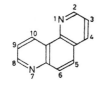

Phenanthroline
(1,7- shown)

Phenanthrolinyl
(1,7-Phenanthrolin-3-yl shown)

(41)

Phenazine

Phenazinyl
(1- shown)

(42)

Phenarsazine

Phenarsazinyl
(2- shown)

(44)

Phenothiazine

Phenothiazinyl
(2- shown)

(46)

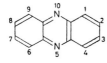

Furazan
(See page 466,
paragraph 2)

Furazanyl
(3- shown)

(47)

Phenoxazine

Phenoxazinyl
(2- shown)

B–2.12—The following trivial and semi-trivial names are retained but are not recommended for use in fusion names. The names of the radicals shown are formed according to Rule **B–5**.

Parent Compound		Radical Name
(1)	Isochroman	Isochromanyl (3- shown)
(2)	Chroman	Chromanyl (7- shown)
(3)	Pyrrolidine	Pyrrolidinyl (2- shown)
(4)	Pyrroline (2- shown*)	Pyrrolinyl (2-Pyrrolin-3-yl*shown)
(5)	Imidazolidine	Imidazolidinyl (2- shown)
(6)	Imidazoline (2- shown*)	Imidazolinyl (2-Imidazolin-4-yl*shown)
(7)	Pyrazolidine	Pyrazolidinyl (2- shown)

* The " 2- " denotes the position of the double bond.

Parent Compound   Radical Name

(8) Pyrazoline
(3- shown*)

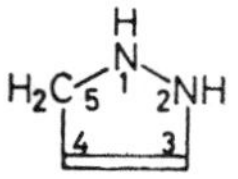

Pyrazolinyl
(3-Pyrazolin-2-yl*shown)

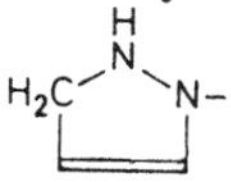

(9) Piperidine

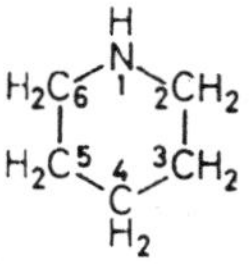

Piperidyl †
(2- shown)

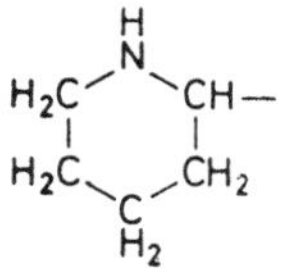

(10) Piperazine

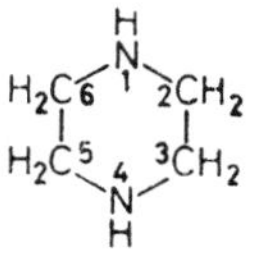

Piperazinyl
(1- shown)

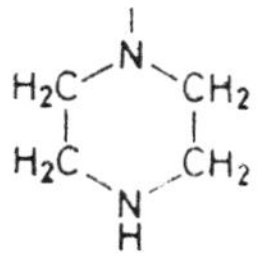

(11) Indoline

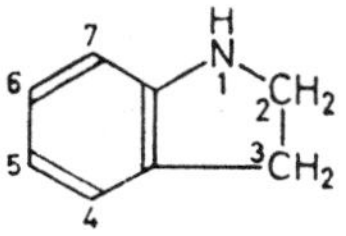

Indolinyl
(1- shown)

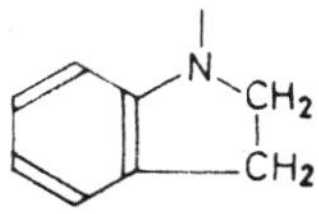

(12) Isoindoline

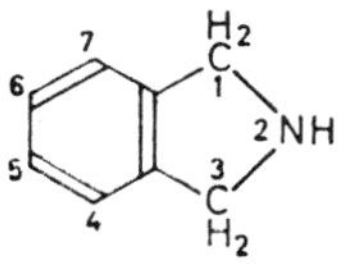

Isoindolinyl
(1- shown)

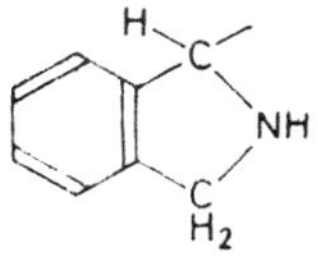

(13) Quinuclidine

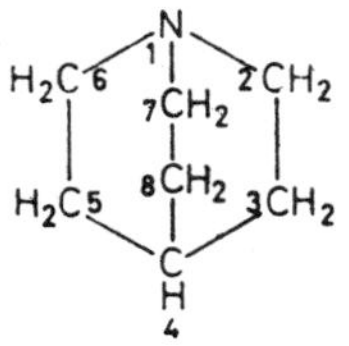

Quinuclidinyl
(2- shown)

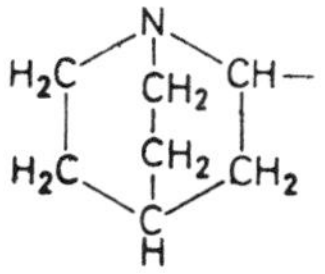

(14) Morpholine

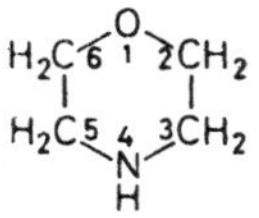

Morpholinyl ‡
(3- shown)

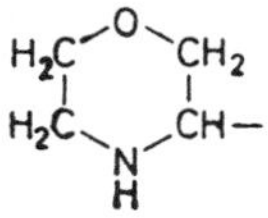

* The " 3- " denotes the position of the double bond.
† For 1-Piperidyl use piperidino.
‡ For 4-Morpholinyl use morpholino.

63

Rule B–3. Fused Heterocyclic Systems

3.1—" *Ortho*-fused " and " *ortho*- and *peri*-fused " ring compounds containing hetero atoms are named according to the fusion principle described in Rule A–21 for hydrocarbons. The components are named according to Rules A–21, B–1 and B–2. When the name of a component in a fusion name contains locants (numerals or letters) that do not apply also to the numbering of the fused system, these locants are placed in square brackets (as are also the locants for fusion positions required by Rule A–21.5). The base component should be a heterocyclic system. If there is a choice, the base component should be, by order of preference:

(*a*) A nitrogen-containing component.

Example:

Benz[*h*]isoquinoline

not Pyrido[3,4-*a*]naphthalene

(*b*) A component containing a heteroatom (in the absence of nitrogen) as high as possible in Table I.

Example:

Thieno[2,3-*b*]furan
not Furo[2,3-*b*]thiophene

(*c*) A component containing the greatest number of rings.

Example:

7*H*-Pyrazino[2,3-*c*]carbazole
not 7*H*-Indolo[3,2-*f*]quinoxaline

(*d*) A component containing the largest possible individual ring.

Example:

2*H*-Furo[3,2-*b*]pyran
not 2*H*-Pyrano[3,2-*b*]furan

(*e*) A component containing the greatest number of hetero atoms of any kind.

Example:

5*H*–Pyrido[2,3–*d*]–*o*–oxazine
not *o*–Oxazino[4,5–*b*]pyridine

(*f*) A component containing the greatest variety of hetero atoms.

Examples:

1*H*-Pyrazolo[4,3-*d*]oxazole
not 1*H*-Oxazolo[5,4-*c*]pyrazole

4*H*-Imidazo[4,5-*d*]thiazole
not 4*H*-Thiazolo[4,5-*d*]imidazole

(*g*) A component containing the greatest number of hetero atoms first listed in Table I.

Example:

Selenazolo[5,4-*f*]benzothiazole *
not Thiazolo[5,4-*f*]benzoselenazole

(*h*) If there is a choice between components of the same size containing the same number and kind of hetero atoms choose as the base component that one with the lower numbers for the hetero atoms before fusion.

Example:

Pyrazino[2,3-*d*]pyridazine

3.2—If a position of fusion is occupied by a hetero atom, the names of the component rings to be fused are so chosen as both to contain the hetero atom.

* In this example the hetero atom first listed in Table I is sulfur and the greatest number of sulfur atoms in a ring is one.

Example:

Imidazo [2,1-*b*] thiazole

3.3—The following contracted fusion prefixes may be used: furo, imidazo, isoquino, pyrido, pyrimido, quino and thieno.

Examples:

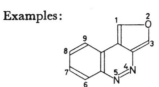

Furo [3,4-*c*]cinnoline

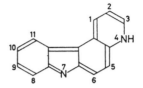

4*H*-Pyrido[2,3-*c*]carbazole

3.4—In peripheral numbering of the complete fused systems, the ring system is oriented and numbered according to the principles of Rule **A–22**. When there is a choice of orientations, it is made in the following sequence in order to:

(*a*) Give low numbers to hetero atoms, thus:

Benzo[*b*]furan   Cyclopenta[*b*]pyran   4*H*-[1,3]Oxathiolo[5,4-*b*]pyrrole
                                                      (note:1,3,4 lower than 1,3,6)

(*b*) Give low numbers to hetero atoms in order of Table I, thus:

Thieno [2,3-*b*] furan

(*c*) Allow carbon atoms common to two or more rings to follow the lowest possible numbers (see Rules **A–22.2** and **A–22.3**). [A hetero atom common to two rings is numbered according to Rule **B–3.4(*e*)**], thus:

not   or

Imidazo[1,2-b][1,2,4]triazine*

In a compound name for a fusion prefix (*i.e.*, when more than one pair of square brackets is required), the points of fusion in the compound prefix

*In Hantzsch-Widman names the locants are inseparable from the name.

are indicated by the use of unprimed and primed numbers, the unprimed numbers being assigned to the ring attached directly to the base component, thus:

Pyrido[1′,2′:1,2]imidazo–[4,5-b]quinoxaline

not

or

or

(d) Give hydrogen atoms lowest numbers possible, thus:

4H-1,3-Dioxolo[4,5-d]imidazole

(e) The ring is numbered as for hydrocarbons but numbers are given to all hetero atoms even when common to two or more rings. Interior hetero atoms are numbered last following the shortest path from the highest previous number.

3.5—As exceptions, two-ring systems in which a benzene ring is fused to a hetero ring may be named by prefixing numbers indicating the positions of the hetero atoms to benzo followed by the name of the heterocyclic component. Numbering is assigned by the principles set forth in Rule B–3.4 (a), (b), and (d). The names provided by this rule may also be used for components of more complex fused systems.

Examples:

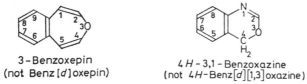

3-Benzoxepin
(not Benz[d]oxepin)

4H-3,1-Benzoxazine
(not 4H-Benz[d][1,3]oxazine)

1H-Pyrrolo[1,2-b][2]benzazepine

(not 1H - Benzo[e]pyrrolo[1,2-a]azepine)

Rule B–4. Replacement Nomenclature (also known as "a" Nomenclature)*

4.1—Names of monocyclic hetero compounds may be formed by prefixing " a " terms (see Table I of Rule **B–1.1**), preceded by their locants, to the name of the corresponding hydrocarbon†. Numbering is assigned so as to give locant 1 to the hetero atom listed earlier in Table I, next to hetero atoms as a complete set, and then to hetero atoms in order of Table I. If a further choice is possible, numbering is assigned according to Rule C-0.15.

Examples:

Sila-2,4-cyclopentadiene

Sila-1,3-cyclopentadiene

Silabenzene

1-Thia-4-aza-2,6-disilacyclohexane

4.2—Fused heterocyclic systems may be named by prefixing " a " terms preceded by their locants, to the name of the corresponding hydrocarbon. The numbering of the corresponding hydrocarbon is retained, irrespective of the position of the hetero atoms; where there is a choice, low numbers are assigned in the following order: first to hetero atoms as a complete set, next to hetero atoms in order of Table I, and then to multiple bonds in the

---

*The Stelzner Method is abandoned.

†The " corresponding hydrocarbon " is obtained from the heterocyclic compound by formally replacing each hetero atom with $>CH_2$, $>CH$, or $—C—$ in accord with the valence 2, 3 or 4 of the hetero atom replaced.

heterocyclic compound according to the principles of Rule **A–11.3**. These principles are applied in one of two ways, as follows:

(*a*) When the corresponding hydrocarbon does not contain the maximum number of non-cumulative double bonds and can be named without the use of hydro prefixes, as for indan, then the hydrocarbon is named in that state of hydrogenation.
Example:

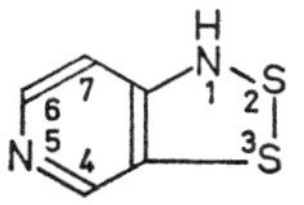

2,3-Dithia-1,5-diazaindan

(*b*) When the two conditions of paragraph (*a*) are not fulfilled, positions in the skeleton of the corresponding hydrocarbon that are occupied by hetero atoms are denoted by " a " prefixes, and the parent heterocyclic compound is considered to be that which contains the maximum number of conjugated or isolated* double bonds, but the corresponding hydrocarbon is named in the form in which it contains the maximum number of non-cumulative double bonds. Hydrogen additional to that present in the parent heterocyclic compound is named by hydro prefixes and/or as *H* in front of the " a " terms.

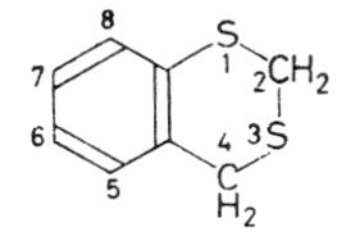

4*H*-1,3- Dithianaphthalene

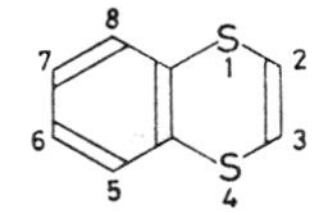

1,4 - Dithianaphthalene

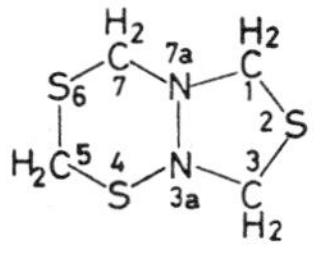

2,4,6 -Trithia -3a,7a- diazaindene

* Isolated double bonds are those which are neither conjugated nor cumulative as in

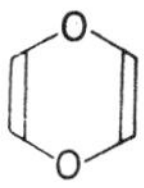

or the B ring of

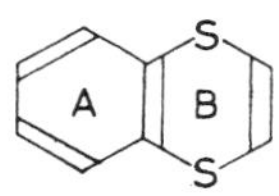

1*H*-2-Oxapyrene

2,7,9-Triazaphenanthrene

4.3—In fusion names, the " a " terms precede the completed name of the parent hydrocarbon. Prefixes denoting ordinary substitution precede the " a " terms.

Example:

3,4-Dimethyl-5-azabenz [*a*] anthracene

Rule B–5.   Radicals

5.11—Univalent radicals derived from heterocyclic compounds by removal of hydrogen from a ring are in principle named by adding " yl " to the names of the parent compounds (with elision of final " e ", if present).

Examples:

Indolyl	from	indole
Pyrrolinyl	from	pyrroline
Triazolyl	from	triazole
Triazinyl	from	triazine

(For further examples see Rule **B–2.11**.)

The following exceptions are retained: furyl, pyridyl, piperidyl, quinolyl, isoquinolyl and thienyl (from thiophene) (see also Rule **B–2.12**). Also retained are furfuryl (for 2-furylmethyl), furfurylidene (for 2-furylmethylene), furfurylidyne (for 2-furylmethylidyne), thenyl (for thienylmethyl), thenylidene (for thienylmethylene) and thenylidyne (for thienylmethylidyne).

As exceptions, the names " piperidino " and " morpholino " are preferred to " 1-piperidyl " and " 4-morpholinyl ".

5.12—Bivalent radicals derived from univalent heterocyclic radicals whose names end in " -yl " by removal of one hydrogen atom from the atom with the free valence are named by adding " -idene " to the name of the corresponding univalent radical.

Example:

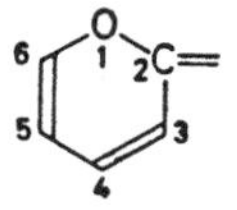

2*H*-Pyran-2-ylidene

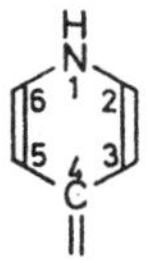

4 (1*H*)-Pyridylidene
or 1,4 – Dihydro – 4 – pyridylidene

5.13—Multivalent radicals derived from heterocyclic compounds by removal of two or more hydrogen atoms from different atoms in the ring are named by adding " -diyl ", " -triyl ", *etc.*, to the name of the ring system.

Example:

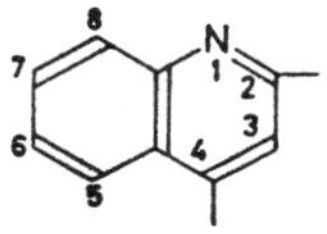

2,4-Quinolinediyl

5.21—The use of " a " terms (Rule B–4) does not affect the formation of radical names. Such names are strictly analogous to those of the hydrocarbon analogs except that the " a " terms establish enumeration in whole or in part.

Examples:

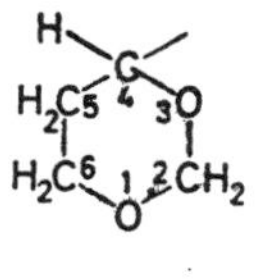

1,3-Dioxa-4-cyclohexyl

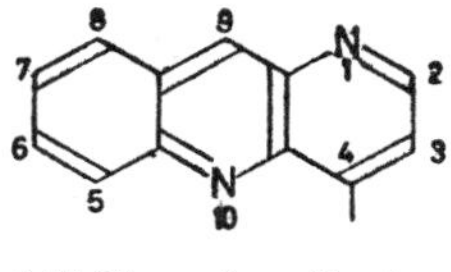

1,10-Diaza-4-anthryl

Rule B–6.   Cationic Hetero Atoms

6.1—According to the " a " nomenclature, heterocyclic compounds containing cationic hetero atoms are named in conformity with the preceding rules by replacing " oxa- ", " thia- ", " aza- ", *etc.*, by " oxonia- ", " thionia- ", " azonia- ", *etc.*, the anion being designated in the usual way. Cationic hetero atoms follow immediately after the corresponding noncationic atoms: oxonia follows oxa, thionia follows thia, azonia follows aza, etc. (cf. Table I of Rule B–1.1). (cf. Rule C-82 and C-83).

Examples:

1-Oxoniaanthracene   chloride

4a-Azoniaanthracene   chloride

1-Thioniabicyclo [2.2.1] heptane   chloride

1-Methyl -1-oxoniacyclohexane   chloride

HETEROCYCLIC SPIRO COMPOUNDS

Rule B–10.   Compounds: Method 1

10.1—Heterocyclic spiro compounds containing single-ring units only may be named by prefixing " a " terms (see Table I, Rule B–1.1) to the names of the spiro hydrocarbons formed according to Rules A–41.1, A–41.2, A–41.3 and A–41.6. The numbering of the spiro hydrocarbon is retained and the hetero atoms as in Rule B-4 are given as low numbers as are consistent with the fixed numbering of the ring. When there is a choice, hetero atoms are given lower numbers than double bonds.

Examples:

1-Oxaspiro [4.5]decane   6,8 – Diazoniadispiro [5.1.6.2] hexadecane
                                            dichloride

10.2—If at least one component of a mono- or poly-spiro compound is a fused polycyclic system, the spiro compound is named according to Rule A–41.4 or A–41.7, giving the spiro atom as low a number as possible consistent with the fixed numberings of the component systems.

Examples:

3,3'-Spirobi[3H-indole]   Spiro[piperidine–4,9'–xanthene]

Rule B–11.   Compounds: Method 2

11.1—Heterocyclic spiro compounds are named according to Rule **A–42**, the following criteria being applied where necessary: (*a*) spiro atoms have numbers as low as consistent with the numbering of the individual component systems; (*b*) heterocyclic components have priority over homocyclic components of the same size; (*c*) priority of heterocyclic components is decided according to Rule **B–3**.   Parentheses are used where necessary for clarity in complex expressions.

Examples:

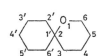

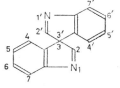

Cyclohexanespiro -2'-(tetra- hydrofuran)

Tetrahydropyran-2-spiro- cyclohexane

3,3'-Spirobi(3H-indole)

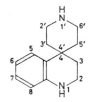

1,2,3,4-Tetrahydroquinoline -4- spiro-4'-piperidine

Hexahydroazepinium –1–spiro–1'–imidazolidine–3'– spiro–1"–piperidinium dibromide

Rule B–12.   Radicals

12.1—Radicals derived from heterocyclic spiro compounds named by Rule **B–10.1** are named according to the principles set forth in Rules **A–11** and **B–5.21**.   Radicals derived from other heterocyclic spiro compounds are named by adding " -yl ", " -diyl ", etc., to the name of the spiro compound (with elision of final " e ", if present, before a vowel).   The numbering of the spiro compound is retained and the point or points of attachment are given numbers as low as is consistent with any fixed numbering of the heterocyclic spiro compound.

Examples:

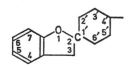

1−Oxaspiro[4.5]dec−2−yl
(cf. Rule B-10.1)
or Cyclohexanespiro−2′−(tetrahydrofuran)−5′−yl
(cf. Rule B−11.1)

Spiro[benzofuran−2(3H),1′−cyclohexan]−4′−yl

Spiro[naphthalene −2(3H),2′−thian]−4′−yl

HETEROCYCLIC RING ASSEMBLIES

Rule B–13

13.1—Assemblies of two or more identical heterocyclic systems are named by placing the prefix " bi- ", " ter- ", " quater- ", etc., before the name of the heterocyclic system or radical. The numbering of the assembly is that of the corresponding heterocyclic systems, one component being assigned unprimed and the others primed, doubly primed, etc., numbers. The points of attachment are indicated by appropriate locants before the name. Other structural features are described as recorded for hydrocarbon ring assemblies in Rules A–52.1 (double bond between two components), A–52.3 (substituents), A–53.3 (hydrogenation), A–54.1 (numerical prefixes), and A–55.1 and A–56.1 (radicals), insofar as fixed numbering of the heterocyclic system allows.

Examples:

2, 3′−Bifuran
or 2, 3′−Bifuryl

2, 2′– Bipyridin – 6 –yl
or 2, 2′– Bipyridyl – 6 –yl
or 6 –(2–Pyridyl) –2–pyridyl

2, 2′: 6′, 2″: 6″, 2‴ – Quaterpyridine

BRIDGED HETEROCYCLIC SYSTEMS

Rule B–14.   Extension of the von Baeyer System

14.1—Bridged heterocyclic systems are named according to the principles of Rules **A–31** and **A–32**, the hetero atoms being indicated according to Rule **B–4.2** and derived radicals by the principles set forth in Rule **A–31.4**.

Examples:

7–Azabicyclo [2.2.1] heptane

3.6.8–Trioxabicyclo [3.2.2] nonane

2.6–Dioxabicyclo [3.2.1] oct–7–yl

Rule B–15.   Hetero Bridges

B–15.1—A hetero polycyclic system that contains an " *ortho*-fused " or an " *ortho*- and *peri*-fused " system according to Rule **A–21** or **B–3** and has one or more atomic bridges is named as an " *ortho*-fused " or " *ortho*- and *peri*-fused " system.   The atomic bridges are then indicated by prefixes as exemplified in the annexed Table or by Rule **A–34.1**.   The name of a bridge containing hetero atoms is constructed from units beginning with the terminal atom that occurs first in Table I of Rule **B–1.1**, the final " o " of a prefix being elided before a vowel in a following prefix; to illustrate this, the formulae in the annexed Table are arranged from left to right in the same order as the prefixes.   If bridges of different types are present, they are cited in alphabetical order.   For examples see Rule **B–15.2**.

Azimino	—N=N—NH—
Azo	—N=N—
Biimino	—NH—NH—
Epidioxy	—O—O—
Epidithio	—S—S—
Epithio	—S—
Epithioximino	—S—O—NH—
Epoxy (see also Rule C–212.2)	—O—
Epoxyimino	—O—NH—
Epoxynitrilo	—O—N=
Epoxythio	—O—S—
Epoxythioxy	—O—S—O—
Furano (usually 3, 4–)	—C₄H₂O—
Imino (see also Rule C–815.2)	—NH—
Nitrilo	—N=

15.2—Systems described in Rule **B–15.1** are numbered according to the principles set forth in Rules **A–34.1** and **A–34.2** for compounds containing hydrocarbon bridges. In the name of the complete compound, the name of the bridge is preceded by two locants, that for the unit cited first in a composite bridge preceding that for the other end of the bridge. Radicals derived from the polycyclic systems described in Rule **B–15.1** are formed by the principles set forth in Rule **B–5**.

1,4-Dihydro-1,4-epoxynaphthalene

Perhydro-1,4-epoxy-4a,8a-(methanoxy-methano)naphthalene

Perhydro-5,3-(epoxymethano) benzofuran

Perhydro-3, 5-(epoxymethano)benzofuran

Definitive Rules for Nomenclature of Organic Chemistry
Section C
Characteristic Groups Containing Carbon, Hydrogen, Oxygen, Nitrogen, Halogen, Sulfur, Selenium, and/or Tellurium

*Issued by the Commission on
the Nomenclature of Organic Chemistry
of the
International Union of Pure and Applied Chemistry*

*Fourth Edition
1979*

PREFACE TO THE SECOND EDITION OF SECTION C

This Second Edition of Section C contains a considerable number of changes from the First Edition, but almost all of them are confined to correction of errors, clarification, or provision of better examples. However, one major change has been made, namely, deletion of the rules for order of complexity of substituents.

The Commission takes this opportunity to thank all those who submitted comments on the Tentative Version (1961) of these Rules and on the First Edition (1965). All these comments have been considered and many of them have been adopted.

PREAMBLE

SCOPE OF SECTION C

Section C deals with the nomenclature of compounds with **characteristic groups** (for the definition of this new terminology see p. 82) containing carbon, hydrogen, oxygen, nitrogen, halogen, sulfur, selenium, and/or tellurium. It thus supersedes almost all of Sections IV, V, and VII of the Liége rules*, and also the relevant parts of the later revision† of Section VI.

USE OF THE RULES

The " General Principles " stated in the Introduction to Sections A and B (see p. xi), have been applied also in Section C. In the present rules the Commission has in general continued its practice of codifying satisfactory existing nomenclature in preference to elaborating novel systems.

Alternative nomenclatures are recorded in many cases in Section C, as in Sections A and B. The Commission has attempted to keep the number of alternatives to a minimum. For certain types of compound, however, *e.g.* azo and hydrazo compounds, Beilstein's *Handbuch der organischen Chemie* and *Chemical Abstracts* have practised different but adequate systems of nomenclature. In such cases rules for the systems used by the two groups are presented as alternatives without prejudice. Occasionally other alternatives are included for what seem to be significant reasons. It is recognized that in chemical discussions it may sometimes be necessary to depart from the present rules in order to provide a name more appropriate to the chemical intent or to avoid obscuring an important feature; nevertheless, it is hoped that such

*†see footnotes on p.83

deviations from the rules will be made only when necessary and that, at least when the divergence is great, authors will give also the names formed according to the systematic rules.

Many trivial names occur in the rules that follow. Needless creation of new trivial names is discouraged (*cf.* the Introduction of Sections A and B), and old trivial names which are seldom used should be replaced by systematic ones. If a new trivial name is created, any syllables in it that are used in systematic nomenclature must bear their systematic meaning and be used by the systematic methods.

CONVENTIONS

(1) As in Sections A and B, the following rules are written, solely for uniformity, in accord with the conventions of *Chemical Abstracts* for spelling, position of numerals, punctuation, italicizing, abbreviations, elision of vowels, and certain terminations, and are not to be considered as recommendations by the Commission in these respects. However, it is hoped that variations therefrom may be reduced so far as practicable.*

(2) Except where specifically stated to the contrary, symbols R, *etc.*, are used to denote univalent radicals attached by means of carbon and derived from aliphatic, carbocyclic, or heterocyclic compounds, which may be saturated or unsaturated, and unsubstituted or substituted, but they are not used for —CN, —CNO, —CNS, or —CNSe groups or for groups attached directly through $>C=X$ where X is O, S, Se, Te, NH, or substituted NH. Up to three such groups may be designated R, R', and R", or R^1, R^2, and R^3; for more than three different groups the sequence R^1, R^2, R^3, R^4 . . . is recommended (where numerals are used, the simple R is not included). R_2, R_3, and R_4 should be used only to denote two, three, or four, respectively, of the same group denoted by simple R. For simplicity, in the rules in this Report only R^1, R^2 . . . *etc.* are used.

Individual Rules of the I.U.P.A.C. 1957 Rules are cited in the present text as Rule A–2, Rule B–3.4, *etc.*

MULTIPLYING AFFIXES

The multiplying affixes, di-, tri-, tetra-, penta-, *etc.*, are used to indicate a set of identical unsubstituted radicals or parent compounds:

Examples:	*Cf. Rules*	*Page*
1,2-Ethanediol	C–10.3	87
2,3-Naphthalenediacetic acid	C–52.2	119
Triethylamine	C–11.41	91
4-Chloro-2,4′-iminodibenzoic acid	C–73.1	132

The forms bis-, tris-, tetrakis-, pentakis-, *etc.*, are used to indicate a set of identical radicals or parent compounds each substituted in the same way (cf. Rule A-2.5):

*Over the period of time that these rules were written, the practices of *Chemical Abstracts* have changed. No attempt has been made to conform to all such changes.

Examples:	*Cf. Rule*	*Page*
Bis(2-chloroethyl)amine	C–814.2	254
2,7-Bis(phenylazo)-1,8-naphthalenediol	C–912.5	282

or to avoid ambiguity:

Examples:	*Cf. Rule*	*Page*
p-Phenylenebisketene	C–321.1	177
7-(Aminomethyl)-2,3-dibenzofuranbis-(ethylamine)	C–813.4	254
Tris(decyl)amine	C–814.2	254

Bi-, ter-, quater-, *etc.* are used to indicate the number of identical rings joined to one another by a link (single or double) (*cf.* note to Rule C–14.11, p. 101):

Examples:	*Cf. Rule*	*Page*
Biphenyl	A–52.4	43
2,1′:5′,2″:6″,2‴-Quaternaphthalene	A–52.2	44
[1,1′-Binaphthalene]-3,3′,4,4′-tetramine	C–71.3	128
3,4′-Bi-2-naphthol	C–71.4	129

Multiplying affixes may be omitted for very common compounds when no ambiguity is caused thereby, for example, phenyl ether for diphenyl ether, malonamide for malondiamide. Such affixes are, however, included throughout the text below for uniformity.

PRIMED NUMBERS OR PARENTHESES

A complete expression denoting a substituted side chain may be enclosed in parentheses, or the carbon atoms in side chains may be indicated by primed numbers (*cf.* Rule A–2.5).

NUMERALS 1 (UNITY)

A numeral 1 (unity) may be omitted from a name when no ambiguity is caused by so doing. However, in the text below this numeral is included, for consistency, when another numerical locant appears in the same name.

GLOSSARY

Section IV of the Liége Rules was headed " Simple Functions ", and Section V " Complex Functions ". Those rules defined compounds of simple function as those containing one kind of function only, and compounds of complex function as those possessing different functions (that is, functions of more than one kind). But the rules did not define or explain the meaning to be attached to the word " function " itself; nor can its precise meaning be inferred uniformly from the usages in the rules themselves. Usage of this term by chemical authors has varied greatly.

To minimize confusion the following terms are used in the present rules:

Functional class name: a word such as ketone, chloride, or alcohol, used in radicofunctional nomenclature (according to the language) as an ending or as a separate word (see Rules C–21 to C–24).

Substituent: any atom or group replacing hydrogen of a parent compound.

Characteristic group: an atom or group that is incorporated into a parent compound otherwise than by a direct carbon–carbon linkage, but including groups —C≡N and >C=X where X is O, S, Se, Te, NH, or substituted NH. (*N.B.* The phrase "characteristic group" includes both groups such as OH, NH₂, COOH, and single atoms such as halogen, =O, and ≡N. It does not apply to substituents such as methyl, phenyl, 2-pyridyl, but does include, for example, piperidino and acetyl.)

Principal group: the characteristic group chosen for expression as suffix in a particular name. (This is equivalent to the "principal function" of the Liége rules.)

SENIORITY; SENIOR

These words are used with reference to priority in a prescribed order, as laid down in the rules that follow or in Section A or B, as noted in each case.

ELISION OF VOWELS

The terminal " e " in names of parent compounds is elided when followed by a suffix beginning with " a ", " i ", or " o ":

Examples:	*Cf. Rule*	*Page*
4-Penten(e)-1-ol	C–10.3	87
2-Hexan(e)amine	C–11.42	91
Anilin(e)ium	C–82.2	135
Butan(e)ide	C–84.3	140
1-Chrysen(e)ol	C–202.1	150
Hexan(e)al	C–302.1	162
2-Butan(e)one	C–312.1	167
Heptan(e)oic acid	C–401.1	182
5-Pentan(e)olide	C–472.2	205
1-Hexan(e)imine	C–815.3	258
Hexan(e)amide	C–822.1	262

The terminal " a " in the names of multiplying affixes is elided when followed by a suffix or ending (not a separate word) beginning with " a " or " o ":

Examples:	*Cf. Rule*	*Page*
[1,1'-Binaphthalene]-3,3',4,4'-tetramine	C–71.3	128
1,3,6,8(2*H*,7*H*)-Pyrenetetrone	C–315.1	172
Benzenehexol		

There is no elision in the following cases:

		Cf. Rule	*Page*
(1) in conjunctive nomenclature,			
e.g., Quinolineacetic acid		C–51.2	118
α,β-Dimethyl-2-naphthaleneethanol		C–52.1	119
(2) in replacement nomenclature,			
e.g., Tetraoxatridecane		C–61.1	123
(3) in multiplying parent compounds,			
e.g., Ethylenediaminetetraacetic acid		C–815.1	257

(4) with multiplying prefixes,
   e.g., 1,4,6,9-Tetraoxo-2-pyrenecarboxylic acid

(5) and (6) See page 84.

ADDITION OF VOWELS

On relatively rare occasions, vowels are inserted between consonants for euphony:

e.g., 2-Naphthalenesulfonodiimidic acid (*cf.* Rule C–642.2, p.240)
   Ethyl ethanesulfonohydroximate (*cf.* Rule C–642.1, p. 239).

TERMINAL CHARACTERISTIC GROUPS

Prefixes that represent complete terminal characteristic groups are preferred to those representing only a portion of a given group.

Examples:	*Cf. Rule*	*Page*
3-(Formylmethyl)heptanedial preferred to		
3-(2-Oxoethyl)heptanedial	C–303.3	163
7-Thioformylheptanoic acid preferred to		
8-Thioxooctanoic acid	C–531.3	218

ORTHO-, META-, PARA-

The positions of substituents are indicated by numbers except that *o*- (*ortho*), *m*- (*meta*), and *p*- (*para*) may be used in place of 1,2-, 1,3-, and l, 4-, respectively (*cf.* Rule A–12.3), for benzene derivatives.

---

* Definitive Report of the Commission for the Reform of Nomenclature in Organic Chemistry of the International Union of Chemistry, which appeared after the Liége Conference (Comptes rendus of the 10th Conference, 1930, pp. 57–64), supplemented by less extensive reports from the meetings of the Commission at Lucerne and at Rome (Comptes rendus of the 12th Conference, 1936, pp. 39–42, and of the 13th Conference, 1938, pp. 36–37).

† Comptes rendus of the 15th Conference of the International Union of Pure and Applied Chemistry, 1949, pp. 127–186, and of the 16th Conference, 1951, pp. 100–104.

(5) The terminal "e" in the names of parent hydrides is elided when adding "-yl" in formation of names of radicals. For example, pyrene, pyrenyl. However, when ambiguity results the "e" is retained, e.g. pentaphene, pentapheneyl to avoid confusion with pentaphenyl (five phenyl radicals).

(6) For elision of vowels in forming names of fused polycyclic compounds see Rule A-21.4 and in Hantzsch-Widman names see Rule B-1.1.

C-0. NOMENCLATURE SYSTEMS

C-0.0. GENERAL PRINCIPLES

(1) The various classes of compound are dealt with in detail in the later Subsections C-1 to C-9.   The following is a summary of general principles.

(2) The formation of a name for a chemical compound usually involves the following steps, to be taken so far as they are applicable in the order given:

(*a*) From the nature of the compound determine the type of nomenclature to be used (substitutive, radicofunctional, additive, subtractive, conjunctive or replacement); or treat as an assembly of identical units.

(*b*) Determine the kind of characteristic group for use as the principal group, if any.   Only one kind of characteristic group should be cited as suffix or functional class name.   All substituents not so cited must be named as prefixes.

(*c*) Determine the parent structure (principal chain, parent ring system, or conjunctive components).

(*d*) Name the parent structure and the principal group(s).

(*e*) Determine and name the infixes and prefixes

(*f*) Complete the numbering.

(*g*) Assemble the partial names into a complete name, using for all the detachable prefixes the alphabetical order.

(3) Details for each of the above steps and for each type of nomenclature are given in the rules which follow.

(4) In substitutive nomenclature some characteristic groups can be denoted either as prefixes or as suffixes, but others only as prefixes.   Radicofunctional nomenclature differs in that separate words (or in some languages suffixes) designating the principal group(s) are associated with radical names designating the remainder of the structure.   The characteristic groups that can be cited as suffixes in substitutive nomenclature are not necessarily identical with the groups designated by the functional class name when radicofunctional nomenclature is used (*e.g.*, butanone and ethyl methyl ketone, where -one denotes $=O$ and ketone denotes $>CO$).

C-0.1. SUBSTITUTIVE NOMENCLATURE

C-0.10. CHARACTERISTIC GROUPS

Compulsory Prefixes

Rule C-10.1

10.1—The characteristic groups listed in Table I are always cited by prefixes, as given in the Table, to the name of the parent compound (for choice of the parent compound see Rule C-12).   Multiplying affixes and locants are added as necessary (for numbering see Subsection C-0.15, p. 105).

Example:

1,2-Dichlorocyclohexane

The complete prefix is 1,2-dichloro-

TABLE I. Characteristic groups cited only as prefixes in substitutive nomenclature
(see Rule C–10.1)

Characteristic group	Prefix	Cf. Rule	Page
—Br	Bromo	C–102.1	144
—Cl	Chloro	C–102.1	144
—ClO	Chlorosyl	C–106.2	146
—ClO$_2$	Chloryl	C–106.2	146
—ClO$_3$	Perchloryl	C–106.2	146
—F	Fluoro	C–102.1	144
—I	Iodo	C–102.1	144
—IO	Iodosyl	C–106.1	146
—IO$_2$	Iodyl (replacing iodoxy)	C–106.1	146
—I(OH)$_2$	Dihydroxyiodo	C–106.3	146
—IX$_2$	X may be halogen or a radical, and the prefix names are dihalogenoiodo, *etc.*, or, for radicals, patterned on diacetoxyiodo	C–106.3	146
=N$_2$	Diazo	C–931.4	290
—N$_3$	Azido	C–941.1	291
—NO	Nitroso	C–851.1	275
—NO$_2$	Nitro	C–852.1	275
=N(O)OH	*aci*-Nitro	C–852.2	275
—OR	R-oxy	C–205.1	154
—SR	R-thio (similarly R-seleno and R-telluro)	C–514.1	213

TABLE II. Some general classes of compound in the order in which the characteristic groups have decreasing priority for citation as principal group (see Rule C–10.3)*

1. 'Onium and similar cations (see Subsection C–0.8)
2. Acids: in the order COOH, C(=O)OOH, then successively their S and Se derivatives, followed by sulfonic, sulfinic acids, *etc.*
3. Derivatives of acids: in the order anhydrides, esters, acyl halides, amides, hydrazides, imides, amidines, *etc.*
4. Nitriles (cyanides), then isocyanides
5. Aldehydes, then successively their S and Se analogues; then their derivatives
6. Ketones, then their analogues and derivatives, in the same order as for aldehydes
7. Alcohols and phenols; then their S, Se and Te analogues; then neutral esters of alcohols and phenols with inorganic acids, except hydrogen halides, in the same order.
8. Hydroperoxides
9. Amines; then imines, hydrazines, *etc.*
10. Ethers; then successively their S and Se analogues
11. Peroxides

*Rule D-1.32 extends this table.

Principal Group for Citation as Suffix

Rule C–10.2

10.2—Characteristic groups other than those listed in Table I may be cited as suffixes or prefixes to the name of the parent compound (for choice of the parent compound see Rule C–12).

Rule C–10.3

10.3—If any characteristic groups other than those in Table I are present, one kind must be cited as suffix, but only one kind. That kind of group is termed the principal group. When a compound contains more than one kind of group not listed in Table I, the principal group is that which characterizes the class occurring as high as possible in Table II; all other characteristic groups are then cited as prefixes. Some suffixes and prefixes to be used with the general classes given in Table II are listed in Table III. Further details will be found in Subsections C–1 to C–9. Multiplying affixes and locants are added as necessary. If, but only if, the complete suffix (that is, the suffix itself plus its multiplying affix, if any) begins with a vowel, a terminal " e " (if any) of the preceding parent name is elided. Elision or retention of the terminal " e " is independent of the presence of numerals between it and the following letter.

Examples:

$$\overset{2}{C}H_3—\overset{1}{C}H_2OH \quad \text{and} \quad HO\overset{2}{C}H_2—\overset{1}{C}H_2OH$$

Suffix for principal group, OH: -ol
Parent name for $CH_3—CH_3$:   Ethane
Full names: Ethanol
                        1,2-Ethanediol

$$HO\overset{7}{C}H_2—\overset{6}{C}H_2—\overset{5}{C}H_2—\overset{4}{C}H_2—\overset{3}{C}H_2—\overset{2}{C}O—\overset{1}{C}H_3$$

Principal group of class higher in Table II: >(C)=O, denoted by suffix -one.

Parent name: Heptane
Name based on: Heptanone
Full name: 7-Hydroxy-2-heptanone

$$\overset{5}{C}H_2=\overset{4}{C}H—\overset{3}{C}H_2—\overset{2}{C}H_2—\overset{1}{C}H_2OH$$

Suffix: -ol
Parent name: Pentane, altered to pentene for unsaturation (see Rule A–3.1)
Full name: 4-Penten-1-ol

Rule C–10.4

10.41—Derivative groups have priority for citation as principal group after the respective parents of their general class, as indicated in Table II.

Example:

2-Hydroxy-1-cyclohexanecarboxamide

The compound is both an amide and an alcohol. Amides, being derivatives of acids, are higher than alcohols in Table II. The amide group is therefore cited as suffix and the hydroxyl group as prefix.

TABLE III. Suffixes and prefixes used for some important groups in substitutive nomenclature (see Rule C–10.3) (for further details see the rules for individual classes)

Class	Formula*	Prefix	Suffix	Cf. Rule
Cations		-onio -onia	-onium	C–82 B–6
Carboxylic acid	—COOH —(C)OOH	Carboxy —	-carboxylic acid -oic acid	C–401
Sulfonic acid	—SO₃H	Sulfo	-sulfonic acid	C–641
Salts	—COOM —(C)OOM	— —	Metal . . .carboxyl-ate Metal . . .oate	C–461
Esters	—COOR —(C)OOR	R-oxycar-bonyl —	R . . .carboxylate R . . .oate	C–463
Acid halides	—CO—Halogen —(C)O—Halogen	Halo-formyl —	-carbonyl halide -oyl halide	C–481
Amides	—CO—NH₂ —(C)O—NH₂	Carbamoyl —	-carboxamide -amide	C–821 C–822
Amidines	—C(=NH)—NH₂ —(C)(=NH)—NH₂	Amidino —	-carboxamidine -amidine	C–951
Nitriles	—C≡N —(C)≡N	Cyano —	-carbonitrile -nitrile	C–832
Aldehydes	—CHO —(C)HO	Formyl Oxo	-carbaldehyde -al	C–301 C–301
Ketones	>(C)=O	Oxo	-one	C–311
Alcohols	—OH	Hydroxy	-ol	C–201
Phenols	—OH	Hydroxy	-ol	C–202
Thiols	—SH	Mercapto	-thiol	C–511
Hydro-peroxides	—O—OH	Hydro-peroxy	—	C–218
Amines	—NH₂	Amino	-amine	C–812
Imines	=NH	Imino	-imine	C–815
Ethers	—OR	R-oxy	—	C–211
Sulfides	—SR	R-thio	—	C–514
Peroxides	—O—OR	R-dioxy	—	C–218

* Carbon atoms enclosed in parentheses are included in the name of the parent compound and not in the suffix or prefix (cf. Rules C–11.11, C–11.31, etc.).

10.42—Groups in which oxygen is replaced by sulfur, selenium, or tellurium have priority, in that descending order, for choice as principal group, after the respective oxygen analogue as indicated in Table II. A more detailed order of priority is exemplified for some sulfur derivatives in Table XII (see p. 234).

Rule C–10.5

10.51—The general method of adding suffixes is modified for a few classes of compound as described in Subsection C–0.11.

C–0.11. EXCEPTIONAL TREATMENT OF SOME CHARACTERISTIC GROUPS

There are two methods of using suffixes in substitutive nomenclature for aliphatic carboxylic acids and their derivatives, for aliphatic nitriles, and for aliphatic aldehydes*; of these methods, one involves modification of the preceding rules, the other does not, as illustrated for carboxylic acids by Rules C–11.11(*b*) and (*a*), respectively, below. However, when the characteristic group concerned is attached to a ring, only the method of Rule C–11.11 (*b*) is used in systematic nomenclature.

A different type of treatment for certain amines is also described below (Rule C–11.4)*.

Carboxylic Acids

Rule C–11.1

11.11—(*a*) The atom =O and the group —OH on one and the same carbon atom are together denoted by a suffix " -oic " attached to the name of the parent aliphatic chain and the word " acid " is added thereafter. Thus the change " -ane " to " -anoic acid " denotes the change of —CH_3 to —COOH†. For numbering see Rule C–401.1.

Example:

$$\overset{6}{CH_3}—\overset{5}{CH_2}—\overset{4}{CH_2}—\overset{3}{CH_2}—\overset{2}{CH_2}—\overset{1}{COOH} \quad \text{Hexanoic acid}$$

(*b*) The group —COOH is treated as a complete substituent, denoted by the ending " carboxylic acid ", which includes the carbon atom of the carboxyl group. For numbering see Rule C–402.2

Examples:

$$HOOC—\overset{5}{CH_2}—\overset{4}{CH_2}—\overset{3}{CH}—\overset{2}{CH_2}—\overset{1}{CH_2}—COOH$$
$$\underset{COOH}{|}$$

1,3,5-Pentanetri-carboxylic acid

—COOH   Cyclohexanecarboxylic acid

---

* For more extensive recommendations (including numbering and guides as to which system to use) see the detailed rules of Subsections C–3 (p.162), C–4 (p.182), and C–8 (p.249).

† In some languages the word corresponding to " acid " is placed first owing to the adjectival nature of the " -ic " ending (for example, in French, *acide hexanoïque*); in others it forms a suffix, the adjectival ending being omitted (for example, in German, *Hexansäure*).

Derivatives and Radicals from Carboxylic Acids

Rule C–11.2

11.21—The two methods described in Rule C–11.1 apply to all derivatives of aliphatic acids and to acyl radicals; choice is made between endings such as —CO—NH$_2$ -amide or -carboxamide, —CO—NH—NH$_2$ -ohydrazide or -carbohydrazide, —COCl -oyl chloride or -carbonyl chloride, —COOR (for esters) -oate or -carboxylate, *etc.*, and for radicals R—CO— -oyl or -carbonyl. For further details (including numbering) see the individual rules for each class of compound.

Examples:

$$\overset{6}{\text{CH}_3}—\overset{5}{\text{CH}_2}—\overset{4}{\text{CH}_2}—\overset{3}{\text{CH}_2}—\overset{2}{\text{CH}_2}—\overset{1}{\text{CO}}—\text{NH}_2$$

Hexanamide

$$\text{NH}_2—\text{CO}—\overset{5}{\text{CH}_2}—\overset{4}{\text{CH}_2}—\overset{3}{\text{CH}}—\overset{2}{\text{CH}_2}—\overset{1}{\text{CH}_2}—\text{CO}—\text{NH}_2$$
$$\underset{\text{CO}—\text{NH}_2}{|}$$

1,3,5-Pentanetricarboxamide

$$\overset{6}{\text{CH}_3}—\overset{5}{\text{CH}_2}—\overset{4}{\text{CH}_2}—\overset{3}{\text{CH}_2}—\overset{2}{\text{CH}_2}—\overset{1}{\text{CO}}—\text{OC}_2\text{H}_5$$

Ethyl hexanoate

$$\text{C}_2\text{H}_5\text{O}—\text{CO}—\overset{5}{\text{CH}_2}—\overset{4}{\text{CH}_2}—\overset{3}{\text{CH}}—\overset{2}{\text{CH}_2}—\overset{1}{\text{CH}_2}—\text{CO}—\text{OC}_2\text{H}_5$$
$$\underset{\text{CO}—\text{OC}_2\text{H}_5}{|}$$

Triethyl 1,3,5-pentanetricarboxylate

1-Hexanoyl-4-carbazolecarbaldehyde

1-Cyclohexylcarbonyl-4-carbazolesulfonic acid
(*cf.* Note to Rule C–403.2)

Aldehydes and Nitriles

Rule C–11.3

11.31—The two methods described in Rule C–11.1 apply in principle also to aliphatic aldehydes and their derivatives and to nitriles; choice is made between the endings:

$$-(C){\overset{\displaystyle O}{\underset{\displaystyle H}{}}} \quad \text{-al} \qquad or \qquad -C{\overset{\displaystyle O}{\underset{\displaystyle H}{}}} \quad \text{-carbaldehyde}$$

$$-(C)\equiv N \quad \text{-nitrile} \qquad or \qquad -C\equiv N \quad \text{-carbonitrile}$$

Examples:

$$\overset{6}{C}H_3-\overset{5}{C}H_2-\overset{4}{C}H_2-\overset{3}{C}H_2-\overset{2}{C}H_2-\overset{1}{C}HO$$
Hexanal

$$OHC-\overset{5}{C}H_2-\overset{4}{C}H_2-\overset{3}{C}H-\overset{2}{C}H_2-\overset{1}{C}H_2-CHO$$
$$\underset{CHO}{|}$$
1,3,5-Pentanetricarbaldehyde

$$\overset{6}{C}H_3-\overset{5}{C}H_2-\overset{4}{C}H_2-\overset{3}{C}H_2-\overset{2}{C}H_2-\overset{1}{C}N$$
Hexanenitrile

$$NC-\overset{5}{C}H_2-\overset{4}{C}H_2-\overset{3}{C}H-\overset{2}{C}H_2-\overset{1}{C}H_2-CN$$
$$\underset{CN}{|}$$
1,3,5-Pentanetricarbonitrile

Amines

Amines are named by traditional methods, details of which are given in Subsection C–8.1; the following are the two main principles.

Rule C–11.4

11.41—For simple amines RNH_2, R^1R^2NH, and $R^1R^2R^3N$, the names of the radicals are attached in front of the word " amine " (see Subsection C–8.1).

Examples:

Butylamine	$CH_3-CH_2-CH_2-CH_2-NH_2$
N-Methylbutylamine	$CH_3-CH_2-CH_2-CH_2-NH-CH_3$
Triethylamine	$(C_2H_5)_3N$

In this method, amine can be considered to replace ammonia as parent compound.

11.42—Primary amines may also be named (see Subsection C–8.1) by adding the suffix "-amine " to the name of the parent compound (with elision of terminal " e ", if present).

Examples:

2-Hexanamine $CH_3—CH_2—CH_2—CH_2—CH(NH_2)—CH_3$

2-Furanamine

C–0.12. GUIDE TO CONSTRUCTION OF THE NAME

Rule C–12

After the principal group(s) have been chosen and named, the parent compound is next chosen by one of the following methods. For details of numbering, see Subsection C–0.15 and for the arrangement of prefixes see Subsection C–0.16.

12.1—If the compound is purely acyclic, the principal chain is chosen as parent for nomenclature by the methods given in Subsection C–0.13 and is named according to the Rules **1.1**, **3.1**, **3.2**, and **3.3** of Section A.

Example:

$$\overset{6}{HO—CH_2}—\overset{5}{CH}—\overset{4}{CH}=\overset{3}{CCl}—\overset{2}{CO}—\overset{1}{CH_3}$$
$$\underset{CH_3}{|}$$

Principal group:	—CO—	-one
Principal chain:	C—C—C—C—C—C	Hexane
Principal chain including principal group:	C—C—C—C—CO—C	2-Hexanone
Subtractive modification:	C—C—C=C—CO—C	3-Hexen-2-one
Prefixes:	Cl—	Chloro-
	HO—	Hydroxy-
	CH_3—	Methyl-

Together with later rules this leads to the name:

3-Chloro-6-hydroxy-5-methyl-3-hexen-2-one

12.2—If the principal group occurs only in a chain which carries a cyclic substituent, the compound is named as an aliphatic compound into which the cyclic component is substituted, by means of a radical prefix. It is not necessary that the chain bearing the cyclic substituent shall be the longest chain.

Example:

Principal group:	—COOH	-oic acid
Principal chain:	C—C—C—C—C—C	Hexane

Principal chain including principal group:

$$C—C—C—C—C—COOH$$

Hexanoic acid

Subtractive modification: $C—C—C{=}C—C—COOH$

3-Hexenoic acid

Prefixes: $Cl—$ Chloro-

Cyclohexyl-

$HO—$ Hydroxy-

$CH_3—$ Methyl-

Together with later rules this leads to the name:

3-Chloro-5-cyclohexyl-6-hydroxy-5-methyl-3-hexenoic acid

12.3—If the principal group occurs in two or more carbon chains that are not attached directly to one another (that is, do not together form a branched chain but are separated by, for instance, a ring or hetero atom), then that chain is chosen as parent for nomenclature which carries the largest number of the principal group; if the numbers of these groups in two or more chains are the same, choice is made by the principles for selection of the principal chain (see Subsection C–0.13); if this does not effect a choice, the compound is named as an assembly of identical units (see Subsection C–0.7).

Example:

$$HO\overset{3}{C}H_2—\overset{2}{C}H_2—\overset{1}{C}H_2— \quad —\overset{1}{C}H—\overset{2}{C}H_2\,OH$$
$$\underset{OH}{}$$

Principal group: —OH   -ol

Principal chain, *i.e.*, that carrying two OH:

HO—C—C—OH   Ethanediol

Prefix: $HOCH_2—CH_2—CH_2—$ *p*-(3-Hydroxypropyl)phenyl

Together with later rules this leads to the name:

1-[*p*-(3-Hydroxypropyl)phenyl]-1,2-ethanediol

12.4—If the principal group occurs only in one cyclic system, that system forms the parent for nomenclature and is named according to the rules of Sections A and B.

Example:

$$—CH_2—CH_3$$
$$—OH$$

Principal group: —OH   -ol

93

Parent: Cyclohexane

Prefix: $CH_3—CH_2—$ Ethyl-

Together with later rules this leads to the name:

2-Ethyl-1-cyclohexanol

12.5—If the principal group occurs in more than one cyclic system, that system is chosen as parent for nomenclature which carries the largest number of the principal group; if the numbers in two or more systems are the same, the senior ring system is chosen as parent according to the rules for seniority of ring systems in Subsection C–0.14; if these rules do not effect a choice, the substance is named as an assembly of identical units (see Subsection C–0.7).

Examples:

Principal group: —COOH -carboxylic acid

Senior ring system for parent: Fluorene

Name of parent including principal group:

Fluorene-2-carboxylic acid

Prefixes: Phenyl-

HOOC— Carboxy-

Together with later rules this leads to the name:

6-(*p*-Carboxyphenyl)fluorene-2-carboxylic acid

12.6—If the principal group occurs both in a chain and in a cyclic system, the parent for nomenclature is that portion in which the principal group occurs in largest number; if the numbers are the same in two or more portions, that portion is chosen as parent for nomenclature which is considered to be the most important (for general considerations see Rule A–61) or is the senior (see Subsection C–0.14).

Examples:

(1)

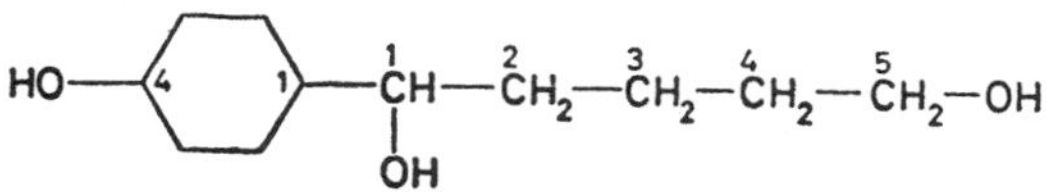

$$HO-\langle\text{cyclohexyl}\rangle-\overset{1}{C}H-\overset{2}{C}H_2-\overset{3}{C}H_2-\overset{4}{C}H_2-\overset{5}{C}H_2-OH$$
$$\overset{|}{OH}$$

Principal group: OH   -ol

Component carrying largest number of principal groups:

        —CH(OH)—CH$_2$—CH$_2$—CH$_2$—CH$_2$—OH

Name of parent including suffix: Pentanediol

Substituent to be named as prefix: HO—⟨ ⟩—   4-Hydroxy·
                                              cyclohexyl

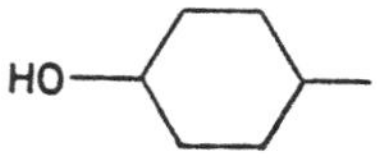

Together with later rules this leads to the name:

        1-(4-Hydroxycyclohexyl)-1,5-pentanediol

(2)

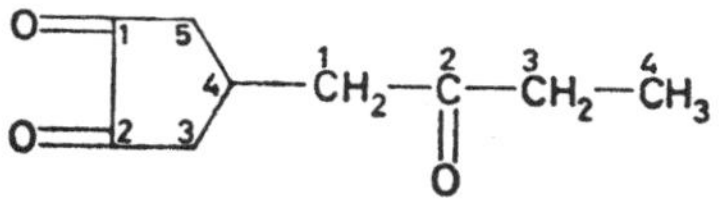

$$O=\square-\overset{1}{C}H_2-\overset{2}{C}-\overset{3}{C}H_2-\overset{4}{C}H_3$$
$$\overset{||}{O}$$

Principal group: =O   -one

Component carrying largest number of principal groups:

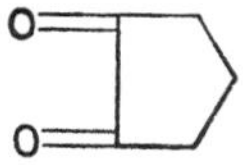

        1,2-Cyclopentanedione

Substituent to be named as prefix:

        —CH$_2$—C—CH$_2$—CH$_3$ 2-Oxobutyl
             ||
             O

Together with later rules this leads to the name:

        4-(2-Oxobutyl)-1,2-cyclopentanedione

(3)

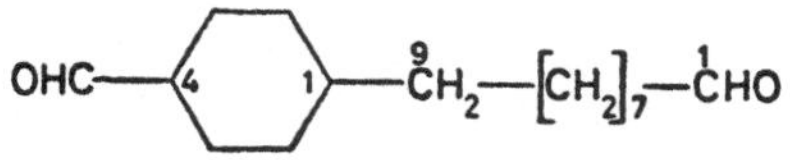

$$OHC-\langle\text{cyclohexyl}\rangle-\overset{9}{C}H_2-[CH_2]_7-\overset{1}{C}HO$$

Principal group: —(C)⟨$\overset{O}{H}$⟩   -al

Parent (Rule A–61.4):   C$_9$ chain   Nonane

Parent including suffix:

C—C—C—C—C—C—C—C—CHO   Nonanal

Substituent to be named as prefix:

4-Formylcyclohexyl

Together with later rules this leads to the name:

9-(4-Formylcyclohexyl)nonanal

12.7—When a substituent is itself substituted, all the subsidiary substituents are named as prefixes. The substituent bearing the subsidiary substituents is regarded as a " parent radical " (analogous to a parent compound). The nomenclature of the whole substituent is subject to all the procedures adopted for compounds (for example, choice of principal chain) with two exceptions, namely (*a*) that no suffix is used and (*b*) that the point of attachment of the radical bears the lowest permissible number which for a chain must be 1 (see Rules **A–1.2**, **A–2.25**, and **A–3.5**).

Example:

Principal group:	—COOH	-carboxylic acid
Parent (Rule C–12.4):		Cyclohexane

Substituent to be named as prefix:

$$-CH_2-CH-CH_2-CHCl-CO-CH_3$$
$$\quad\quad\quad | $$
$$\quad\quad CH_2-OH$$

Parent substituent chain:	C—C—C—C—C—C	Hexyl-
Subsidiary prefixes:	—Cl	Chloro-
	=O	Oxo-
	—CH₂—OH name composed of:	
	—CH₃	Methyl-
	—OH	Hydroxy-
	giving Hydroxymethyl-	

Other substituent in cyclohexane ring:

	—Cl	Chloro-

Together with later rules this gives the name:

4,5-Dichloro-2-[4-chloro-2-(hydroxymethyl)-5-oxo-
hexyl]-1-cyclohexanecarboxylic acid

Rule C–12.8

12.81—When the parent compound (principal chain, ring system), principal group, and other substituents have been selected and named, the numbering of the complete compound is allocated. This depends in the first place on the numbering laid down in Sections A and B for the principal chain and ring systems, but in so far as choice remains the numbering is carried out in accordance with the Subsection on numbering (C–0.15).

Example:

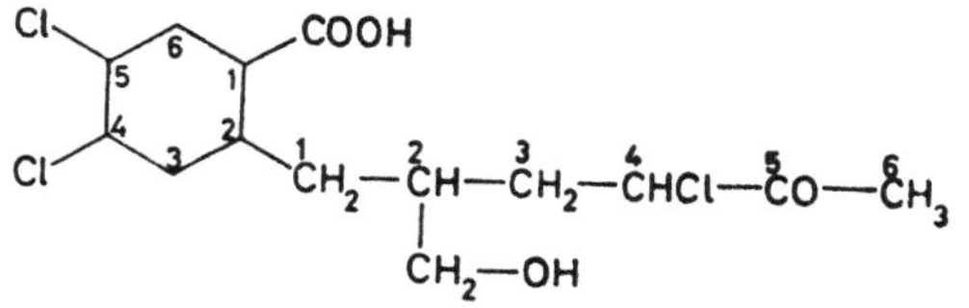

The preceding rules and the rules for numbering given in Subsection C–0.15 lead to the numbering shown in the formula and used in the example to Rule C–12.7.

Rule C–12.9

12.91—The various components having been selected, named, and numbered, any necessary additive and subtractive modifications are made (see Subsections C–0.3 and C–0.4), and the complete name is assembled, prefixes being arranged in order as laid down in Subsection C–0.16 for order of prefixes.

Example:

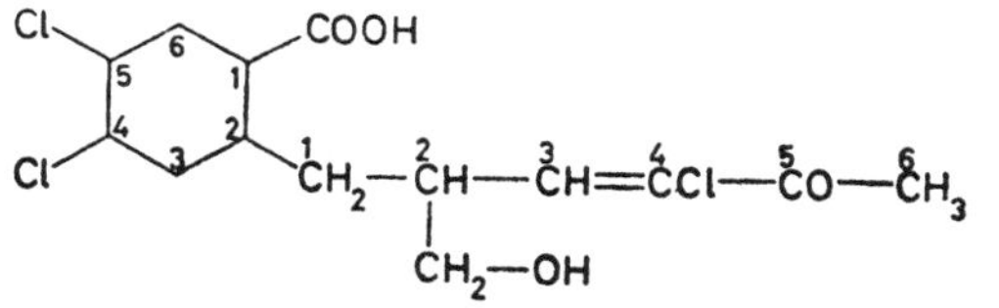

Principal group:	Carboxylic acid
Parent:	Cyclohexane
Prefixes, main substituents:	4,5-Dichloro, 2-Hexyl
subtractive modification:	Hexyl to 3-Hexenyl
subsidiary substituents:	4-Chloro, 5-Oxo,
	2-(Hydroxymethyl)

Assembly of name: 4,5-Dichloro-2-[4-chloro-2-(hydroxymethyl)-5-oxo-3-hexenyl]-1-cyclohexanecarboxylic acid

C–0.13. SENIORITY OF CHAINS (THE PRINCIPAL CHAIN)*

Rule C–13.1

13.11—In an acyclic compound, that chain upon which the nomenclature and numbering are based is called the " principal chain ". When

* Compare Rule C–12.

in an acyclic compound there is a choice for principal chain, the following criteria are applied successively, in the order listed, until a decision is reached:

(a) Maximum number of substituents corresponding to the principal group.

Example:

$$
\begin{array}{c}
\overset{\displaystyle OH}{\underset{\displaystyle 4|}{}} \\
\overset{5}{CH_3}\!-\!\overset{4}{CH} \\
\end{array}
$$

$$
\underset{\overset{4}{Cl}-\overset{3}{CH_2}-\overset{2}{CH_2}-\overset{1}{CH_2}-\overset{}{CH_2}}{\overset{3}{\diagup}}\!\!\!\!\diagdown\!\!\overset{3}{CH}\!-\!\overset{2}{CH_2}\!-\!\overset{1}{CH_2}\!-\!OH
$$

3-(4-Chlorobutyl)-1,4-pentanediol

(b) Maximum number of double and triple bonds considered together

Example:

$$
\overset{7}{C}\!\equiv\!\overset{8}{C}\!-\!\overset{9}{C}\!\equiv\!\overset{10}{C}\!-\!\overset{11}{CH_2}\!-\!OH
$$

$$
\overset{6}{CH}\!-\!\overset{5}{CH}\!=\!\overset{4}{CH}\!-\!\overset{3}{CH}\!=\!\overset{2}{CH}\!-\!\overset{1}{CH_2}\!-\!OH
$$

$$
\overset{1}{CH}\!=\!\overset{2}{CH}\!-\!\overset{3}{CH_2}\!-\!\overset{4}{CH}\!-\!\overset{5}{CH_2}\!-\!\overset{6}{CH_3}
$$

$$
\underset{OH}{|}
$$

6-(4-Hydroxy-1-hexenyl)-2,4-undecadiene-7,9-diyne-1,11-diol

(c) Maximum length.

Example:

$$
HO\!-\!\overset{9}{CH_2}\!-\!\overset{8}{CH_2}\!-\!\overset{7}{C}\!\equiv\!\overset{6}{C}
$$

$$
\overset{5}{CH}\!-\!\overset{4}{CH}\!=\!\overset{3}{CH}\!-\!\overset{2}{CH_2}\!-\!\overset{1}{CH_2}\!-\!OH
$$

$$
HO\!-\!\overset{3}{CH_2}\!-\!\overset{2}{CH}\!=\!\overset{1}{CH}
$$

5-(3-Hydroxypropenyl)-3-nonen-6-yne-1,9-diol

(d) Maximum number of double bonds.

Example:

$$
HO\!-\!\overset{9}{CH_2}\!-\!\overset{8}{CH_2}\!-\!\overset{7}{CH}\!=\!\overset{6}{CH}
$$

$$
\overset{5}{CH}\!-\!\overset{4}{CH_2}\!-\!\overset{3}{CH}\!=\!\overset{2}{CH}\!-\!\overset{1}{CH_2}\!-\!OH
$$

$$
HO\!-\!\overset{4}{CH_2}\!-\!\overset{3}{CH_2}\!-\!\overset{2}{C}\!\equiv\!\overset{1}{C}
$$

5-(4-Hydroxy-1-butynyl)-2,6-nonadiene-1,9-diol

(e) Lowest locants* for the principal groups (that is, for the suffix).
Example:

$$\begin{array}{c}
\overset{OH}{\underset{8}{Cl}-\overset{7|}{CH_2}-\overset{}{CH}-\overset{6}{CH_2}}\\
\overset{3}{HO}-\overset{2}{CH_2}-\overset{1}{CH_2}-\overset{}{CH}
\end{array} \hspace{-0.5em}\begin{array}{c}\\ \diagdown \\ \diagup \\ | \\ Cl\end{array}\hspace{-0.5em}\overset{5}{CH}-\overset{4}{CH_2}-\overset{3}{CH_2}-\overset{2}{CH_2}-\overset{1}{CH_2}-OH$$

8-Chloro-5-(1-chloro-3-hydroxypropyl)-1,7-octanediol

(f) Lowest locants for multiple bonds.
Example:

$$\begin{array}{c}
\overset{9}{HO}-\overset{8}{CH_2}-\overset{7}{CH_2}-\overset{6}{CH}=\overset{}{CH}\\
\overset{4}{HO}-\overset{3}{CH_2}-\overset{2}{CH}=\overset{1}{CH}-\overset{}{CH_2}
\end{array}\hspace{-0.5em}\begin{array}{c}\\ \diagdown \\ \diagup \end{array}\hspace{-0.5em}\overset{5}{CH}-\overset{4}{CH_2}-\overset{3}{CH}=\overset{2}{CH}-\overset{1}{CH_2}-OH$$

5-(4-Hydroxy-2-butenyl)-2,6-nonadiene-1,9-diol

(g) Lowest locants for double bonds.
Example:

$$\begin{array}{c}
\overset{11}{HO}-\overset{10}{CH_2}-\overset{9}{C}\equiv\overset{8}{C}-\overset{7}{CH}=\overset{}{CH}\\
\overset{5}{HO}-\overset{4}{CH_2}-\overset{3}{CH}=\overset{2}{CH}-\overset{1}{C}\equiv\overset{}{C}
\end{array}\hspace{-0.5em}\begin{array}{c}\\ \diagdown \\ \diagup \end{array}\hspace{-0.5em}\overset{6}{CH}-\overset{5}{CH}=\overset{4}{CH}-\overset{3}{CH}=\overset{2}{CH}-\overset{1}{CH_2}-OH$$

6-(5-Hydroxy-3-penten-1-ynyl)-2,4,7-undecatrien-9-yne-1,11-diol

(h) Maximum number of substituents cited as prefixes.
Example:

$$\begin{array}{c}
\overset{9}{CH_3}-\overset{8|}{\overset{OH}{CH}}-\overset{7}{CH_2}-\overset{6|}{\overset{CH_3}{CH}}\\
\overset{4}{CH_3}-\overset{3}{CH}-\overset{2}{CH_2}-\overset{1}{CH_2}\\
\hspace{2em}|\\ \hspace{2em}OH
\end{array}\hspace{-0.5em}\begin{array}{c}\\ \diagdown \\ \diagup \end{array}\hspace{-0.5em}\overset{5}{CH}-\overset{4|}{\overset{H_3C}{CH}}-\overset{3|}{\overset{Cl}{CH}}-\overset{2|}{\overset{OH}{CH}}-\overset{1}{CH_3}$$

3-Chloro-5-(3-hydroxybutyl)-4,6-dimethyl-2,8-nonanediol

(i) Lowest locants for all substituents in the principal chain cited as prefixes.

Example:

$$\begin{array}{c}
\overset{9}{HO}-\overset{8}{CH_2}-\overset{7}{CH_2}-\overset{6|}{\overset{CH_3}{CH}}\\
\overset{4}{HO}-\overset{3}{CH_2}-\overset{2}{CH_2}-\overset{1}{CH}-\overset{}{CH_2}\\
\hspace{4em}|\\ \hspace{4em}Cl
\end{array}\hspace{-0.5em}\begin{array}{c}\\ \diagdown \\ \diagup \end{array}\hspace{-0.5em}\overset{5}{CH}-\overset{4}{CH_2}-\overset{3}{CH_2}-\overset{2|}{\overset{NH_2}{CH}}-\overset{1}{CH_2}-OH$$

2-Amino-5-(2-chloro-4-hydroxybutyl)-6-methyl-1,9-nonanediol

*When series of locants containing the same number of terms are compared term by term, that series is "lowest" which contains the lowest number on the occasion of the first difference.

(*j*) The substituent first cited in alphabetical order.

Example:

$$\begin{array}{c} ClCH_2 \\ \diagdown \\ \\ BrCH_2 \diagup \end{array} CH-[CH_2]_5-CHClBr$$

1,8-Dibromo-1-chloro-7-(chloromethyl)octane

(*k*) Lowest locants for the substituent cited first as prefix in alphabetical order (see Subsection C–0.16).

Example:

$$\overset{CH_3NO_2}{\underset{}{HOOC-CH_2-CH_2-\overset{|}{CH}-\overset{|}{CH}-CH_2-CH_2-COOH}}$$

4-Methyl-5-nitrooctanedioic acid

Rule C–13.2

13.21—For choice of the principal chain, an acid is normally named by method (*a*) of Rule **C–11.11** (-oic name) or an existing trivial name is used, unless use of the ending carboxylic acid (method (*b*) of Rule **C–11.11**) leads to citation of a larger number of carboxyl groups as ending of the name.

Examples:

Pentanoic acid $CH_3-CH_2-CH_2-CH_2-COOH$
or Valeric acid
(preferred to 1-Butanecarboxylic acid)

2-Ethylbutanoic acid $CH_3-CH_2-CH-CO_2H$
or 2-Ethylbutyric acid $|$
(preferred to 3-Pentanecarboxylic acid) CH_3-CH_2

Propylmalonic acid $CH_3-CH_2-CH_2-CH(COOH)_2$
(preferred to 1,1-Butanedicarboxylic acid)

1,1,4-Butanetri- $\overset{COOH}{\overset{|}{}}$
  carboxylic acid $HOOC-CH_2-CH_2-CH_2-CH-COOH$
(preferred to 2-Carboxyadipic acid)
(not 3-Carboxypropylmalonic acid)

Rule C–13.3

13.31—Methods similar to those described in Rule **C–13.2** are used for choice between alternative names for aldehydes, nitriles, and derivatives of carboxylic acids and aldehydes (see Subsection C–0.11).

Examples:

Hexanal $CH_3—CH_2—CH_2—CH_2—CH_2—CHO$

(preferred to 1-Pentanecarbaldehyde)

$$\begin{array}{l} CH_3O—OC \qquad\qquad\qquad\qquad\qquad CO—OCH_3 \\ \qquad\qquad \rangle CH—CH_2—CH—CH \langle \\ CH_3O—OC \qquad\qquad\quad | \qquad\qquad CO—OCH_3 \\ \qquad\qquad\qquad\qquad CH_2—CO—OCH_3 \end{array}$$

Tetramethyl 2-(methoxycarbonylmethyl)-1,1,4,4-butanetetracarboxylate

(preferred to Dimethyl 2,5-bis(methoxycarbonyl)-3-(methoxycarbonylmethyl)adipate)

(not Trimethyl 3-[bis(methoxycarbonyl)methyl]-1,1,4-butanetricarboxylate)

C–0.14. SENIORITY OF RING SYSTEMS*

Rule C–14.1

14.11—Seniority of ring systems is decided by applying the following criteria, successively in the order given, until a decision is reached:

(*a*) All heterocycles are senior to all carbocycles.

Example:

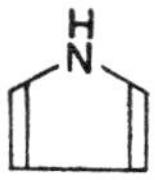

senior to

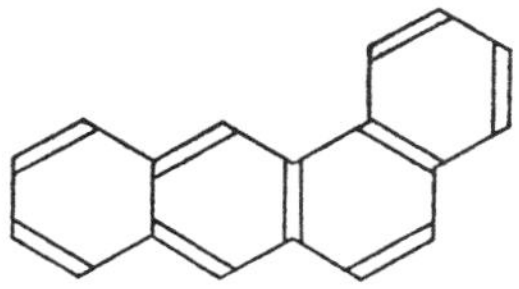

(*b*) For heterocycles the criteria based on the nature and position of the hetero atoms set out in Rule **B–3.1.**

Examples:

See Rule **B–3.1**, such as:

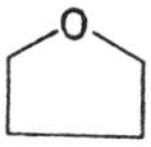

senior to

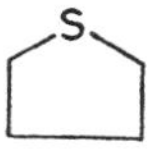

(*c*) Largest number of rings.

Example:

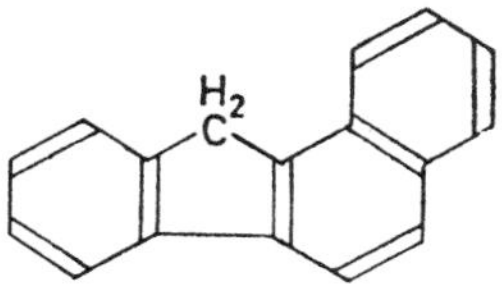

senior to

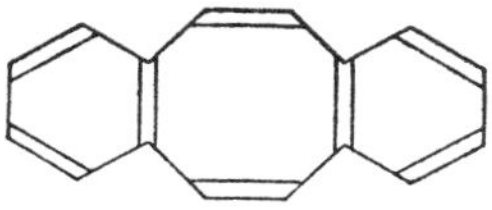

* Compare Rule **C–12.5.**

(*d*) Largest individual ring at first point of difference.

Examples:

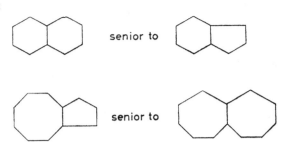

(*e*) Largest number of atoms in common among rings.

Examples:

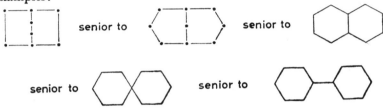

Note: Rings joined by a link (single or double) are included in this choice only when identical and named by the bi-, ter-, quater-, *etc.*, system (see Rule **A–54.1**).

(*f*) Lowest letters* (*a*, *b*, *etc.*, see Rule **A–21.5**) in the expression for ring junctions.

Example:

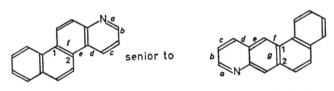

Naphtho [2,1– *f*]quinoline   Naphtho [1,2– *g*]quinoline

(*g*) Lowest numbers at the first point of difference in the expression for ring junctions. (See Rules: **A–21.5** for *ortho*-fusion and *ortho-peri*-fusion; **A–32** for tricyclo, *etc.*, systems; **A–41**, **A–42**, **B–10**, and **B–11** for spirans; **A–52** for assemblies of identical units.)

* Lowest means *a* before *b* before *c*, *etc.*

Examples:

Naphtho[1,2-*f*]quinoline   Naphtho[2,1-*f*]quinoline

senior to

Naphtho[2,3-*f*]quinoline

Tricyclo[5.3.1.0²,⁴]undecane   Tricyclo[5.3.1.0³,⁵]undecane

senior to

Spiro[cyclopentane-1,1'-indene]   Spiro[cyclopentane-1,2'-2'*H*-indene]
or Indene-1-spiro-1'-cyclopentane   or 2*H*-Indene-2-spiro-1'-cyclopentane

2,3'-Bipyridine   3,3'-Bipyridine

(*h*) Lowest state of hydrogenation.

Example:

senior to   senior to

(*i*) Lowest locant for indicated hydrogen.

Example:

Indole*   3*H*-Indole   3a*H*-Indole

(*j*) Lowest locant for point of attachment (if a radical).

Example:

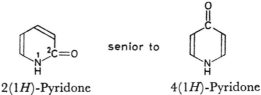

2-Pyridyl   3-Pyridyl

(*k*) Lowest locant for an attached group expressed as suffix.

Example:

2(1*H*)-Pyridone   4(1*H*)-Pyridone

(k₁) Maximum number of substituents cited as prefixes.

(*l*) Lowest locant for substituents named as prefixes, hydro prefixes, -ene, and -yne, all considered together in one series in ascending numerical order independently of their nature.

Examples:

2-Chloro-1-methyl   2-Chloro-3-methyl

3-Chloro-1,2-dihydro-2-methyl   2-Chloro-2,3-dihydro-3-methyl

* 1*H*- is understood to be present.

104

(*m*) Lowest locant for that substituent named as prefix which is cited first in the name (see Subsection C–0.16).

Example:

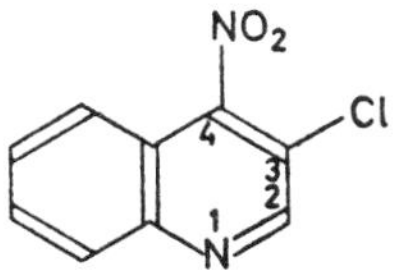

| 3-Chloro-4-nitro | senior to | 4-Chloro-3-nitro |

| 1-Ethyl-2-methyl | senior to | 2-Ethyl-1-methyl |

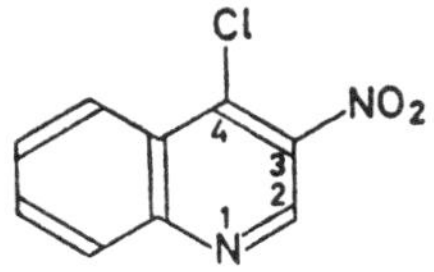

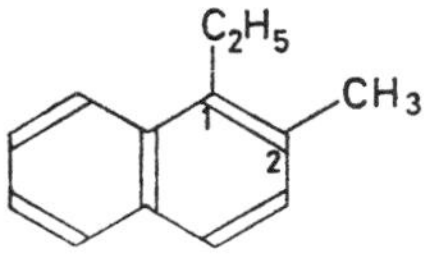

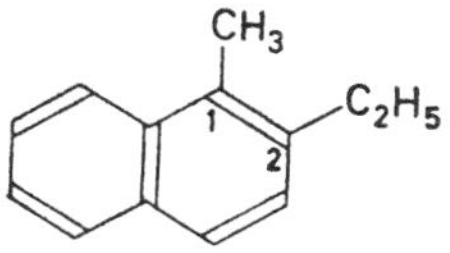

Note: Hydro and dehydro prefixes, if treated as detachable (see Rule C–16.11), are considered along with prefixes for substituents when this criterion is applied.

C–0.15.   NUMBERING OF COMPOUNDS*

Rule C–15.1

15.11—In so far as Sections A and B of the I.U.P.A.C. rules leave a choice, the starting point and direction of numbering of a compound are chosen so as to give lowest locants to the following structural factors (if present), considered successively in the order listed until a decision is reached; however principal groups take precedence over multiple bonds.

(*a*) Indicated hydrogen (whether cited in the name or omitted as being conventional).

Examples:

1*H*-Phenalene-4-carboxylic acid

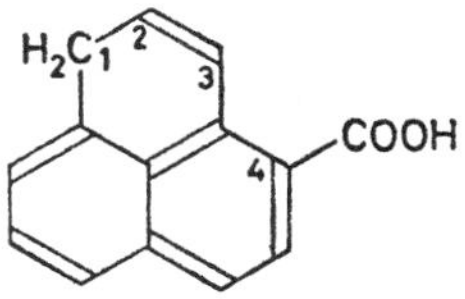

2*H*-Pyran-6-carboxylic acid

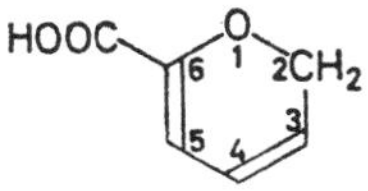

* *Cf.* Rule C–12.8.

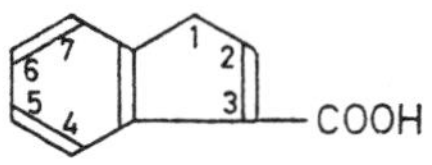

Indene-3-carboxylic acid

(b) Principal groups named as suffix.

Examples:

3,4-Dichloro-1,6-naphthalene-
   dicarboxylic acid

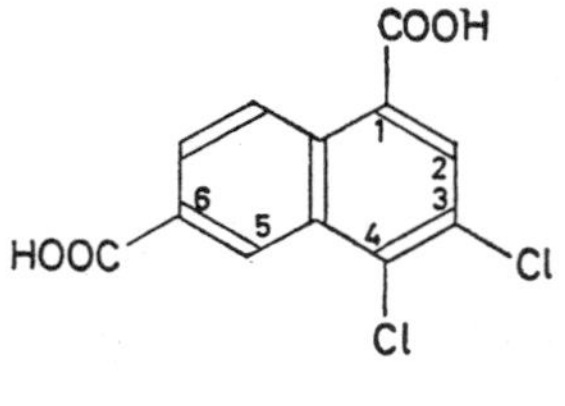

2-Cyclohexen-1-ol

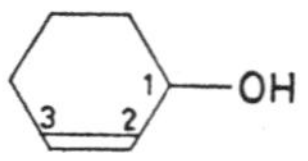

(c) Multiple bonds in acyclic compounds (see Rules **A–3.1** to **A–3.4**), in cycloalkanes (see Rule **A–11**), and in bi-, tri-, and poly-cycloalkanes (see Rules **A–31** and **A–32**), double having priority over triple bonds, and in heterocyclic systems whose names end in -etine, -oline, or -olene (see Rules **B–1.1** and **B–1.2**).

Examples:

3,4-Dichloro-1-cyclohexene

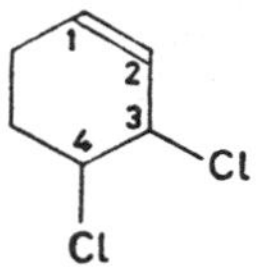

5-Chloro-2-pyrroline

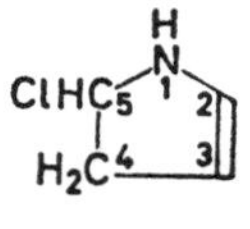

(d) Lowest locant for substituents named as prefixes, hydro prefixes, -ene and -yne, all considered together in one series in ascending numerical order.

Examples:

5,6-Dichloro-1,2,3,4-tetrahydronaphthalene

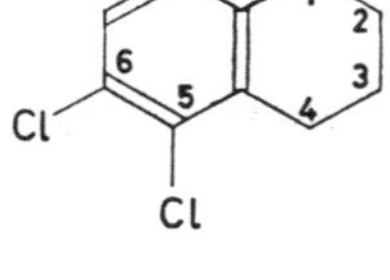

8-Hydroxy-4,5-dimethyl-2-azulenecarboxylic
                                  acid

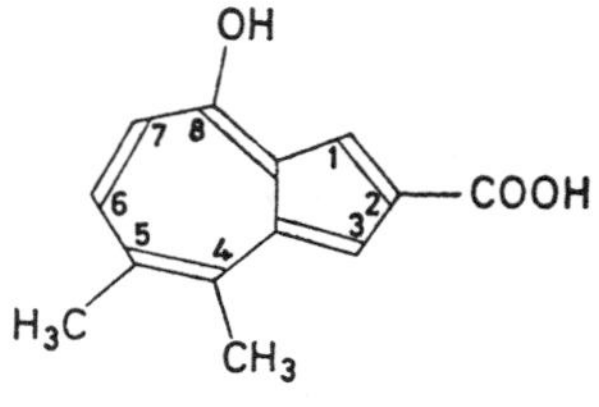

(*e*) Lowest locant for that substituent named as prefix which is cited first in the name (see Subsection C–0.16).

Examples:

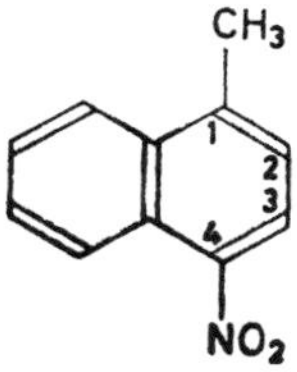

1-Methyl-4-nitronaphthalene 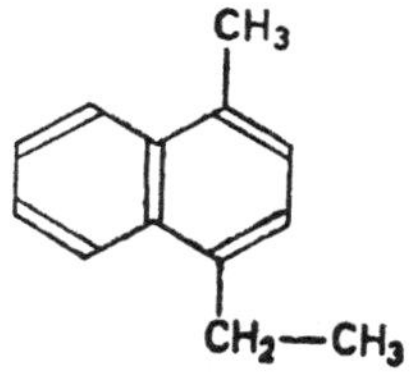   1-Ethyl-4-methylnaphthalene

Note: Hydro and dehydro prefixes, if treated as detachable (see Rule C–16.11), are considered along with prefixes for substituents when this criterion is applied.

Rule C–15.2

15.21—For cyclic radicals, indicated hydrogen and thereafter the point of attachment (free valency) have priority for lowest available number, the criteria of Rule C–15.1 being applied only if a choice remains after the requirements of the rules in Sections A and B have been satisfied. For acyclic radicals see Rules A–2.25, A–2.3, and A–3.5 to A–4.4.

Examples:

1*H*-Phenalen-7-yl

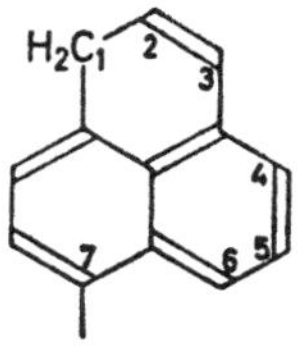

5-Carboxy-2-chlorophenyl

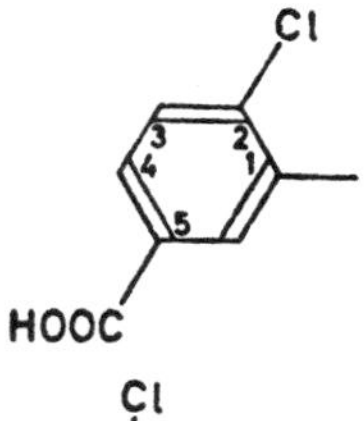

3,5,8-Trichloro-2-naphthyl

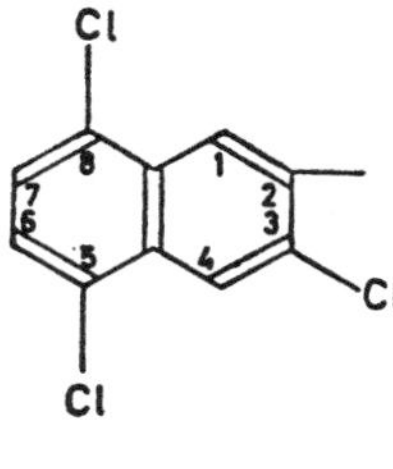

C–0.16.   ORDER OF PREFIXES

The alphabetical order of prefixes is used in this document. Examples showing the alphabetical arrangement of prefixes are given in Subsection C–0.12. Guidance for arranging the radicals according to alphabetical order is given in Rule **C–16.3**, this being an extension of the Rule **A–2.3** for acyclic hydrocarbon radicals.

General Procedure

Rule C–16.1

16.11—In the complete name of a compound or radical, the prefixes are arranged in order, as described in the following rules, in front of the parent name. For this purpose the term " parent name " includes any syllables denoting a change of ring members or relating to the structure of a carbon chain; modifying syllables of this kind always stay directly attached to the remainder of the parent name, and are termed " non-detachable ".

Examples:

(*a*) Non-detachable parts of parent names:

   (*i*) forming rings: cyclo, bicyclo, tricyclo, *etc.*, spiro.
   (*ii*) breaking rings: seco.
   (*iii*) changing size: nor, homo.
   (*iv*) fusing two or more rings: benzo, naphtho, imidazo, *etc.*
   (*v*) substituting one ring-member or chain-member for another: oxa, aza, azonia, *etc.*
   (*vi*) changing positions of ring-members or chain-members: iso, *sec-, tert-.*
   (*vii*) indicated hydrogen.
   (*viii*) forming bridges: ethano, *etc.*, epoxy, *etc.*
   (*ix*) hydro (see note).

(*b*) Detachable prefixes:

   (*i*) denoting substitution: chloro, amino, acetylamino (acetamido), epoxy, methyl, phenyl, imidazolyl, cyclohexyl, imidazotriazolyl, triazaphenanthryl, *etc.*
   (*ii*) forming hetero bridges: epoxy, *etc.*
   (*iii*) hydro (see note).

Note: Hydro and syllables denoting subtraction may be treated as detachable or non-detachable.

Rule C–16.2

16.21—Prefixes for sub-substituents (that is, substituents of substituents) are arranged in order amongst themselves in the complete name of the substituted substituent in the same way as are the prefixes for substituents of a parent compound; then the complete names of the substituted sub-stituents are treated as single (complex) prefixes to the name of the parent compound.

Examples:

(*a*) If isoquinoline is substituted by a chloronitrophenyl group and by a chlorofluoroethyl group, then

isoquinoline   chlorofluoroethyl   chloronitrophenyl

(*i*) Chloro and nitro are arranged amongst themselves and placed immediately in front of phenyl.

(*ii*) Chloro and fluoro are arranged amongst themselves and placed immediately in front of ethyl.

(*iii*) Chloronitrophenyl and chlorofluoroethyl are arranged amongst themselves in front of isoquinoline.

Together with locants, the following name is thus derived:

3-(2-Chloro-1-fluoroethyl)-5-(3-chloro-4-nitrophenyl)isoquinoline

(*b*) If benzoic acid is substituted by a group $\begin{array}{c} C_2H_5 \\ \diagdown \\ \diagup \\ CH_3 \end{array}$ N— in the *para*-position, then

(*i*) benzoic acid is selected as the parent name,

(*ii*) the ethyl and methyl radicals are arranged in order immediately in front of amino, and

(*iii*) ethylmethylamino is then placed in front of benzoic acid.

Together with the locant, the following name is thus derived:

p-(Ethylmethylamino)benzoic acid

Rule C–16.3

16.31—Simple prefixes (that is, those for unsubstituted substituents and for atoms) are first arranged alphabetically*; multiplying affixes (if necessary) are then inserted and do not alter the alphabetical order already attained.

* Use of an order of complexity given as alternative in the First Edition is abandoned

Examples:

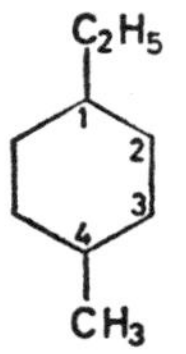

1-Ethyl-4-methylcyclohexane

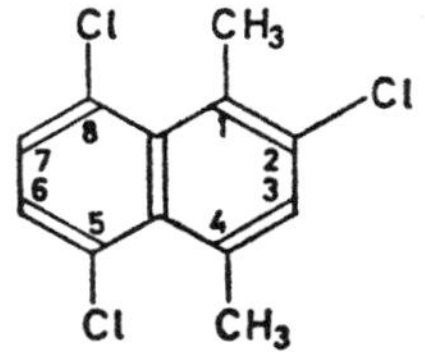

2,5,8-Trichloro-1,4-dimethylnaphthalene

(Letters causing the alphabetical arrangement in this and similar names are printed in bold type.)

16.32—For this method subtractive prefixes[*](such as anhydro, dehydro, demethyl: see Subsection C–0.4) may be alphabetized among those arising from substitutive nomenclature or treated as non-detachable. They are considered to begin with the first letter of the complete prefix.

Example:

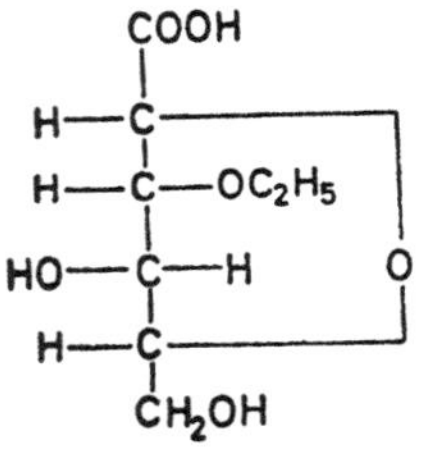

**2,5-Anhydro-3-*O*-ethyl-D-gulonic acid
or 3-*O*-Ethyl-2,5-anhydro-D-gulonic acid**

[*]Steroid Rule 2S-7.1 specifies alphabetic order for non-detachable prefixes when more than one is present.

16.33—Hydro prefixes may be treated as detachable and arranged in alphabetical order among other detachable prefixes or they may be treated as non-detachable from the name of the parent compound or radical.

Example:

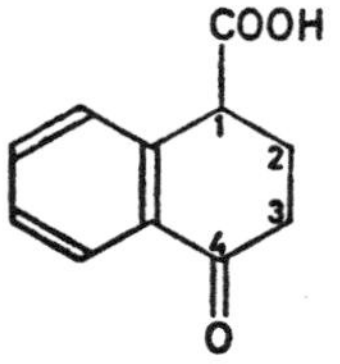

Detachable: 1,2,3,4-Tetrahydro-4-oxo-1-naphthoic acid
Non-detachable: 4-Oxo-1,2,3,4-tetrahydro-1-naphthoic acid

16.34—The name of a prefix for a substituted substituent is considered to begin with the first letter of its complete name.

Examples:

$$\overset{13}{CH_3}-[CH_2]_4-\overset{8}{CH_2}-\overset{7}{CH}-\overset{6}{CH_2}-\overset{5}{CH}-\overset{4}{CH_2}-\overset{3}{CH_2}-\overset{2}{CH_2}-\overset{1}{CH_3}$$

$$\overset{1}{CH_3}-\overset{2}{CH}-\overset{3}{CH}-\overset{4}{CH_2}-\overset{5}{CH_3}\quad\quad\overset{}{C_2H_5}$$

$$CH_3$$

7-(1,2-Dimethylpentyl)-5-ethyltridecane
(Dimethylpentyl as a complete substituent is alphabetized under " d ".)

1-Aminomethyl-2-**d**imethyl-
amino-3-**m**ethoxy-9,9-dimethyl-
fluorene-4-carboxylic acid

2-Cyano-4-(3-ethyl-1-methylureido)-
3,6-dihydroxybenzoic acid
(In the substituted radical, ethyl and methyl are first alphabetized before ureido, and the whole is then considered to begin with its first letter " e ".)

1-(2-Hydroxyethyl)-8-(2-hydroxy-propyl)-2-naphthoic acid

(The first point of difference, printed bold, is decisive.)

16.35—When two or more prefixes are composed of identical words, priority for citation is given to that radical which contains the lowest locant at the first-cited point of difference.

Example:

6-(**1**-Chloroethyl)-5-(**2**-chloroethyl)-
1-ethylbenzimidazole

16.36—*o-* (*ortho-*), *m-* (*meta-*), and *p-* (*para-*) Substituted substituents, if otherwise identical, are arranged in that order (the same as that of their numerical equivalents, 2-, 3-, and 4-, respectively).

Example:

2-(o-Nitrophenyl)-7-(m-nitrophenyl)naphthalene

C–0.2. RADICOFUNCTIONAL NOMENCLATURE

Although radicofunctional nomenclature is described in this Subsection, it is to be understood that substitutive nomenclature is, in general, to be preferred.

The procedures of radicofunctional nomenclature are identical with those of substitutive nomenclature except that suffixes are never used. Instead of the principal group being named as suffix, the functional class name (characteristic activity) of the compound is expressed as one word and the remainder of the molecule as another.

Acids and their derivatives, aldehydes and their derivatives, and amines are named as described in Subsection C–0.11 for substitutive nomenclature.

Rule C–21

21.1—One characteristic group in the compound is expressed as one word (called the " functional class name "), as shown in Table IV. Provided that this group is univalent, the remainder of the molecule attached to that group is expressed in its radical form as another word which precedes the functional class name.

Example:

$$CH_3—OH$$

Characteristic group, OH:	Alcohol (functional class name)
Radical, CH_3:	Methyl
Complete name:	Methyl alcohol

Note: In English and French the functional name and the radical are named as separate words. The radical name has adjectival significance; so in English the radical name precedes the name of the functional class (methyl alcohol) but in French it may follow it (*alcool méthylique*). In some other languages the name of the radical precedes that of the functional class name but the whole is written as one word (*Methylalkohol*).

Rule C–22

22.1—When the functional class name refers to a characteristic group that is bivalent, the two radicals attached to it are each named; when different, they are written as separate words, being arranged in alphabetical order (see Subsection C–0.16).

Examples:

$$CH_3—CH_2—CO—CH_3$$

Functional class name for CO:	Ketone
Radicals, $CH_3—CH_2$:	Ethyl
CH_3:	Methyl
Complete name:	

Ethyl methyl ketone

$$CH_3—CH_2—CH_2—CO—CH_2—CH_2—CH_3$$

Functional class name for CO:	Ketone
Radicals, $CH_3—CH_2—CH_2$:	Propyl
Complete name:	Dipropyl ketone

TABLE IV. Some functional class names used in radicofunctional nomenclature, in order of decreasing priority for choice as such (see Rule C–23)

Group	Functional class name
X in acid derivatives $RCO—X$, $RSO_2—X$ etc.	Name of X; in the order fluoride, chloride, bromide iodide; cyanide, azide, etc.; then their S, followed by their Se analogues
—CN, —NC	Cyanide, isocyanide
>CO	Ketone, then S, then Se analogues
—OH	Alcohol; followed by S and then Se analogues
—O—OH	Hydroperoxide
>O	Ether or oxide
>S, >SO, >SO$_2$	Sulfide, sulfoxide, sulfone
>Se, >SeO, >SeO$_2$	Selenide, selenoxide, selenone
—F, —Cl, —Br, —I	Fluoride, chloride, bromide, iodide
—N$_3$	Azide

Rule C–23

23.1—When a compound contains more than one kind of group listed in Table IV, that kind is cited as functional class name which occurs higher in the Table, all others being expressed as prefixes.

Example:

2-Hydroxyethyl methyl ketone $HO—CH_2—CH_2—CO—CH_3$

Rule C–24

24.1—The procedures described for completing substitutive names in Subsections C–0.12 to C–0.16 are applied also in radicofunctional nomenclature, consideration of the functional class name replacing that of the suffix (for instance, for choice of the principal chain).

C–0.3. ADDITIVE NOMENCLATURE

Rule C–31

31.1—Additive nomenclature involves naming atoms added to the structure denoted by the rest of the name. (See also Rule D-1.5)

Rule C–32

32.1—The prefix " hydro- ", to denote added hydrogen atoms, is the only case where addition is expressed by a prefix; its use is described in Rules **A–23.1, B–1.1, B–1.2** and **C–16,** but is extended also to trivial names.

Examples:

1,2,3,4-Tetrahydronaphthalene   1,2-Dihydroazocine

Dihydroquinine
(asterisks denote the
  positions to which
hydrogen has been
added).

In spite of precedent it is better, even when devising trivial names for a series of naturally occurring or other related products, to choose as parent either the fully saturated compound or one with the maximum number of non-cumulative double bonds, so that " ene " syllables or " hydro " prefixes can be used with their normal connotation.

Rule C–33

33.1—In additive nomenclature, except for hydro prefixes, the name of the added atoms is placed, in its ion form, after the name of the parent compound. This method of nomenclature is in general discouraged, although it is useful in a few specialized cases, particularly when the stereochemistry of addition is unknown or when the systematic name of the adduct might obscure the relation to the parent compound; examples are:

Styrene oxide
[2-**Phenyloxirane** or
(1,2-**Epoxyethyl**)benzene]

2-**Hexene ozonide** $C_6H_{12}O_3$

Cholesterol dibromide
(5,6ξ-Dibromo-5ξ-cholestan-
3β-ol)

Rule C-34

34.1—The use of the prefix " homo- " to denote incorporation of CH_2 as an additional member into a ring is described in Steroid Rules* **2S-7.1** and **2S-7.3**. Use of the prefix " seco- " to denote ring fission, with addition of a hydrogen atom at each terminal group thus created, is described in Rule **2S-8.1**. Wherever feasible, however, systematic names are preferred; for example (3,4-methylenedioxyphenyl)acetic acid is preferred to homo-piperonylic acid.

C-0.4. SUBTRACTIVE NOMENCLATURE

Subtractive prefixes may be used to express removal of atoms or groups from a compound denoted by a systematic or trivial name.

In a few cases its use is widely recognized: unsaturation in aliphatic compounds and cycloalkanes, bicycloalkanes, *etc.*, is specified by endings " -ene " and " -yne " denoting loss of hydrogen (Rules **A-3** and **A-11.3**); loss of water from two molecules of a monobasic acid or internal loss of water from a dibasic acid is described by the functional class name " anhydride " (Subsection C-4.9); other uses are for naming lactones and related compounds (Subsection C-4.7) and their sulfur analogues.

Three further methods of subtractive nomenclature are in use, namely, the use of the prefixes " de- " (Rule **C-41**), " nor- " (Rules **C-42** and **C-43**) and " anhydro- " (Rule **C-44**).

Rule C-41

41.1—The prefix " de- " (not " des- ")† , followed by the name of a group or atom (other than hydrogen), denotes replacement of that group or atom by hydrogen.

Example:

De-*N*-methylmorphine

41.2—Loss of two hydrogen atoms from a compound designated by a trivial name is denoted by a prefix " didehydro- ".

Example:

7,8-Didehydrocholesterol

---

*Rules for Nomenclature of Steroids, *Pure and Applied Chemistry,* 1972, Vol. 30, Nos. 1-2.
† In some languages (for example, French and German) " des " is, however, customarily preferred.

Note: In common usage " dehydro- " is often used in place of " didehydro- ". For example the above compound is often termed dehydrocholesterol; another such example is dehydroascorbic acid, where " dehydro " denotes loss of hydrogen from two hydroxyl groups in the conversion of HO—C═C—OH into CO—CO at positions 2 and 3:

Didehydro-L-ascorbic acid
(commonly called
Dehydro-L-ascorbic acid)

$$
\begin{array}{c}
\text{CO} \\
|\\
\text{CO} \\
|\\
\text{CO} \quad\quad \text{O}\\
|\\
\text{H—C} \\
|\\
\text{HO—C—H} \\
|\\
\text{CH}_2\text{OH}
\end{array}
$$

41.3—As an exception, " deoxy- ", when applied to hydroxy compounds, denotes replacement of a hydroxyl group by a hydrogen atom.

Example:

6-Deoxy-α-D-glucopyranose

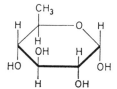

Rule C–42

42.1—By a special convention, which should not be extended but has been used particularly for terpenes (compare Rules A–72.1 and A–74.2), " nor- " denotes replacement by hydrogen of all methyl groups attached to a ring system.

Example:

Pinane   Norpinane

Rule C–43

43.1—In more recent use, particularly for steroids* (see Rules 2S–6.1 and 2S–7.1) and higher terpenes, " nor- " denotes (a) elimination of one methylene group from a chain, the highest permissible locant being used, or (b) contraction of a ring by one CH_2 unit, the locant being the capital letter identifying the ring. Elimination of two or more methylene groups may be denoted by the prefixes " dinor-", " trinor- ", and so on.

* See footnote on p. 115.

Examples:

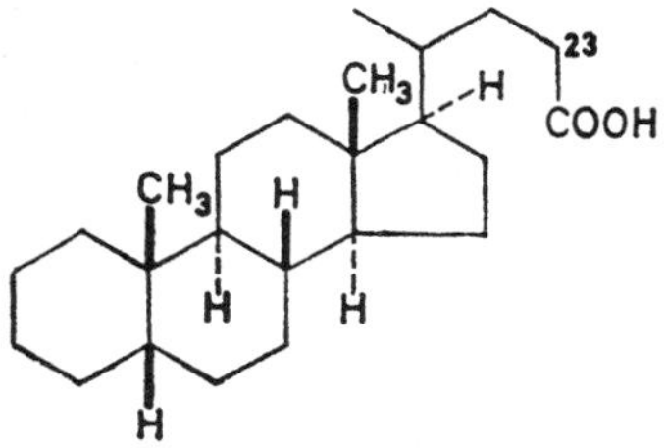

5β-Cholanoic acid

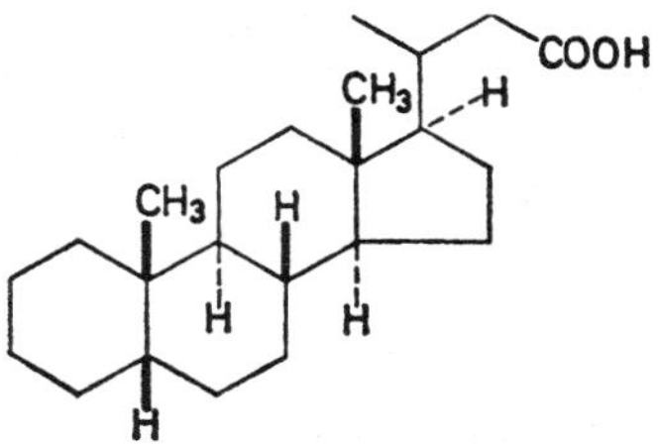

23-Nor-5β-cholanoic acid

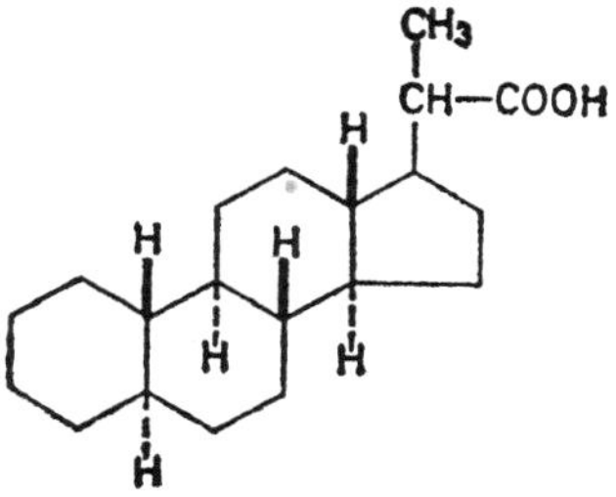

18,19-Dinor-5α-pregnane-20-carboxylic acid

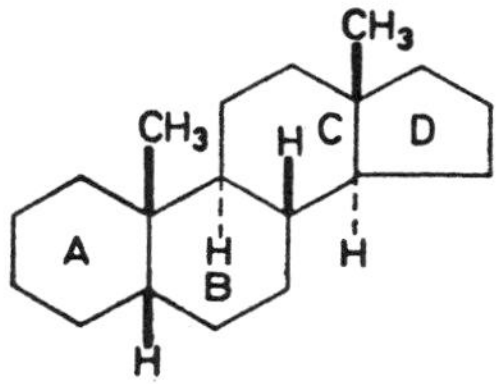

5β-Androstane

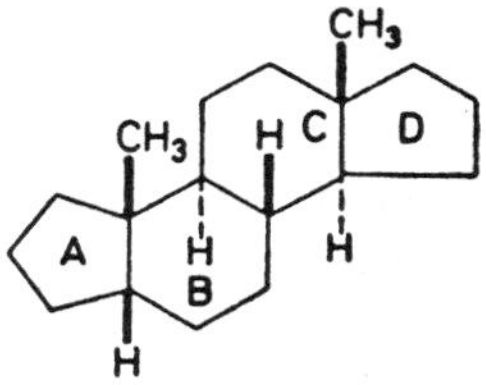

A-Nor-5β-androstane

Rule C-44

44.1—The prefix "anhydro-" denotes the loss of the elements of water from within one molecule, the positions from which removal has occurred being indicated by locants.

Example:

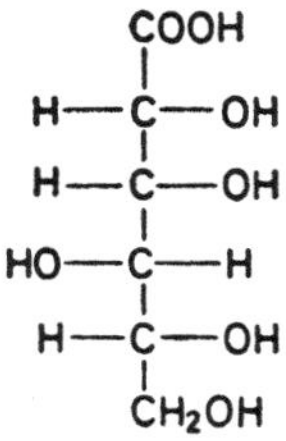

D-Gulonic acid

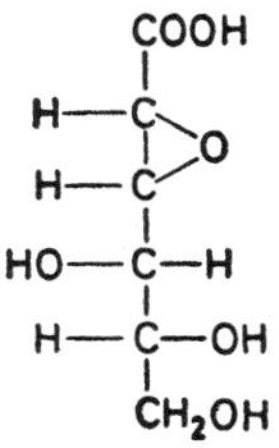

2,3-Anhydro-D-gulonic acid

C–0.5. CONJUNCTIVE NOMENCLATURE

The conjunctive nomenclature described in this Subsection is used especially in *Chemical Abstracts*, where it was formerly termed " additive nomenclature ". It is particularly useful in inverted indexes.

Conjunctive nomenclature may be applied when a principal group is attached to an acyclic component that is directly attached by a carbon–carbon bond to a cyclic component.   The name of the cyclic component is attached directly in front of that of the acyclic component, each name being that of the component as a molecule, that is as if the other component were replaced by hydrogen. This juxtaposition implies that hydrogen has been eliminated from each component by a process of mutual substitution, as in coroneneacetic acid (which would be named coronenylacetic acid by substitutive nomenclature).   For exceptional treatment of names of parent compounds requiring indicated hydrogen see Rule C–56.1.

Conjunctive nomenclature may be used when the acyclic component is an amine named by Rule C–812.1(*a*), for example ethylamine, and when the acyclic component is a monocarboxylic acid having a trivial name; it is, however, not used when an unsaturated side chain is named systematically, owing to difficulty in specifying the site of unsaturation.

The precise procedures are governed by the following rules:

Rule C–51

51.1—When a side chain attached to a cyclic component of a structure carries one principal group, the name of the molecule corresponding to the cyclic component is placed before the substitutive name of the molecule corresponding to the side chain.

Example:

Cyclohexanemethanol

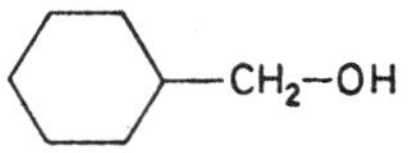

51.2—When necessary, the position of the side chain is indicated by a numeral placed before the name of the cyclic component.   Carbon atoms of the side chain are indicated by Greek letters proceeding from the principal group to the cyclic component, but these are used in the name only to locate substituents in the side chain.   The terminal carbon atom of acids, aldehydes, and nitriles is omitted when allocating Greek positional letters.

Examples:

2-Naphthalenepropanol

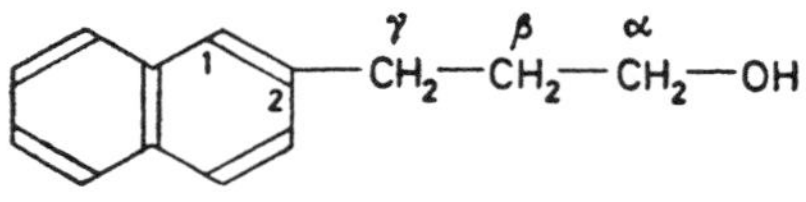

3-Quinolineacetic acid

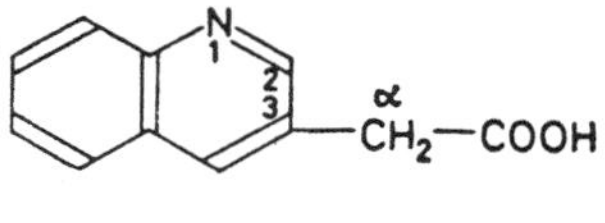

α-Chloro-3-quinoline-
acetic acid

Note: Conjunctive nomenclature is not used when the side chain carries more than one of the principal group (see Rule **C–51.1**). As exceptions to this note, conjunctive nomenclature may, however, be used with the names malonic and succinic acid.

Examples:

2-Naphthalenemalonic acid

β-Chloro-3-quinolinesuccinic
acid

Rule C–52

52.1—For all purposes in conjunctive nomenclature the side chain is considered to extend only from the principal group to the cyclic component. Any other chain members, even those extending this side chain terminally, are named as substituents, appropriate prefixes being placed before the name of the cyclic component.

Example:

α,β-Dimethyl-2-naphthaleneethanol

52.2—When a cyclic component carries more than one identical side chain*, the name of the cyclic component is followed by di-, tri-, *etc.* (to indicate the number of such side chains), and then by the name of the acyclic component, and it is preceded by the locants for the side chains.**

* Throughout this Subsection the phrase " identical side chains " denotes side chains of the same type whose substituents are identical in type, number, and position.
**Bis-, tris-, *etc.* are used with alkylamine side chains to avoid ambiguity, e.g. 1,2-cyclo-hexanebis(ethylamine), not 1,2-cyclohexanediethylamine.

119

Example:

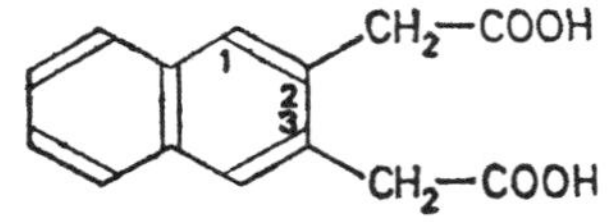

2,3-Naphthalenediacetic acid

52.3—When side chains of two or more different kinds are attached to a cyclic component, only the senior side chain (see Subsection C–0.13) is named by the conjunctive method. The remaining side chains are named as prefixes.

Examples:

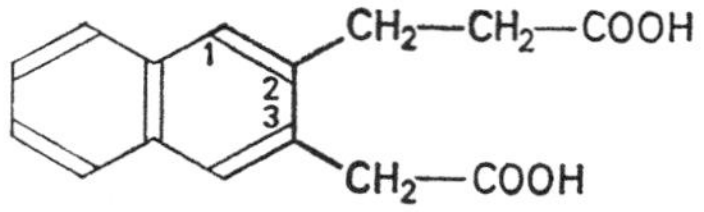

3-(Carboxymethyl)-2-naphthalenepropionic acid

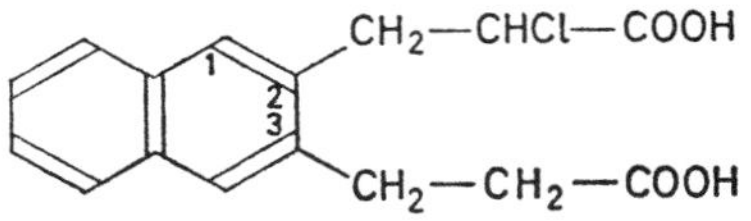

3-(2-Carboxyethyl)-α-chloro-2-naphthalenepropionic acid

Rule C–53

53.1—Benzene derivatives may be named by conjunctive nomenclature only when two or more identical side chains are present.

Example:

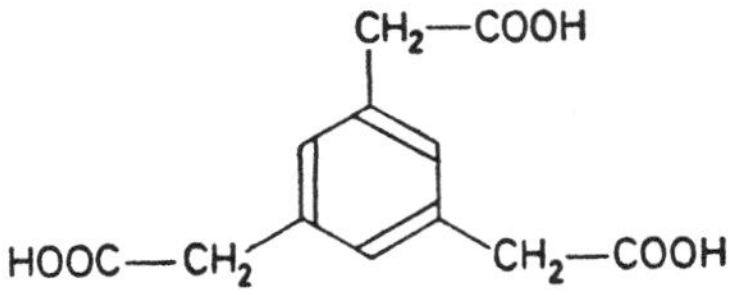

1,3,5-Benzenetriacetic acid

53.2—Trivial names for oxo carboxylic acids (see Rule C–416.3) may be used for the acyclic component.

Examples:

2-Naphthaleneglyoxylic acid

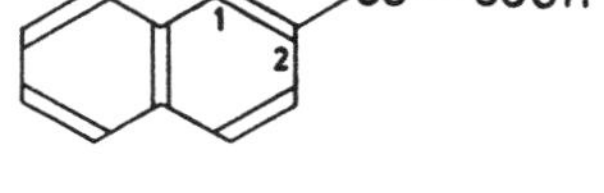

1-Indeneacetoacetic acid

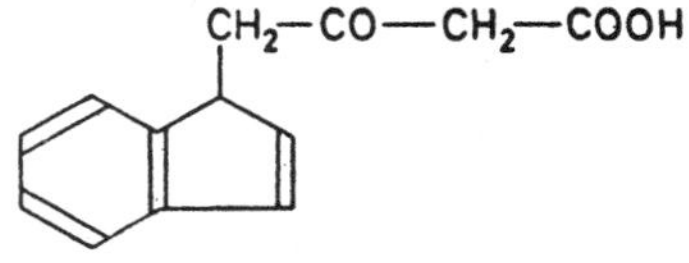

Rule C–54

54.1—When different side chains are attached to a cyclic component, (*a*) that type is chosen for naming by the conjunctive method which contains the principal group, or (*b*), if there is more than one side chain containing this group, that type which is present in greatest number, or (*c*), if still no decision has been reached, that which has the requirements for choice as principal chain. (For the last-mentioned criterion it must be remembered that the " side chain " does not extend beyond the cyclic component and most remote principal group.) All other side chains are named as substituents, by prefixes placed before the name of the cyclic component.

Examples:

(*a*) 3-(2,3-Dihydroxypropyl)-2-quinolinevaleric acid

(*b*) 1-(2-Carboxyethyl)-2,3-naphthalenediacetic acid

(*c*) 3-(Carboxymethyl)-2-quinolinepropionic acid

Note: Conjunctive nomenclature is not used when numbering difficulties might arise, as with, for example, unsaturated acids named systematically (names ending in " -enoic ", " -dienoic acid ", *etc.*; see **Rule C–401**). In such cases it would become necessary to denote unsaturation by Greek letters.

54.2—If there is a choice for cyclic component the senior is chosen (see Subsection C–0.14 on seniority of ring systems).

Examples:

6-{[6-(Carboxymethyl)-2-naphthyl]methyl}-2-quinolineacetic acid β-Cyclohexyl-1-naphthalenepropionic acid

Rule C–55

55.1—If the cyclic and the acyclic component are joined by a double bond, the locants of this bond are placed as superscripts to a Greek capital delta that is inserted between the names of the two components. The locant for the cyclic precedes that for the acyclic component.

Example:

Indene-$\Delta^{1,\alpha}$-acetic acid

Rule C–56

56.1—If the insertion of a side chain into a cyclic component has necessitated also addition of a hydrogen atom at another position, then this hydrogen atom is specified in the name by H (indicated hydrogen, see Rule A–21.6).

Example:

4a($2H$)-Naphthaleneacetic acid

Rule C–57

57.1—Radicals with the free valency in the acyclic component are named by the customary modification of the name of the latter.

Example:

2-Naphthaleneacetyl or 2-Naphthylacetyl

Rule C-58

58.1—As an exception, compounds containing carbamic acid side chains joined by nitrogen to the cyclic component may be designated by conjunctive nomenclature, the NH—COOH chain being considered as the acyclic component.

Example:

N-Methylindene-2-carbamic acid

58.2—As a further exception, heterocycles having a side chain attached to a ring-nitrogen atom may be designated by conjunctive nomenclature.

Example:

Indole-1-propionic acid

Note: For names such as ethylenediaminetetraacetic acid see note to Rule C-815.1.

C-0.6. REPLACEMENT NOMENCLATURE

The system of naming heterocyclic compounds whereby carbon atoms of their hydrocarbon analogues are replaced by hetero atoms has been described in Section B under the title of " a " nomenclature. The following Rules embody an extension of that method to be used for naming chains; namely: the chain is considered to be that of an acyclic hydrocarbon in which some CH_2 groups are replaced by hetero atoms. This system, as applied to cyclic and acyclic compounds, is now termed " replacement nomenclature ".

Replacement nomenclature can be applied to acyclic structures whether or not cyclic components are attached to them. It is intended for use only when other nomenclature systems are difficult to apply in the naming of chains containing hetero atoms; it is not expected to be of value when the degree of complexity is notably less than in the examples given below, particularly since it sometimes leads to names which are of unfamiliar type.*

Rule C-61

61.1—In an unbranched structure not containing a group that can be named as principal group, the longest chain of carbon and hetero atoms terminating with carbon is chosen and named as though the entire chain were that of an acyclic hydrocarbon. The hetero atoms present within this

*(See also Rule D-1.6)

123

chain are then named by means of prefixes " aza- ", " oxa- ", " thia- ", *etc.*, as listed in Table I of Rule **B–1**, in the order stated in Rule **B–1.4**, with locants to indicate their positions in the chain. Methods described in Section A are then used for multiplying prefixes and for denoting unsaturation. The chain is numbered from one end to the other, excluding terminal hetero atoms, according to precedences described in Rule **C–62**.

Examples:

$$\overset{13}{C}H_3-\overset{12}{C}H_2-\overset{11}{O}-\overset{10}{C}H_2-\overset{9}{C}H_2-\overset{8}{O}-\overset{7}{C}H_2-\overset{6}{C}H_2-\overset{5}{O}-\overset{4}{C}H_2-\overset{3}{C}H_2-\overset{2}{O}-\overset{1}{C}H_3$$

2,5,8,11-Tetraoxatridecane

$$\overset{8}{N}=\overset{7}{C}H=\overset{6}{C}H-\overset{5}{C}H_2-\overset{4}{O}-\overset{3}{C}H_2-\overset{2}{C}H=\overset{1}{C}H_2$$
$$\overset{}{N}=\overset{}{C}H=\overset{}{C}H-\overset{}{C}H_2-\overset{}{S}-\overset{}{C}H_2-\overset{}{C}H=\overset{}{C}H_2$$
$$\quad 9 \quad 10 \quad 11 \quad 12 \quad 13\ 14 \quad 15 \quad 16$$

4-Oxa-13-thia-8,9-diaza-1,6,8,10,15-hexadecapentaene

Rule C–62

62.1—In an unbranched structure containing one or more groups that can be named as principal groups the latter are selected in the same way as for carbon chains and are similarly named; substitutive nomenclature is used. Locants are applied to the chain, excluding terminal hetero atoms, in accordance with the following criteria, applied successively so far as necessary:

(*i*) lowest locants to the principal groups;

(*ii*) lowest locants to the hetero atoms considered together, and if there is a choice, to the hetero atoms cited first in Table I of Rule **B–1**. In the order of precedence any "onia" prefix follows the corresponding "a" prefix.

(*iii*) according to the principles stated in Rules **A–3.1, A–3.2, A–3.3,** and **A-3.4.**

Examples:

$$\overset{10}{C}H_2-\overset{9}{O}-[CH_2]_2-\overset{6}{O}-[\overset{5,4}{C}H_2]_2-\overset{3}{O}-\overset{2}{C}H_2-\overset{1}{C}H_2-OH$$
$$\overset{}{C}H_2-\overset{}{O}-[CH_2]_2-\overset{}{O}-[CH_2]_2-\overset{}{S}-[CH_2]_{11}-\overset{}{C}H_3$$
$$11 \quad 12 \qquad 15 \qquad 18 \qquad 30$$

3,6,9,12,15-Pentaoxa-18-thiatriacontan-1-ol

$$\overset{11}{H_2N}-\overset{10}{C}H_2-\overset{9}{C}H_2-\overset{8}{N}H-\overset{7}{C}H_2-\overset{6}{C}H_2-\overset{5}{N}H-\overset{4}{C}H_2-\overset{3}{C}H_2-\overset{2}{N}H-\overset{1}{C}H_2-CH_2-NH_2$$

3,6,9-Triazaundecane-1,11-diamine

$$\overset{15}{C}H_3-\overset{14}{C}H_2-\overset{13}{O}-[CH_2]_2-\overset{10}{O}-[CH_2]_2-\overset{7}{O}-[CH_2]_2-\overset{4}{O}-[CH_2]_2-\overset{1}{C}OOH$$

4,7,10,13-Tetraoxapentadecanoic acid

$$\overset{10}{H_2N}-\overset{9}{CO}-\overset{8}{NH}-\overset{7}{CH_2}-\overset{6}{CH_2}-\overset{5}{O}-\overset{4}{CH_2}-\overset{3}{CH_2}-\overset{2}{O}-\overset{1}{CH_2}-COOH$$

10-Amino-10-oxo-3,6-dioxa-9-azadecanoic acid

$$\overset{14}{CH_3}-\overset{13}{CHCl}-\overset{12}{O}-[CH_2]_2-\overset{9}{NH}-[CH_2]_2-\overset{6}{O}-[CH_2]_2-\overset{3}{S}-CH_2-\overset{1}{C}{\equiv}N$$

13-Chloro-6,12-dioxa-3-thia-9-azatetradecanenitrile

$$\overset{17}{H_2N}-CO-[CH_2]_3-\overset{13}{S}-[CH_2]_3-\overset{9}{NH}-[CH_2]_3-\overset{5}{O}-[CH_2]_3-\overset{1}{CO}-NH_2$$

5-Oxa-13-thia-9-azaheptadecanediamide

$$\overset{25}{CH_3}-O-CO-[CH_2]_2-\overset{19}{O}-CO-[CH_2]_2-\overset{15}{O}-\overset{23}{CO}-CH_2\diagdown$$
$$\underset{1}{CH_3}-O-\underset{3}{CO}-[CH_2]_2-\underset{6}{O}-CO-[CH_2]_2-\underset{10}{O}-CO-CH_2\diagup{}^{S^{13}}$$

2,6,10,16,20,24-Hexaoxa-13-thiapentacosane-3,7,11,15,19,23-hexone

(Note: In polyesters named by this method the locants of carbonyl groups differ by unity from the locants of oxa groups. Similar differences apply in the names of polyamides.)

$$\overset{16}{NH_2}-CH_2-\overset{15}{CH}-\overset{14}{CH_2}-[NH-CH_2-CH]_4-\overset{1}{CH_2}-NH_2$$
$$\qquad\quad\underset{NH_2}{|}\qquad\qquad\qquad\underset{NH_2}{|}$$

4,7,10,13-Tetraazahexadecane-1,2,5,8,11,15,16-heptamine

$$\overset{11}{NO_2}-\overset{10}{CH_2}-\overset{9}{CH_2}-\overset{8}{SO_2}-\overset{7}{CH_2}-\overset{6}{CH_2}-\overset{5}{O}-\overset{4}{CH_2}-\overset{3}{CH_2}-\overset{2}{S}-\overset{}{CH_2}-\overset{1}{CH_2}-OH$$

11-Nitro-6-oxa-3,9-dithiaundecan-1-ol 9,9-dioxide

(cf. Rule D-1.5)

Rule C–63

63.1—(*a*) In a branched chain to which replacement nomenclature is applied, the principal chain is chosen in accordance with the criteria, applied successively so far as necessary, that are listed in Rule **C–13.11**, terminal non-carbon atoms being excluded except insofar as choice of the principal group is concerned.

However, the following additional criteria are then inserted in that list:

(*i*) maximum number of hetero atoms: to be considered immediately after the maximum number of principal groups [**C–13.11** (*a*)];

(*ii*) maximum number of hetero atoms first cited in Rule **B–1** (Table I): to be considered immediately after the maximum length [**C–13.11** (*c*)], terminal non-carbon atoms being excluded;

(*iii*) lowest locants for hetero atoms all together, and then for those first cited in Rule **B–1** (Table I): to be considered immediately after the lowest locants for principal groups [**C–13.11** (*e*)].

(*b*) When replacement nomenclature is used for a principal chain it is used also for all the side chains provided that there is more than one hetero atom in any one of the side chains, but in other cases the customary names* are used for side chains.

(*c*) Locants for the main chain are assigned as described in Rule **C–62** and, so far as necessary thereafter, as described in Subsection C–0.13 excepting Rule **C–13.11**(a). Side chains are numbered from the point of attachment, which has locant 1.

Examples:

$$\underset{9}{\text{HO–[CH}_2]_2}-\underset{}{\text{O}}-\underset{6}{\text{[CH}_2]_2}-\underset{}{\text{O}}-\underset{3}{\text{[CH}_2]_2}-\text{CH} \underset{\text{O–CH}_3}{\overset{\overset{2\quad 1}{\text{CH}_2\text{–COOH}}}{<}}$$

11-Hydroxy-3-methoxy-6,9-dioxaundecanoic acid

$$\underset{10\ 9}{\text{HOOC–}}\underset{8\ 7}{\text{CH}_2\text{–O–CH}_2} \atop \text{CH}_3\text{–O–[CH}_2]_4\text{–O}} \overset{6\ 5}{C{=}C} \underset{\text{O–[CH}_2]_4\text{–O–CH}_3}{\overset{\text{CH}_2\text{–O–CH}_2\text{COOH}}{}}$$

5,6-Bis(1,6-dioxaheptyl)-3,8-dioxa-5-decenedioic acid

$$\underset{\text{CH}_3\text{–CH}_2\text{–CH}_2}{\overset{11\quad 10}{\text{CH}_3\text{–O}}}{>}\underset{9}{\text{CH}}-\underset{6}{\text{[CH}_2]_2}-\text{S}-\underset{3}{\text{[CH}_2]_2}-\text{NH}-\underset{2,1}{\text{[CH}_2]_2}-\text{OH}$$

9-Propyl-10-oxa-6-thia-3-aza-1-undecanol

$$\underset{\text{CH}_3\text{–O–CH}_2}{\overset{9}{\text{CH}_3\text{–O–[CH}_2]_2}}-\underset{}{\overset{6\ 5}{\text{S–CH}_2}}{>}\underset{4}{C}\underset{\text{CH}_2\text{–NH–CH}_3}{\overset{\overset{3\quad 2\ 1}{\text{NH–O–CH}_3}}{<}}$$

4-(Methoxymethyl)-4-[(methylamino)methyl]-2,9-dioxa-6-thia-3-azadecane

Rule C–64

64.1—A chain terminated by a cyclic system at one or at all ends is treated as in Rules **C–61**, **C–62**, and **C–63**, the cyclic system(s) being represented by names described in Section A or B.

* That is, not by replacement nomenclature.

Examples:

$$\overset{13}{C_6H_5}-\overset{12}{CH_2}-O-[CH_2]_2-\overset{9}{O}-[CH_2-CH_2-NH]_2-\overset{2}{CH_2}-\overset{1}{COOH}$$

13-Phenyl-9,12-dioxa-3,6-diazatridecanoic acid

2,2'-(2,4,6,8-Tetraazanonamethylene)bis(1-methyl-
quinolinium iodide)

Rule C–65 (Alternative to parts of Rules C–61 to C–64)

65.1—When the main chain of a compound that is to be treated by replacement nomenclature ends in —OR, —SR, —COOR, —CONHR, —CONR¹R², —NHR, or NR¹R², that group may be named as an ether, sulfide, ester, substituted amide, or substituted amine group (as appropriate), provided that the chain of the group R, R¹, or R² does not contain a hetero atom.

Examples:

$$C_6H_5-CO-NH-[\overset{10,9}{CH_2}]_2-\overset{8}{S}-[CH_2]_2-\overset{5}{O}-[CH_2]_2-\overset{2}{NH}-\overset{1}{COOH}$$

10-Benzamido-5-oxa-8-thia-2-azadecanoic acid

$$\overset{11}{CH_3}-\overset{10}{CO}-\overset{9}{O}-[\overset{8}{CH_2}-\overset{7}{CO}-\overset{6}{O}]_2-\overset{2}{CH_2}-\overset{1}{COOCH_3}$$

Methyl 4,7,10-trioxo-3,6,9-trioxaundecanoate

$$\begin{matrix} & & & CH_3 & & & CH_3 & \\ \overset{8}{CH_2}-\overset{7}{CO}-\overset{6}{N}-\overset{5}{CH_2}-\overset{4}{CO}-\overset{3}{N}-\overset{2}{CH_2}-\overset{1}{CO}-NH-CH_3 \\ \overset{9}{N}-\overset{10}{CO}-\overset{11}{CH_2}-N-\overset{12}{CO}-\overset{13}{CH_2}-\overset{14}{N}-CO-CH_3 \\ CH_3 \quad\quad CH_3 \quad\quad CH_3 \end{matrix}$$

N,3,6,9,12-Pentamethyl-14-(*N*-methylacetamido)-4,7,10,13-
tetraoxo-3,6,9,12-tetraazatetradecanamide

C–0.7. NOMENCLATURE OF ASSEMBLIES OF IDENTICAL UNITS

For the purposes of this Subsection, " assemblies of identical units " comprise two or more identical components and are of two kinds: (*a*) those in which the components are joined by a double or single bond, not necessarily at the same position, with loss of a hydrogen atom from each component at each such bond; and (*b*) those in which the components are substitutively linked, not necessarily at the same position, by a symmetrical bivalent, tervalent, or quadrivalent radical. The resulting compounds may also be termed doubled, tripled, *etc.*, compounds.

Rule C-71.   Ring assemblies involving valence bonds

71.1—A molecule in which two or more identical cyclic components are linked to one another by a double or single bond constitutes a ring assembly which is named in accordance with Rules **A-52** and **A-54**.

Examples:

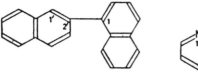

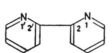

1,2'-Binaphthalene
or   1,2'-Binaphthyl

2,2'-Bipyridine
or   2,2'-Bipyridyl

1,1'-Biimidazole
or   1,1'-Biimidazolyl

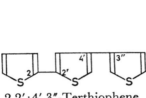

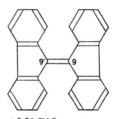

2,2':4',3''-Terthiophene

Δ⁹,⁹'-Bifluorene
or   9,9'-Bifluorenylidene

71.2—Radicals derived from unsubstituted ring assemblies are named by adding " -yl ", " -ylene " or " -diyl ", or " -triyl ", *etc.*, to the name of the assembly, with elision of a terminal "e" (if present) before " yl " or ," ylene ". Radical positions have numbers as low as permitted by the fixed numbering of the assembly. (cf. Rules A-55.1 and B-13).

Examples:

2,2'-Bipyridin-6-yl
or   2,2'-Bipyridyl-6-yl

6,6'-Dichloro-2,2'-
biphenylylene
or   6,6'-Dichloro-2,2'-
biphenyldiyl

71.3 (Partially alternative to Rule C-71.4)—Derivatives of ring assemblies are named systematically with the principal group added as suffix to the name of the ring assembly and any other substituents are named as prefixes. The points of union of the cyclic components are given the lowest possible numbers consistent with the numbering of the ring systems; if a choice remains, the principal groups, and thereafter the other substituents, are given the next lowest permissible numbers. When necessary, the name and the prefixed numbers of the ring assembly are placed in brackets, to denote that the syllables " bi- ", " ter- ", etc., do not multiply the suffixes.

Examples:

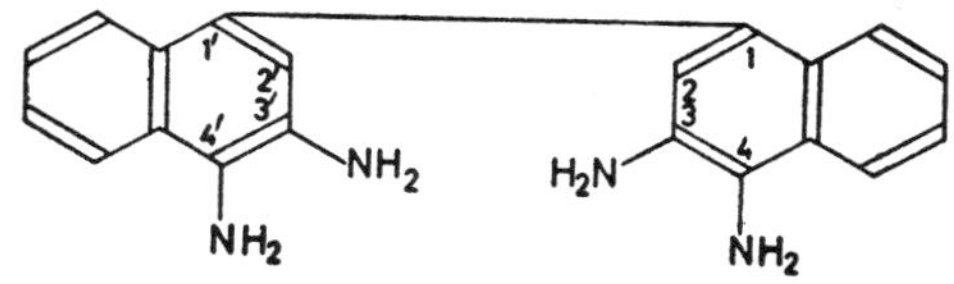

[1,1'-Binaphthalene]-3,3',4,4'-tetramine
or [1,1'-Binaphthyl]-3,3',4,4'-tetramine

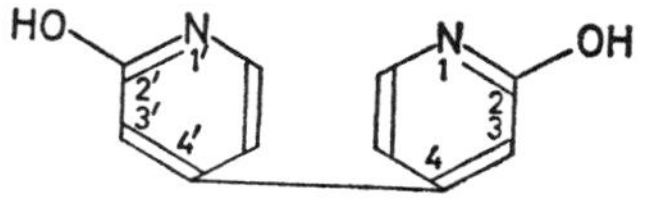

[4,4'-Bipyridine]-2,2'-diol
or [4,4'-Bipyridyl]-2,2'-diol

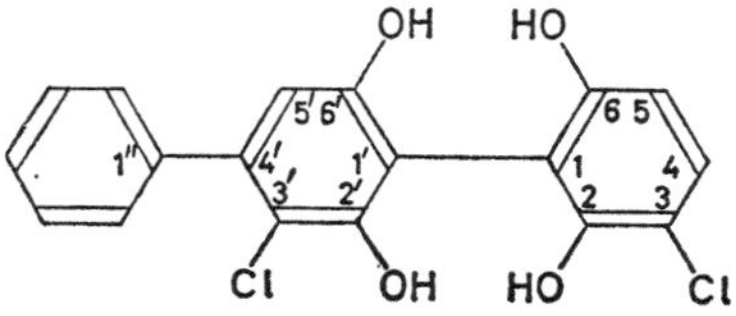

3,3'-Dichloro-1,1':4',1"-terphenyl-
2,2',6,6'-tetrol

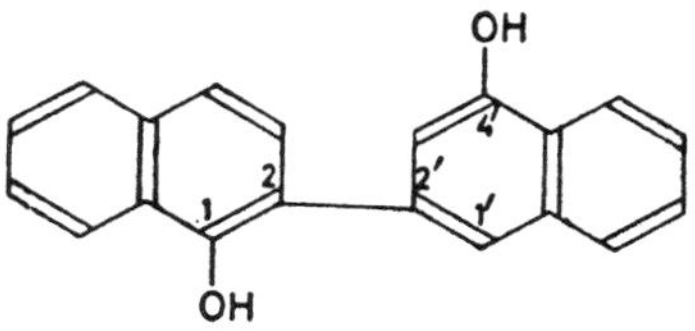

[2,2'-Binaphthalene]-1,4'-diol
or [2,2'-Binaphthyl]-1,4'-diol

71.4 (Alternative to part of Rule **C–71.3**)—When identical component molecules have trivial names that include the principal groups, their union by a single or double bond is denoted by placing the prefix " bi- " in front of the trivial name. The numbering of the individual component is retained and the point of union has the lowest number that is permissible thereafter. Locants for the points of union are placed before " bi- ", and one is primed (the higher if they are different).

Examples:

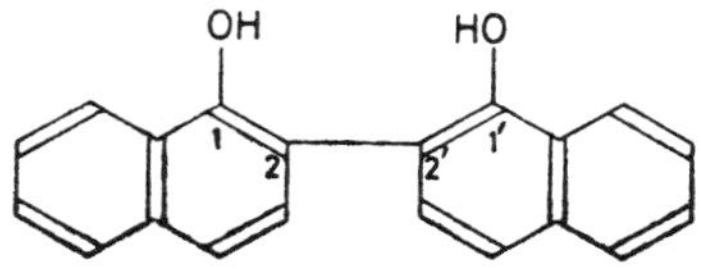

2,2'-Bi-1-naphthol

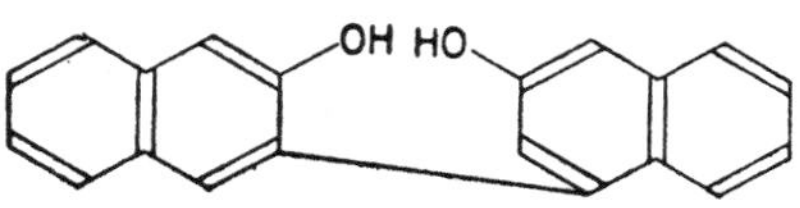

3,4'-Bi-2-naphthol

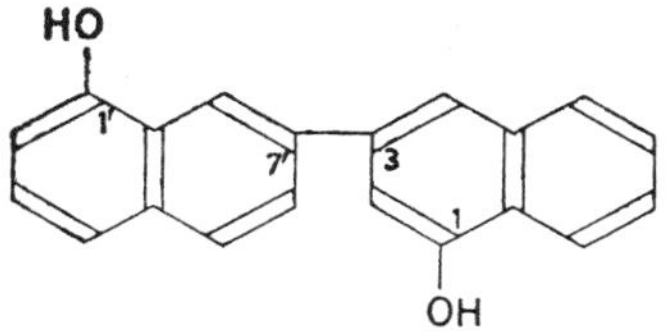

3,7'-Bi-1-naphthol

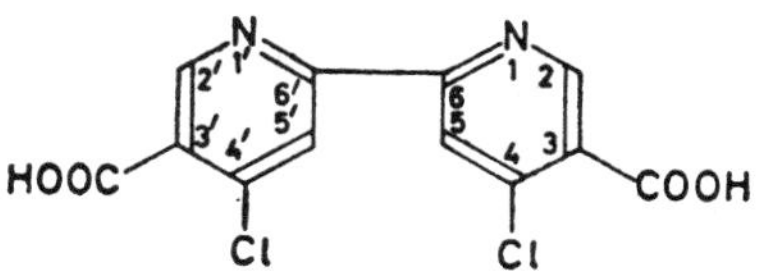

4,4'-Dichloro-6,6'-binicotinic acid

1,1'-Bianthraquinone

Exceptions: This rule is not applied to biphenyl derivatives. Thus

2,2'-Biphenyldiol
(not 2,2'-Biphenol)

71.5—When directly linked units contain principal groups in unequal numbers, all the principal groups are cited (with locants) as suffixes to the name of the assembly. The name of the assembly is placed in square brackets when ambiguity might otherwise arise.

Examples:

[9,9'-Bifluorene]-9-carboxylic acid

3,3',5-Biphenyltriol

Rule C-72. Assemblies involving bi- or multi-valent radicals

72.1—When a compound contains identical units whose only substituents are the principal groups, and when these identical units are linked by a symmetrical bi- or multi-valent radical, it is named by stating successively (*a*) the locants for the positions of substitution of the radical into the units, (*b*) the name of the substituting radical, (*c*) the prefix " di " or " tri ", *etc.*, and (*d*) the name of the unit including the principal groups. The numbering of the unit and the principal group is retained, and when there is a choice the points of substitution by the radical are numbered as low as permissible therewith. Primes, double primes, *etc.*, are used to distinguish the locants, the highest receiving the largest number of primes.

Examples:

3,3'-Thiodipropionic acid $S(CH_2—CH_2—COOH)_2$

Nitrilotriacetic acid $N(CH_2—COOH)_3$

4,4'-Methylenedibenzoic acid $HOOC-\langle\rangle-CH_2-\langle\rangle-COOH$

72.2—Names of symmetrical compound radicals are formed by juxtaposing the names of the individual radicals starting with the central radical *e.g.*, methylenedioxy, oxydiethylene; or the trivial name, if any, for the compound radical may be used.

Examples:

$HOOC-\langle\rangle-O-CH_2-O-\langle\rangle-COOH$

4,4'-Methylenedioxydibenzoic acid

1,6'-Ureylenedi-2-naphthalenesulfonic acid

$HO-CH_2-CH_2-\langle\rangle-O-CH_2-CH_2-O-\langle\rangle-CH_2-CH_2-OH$

4,4'-(Ethylenedioxy)dicyclohexaneethanol

$$O\big\langle\begin{array}{l}CH_2-CH_2-N(CH_2-CH_2-COOH)_2\\CH_2-CH_2-N(CH_2-CH_2-COOH)_2\end{array}$$

3,3',3'',3'''-[Oxybis(ethylenenitrilo)]tetrapropionic acid

(In the last example the compound radical name is formed by starting with the central radical " oxy- ", followed by " -bis- ", and adding successively the names of the radicals " ethylene " ($-CH_2-CH_2-$) and " nitrilo " ($-N=$), and finally the name of the unit " propionic acid " is added.)

Note: On this nomenclature system, unsymmetrical compound radicals are not used for linking identical units because of difficulty in assigning unambiguous numbering to the complete structure. Instead, simple substitutive nomenclature is used, *e.g.*:

$\langle\rangle-O-CH_2-CH_2-\langle\rangle-COOH$
$HOOC$

p-[2-(*m*-Carboxyphenoxy)ethyl]benzoic acid

Rule C–73.   Derivatives of assemblies of identical units

73.1—When assemblies named in accordance with Rule C–72 contain substituents in the identical units in addition to principal groups, these substituents are named by use of prefixes.   These prefixes are assigned the lowest locants available after priority has been given to the principal groups and the linking radical.

Examples:

4-Chloro-2,4′-iminodibenzoic acid

$$N(CHCl-COOH)_3$$

2,2′,2″-Trichloronitrilotriacetic acid

4,4′-Dinitro-2,3′-dithio-
dibenzaldehyde

6,6′-Dibromo-3,3′-oxy-
dibenzoic acid

Note: Assemblies of identical units linked by bi- or multi-valent radicals where the individual units contain different numbers of the principal group are named by the procedure given in Rule C-12.5.

Rule C–74.   Assemblies of identical acyclic units

74.1—Some directly linked, identical, acyclic units are conveniently named as assemblies of units, although in general the use of rules similar to C–71 and C–73 is precluded by the rules of Subsection C–0.13 governing the choice of the principal chain.

Examples:

Biacetyl (unsubstituted compound only) $CH_3-CO-CO-CH_3$
(or 2,3-Butanedione)

Bibenzyl $C_6H_5-\overset{\alpha'}{C}H_2-\overset{\alpha}{C}H_2-C_6H_5$
(or 1,2-Diphenylethane)

However,

1,1,2,2-Ethanetetracarboxylic acid $(HOOC)_2CH-CH(COOH)_2$
(not Bimalonic acid)

74.2—The following acyclic nitrogen compounds, and their derivatives and substitution products, may be named as assemblies of identical units, with the numbering shown (see also Rule C–72.1):

Bicarbamic acid $\qquad$ HOOC—NH—NH—COOH

4,4′-Bisemicarbazide $\quad$ $\overset{1'}{H_2N}$—$\overset{2'}{NH}$—$\overset{3'}{CO}$—$\overset{4'}{NH}$—$\overset{4}{NH}$—$\overset{3}{CO}$—$\overset{2}{NH}$—$\overset{1}{NH_2}$

Biguanidine $\qquad$ $\overset{6}{H_2N}$—$\overset{5}{C}(=NH)$—$\overset{4}{NH}$—$\overset{3}{NH}$—$\overset{2}{C}(=NH)$—$\overset{1}{NH_2}$

(However, for Biguanide, see Rule C–962.1)

Biurea $\qquad$ $\overset{6}{H_2N}$—$\overset{5}{CO}$—$\overset{4}{NH}$—$\overset{3}{NH}$—$\overset{2}{CO}$—$\overset{1}{NH_2}$

C–0.8   FREE RADICALS, IONS, AND RADICAL IONS

FREE RADICALS

Rule C–81

81.1—Free radicals are named as described for the corresponding radicals (groups) when contained in a molecular structure, according to the rules of Sections A, B, and C.   Radicals $X_2C{:}$, where X is an atom or group not containing carbon, but including cyanide, cyanate, isocyanide, isocyanate, and their sulfur, selenium, and tellurium analogues, have the generic name " carbenes ".   When the name of a radical ends in " y ", this letter is changed to " yl ".

Examples:

Methyl	$\cdot CH_3$
Trifluoromethyl	$\cdot CF_3$
Hydroxymethyl	$\cdot CH_2{-}OH$
1-Hydroxyethyl	$CH_3{-}\overset{\cdot}{C}H{-}OH$
1-Cyano-1-methylethyl	$(CH_3)_2\overset{\cdot}{C}{-}CN$
2-Methylbenzyl	
α,α-Dichlorobenzyl	$\cdot CCl_2{-}C_6H_5$
Triphenylmethyl or Trityl	$\cdot C(C_6H_5)_3$
Ethylene	$\overset{\cdot}{C}H_2{-}\overset{\cdot}{C}H_2$
1,3-Propanediyl (or trimethylene)	$\cdot CH_2{-}CH_2{-}CH_2\cdot$
Methylene or Carbene	$:CH_2$
Dichloromethylene or Dichlorocarbene	$:CCl_2$
Acetyl	$CH_3{-}\overset{\cdot}{C}{=}O$

Acetoxyl	$CH_3\text{—}CO\text{—}O\cdot$
Acetyldioxyl	$CH_3\text{—}CO\text{—}O\text{—}O\cdot$
Methoxyl	$CH_3\text{—}O\cdot$
tert-Butyldioxyl	$(CH_3)_3C\text{—}O\text{—}O\cdot$
Methanesulfonyl	$CH_3\text{—}\overset{\cdot}{S}O_2$

Methylsulfanyl
or   Methanesulfenyl   } $\quad CH_3\text{—}S\cdot$
or   Methylthio

$$(C_6H_5)_2\overset{\cdot}{C}\text{—}\boxed{}\text{—}O\text{—}\boxed{}\text{—}\overset{\cdot}{C}(C_6H_5)_2$$

4,4′-Oxybis(α,α-diphenylbenzyl)

81.2—Free radicals formed by loss of a hydrogen atom from bases whose names end in " -amine " are named by changing that ending to " -aminyl ”; for triplet diradicals the ending is " -aminylene ". Free radicals formed from bases with endings other than " -amine " are named in accordance with Rule **B–5**.

Examples:

Difluoroaminyl	$F_2N\cdot$
Dimethylaminyl	$(CH_3)_2N\cdot$
Phenylaminyl	$C_6H_5\text{—}\overset{\cdot}{N}H$
Bis(*p*-methoxyphenyl)aminyl oxide	$(p\text{-}CH_3O\text{—}C_6H_4)_2\overset{\cdot}{N}O$

Pyrrol-l-yl

1-Piperidyl

Methylaminylene	$CH_3\text{—}\overset{\cdot\cdot}{N}\cdot$
Phenylaminylene	$C_6H_5\text{—}\overset{\cdot\cdot}{N}\cdot$

Exception:

Hydrazyl	$H_2N\text{—}\overset{\cdot}{N}H$

(and its substitution products)

CATIONS

Rule C–82

82.1—If the positive charge of a parent organic cation is derived by fixation of a proton on a hetero atom that does not form part of a ring, the ion is named as a substitution product of the parent cation named in Table V. When a halogen cation is present as a ring member, the remainder of the ring may be named as a bivalent substituent therein.

Examples:

Tetramethylammonium	$(CH_3)_4N^+$
Dimethyloxonium	$(CH_3)_2OH^+$
Diphenyliodonium	$(C_6H_5)_2I^+$
Ethylenebromonium	

$$\begin{array}{c} H_2C \\ \quad \diagdown \\ \qquad \Big\rangle Br^+ \\ \quad \diagup \\ H_2C \end{array}$$

Exceptions: Names of the type benzenediazonium (Rule C–931), uronium (Rule C–973), and thiouronium (Rule C–974) are used as laid down in the Rules cited.

TABLE V.   Parent 'onium ions (not necessarily capable of independent existence), in order of decreasing preference for citation at the end of a name

Ion	Parent cation	Cation as prefix*
H_4N^+	Ammonium	Ammonio-
H_3O^+	Oxonium	Oxonio-
H_3S^+	Sulfonium	Sulfonio-
H_2Se^+	Selenonium	Selenonio-
H_2Cl^+	Chloronium	Chloronio-
H_2Br^+	Bromonium	Bromonio-
H_2I^+	Iodonium	Iodonio-

Note: This Table lists also the prefixes required for Rules C–85.1 and C–87.1.

Example:

Trimethylammonio $—^+N(CH_3)_3$

82.2—If the positive charge of an organic cation is derived by fixation of a proton on a hetero atom of a parent compound whose name does not end in amine, the name of the ion is formed by adding "-ium" to the name of the organic parent compound, with elision of a terminal " e " (if present). When there is more than one hetero atom able to fix the proton, the point of attachment of the proton may be shown by the " indicated hydrogen " method.

Examples:

Anilinium	$C_6H_5—NH_3^+$
Guanidinium	$C(NH_2)_3^+$
2-Phenylhydrazinium	$C_6H_5—NH—NH_3^+$
1-Methylhydrazinium	$CH_3—NH_2^+—NH_2$

* In a similar way, prefixes can be derived from other cation names by changing an ending "-ium", formed according to Rule C–82.2, C–82.4, C–83.1, or C–83.2, to "-io", for example pyridinium to pyridinio-.

2'H^+-Acetohydrazidium (for numbering see Rule C–921.5) $\qquad$ CH_3—CO—NH—NH_3^+

NH^+-Benzamidium $\qquad$ C_6H_5—CO—NH_3^+

Acetonium $\qquad$ $(CH_3)_2COH^+$

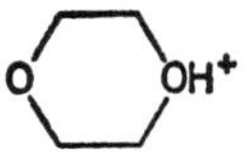

1,4-Dioxanium

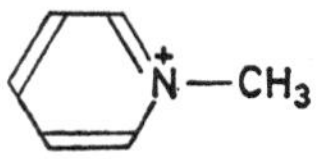

1-Methylpyridinium

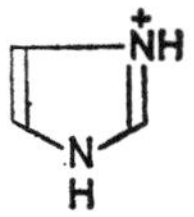

Imidazolium

$^+NH_3$—CH_2—COOH

Glycinium

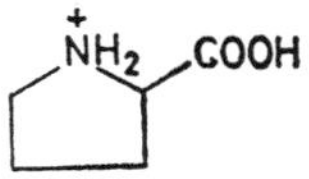

Prolinium

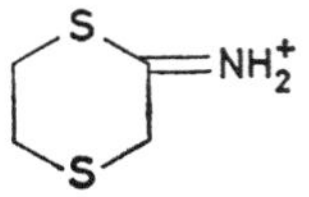

1,4-Dithian-2-iminium

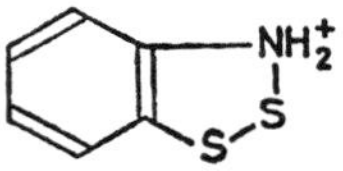

3H,3H^+-Benzo-1,2,3-dithiazolium

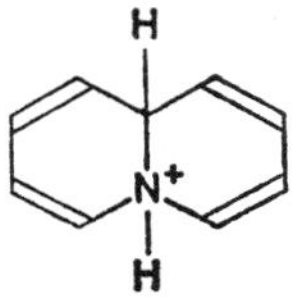

9aH-Quinolizinium

82.3—When replacement ("a") nomenclature is used (see Rule B–6) the ending "-a" of an affix oxa, thia, aza, *etc.*, is changed to "-onia".

Example:

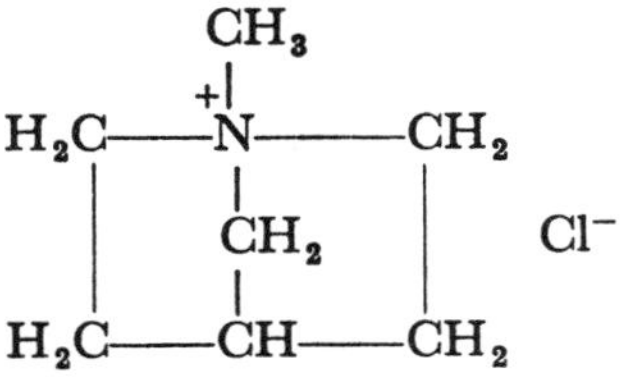

1-Methyl-1-azoniabicyclo[2.2.1]heptane chloride

82.4—An organic cation formed by addition of a proton to an unsaturated hydrocarbon is named by adding "-ium" to the name of the hydrocarbon, with elision of terminal "e" if present. If desired the point of attachment of the proton is shown by the "indicated hydrogen" method. If a π-complex is to be specified, the prefix πH^+ is used. If the position of the charge is to be specified, the appropriate locant is placed before the ending "-ium".

Examples:

Ethenium $CH_3—CH_2^+$ Benzenium $C_6H_7^+$

Allenium (position
of H+ unknown)   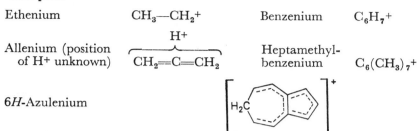   Heptamethyl-
benzenium $C_6(CH_3)_7^+$

6*H*-Azulenium

Rule C–83

83.1—When a cation can be considered as formed by loss of an electron or electrons from a radical at the free valence position(s) it may be named (*a*) by adding the word " cation " to the radical name or (*b*) by replacing the suffix " -yl " of the univalent radical by " -ylium ". Method (*b*) can be extended to multivalent cations by using the suffixes " -diylium ", " -triylium ", *etc.*, attached to the name of a hydrocarbon or heterocyclic system. The name " carbenium " may be used for CH_3^+.

Examples:

$CH_3—CH_2^+$

(*a*) Ethyl cation
(*b*) Ethylium
or   Ethenium (Rule **C–82.4**)

$C_6H_5^+$

(*a*) Phenyl cation
(*b*) Phenylium

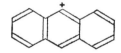

(*a*) 2-Cyclohexen-1-yl cation
(*b*) 2-Cyclohexen-1-ylium

(*a*) 9-Anthryl cation
(*b*) 9-Anthrylium

$(CH_3)_3C—CH_2^+$

(*a*) Neopentyl cation
(*b*) Neopentylium

$(C_6H_5)_3C^+$

(*a*) Triphenylmethyl cation
(*b*) Triphenylmethylium
or Triphenylcarbenium

$C_3H_3^+$

(*a*)· Cyclopropenyl cation
(*b*) Cyclopropenylium

CH_2^{2+}

(*a*) Methylene dication
(*b*) Methanediylium

$CH_3—CH^{2+}$

(*a*) Ethylidene dication
(*b*) Ethane-1,1-diylium

$^+CH_2—CH_2^+$

(*a*) Ethylene dication
(*b*) Ethane-1,2-diylium

$$CH_3—CO^+$$

(*a*) Acetyl cation
(*b*) Acetylium

$$CH_3—S^+$$

(*a*) Methylsulfanyl cation
or   Methanesulfenyl cation

(*b*) Methylsulfanylium
or   Methanesulfenylium

$$CH_3—SO_2^+$$

(*a*) Methanesulfonyl cation
(*b*) Methanesulfonylium

$$C_6H_5—CH{=}N^+$$

(*a*) Benzylideneaminyl cation
(*b*) Benzylideneaminylium

(*a*) Quinolizinyl cation;   (*b*) Quinolizinylium

(*a*) 1,2,3-Benzodithiazolyl cation;   (*b*) 1,2,3-Benzodithiazolylium

Note 1: Method (*a*) may be used for the isolated species, and method (*b*) for salts and for the corresponding substituents with names ending in " -io ". Greater precision may be necessary for resonating structures; *e.g.*, for univalent cations, two types of case may occur.

(*i*) Neither double-bond specification nor indication of hydrogen is necessary for the parent compound, as, for instance, for benzene or anthracene; but it may be necessary to indicate which hydrogen atom has been removed and this is done by the locant in names such as 9-anthryl cation and 9-anthrylium.

(*ii*) Double-bond specification or indication of hydrogen is necessary for the parent compound. Then use of the " -ylium " ending, without a locant, shows that such specification is not needed for the ionic species because the hydrogen atom removed is that which would have been specified as " indicated hydrogen ". This applies also when a CH_2 group in conjugated polyenes is located indirectly by locants for double bonds. The following names, for instance, are thus unambiguous:

Cyclopropenylium   Cycloheptatrienylium
                          or Tropylium

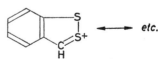

1,2-Benzodithiolylium

Note 2: According to Rules A–21 and B–3, parent compounds for *ortho*-fused, or *ortho*- and *peri*-fused, hydrocarbons and heterocycles are chosen so as to contain the maximum number of non-cumulative double bonds. This principle has sometimes been applied to heterocyclic ions and leads, for instance, to the following names (contrast example above):

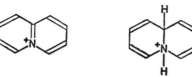

Quinolizinium   5,9a-Dihydroquinolizinium

83.2—The following are examples of well-established trivial names of cations that are retained as exceptions:

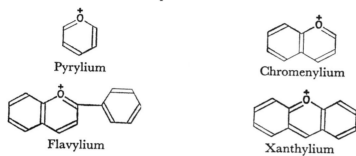

Pyrylium   Chromenylium

Flavylium   Xanthylium

83.3—Cation radicals that can be considered to be formed by addition of a proton to a radical may be named (a) by adding the word " cation " to the name of the compound having the same formula or (b) by adding the suffix " -yl " to the name of the cation.
Examples:

$C_2H_6^{\cdot+}$

(a) Ethane cation
(b) Ethaniumyl

$C_6H_6^{\cdot+}$

(a) Benzene cation
(b) Benzeniumyl

(a) Pyrazine 1-cation
(b) 1-Pyraziniumyl

(a) Quinoline 1-cation
(b) 1-Quinoliniumyl

$(p\text{-}CH_3\text{—}C_6H_4)_3N^{\cdot+}$

(a) Tri-p-tolylamine cation
(b) Tri-p-tolylammoniumyl

$(CH_3)_2S^{\cdot+}$

(a) Dimethylsulfane cation
(b) Dimethylsulfoniumyl

Rule C-84

84.1—In names of anions formed by loss of a proton from an acid, " -ate " replaces " -ic acid ".

Examples:

Hexanoate $C_5H_{11}—COO^-$

Cyclohexanecarboxylate $—COO^-$

Benzenesulfonate $C_6H_5—SO_3^-$

84.2—Anions formed by loss of a proton from an alcohol or phenol have names ending in " olate " or " oxide ", as described in Rule **C–206**. Those formed from a thiol have names ending in " thiolate " or " sulfide ", as described in Rule **C–511.3**.

84.3—Anions formed by removal of a proton or protons from a carbon atom are named by adding " -ide ", " -diide ", *etc.*, to the name of the parent compound, with elision of terminal " e " (if present) before a vowel. Such ions are termed generically " carbanions ".

Examples:

1-Butanide	$CH_3—CH_2—CH_2—CH_2^-$
1-Butyn-1-ide	$CH_3—CH_2—C{\equiv}C^-$
Benzenide	$C_6H_5^-$
Triphenylmethanide	$(C_6H_5)_3C^-$
Diphenylmethanediide	$(C_6H_5)_2C^{2-}$
Cyclopentadienide	$C_5H_5^-$

1,4-Dihydro-1,4-naphthalenediide

Exception: Metal derivatives of acetylene are named " acetylides ", as, for instance:

Monosodium acetylide	$NaC{\equiv}CH$
Disodium acetylide	$NaC{\equiv}CNa$

Note: Names such as butyllithium and phenylsodium are commonly used and are retained. (See Subsection D-3)

84.4—Radical anions that may be considered to be formed by loss of a proton from a radical are named by changing the ending " -yl " of the name of the radical to " -ylide " or by adding the word " anion " to the name of the corresponding substance.

Examples:

Ethylide $C_2H_4{}^{\cdot-}$ Dihydronaphthylide $C_{10}H_8{}^{\cdot-}$
or Ethene anion   or Naphthalene anion

The following names are retained:

Ketyl as in Potassium diphenylketyl $(C_6H_5)_2\dot{C}\!-\!O^-\,K^+$

p-Benzosemiquinone anion

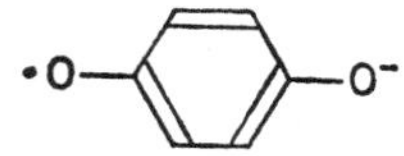

TWO OR MORE KINDS OF IONIC CENTRE WITH THE SAME KIND OF CHARGE IN A SINGLE STRUCTURE

Rule C–85

85.1—When two or more cationic centres named in Table V are present in a single structure, that occurring highest in the Table is designated as suffix and the endings " -onium " of the others are changed to " -onio- " for citation as prefixes. Alternatively replacement nomenclature may be used.

Examples:

$$(CH_3)_2\overset{+}{S}\!-\![CH_2]_6\!-\!\overset{+}{N}(CH_3)_3 \quad 2Cl^-$$

[6-(Dimethylsulfonio)hexyl]trimethylammonium dichloride

8,8,13-Trimethyl-3,6,15,18-tetraoxa-13-thionia-8 -azoniaicosane dichloride

85.2—When two or more cationic centres that cannot all be named as in Table V are present in a single structure, one of them is chosen as senior and its name, or the name of the component containing it, is used as parent; the other cationic centres are named by use of the prefixes designating either the centres themselves or the components in which they occur. The senior cationic centre is chosen according to the nature of its central atom, with the priorities (*a*) carbon and (*b*) then hetero atoms in the order of Table V. If a choice remains, then within each type having the same central atom, cyclic structures take precedence over acyclic, and the order of precedence of cyclic structures is that given in Subsection C–0.14.

Examples:

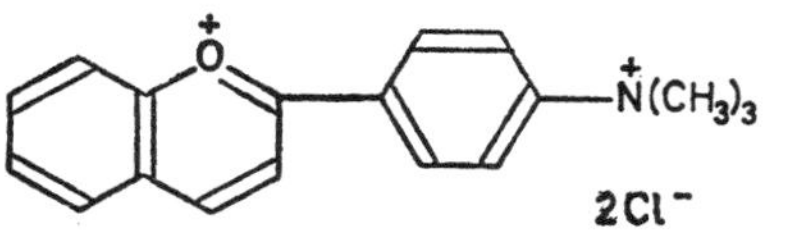

4'-Trimethylammonio-
flavylium dichloride

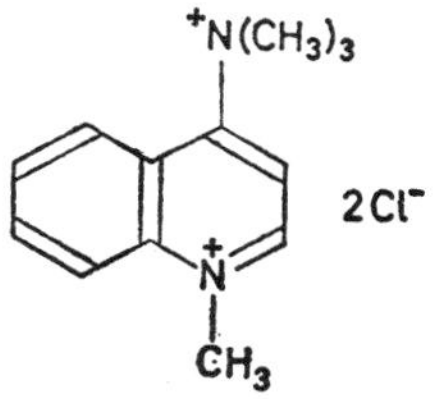

1-Methyl-4-trimethylammonio-
quinolinium dichloride

(1-Methyl-3-pyridinio)tropylium
dibromide

Rule C–86

86.1—When two or more different anionic centres are present in one structure, the ion corresponding to the acid that occurs first in Table II is designated by a suffix; for the other anionic groups the ending " -ate " or " -ide " is changed to " ato- " or " ido- ", respectively, and these groups are named as prefixes

Example:

Disodium 4'-sulfonatoazobenzene-4-carboxylate
or Disodium *p*-(*p*-sulfonatophenylazo)benzoate

86.2—When named as prefix, the anionic substituents —O⁻ and —S⁻ are termed " oxido- " and " sulfido- ", respectively.

Example:

Disodium 3-oxido-2-naphthoate

POSITIVE AND NEGATIVE IONIC CENTRES IN A SINGLE STRUCTURE

Rule C–87

87.1—When positive and negative ionic centres are present in a single structure, the compound is named by giving the name of the cationic group directly before that of the anionic group. This can be done (*a*) by considering that a cationic group, designated by a prefix, is substituted into an anion, the names of cationic substituents being formed by changing the ending " -ium " of the cation name into " -io ", or (*b*) by considering that an anionic substituent, designated by a suffix, is substituted into a cation.

Examples:

$$(CH_3)_3N^+—CH_2—COO^-$$

(a) Trimethylammonioacetate (trivial name, betaine, see Rule C–816.1)

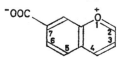

(b) Phenalenylium-1-sulfonate

(b) 1-Oxonianaphthalene-7-
carboxylate

(a) (1-Pyridinio)acetate

(a) (1-Methyl-4-pyridinio)acetate

(a) (1-Pyridinio)sulfate

$$(CH_3)_3N^+—O—SO_3{}^-$$

(a) Trimethylammoniosulfate

(a) (1-Pyridinio)formate
(b) Pyridinium-1-carboxylate

(b) Pyridinium-1-sulfonate

(a) N-(1-Pyridinio)-p-toluene
sulfonamidate

(a) Triethylphosphonio-
cyclopentadienide

(a) N-[1,4-Diphenyl-3-
(1,2,4-triazolio)]anilide

(a) 1,2-Dithiol-3-ylio sulfide
(b) 1,2-Dithiolylium-3-thiolate

(a) 2-(1,2-Dithiol-3-ylio)-
1-naphtholate

PREAMBLE TO SUBSECTIONS C–1 TO C–9

The following Subsections C–1 to C–9 illustrate how the general principles set out in Subsection C–0 are applied to various types of compound. These illustrations are confined in these Subsections to compounds containing only carbon, hydrogen, halogen, oxygen, sulfur, selenium, tellurium, and/or nitrogen.

C–1. HALOGEN DERIVATIVES

Rule C–101

101.1—Halogen derivatives may be named according to three basic systems, namely, by substitutive nomenclature, by radicofunctional nomenclature, or by additive nomenclature. The substitutive names are recommended for general use as the other two types of nomenclature, particularly additive nomenclature, are limited in their application.

Rule C–102

102.1—Substitutive names of halogen derivatives are formed by adding prefixes " fluoro- ", " chloro- ", " bromo- ", or " iodo- " to the name of the parent compound.

Examples:

2-Chlorohexane CH_3—CH_2—CH_2—CH_2—$CHCl$—CH_3

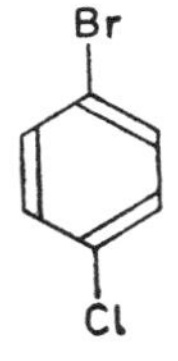

1-Bromo-4-chlorobenzene

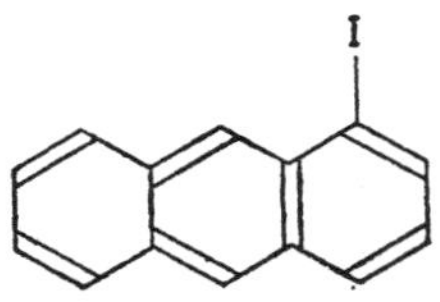

1-Iodoanthracene

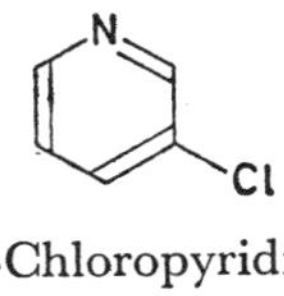

3-Chloropyridine

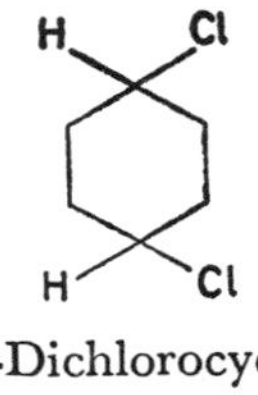

1,4-Dichlorocyclohexane

1,2,3,4,5,6-Hexachlorocyclohexane   4′-Chloro-*m*-terphenyl

Rule C–103

103.1—Radicofunctional names are formed by naming the organic radical followed by the word " fluoride ", " chloride ", " bromide ", or " iodide ".

Examples:

Methyl chloride	CH_3Cl
tert-Butyl chloride	$(CH_3)_3CCl$
Benzyl iodide	C_6H_5—CH_2I
Ethylene dibromide	CH_2Br—CH_2Br
Benzylidene dichloride	C_6H_5—$CHCl_2$

Rule C–104

104.1—Additive names for halogen derivatives are formed from the name of the unsaturated parent compound and an expression showing the number and the kind of halogen atoms added, the latter being named in the ion form.

Examples:

Stilbene dibromide

p-Mentha-1,4(8)-diene tetrabromide

Rule C–105

105.1—Halogen-containing compounds or radicals in which all hydrogen atoms, except those whose replacement would affect the nature of characteristic groups (see p. 82) present, have been replaced by halogen atoms of the same kind may be named by adding the prefixes " perfluoro- ", " perchloro- ", " perbromo- ", or " periodo- " to the name of the corresponding non-halogenated compound or radical. Parentheses are used to indicate the part of the molecule so substituted.

Examples:

Perfluoropentane $CF_3—CF_2—CF_2—CF_2—CF_3$

Perfluoro(decahydro-1-methylnaphthalene)

But:

2-Chloroheptafluoropropane $CF_3—CClF—CF_3$
(not 2-Chloroperfluoropropane)

Rule C-106

106.1—Compounds containing the group —IO or —IO$_2$*, are named by adding the prefix " iodosyl- " (replacing " iodoso- ") or " iodyl- " (replacing " iodoxy- ")*, respectively, to the name of the parent compound.

Examples:

Iodosylbenzene	C_6H_5IO
Iodylbenzene	$C_6H_5IO_2$

106.2—Compounds containing the group —ClO, —ClO$_2$, or —ClO$_3$ are named by adding the prefix " chlorosyl- ", " chloryl- ", or " perchloryl- ", respectively*, to the name of the parent compound.

Example:

     Perchlorylbenzene $C_6H_5ClO_3$

106.3—Compounds containing the group —I(OH)$_2$ or derivatives of this group are named by adding the prefixes " dihydroxyiodo- ", " dichloroiodo- ", " diacetoxyiodo- ", *etc.* to the name of the parent compound.

Examples:

(Dihydroxyiodo)benzene	$C_6H_5I(OH)_2$
(Dichloroiodo)benzene	$C_6H_5ICl_2$
(Diacetoxyiodo)benzene	$C_6H_5I(O—CO—CH_3)_2$

Rule C-107

107.1—Cations of the type $R^1R^2I^+$ are given names derived from the iodonium ion H_2I^+ by substitution. Other halonium ions are named similarly.

---

* See list of names for ions and radicals in I.U.P.A.C. *Nomenclature of Inorganic Chemistry*, 1970, p. 99 (Butterworths Scientific Publications, London, 1971).

Examples:

Diphenyliodonium chloride	$(C_6H_5)_2I^+\,Cl^-$
Diphenyliodonium hydroxide	$(C_6H_5)_2I^+\,OH^-$
Diphenyliodonium iodide	$(C_6H_5)_2I^+\,I^-$

2,2'-Biphenylylenechloronium hydroxide

Rule C–108

108.1—The following trivial names are retained:

CHF_3	Fluoroform
$CHCl_3$	Chloroform
$CHBr_3$	Bromoform
CHI_3	Iodoform
$COCl_2$	Phosgene (alternative to carbonyl dichloride)
$CSCl_2$	Thiophosgene (alternative to thiocarbonyl dichloride)
$:CCl_2$	Dichlorocarbene (alternative to dichloromethylene; see Rule C–81.1)

108.2—Use of customary inorganic nomenclature* leads to names such as the following:

$COCl_2$	Carbonyl dichloride (alternative to phosgene) (also other carbonyl and thiocarbonyl halides)
CCl_4	Carbon tetrachloride (also other carbon tetrahalides)

* See list of names for ions and radicals in I.U.P.A.C. *Nomenclature of Inorganic Chemistry*, 1970, p. 99 (Butterworths Scientific Publications, London, 1971).

C-2. ALCOHOLS, PHENOLS, AND THEIR DERIVATIVES

ALCOHOLS

Rule C-201

201.1—In substitutive (see Subsection C–0.1) and conjunctive nomenclature (see Subsection C–0.5) of alcohols the hydroxyl group (OH) as principal group is indicated by a suffix " -ol ", with elision of terminal " e " (if present) from the name of the parent compound.

Examples:

Methanol	CH_3OH
2-Propanol*	CH_3—$CH(OH)$—CH_3
1,4-Butanediol	$HOCH_2$—CH_2—CH_2—CH_2OH
2-Ethyl-2-buten-1-ol	CH_3—CH=C—CH_2OH
	$\quad\quad\quad\quad \mid$
	$\quad\quad\quad\quad CH_2$—$CH_3$

2-Cyclohexen-1-ol

Bicyclo[3.2.0]heptan-2-ol

Triphenylmethanol†	$(C_6H_5)_3COH$

2-Naphthaleneethanol
or 2-(2-Naphthyl)ethanol

---

* Designations such as isopropanol, *sec*-butanol, and *tert*-butanol are incorrect because there are no hydrocarbons isopropane, *sec*-butane, and *tert*-butane to which the suffix " -ol " can be added; such names should be abandoned. Isopropyl alcohol, *sec*-butyl alcohol, and *tert*-butyl alcohol are, however, permissible (see Rule C–201.3) because the radicals isopropyl, *sec*-butyl, and *tert*-butyl do exist.

† The use of the " carbinol " nomenclature (whereby, for instance, this compound would be named triphenylcarbinol) should be abandoned.

201.2—Hydroxyl groups are indicated by the prefix " hydroxy- " when a group having priority for citation as principal group is also present or when the hydroxyl group is present in a side chain.

Examples:

5-Hydroxyhexanal

$$CH_3—CH—CH_2—CH_2—CH_2—CHO$$
$$\underset{OH}{|}$$

3-Hydroxy-1-cyclohexanecarboxylic acid

2-Hydroxymethyl-1,4-butanediol

$$HO—CH_2—CH_2—CH—CH_2OH$$
$$\underset{CH_2OH}{|}$$

201.3—Radicofunctional names (see Subsection C–0.2) of alcohols consist of the radical name derived from that of the parent compound, followed by the word " alcohol ".

Examples:

Methyl alcohol CH_3OH

Hexyl alcohol $CH_3—CH_2—CH_2—CH_2—CH_2—CH_2OH$

201.4—The following are examples of trivial names which are retained:

Allyl alcohol $CH_2{=}CH—CH_2OH$

tert-Butyl alcohol $(CH_3)_3COH$

Benzyl alcohol $C_6H_5—\overset{\alpha}{C}H_2OH$

Phenethyl alcohol $C_6H_5—\overset{\beta}{C}H_2—\overset{\alpha}{C}H_2OH$

Salicyl alcohol

Ethylene glycol* $HOCH_2—CH_2OH$

Propylene glycol* $CH_3—CH(OH)—CH_2OH$

Glycerol $HOCH_2—CH(OH)—CH_2OH$

Pentaerythritol $C(CH_2OH)_4$

Pinacol $(CH_3)_2C—C(CH_3)_2$
$$\underset{HO}{|}\quad\underset{OH}{|}$$

* The glycol nomenclature should not be applied to other dihydric alcohols.

Geraniol

$$\begin{array}{l}(CH_3)_2C{=}CH{-}CH_2{-}CH_2{-}\underset{\displaystyle CH_3}{\underset{|}{C}}{=}CH{-}CH_2OH\end{array}$$

Phytol

$$\begin{array}{l}\qquad\qquad\quad CH_3\\ \qquad\qquad\quad\ |\\ CH_2{-}CH_2{-}CH{-}[CH_2]_3{-}CH(CH_3)_2\\ |\\ CH_2{-}\underset{\displaystyle CH_3}{\underset{|}{CH}}{-}[CH_2]_3{-}\underset{\displaystyle CH_3}{\underset{|}{C}}{=}CH{-}CH_2OH\end{array}$$

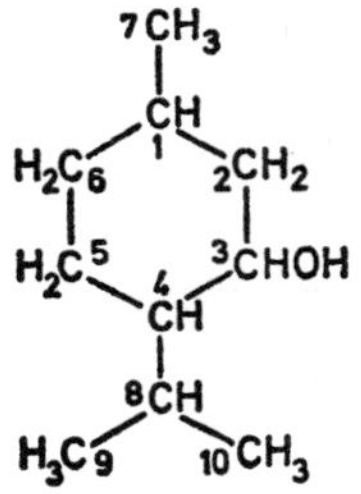

Menthol

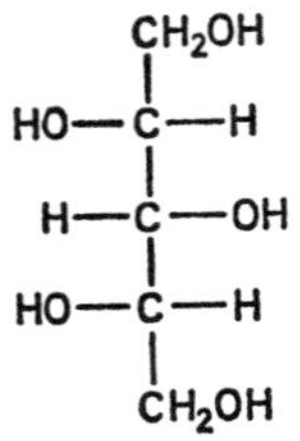

Xylitol

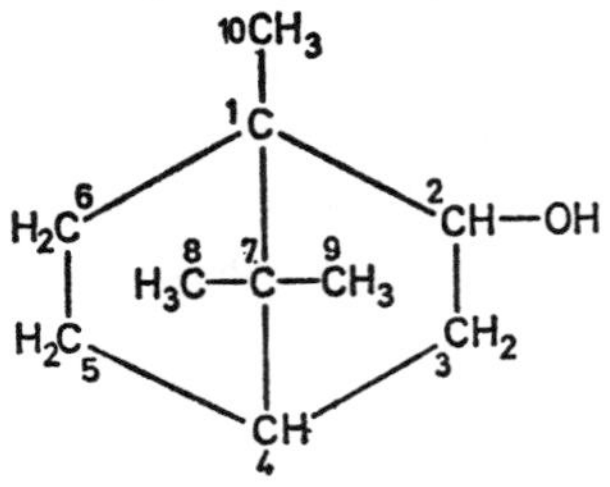

Borneol

PHENOLS

Rule C–202

202.1—Hydroxy derivatives of benzene and other aromatic carbocyclic systems are named by adding the suffix " -ol ", " -diol ", *etc.*, to the name of the hydrocarbon, with elision of terminal " e " (if present) before " ol ".

Examples:

1-Chrysenol

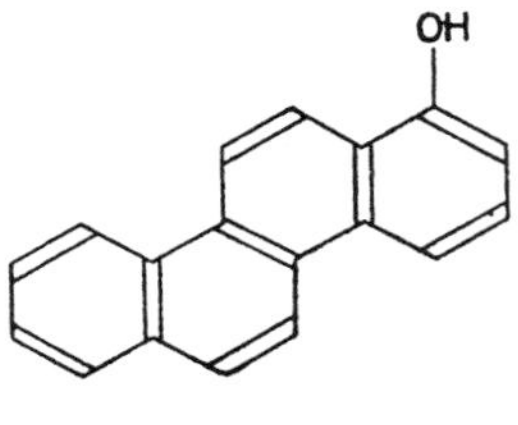

1,2,4-Benzenetriol

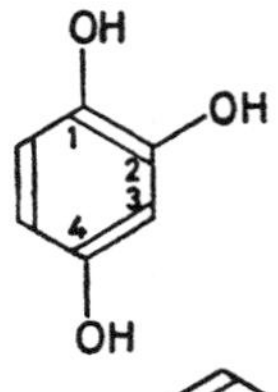

Benz[*a*]anthracen-7-ol

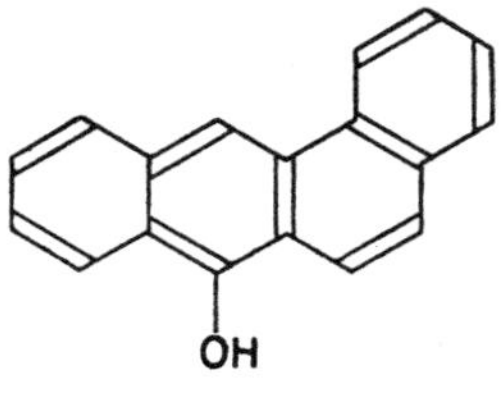

202.2—The following are examples of trivial names of aromatic hydroxy compounds which are retained:

Phenol

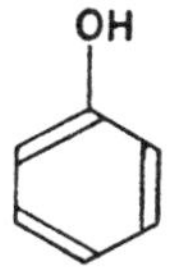

Cresol (*m-* shown)
(note the numbering)

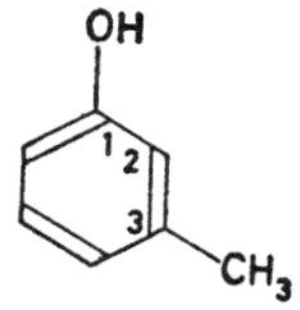

2,3-Xylenol

(note the numbering and that the numeral 1 for the hydroxyl group is omitted from the name)

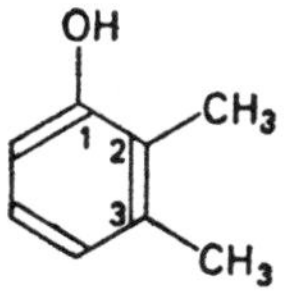

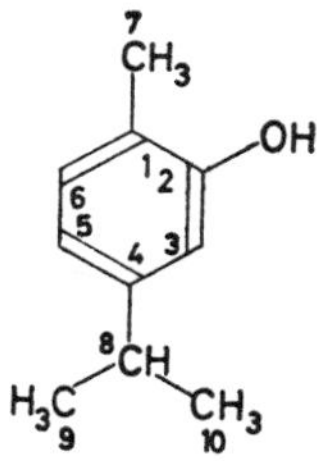

Carvacrol

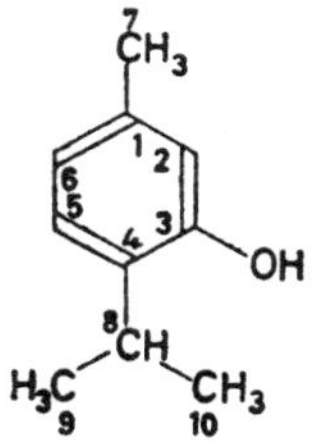

Thymol

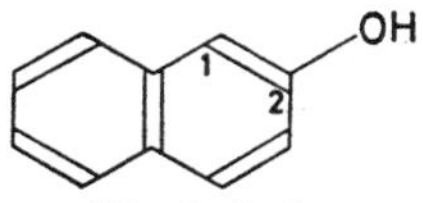

Naphthol
(2- shown)

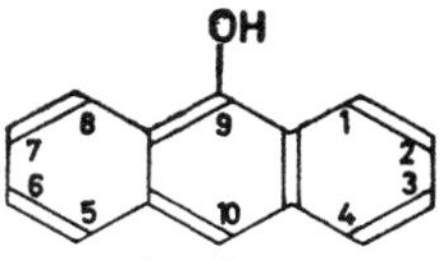

Anthrol
(9- shown)

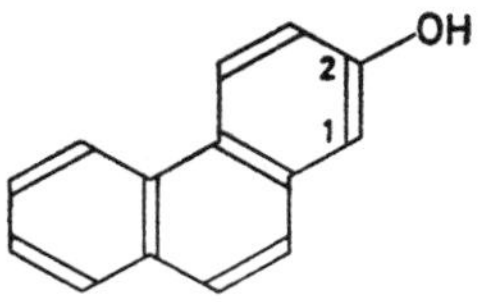

Phenanthrol
(2- shown)

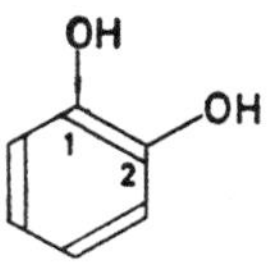

Pyrocatechol

151

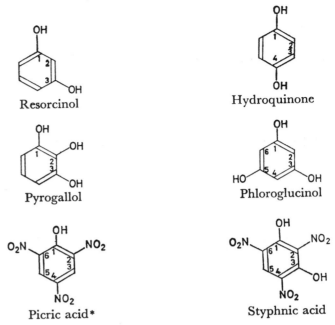

Resorcinol

Hydroquinone

Pyrogallol

Phloroglucinol

Picric acid*

Styphnic acid

Rule C–203

203.1—Hydroxy derivatives of ring assemblies are named (*a*) by adding the suffix " -ol " to the name of the corresponding hydrocarbon or (*b*) as substituted derivatives of the component present that contains the largest number of hydroxyl groups.

Examples:

(*a*) *p*-Terphenyl-2′-ol

or   (*b*) 2,5-Diphenylphenol

(*a*) 4,4′-Biphenyldiol
     (see exception to
     Rule C – **71.4**)

HO— —OH

* Gives the radical " picryl " or 2,4,6-trinitrophenyl

(*a*) 2,4,4′,6-Biphenyltetrol

or   (*b*) (*p*-Hydroxyphenyl)phloroglucinol

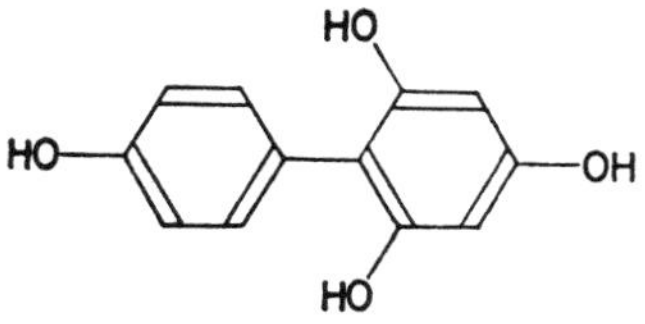

(*b*) 3-(2-Naphthyl)phenol

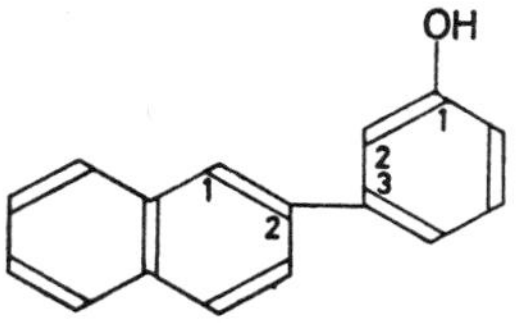

(*b*) 4-Phenyl-1-naphthol

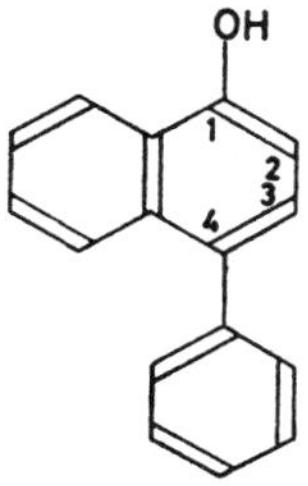

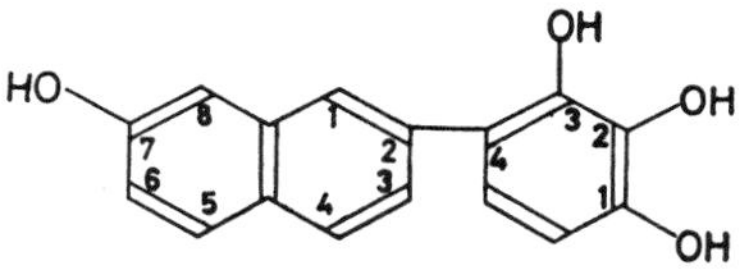

(*b*) 4-(7-Hydroxy-2-naphthyl)pyrogallol

HETEROCYCLIC COMPOUNDS

Rule C–204

204.1—Hydroxy derivatives of heterocyclic compounds in which the hydroxyl group is attached to a carbon atom are named by adding the suffix " -ol ", " -diol ", etc., to the name of the corresponding parent compound, with elision of terminal " e " (if present) before " ol ".

Examples:

8-Quinolinol

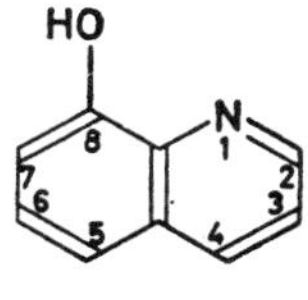

Indol-5-ol

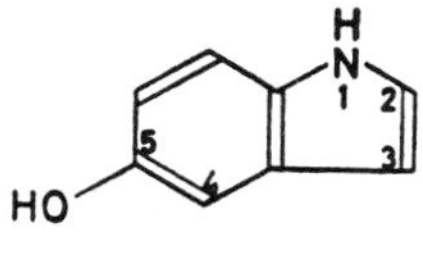

When confusion arises, the hydroxyl group is indicated as a prefix:

Example:
3-Hydroxythiophene
(not thiophenol, which is C_6H_5—SH)

204.2—The presence of a hydroxyl group attached to a hetero atom in a heterocyclic compound may be indicated by the prefix " hydroxy- ".

Example:

1-Hydroxypiperidine

RADICALS

Rule C–205

205.1—Radicals RO— are named by adding " oxy " as a suffix to the name of the radical R.

Examples:

Pentyloxy	CH_3—CH_2—CH_2—CH_2—CH_2—O—
Hexyloxy	CH_3—CH_2—CH_2—CH_2—CH_2—CH_2—O—
Allyloxy	CH_2=CH—CH_2—O—
Isopentyloxy	$(CH_3)_2$CH—CH_2—CH_2—O—
Benzyloxy	C_6H_5—CH_2—O—

2-Naphthyloxy

2-Pyridyloxy

Only the following contractions for oxygen-containing radical names are recommended as exceptions to this rule:

Methoxy	CH_3—O—
Ethoxy	CH_3—CH_2—O—
Propoxy	CH_3—CH_2—CH_2—O—
Butoxy	CH_3—CH_2—CH_2—CH_2—O—
Phenoxy	C_6H_5—O—
Isopropoxy*	$(CH_3)_2$CH—O—
Isobutoxy*	$(CH_3)_2$CH—CH_2—O—
sec-Butoxy*	CH_3—CH_2—CH(CH_3)—O—
tert-Butoxy*	$(CH_3)_3$C—O—

* For the unsubstituted radical only.

205.2—Except when forming part of a ring system, bivalent radicals of the form —O—X—O— are named by adding " dioxy " to the name of the bivalent radical —X— (see Rules C–72.2 and C–331). (cf. Rule B-15)

Examples:*

Methylenedioxy	—O—CH$_2$—O—
Ethylenedioxy	—O—CH$_2$—CH$_2$—O—
Trimethylenedioxy	—O—CH$_2$—CH$_2$—CH$_2$—O—
Carbonyldioxy	—O—CO—O—
Sulfonyldioxy	—O—SO$_2$—O—

3,3'-(Sulfonyldioxy)dipropionic acid
$$O_2S\big\langle{}^{O—CH_2—CH_2—CO_2H}_{O—CH_2—CH_2—CO_2H}$$

SALTS

Rule C–206

206.1 (Alternative to Rule C–206.2)—Anions derived from alcohols or phenols are named by changing the final " -ol " of the name of the alcohol or phenol to " -olate " (compare Rule C–84.2). This applies to substitutive, radicofunctional, and trivial names.

Examples:

Sodium methanolate	CH$_3$ONa
Magnesium di(1-propanolate)	(CH$_3$—CH$_2$—CH$_2$O)$_2$Mg
Aluminium tri(2-propanolate)	[(CH$_3$)$_2$CHO]$_3$Al
Sodium benzyl alcoholate	C$_6$H$_5$—CH$_2$ONa

Monosodium pinacolate	(CH$_3$)$_2$C——C(CH$_3$)$_2$
	    $\mid$ $\mid$
	   ONa OH

Sodium phenolate	C$_6$H$_5$ONa

Calcium di(thymolate)

---

* Although the radicals listed in the examples are not all alcohol derivatives it is convenient to list them here.

Sodium 1-chrysenolate

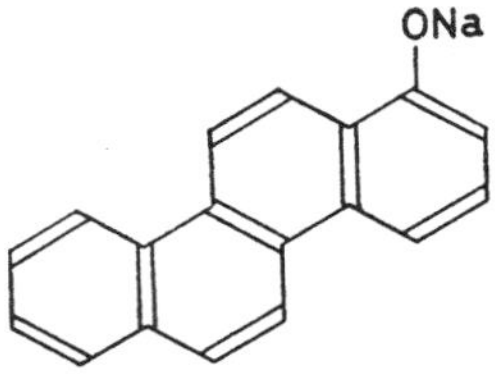

206.2 (Alternative to Rule C–206.1)—Salts composed of an anion RO and a cation (usually a metal) are named by citing, first, the cation and then the RO anion; the latter has the unabbreviated name used for the RO— radical but with the ending " -yloxy- " changed to " -yl oxide " (compare Rule C–84.2).

Examples:
Sodium benzyl oxide $\qquad$ C_6H_5—CH_2ONa
Calcium bis(pentyl oxide) $\qquad$ $(C_5H_{11}O)_2Ca$

Exceptions: When the radical RO— has an abbreviated name listed in the exceptions to Rule C–205.1, the ending " -oxy- " of this name is changed to " -oxide ".

Examples:
Sodium methoxide $\qquad$ CH_3ONa
Thorium tetraisopropoxide $\qquad$ $[(CH_3)_2CHO]_4Th$
Potassium phenoxide $\qquad$ C_6H_5OK

Sodium o-fluorophenoxide

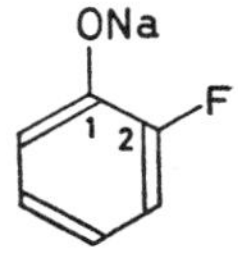

Potassium *tert*-butoxide $\qquad$ $(CH_3)_3COK$

Notes: (1) Radicofunctional names for salts of alcohols and phenols may, however, also be used in some languages (not English), whereby the ending " -yl alcohol " is changed to " -ylate ", as in:

Aluminium tri(isopropylate) $\qquad$ $[(CH_3)_2CHO]_3Al$
Sodium *tert*-butylate $\qquad$ $(CH_3)_3CONa$
Potassium benzylate $\qquad$ C_6H_5—CH_2OK

(2) For designation of O^- as a prefix, see Rule C–86.2.

ETHERS

Rule C–211

211.1—Compounds R^1—O—R^2 have the generic name " ethers "* and may be named by either the substitutive or the radicofunctional method.

* The term " oxide " is used in place of ether in some languages.

211.2—Substitutive names of unsymmetrical ethers are formed by using names of radicals R^1O— as prefixes to the names of the hydrocarbons corresponding to the second radical R^2. The senior component is selected as the parent compound.

Examples:

1-Isopropoxypropane $(CH_3)_2CH—O—CH_2—CH_2—CH_3$

Ethoxyethylene $CH_3—CH_2—O—CH=CH_2$

1-Chloro-2-ethoxyethane $CH_3—CH_2—O—CH_2—CH_2Cl$

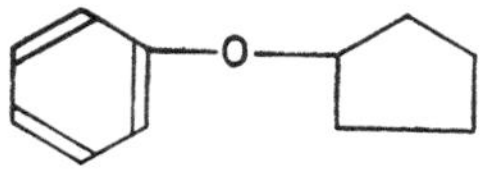

(Cyclopentyloxy)benzene 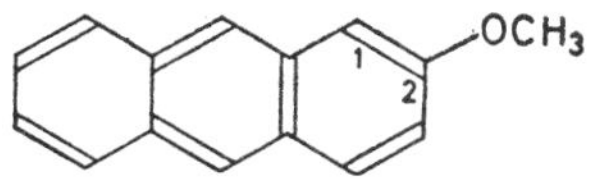   2-Methoxyanthracene

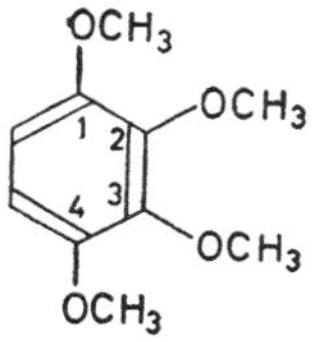

1,2,3,4-Tetramethoxybenzene 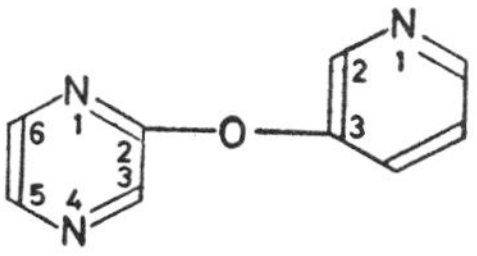   (3-Pyridyloxy)pyrazine

211.3—Radicofunctional names of ethers are formed by citing the names of the radicals R^1 and R^2 followed by the word " ether ".

Examples:

Ethyl methyl ether $CH_3—O—CH_2—CH_3$

Diethyl ether $CH_3—CH_2—O—CH_2—CH_3$

Bis(2-chloroethyl) ether $ClCH_2—CH_2—O—CH_2—CH_2Cl$

Cyclopentyl phenyl ether

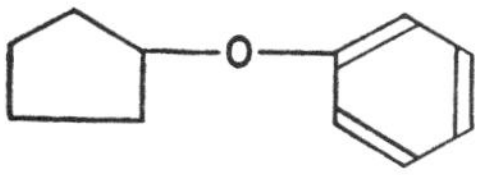

Ethyl vinyl ether $CH_3—CH_2—O—CH=CH_2$

2-Pyrazinyl 3-pyridyl ether

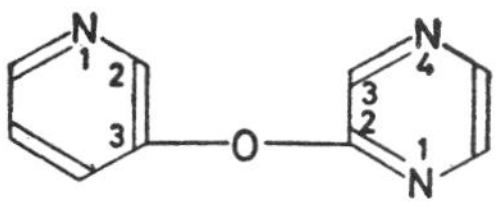

Rule C–212

212.1—Oxygen linking two identical parent compounds may be indicated by the prefix " oxy " when a group having priority for citation as suffix is present, the nomenclature for assemblies of identical units (compare Subsection C–0.7) being used.

Examples:

2,2'-Oxydiethanol

$$HOCH_2\text{—}CH_2\text{—}O\text{—}CH_2\text{—}CH_2OH$$

3,3'-Oxydipropionic acid

$$HOOC\text{—}CH_2\text{—}CH_2\text{—}O\text{—}CH_2\text{—}CH_2\text{—}COOH$$

3,3'-Oxydiphenol

212.2—An oxygen atom directly attached to two carbon atoms already forming part of a ring system or to two carbon atoms of a chain may be indicated by the prefix "epoxy-", particularly when it is desired to preserve the name of a specific complex structure, as, for example, steroids or carotenoids. (see also Rule **B-15**).

Examples:

1-Chloro-2,3-epoxypropane

$$CH_2Cl\text{—}CH\text{—}CH_2$$

α,α'-Epoxybibenzyl

$$C_6H_5\overset{\alpha'}{CH}\text{—}\overset{\alpha}{CH}C_6H_5$$

2-Methyl-1, 3-epoxyhexane
[see Rule **C-16.11**, (a) (viii)]
or 1,3-Epoxy-2-methylhexane
[See Rule **C-16.11**, (b) (ii)]

$$CH_2\text{—}CH\text{—}CH\text{—}CH_2\text{—}CH_2\text{—}CH_3$$

8,13-Dihydro-8,13-epoxy-
benzo[a]naphthacene

[Note: epoxy is not alphabetized; see Rule **C-16.11**, (a) (viii)].

212.3—Compounds RO—X—OR, where the two parent compounds RH are identical and contain a group having priority over ether for citation as suffix, and —X— is any bivalent radical, are named by the method for assemblies of identical units (Subsection C-0.7).

Example:

4,4'-(Ethylenedioxy)dibenzoic acid

212.4—Linear polyethers derived from three or more molecules of aliphatic dihydroxy compounds are most conveniently named by the open-chain replacement nomenclature (see Subsection C–0.6).

Examples:

$$\overset{8}{\text{HOCH}_2}-\overset{7}{\text{CH}_2}-\overset{6}{\text{O}}-\overset{5}{\text{CH}_2}-\overset{4}{\text{CH}_2}-\overset{3}{\text{O}}-\overset{2}{\text{CH}_2}-\overset{1}{\text{CH}_2\text{OH}}$$

3,6-Dioxaoctane-1,8-diol

$$\overset{12}{\text{O}}-\overset{11,\,10}{(\text{CH}_2)_2}-\overset{9}{\text{O}}-\overset{8-6}{(\text{CH}_2)_3}-\overset{5}{\text{O}}-\overset{4-1}{(\text{CH}_2)_4}-\text{OH}$$

$$\underset{13,\,14}{[\text{CH}_2]_2}-\underset{15}{\text{O}}-\underset{16-20}{[\text{CH}_2]_5}-\underset{21}{\text{O}}-\underset{22-27}{[\text{CH}_2]_6}-\text{OH}$$

5,9,12,15,21-Pentaoxaheptacosane-1,27-diol

212.5—Symmetrical linear polyethers may be named in terms of the central oxygen atom when there is an odd number of ether oxygen atoms, or in terms of the central hydrocarbon group when there is an even number of ether oxygen atoms.*

Examples:

$$\text{C}_6\text{H}_5-\text{O}-\text{C}_6\text{H}_4-\text{O}-\text{C}_6\text{H}_4-\text{O}-\text{C}_6\text{H}_4-\text{O}-\text{C}_6\text{H}_4-\text{O}-\text{C}_6\text{H}_5$$

Bis[*p*-(*p*-phenoxyphenoxy)phenyl] ether

$$\text{C}_6\text{H}_5-\text{O}-\text{C}_6\text{H}_4-\text{O}-\text{C}_6\text{H}_4-\text{O}-\text{C}_6\text{H}_4-\text{O}-\text{C}_6\text{H}_5$$

p-Bis(*p*-phenoxyphenoxy)benzene

Rule C–213

213.1—Partial ethers of polyhydroxy compounds may be named either by substitutive nomenclature or by stating the name of the polyhydroxy compound followed, first, by the name of the etherifying radical(s) with any required multiplying prefix, and then by the word " ether ".

Examples:

$$\text{HO}-\text{CH}_2-\text{CH}_2-\text{O}-\text{CH}_2-\text{CH}_3$$

2-Ethoxyethanol

or   Ethylene glycol monoethyl ether

$$\text{CH}_2\text{OCH}_3$$
$$|$$
$$\text{CHOH}$$
$$|$$
$$\text{CH}_2\text{OH}$$

3-Methoxy-1,2-propanediol
or   Glycerol 1-methyl ether
or   1-*O*-Methylglycerol

m-Methoxyphenol
or   Resorcinol monomethyl ether
or   1-*O*-Methylresorcinol

*For a general method for naming polymers see Nomenclature of Regular Single-strand Organic Polymers, IUPAC Information Bulletin Appendix No. 29, November 1972.

Rule C–214

214.1—The following are examples of trivial names retained for ethers:

Anisole C_6H_5—O—CH_3

Phenetole C_6H_5—O—CH_2—CH_3

Anethole

Guaiacol

Veratrole

Eugenol

Rule C–215

215.1—The names of cyclic ethers are preferably formed according to the rules of Section B (but see also Rule **C–331.2** and the exceptions thereto).

Examples:

Oxolane
or   Tetrahydrofuran

Oxirane
(also known as ethylene oxide)

Rule C–216

216.1—Cyclic ethers derived from dihydroxy compounds are named as heterocyclic compounds (see Section B of these Rules) except as laid down in Rule **C–331.2**.

Example:

Dibenzo[e,g][1,4]dioxocin

PEROXIDES

Rule C–218

218.1—Compounds RO—OH are named (*a*) by placing the name of the radical R before the word " hydroperoxide " or (*b*) by use of the prefix " hydroperoxy- ".

Examples:

(*a*) *p*-Cumenyl hydroperoxide

(*b*) Ethyl 4-hydroperoxy-2,5-cyclohexadiene-1-carboxylate

218.2—Compounds R^1O—OR^2 are named (*a*) by placing the names of the radicals R^1 and R^2, in alphabetical order, before the word " peroxide ", when the group —O—O— links two chains, or two rings, or a ring and a chain, (*b*) by use of the affix " dioxy- " to denote the bivalent group —O—O— for naming assemblies of identical units or to form part of a prefix, or (*c*) by use of the prefix " epidioxy- " when the —O—O— group forms a bridge between two carbon atoms of a ring or a ring system.

Examples:

(*a*) Ethyl phenyl peroxide C_6H_5O—OC_2H_5

(*a*) Dibenzoyl peroxide
or Benzoyl peroxide C_6H_5—CO—O—O—CO—C_6H_5

(*b*) 4,4'-Dioxydibenzoic acid $(O$—C_6H_4—$COOH)_2$

(*b*) Methyl *p*-(ethyldioxy)benzoate   *p*-C_2H_5O—O—C_6H_4—$COOCH_3$

(*c*) 3β,6β-Epidioxy-5α-androstane

C-3. ALDEHYDES, KETONES, AND THEIR DERIVATIVES

C-3.0. ALDEHYDES

Rule C-301

301.1—The term " aldehyde " is applied to compounds which contain the group —C(=O)H attached to carbon. Aldehydes are named by means of the suffixes " -al ", " -aldehyde ", or " -carbaldehyde ", or by the prefix " formyl- " [representing the —C(=O)H group when present as the terminal group of a carbon chain], or, in connexion with trivial names, by the prefix " oxo- " (representing =O).

ACYCLIC ALDEHYDES

Rule C-302

302.1—The name of an unbranched acyclic mono- or di-aldehyde is formed by adding the suffix " -al " (for a monoaldehyde), with elision of a terminal " e " (if present), or " -dial " (for a dialdehyde) to the name of the hydrocarbon containing the same number of carbon atoms.

Examples:

Ethanal	CH_3—CHO
Hexanal	CH_3—CH_2—CH_2—CH_2—CH_2—CHO
Undecanedial	OHC—$[CH_2]_9$—CHO

$$CH_2{=}CH$$
$$|$$
3-Vinyl-2-hepten-6-ynal $\quad$ HC≡C—CH_2—CH_2—C=CH—CHO

2-Hexenedial $\quad$ OHC—CH_2—CH_2—CH=CH—CHO

$$CH_2{=}CH—CH{=}CH$$
$$|$$
$$OHC—CH_2—CH—CH_2—CH{=}CH—CHO$$
5-(1,3-Butadienyl)-2-heptenedial

Rule C-303

303.1—(a) The name of an acyclic polyaldehyde in which more than two aldehyde groups are attached to an unbranched chain is formed by adding " -tricarbaldehyde ", " -tetracarbaldehyde ", etc., to the name of the longest chain carrying the maximum number of aldehyde groups. The name and numbering of the main chain do not include the aldehyde groups, and numbering follows the general principles for unsaturation and substituents. (b) Alternatively, the name is formed by adding the prefix " formyl- " to the name of the dial incorporating the principal chain.

Examples:

$$\text{(a)} \quad \overset{\displaystyle \text{CHO}}{\underset{\displaystyle \text{1,2,4-Butanetricarbaldehyde}}{\overset{4 \quad 3 \quad \overset{2|}{} \quad 1}{\text{OHC}-\text{CH}_2-\text{CH}_2-\text{CH}-\text{CH}_2-\text{CHO}}}}$$

$$\text{or} \quad \text{(b)} \quad \overset{\displaystyle \text{CHO}}{\underset{\displaystyle \text{3-Formylhexanedial}}{\overset{6 \quad 5 \quad 4 \quad \overset{3|}{} \quad 2 \quad 1}{\text{OHC}-\text{CH}_2-\text{CH}_2-\text{CH}-\text{CH}_2-\text{CHO}}}}$$

303.2—An aldehyde group in an acyclic compound is named by the prefix " formyl- " when a group having priority for citation as principal group is also present. (See, however, also Rules C-303.4, C-415, and C-416.)

Example:

$$\text{OHC}-[\text{CH}_2]_6-\text{N}(\text{CH}_3)_3{}^+ \text{ I}^-$$
(6-Formylhexyl)trimethylammonium iodide

303.3—For an acyclic polyaldehyde, in which the aldehyde groups —CHO are attached to more than one branch of a branched chain, the name of the longest chain carrying the greatest number of aldehyde groups is used together with a suffix " -dial " (see Rule C-302.1), " -tricarbalde-hyde ", etc. (see Rule C-303.1), and other chains carrying aldehyde groups are named by use of " formylalkyl- " prefixes.

Examples:

$$\overset{\displaystyle \text{CH}_2-\text{CHO}}{\underset{\displaystyle \text{3-(Formylmethyl)heptanedial}}{\overset{7 \quad 6 \quad 5 \quad 4 \quad \overset{3|}{} \quad 2 \quad 1}{\text{OHC}-\text{CH}_2-\text{CH}_2-\text{CH}_2-\text{CH}-\text{CH}_2-\text{CHO}}}}$$

3-(2-Formylethyl)-1,2,6-hexanetricarbaldehyde

Rule C-304

304.1—The name of an aldehyde in which the aldehyde group is directly attached to a carbon atom of a ring system is formed by adding the suffix " -carbaldehyde ", " -dicarbaldehyde ", etc., to the name of the ring system.

Examples:

Cyclohexanecarbaldehyde

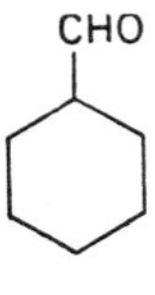

1,2-Naphthalenedicarbaldehyde

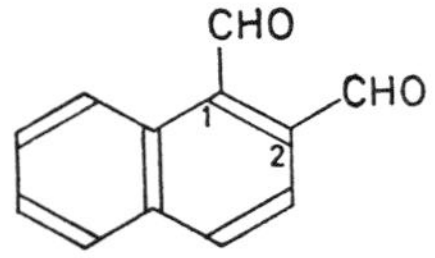

2-Thiazolidinecarbaldehyde

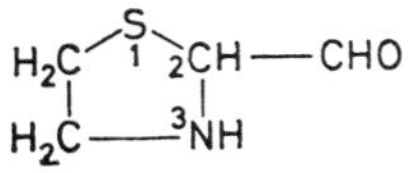

304.2—An aldehyde group in a cyclic compound is named by the prefix " formyl- " when a group having priority for citation as principal group is also present.

Example:

p-Formylphenylacetic acid

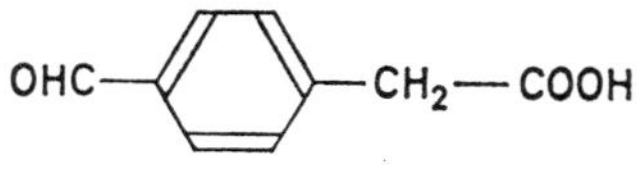

304.3—A —CHO group attached to a nitrogen atom in a ring system is denoted by the prefix " formyl- " (see Rule C–824.2) or according to Rule C–304.1.

Example:

    3-Formylthiazolidine
or   3-Thiazolidinecarbaldehyde

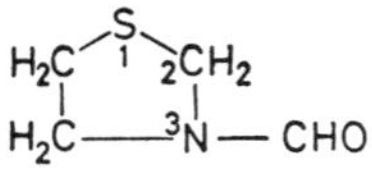

304.4—When the aldehyde group is separated from the ring by a chain of carbon atoms, the compound is named (a) as a derivative of the acyclic system or (b) by conjunctive nomenclature.

Examples:

    (a) (1-Naphthyl)acetaldehyde
or   (b) 1-Naphthaleneacetaldehyde

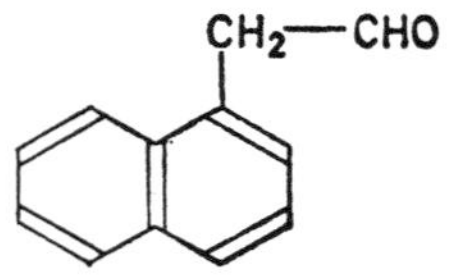

Examples:

    (a) 6,6'-(1,4-Cyclohexylene)dihexanal
or (b) 1,4-Cyclohexanedihexanal

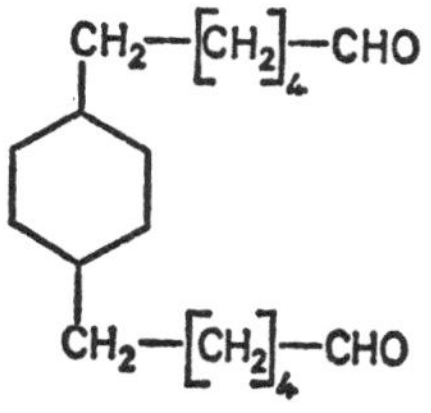

    (b) 1,4,5,8-Naphthalenetetraacetaldehyde

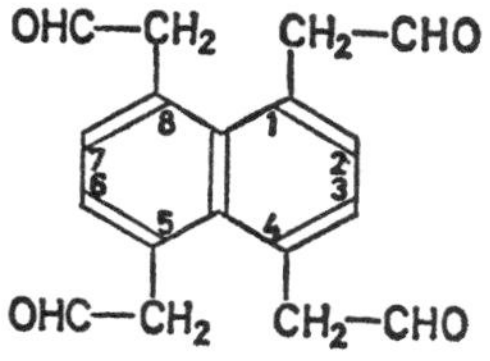

    (a) (1-Naphthylmethyl)malonaldehyde

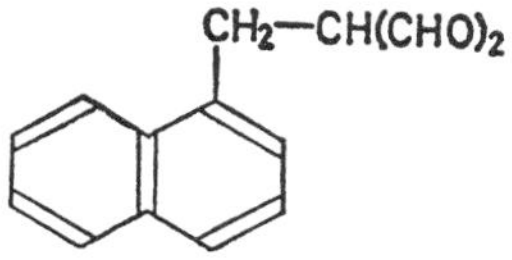

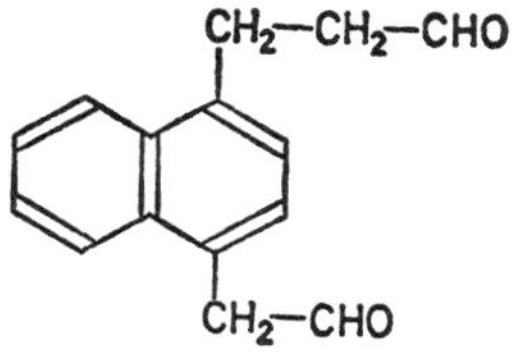

    (a) 3-[4-(Formylmethyl)-1-naphthyl]propionaldehyde
or (b) 4-(Formylmethyl)-1-naphthalenepropionaldehyde

304.5—When a compound RCH_3 has a trivial name, the related aldehyde RCHO may be named as the product of its substitution by $O=$; the prefix " oxo- " is used to denote this substitution (see Rules C–411.1 and C–415.1).

Example:

    19-Oxo-5β-etianic acid

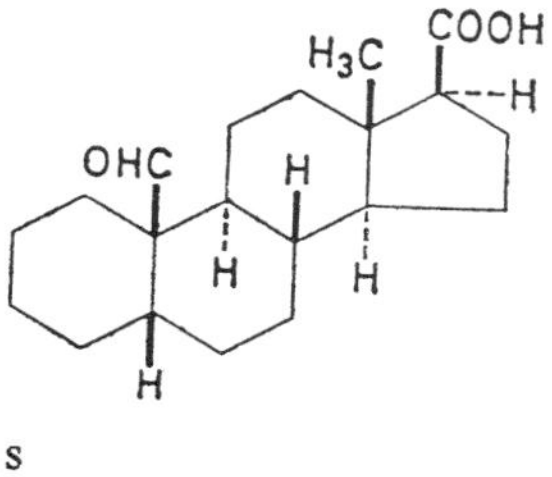

TRIVIAL NAMES

Rule C–305

305.1—When the corresponding monobasic acid has a trivial name, the name of the aldehyde may be formed from the trivial name of the acid by changing the ending " -ic acid " or " -oic acid " to " -aldehyde ".

Examples:

Formaldehyde	HCHO
Acetaldehyde	CH_3—CHO
Propionaldehyde	CH_3—CH_2—CHO
Butyraldehyde	CH_3—CH_2—CH_2—CHO
Isobutyraldehyde	$(CH_3)_2CH$—CHO
Valeraldehyde	CH_3—CH_2—CH_2—CH_2—CHO
Isovaleraldehyde	$(CH_3)_2CH$—CH_2—CHO
Acrylaldehyde (preferred to Acrolein)	CH_2=CH—CHO
Benzaldehyde	C_6H_5—CHO
Cinnamaldehyde	C_6H_5—CH=CH—CHO

Anisaldehyde (*p*- shown)

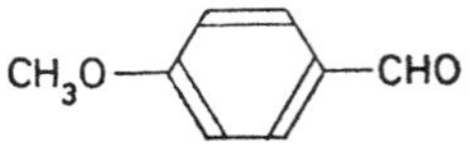

Nicotinaldehyde

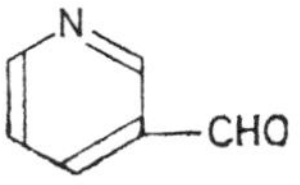

2-Furaldehyde (preferred to Furfural)

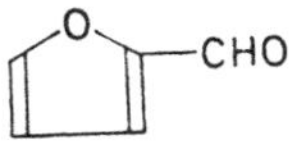

Glyceraldehyde	$HOCH_2$—CH(OH)—CHO
Glycolaldehyde	$HOCH_2$—CHO

305.2—The following are examples of trivial names retained for aldehydes:

$$(CH_3)_2C=CH-CH_2-CH_2-C(CH_3)=CH-CHO$$
Citral

Vanillin
(preferred to Vanillaldehyde)

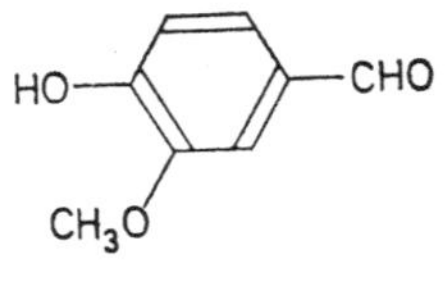

Piperonal
(preferred to Piperonylaldehyde)

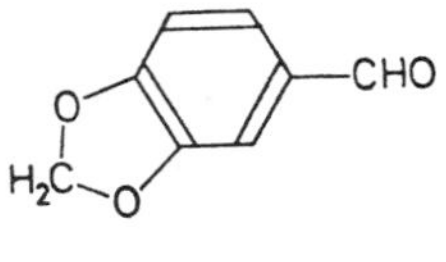

305.3—The names of aldehydes corresponding to trivially named polybasic acids in which all of the carboxyl groups are changed to aldehyde groups are formed from the names of the acids by changing " -ic acid " to " -aldehyde ".

Examples:

Malonaldehyde	$CH_2(CHO)_2$
Succinaldehyde	$OHC—[CH_2]_2—CHO$
Glutaraldehyde	$OHC—[CH_2]_3—CHO$
Adipaldehyde	$OHC—[CH_2]_4—CHO$
Phthalaldehyde	$o\text{-}C_6H_4(CHO)_2$
Isophthalaldehyde	$m\text{-}C_6H_4(CHO)_2$
Terephthalaldehyde	$p\text{-}C_6H_4(CHO)_2$

Exception:

Glyoxal	$OHC—CHO$

Note: Since all the carboxyl groups of a polycarboxylic acid must be changed to aldehyde groups in order that the compound be named according to this Rule it is not necessary to introduce " di " or " tri " into such aldehyde names. (For aldehydic acids, see Rule C–415 and C–416.)

305.4—The names of amino aldehydes are formed from the names of the corresponding amino-free aldehydes. They are not formed from the trivial names of the amino acids unless the latter end in " -ic acid ".*

Examples:

Aminoacetaldehyde (not Glycinaldehyde)	$H_2N—CH_2—CHO$
2-Aminopropionaldehyde (not Alaninaldehyde)	$CH_3—CH(NH_2)—CHO$
2-Aminobutyraldehyde	$CH_3—CH_2—CH(NH_2)—CHO$

However:

Aspartaldehyde (compare Rule C–305.3)	$OHC—CH_2—CH(NH_2)—CHO$

C–3.1. KETONES

GENERAL

Rule C–311

311.1—The generic name " ketone " is given to compounds containing an oxygen atom doubly bound to a single carbon atom with the carbonyl group $>C{=}O$ joined to two carbon atoms. Ketones are named by means of the suffix " -one ", the prefix " oxo- ", the functional class name " ketone ", or, in special cases, the suffix " -quinone ".

Rule C–312

312.1—In substitutive nomenclature the name of an acyclic ketone is formed by adding the suffix " -one " or " -dione ", etc., to the name of the hydrocarbon corresponding to the principal chain, with elision of terminal " e " (if present) before " one ".

*An alternative method is given in Nomenclature of α-Amino Acids, *IUPAC Information Bulletin* Appendix No. 46, September 1975.

Examples:

2-Butanone

$$CH_3—CH_2—CO—CH_3$$

2,4-Hexanedione

$$\overset{6}{C}H_3—\overset{5}{C}H_2—\overset{4}{C}O—\overset{3}{C}H_2—\overset{2}{C}O—\overset{1}{C}H_3$$

4-Penten-2-one

$$\overset{5}{C}H_2{=}\overset{4}{C}H—\overset{3}{C}H_2—\overset{2}{C}O—\overset{1}{C}H_3$$

3-Allyl-2,4-pentanedione

$$\overset{5}{C}H_3—\overset{4}{C}O—\overset{3}{C}H—\overset{2}{C}O—\overset{1}{C}H_3$$
$$\qquad\qquad\;\; |$$
$$\qquad\qquad CH_2—CH{=}CH_2$$

5-Propyl-5-hexene-2,4-dione

$$CH_2{=}C—CO—CH_2—CO—CH_3$$
$$\qquad\; |$$
$$\quad CH_2—CH_2—CH_3$$

Note: The following names are retained:

Acetone $\qquad\qquad CH_3—CO—CH_3$

Biacetyl (see Rule C–74.1) $\quad CH_3—CO—CO—CH_3$

However, the use of other trivial names, such as butyrone, valerone, stearone, formed from acid names, is discouraged.

312.2—Radicofunctional names of ketones, $R^1—CO—R^2$, are formed by citing the names of the radicals R^1 and R^2 followed by the word " ketone ".

Examples:

Ethyl methyl ketone

$$CH_3—CO—\overset{1}{C}H_2—\overset{2}{C}H_3$$

Diethyl ketone

$$\overset{2'}{C}H_3—\overset{1'}{C}H_2—CO—\overset{1}{C}H_2—\overset{2}{C}H_3$$

Rule C–313

313.1—The name of a ketone having a ring or rings attached to a chain which contains the carbonyl group or groups is formed from the name of the corresponding acyclic hydrocarbon, with the radical name of the ring system as a substituent, the ketone group or groups being indicated by a suffix " -one ", " -dione ", etc.

Examples:

Cyclohexyl-2-propanone

or   Cyclohexylacetone

$$C_6H_5—\overset{6}{C}H_2—\overset{5}{C}O—\overset{4}{C}H_2—\overset{3}{C}H_2—\overset{2}{C}O—\overset{1}{C}H_2—$$

1-(2-Naphthyl)-6-phenyl-2,5-hexanedione

$$—CO—CH_2—CH_2—CH_2—CO—$$

1,5-Di(2-furyl)-1,5-pentanedione

313.2—Acyclic monoacyl derivatives of cyclic compounds are named (a) by the method of Rule C–313.1, (b) by the radicofunctional method of Rule C–312.2, (c) by prefixing the name of the acyl group to the name of the cyclic component, or (d) if the cyclic component is benzene or naphthalene by changing the ending " -ic acid " or " -oic acid " of the name of the acid corresponding to the acyl group to " -ophenone " or " -onaphthone ", respectively.

Examples:

(a)   1-(2-Pyridyl)-1-butanone
(b)   Propyl 2-pyridyl ketone
(c)   2-Butyrylpyridine

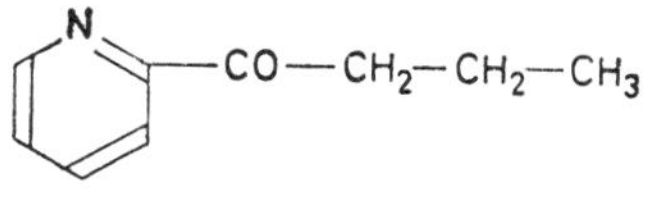

(d)   Acetophenone

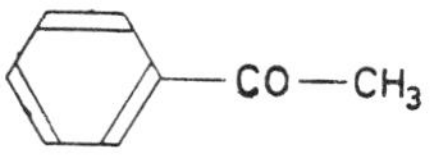

(d)   1'-Butyronaphthone

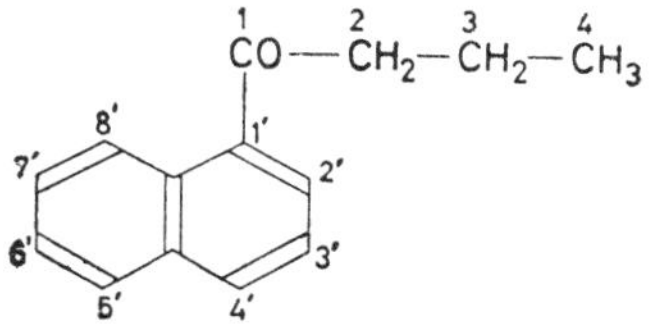

(d)   4'-Chloroundecanophenone

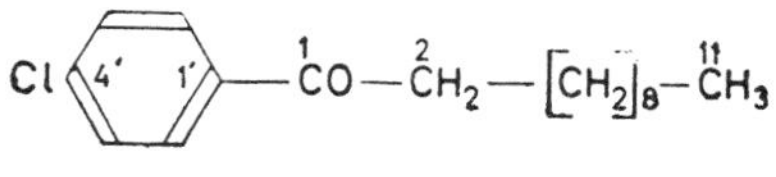

Exceptions:

The following are examples of trivial names retained for ketones:

Propiophenone
   (not Propionophenone)

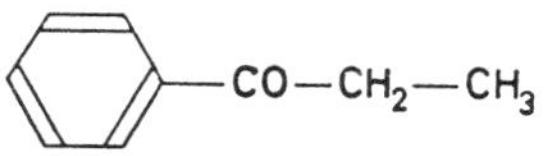

Chalcone

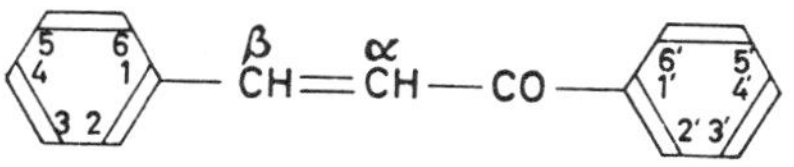

(preferred to Cinnamophenone, Benzylideneacetophenone, or 3-Phenylacrylophenone)

Deoxybenzoin

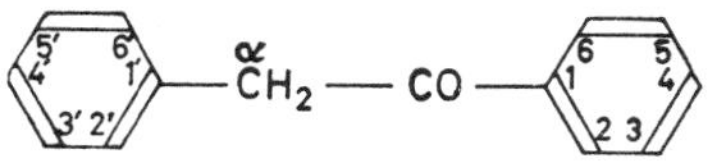

313.3—When a carbonyl group is attached directly to carbon atoms in two ring systems, the compound is named by the radicofunctional method provided that no other substituent is present having priority for citation as principal group (compare Rule C–316.2).

Examples:

Di-2-furyl ketone

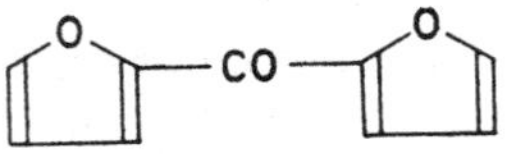

2-Furyl 2-pyridyl ketone

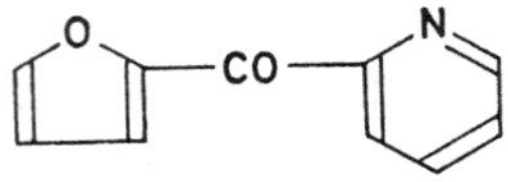

2-Bromo-1-naphthyl
  1-chloro-2-naphthyl ketone

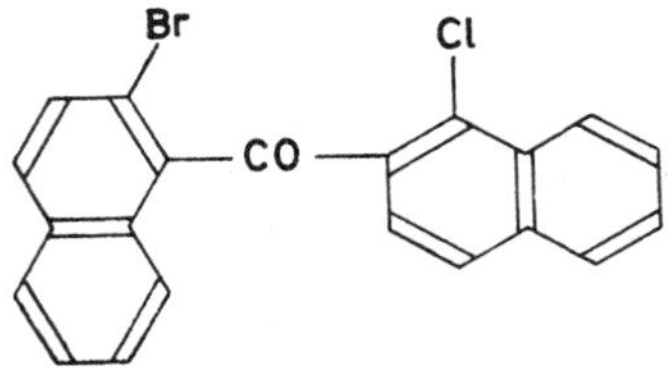

Bis(5-methyl-2-thienyl) ketone

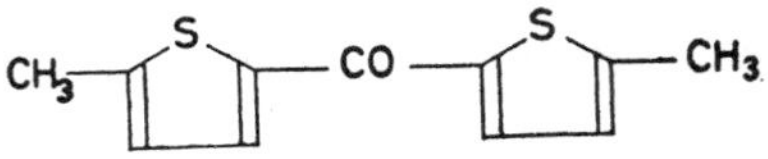

Exception:
Benzophenone C_6H_5—CO—C_6H_5

313.4—The name of a polyketone in which two or more contiguous carbonyl groups have rings attached at each end is formed (*a*) by the radico-functional method with the functional class name " diketone ", " triketone ", *etc.*, or (*b*) by substitutive nomenclature with " -one " suffixes.

Examples:
  (*a*) Di-2-naphthyl
                    diketone
or (*b*) Di-2-naphthyl-
                    ethanedione

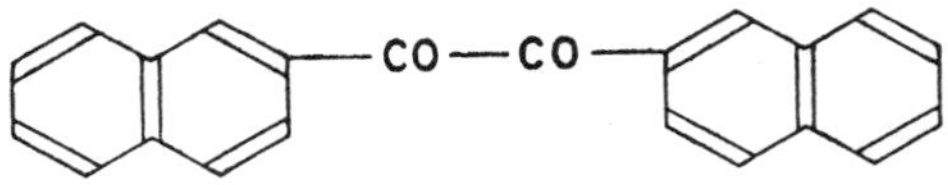

  (*a*) Di-3-pyridyl triketone
or (*b*) Di-3-pyridyl-
                    propanetrione

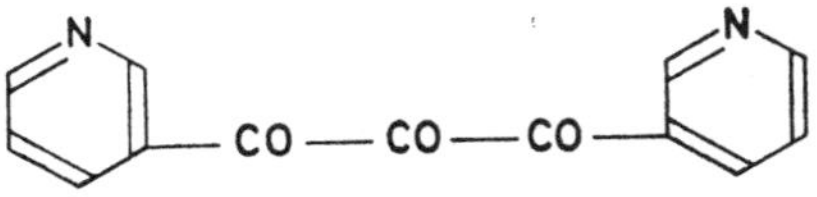

  (*a*) 2-Furyl 2-pyrrolyl diketone
or (*b*) 1-(2-Furyl)-2-(2-pyrrolyl)-
                    ethanedione

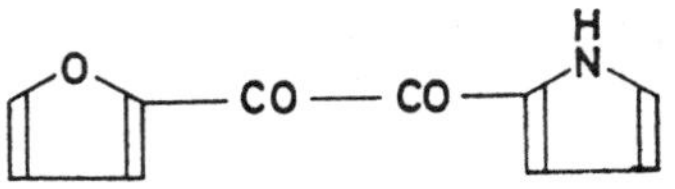

Exceptions:
Benzil C_6H_5—CO—CO—C_6H_5

2,2'-Furil

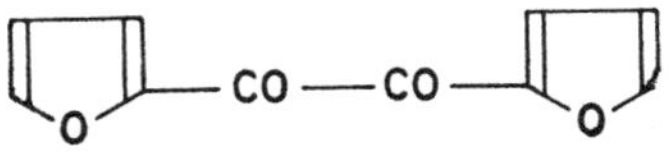

CARBOCYCLIC AND HETEROCYCLIC KETONES

The name of a carbocyclic or heterocyclic ketone which contains, in addition to the carbonyl group as a ring member, the maximum number of non-cumulative double bonds in the ring system, is formed by adding the suffix "-one", "-dione", "-trione", etc., to the name of the parent compound, or, in the case of certain diones, tetrones, etc., the name is formed by adding the suffix "-quinone", "-diquinone", etc., to the name (sometimes in modified form) of the parent compound. The ketonic oxygen atom may also be expressed by the prefix "oxo-".

Rule C–314

314.1 (Alternative to Rules C–315.1 and C–316.1)—Classical procedure: When in a cyclic compound a group >CH$_2$ is replaced by >CO, but not otherwise, this change is denoted by attaching the suffix "-one", with elision of terminal "e" (if present), to the name of the ring system. Suffixes "-dione", "-trione", etc., denote similar replacement of two, three, etc., >CH$_2$ groups by >CO.

Examples:

Cyclohexanone

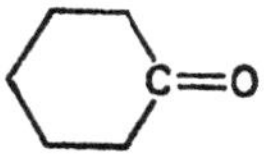

Inden-1-one

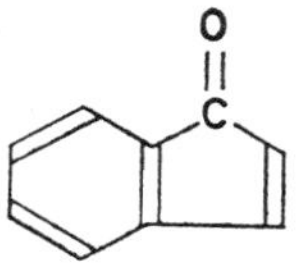

1,2,3-Indantrione

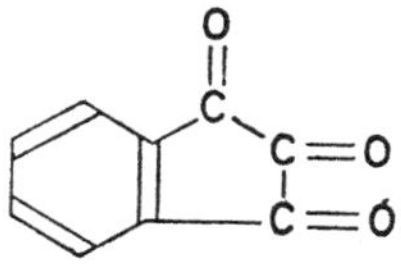

Fluoren-9-one

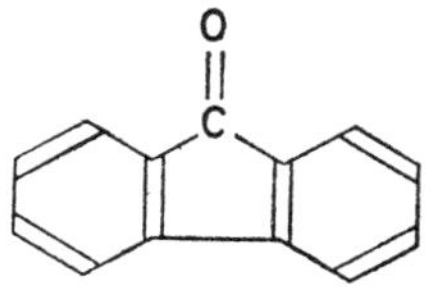

2,5-Cyclohexadien-1-one

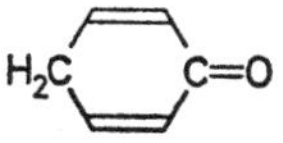

4H-Pyran-4-one

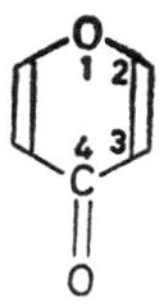

Note: The *H* for indicated hydrogen and its locant may be omitted when no ambiguity results; for example, 4*H*-Pyran-4-one may be named 4-Pyranone, or its contraction 4-Pyrone may be used.

Rule C–315

315.1 (Alternative to Rules C–314.1 and C–316.1)—Polycarbocyclic and heterocyclic ketones may be named by use of a suffix " -one " to denote replacement of either $>CH_2$ or $\geqslant CH$ in an unsaturated or aromatic system by $>CO$; when the ring system is aromatic the maximum number of non-cumulative double bonds is added after introduction of the $>CO$ group(s) and any hydrogen that remains to be added is denoted as indicated hydrogen. As an exception to Rule C–15.11 (*a*), the carbonyl group has then priority over indicated hydrogen for lower number.

Examples:

2(3*H*)-Pyrazinone*

1(2*H*)-Naphthalenone

5,8-Quinolinedione

1,3,6,8(2*H*,7*H*)-Pyrenetetrone

315.2—The following modifications (but not others) of names of carbocyclic ketones formed according to Rule C–315.1 may be used: anthrone [instead of 9(10*H*)-anthracenone], phenanthrone (instead of 9(10*H*)-phenanthrenone).

315.3—The following contracted names for oxo-derivatives of heterocyclic nitrogen compounds are retained as alternatives for the systematic names,

---

* This and similar compounds are classified as lactams (see Rule **C–475.1**), but are included here and in Rule **C–315.3** because they may be named in the same way as ketones.

but further hydrogenated derivatives must not be named as hydro derivatives of these shorter forms:

2-Pyridone   4-Pyridone   2-Quinolone   4-Quinolone

1-Isoquinolone   4-Oxazolone   4-Pyrazolone*   5-Pyrazolone*

4-Isoxazolone   4-Thiazolone   9-Acridone

(Note: Indicated hydrogen is sometimes used with the above names.)

* For some compounds, such as these, the names given have been widely used although the systematic names end in " -olinone ". In view of this, the least hydrogenated forms should be named as " 4-oxopyrazole ", " 5-oxopyrazole ", *etc*. The locant designates the position of the keto group.

In addition, names of oxo-derivatives of fully saturated nitrogen heterocycles, which systematically end in " -idinone ", are often contracted to end in " idone ", and such abbreviated names may also be used when no ambiguity thereby results.

Examples:

4-Piperidone   2-Pyrrolidone   4-Thiazolidone
                 (2-Pyrrolidinone) (4-Thiazolidinone)

Rule C–316

316.1 (Alternative to Rules **C–314.1** and **C–315.1**)—Heterocyclic and di- and poly-carbocyclic ketones may also be named by means of the prefix " oxo- " with the hydrogenation indicated by hydro prefixes; in this case the hydrogenation is considered to have occurred before the introduction of the keto group.

Examples:

1-Oxo-1,2-dihydronaphthalene

5,8-Dioxo-5,8-dihydroquinoline

1,3,6,8-Tetraoxo-1,2,3,6,7,8-hexahydropyrene

316.2—The prefix " oxo- " is used when another group having priority for citation as principal group is also present.

Examples:

4-Oxo-1-cyclohexanecarboxylic acid

5-Oxo-2-pyrrolidinecarboxylic acid

Rule C–317

317.1—Diketones and tetraketones derived from aromatic compounds by conversion of two or four CH groups into CO groups with any necessary

rearrangement of double bonds to a quinonoid structure are named by adding the suffix " -quinone " or " -diquinone " to the name of the aromatic
compound (this name sometimes being modified).

Examples:

p-Benzoquinone

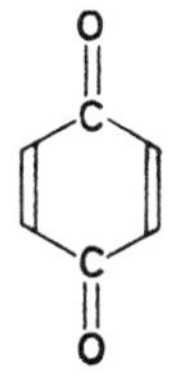

Anthraquinone

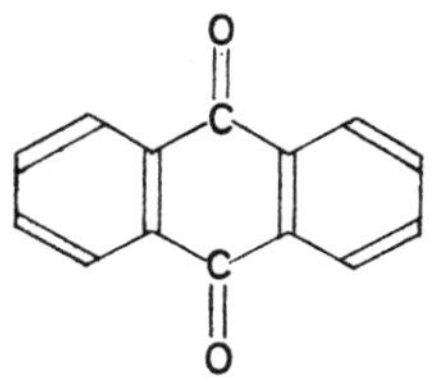

1,4-Naphthoquinone

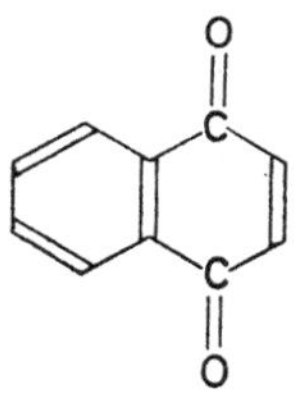

5,6-Chrysenequinone

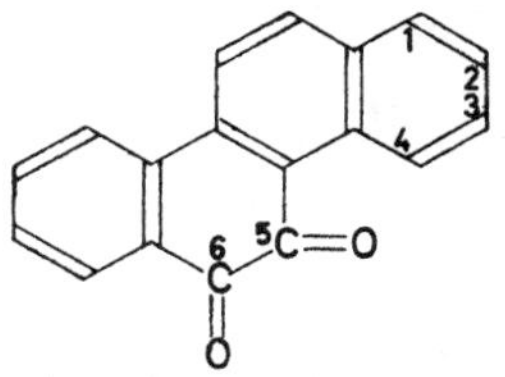

6,12-Chrysenequinone

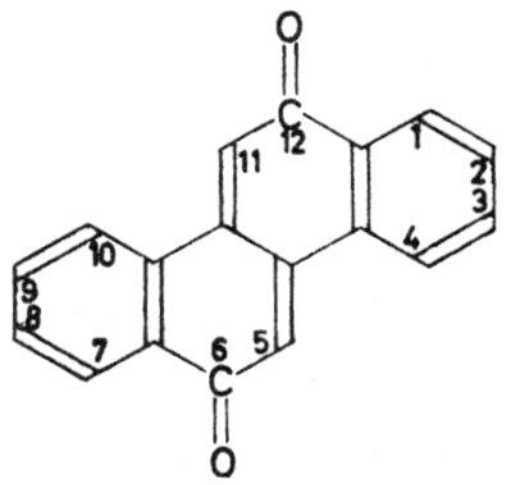

The following names are retained:

Acenaphthoquinone
or Acenaphthenequinone

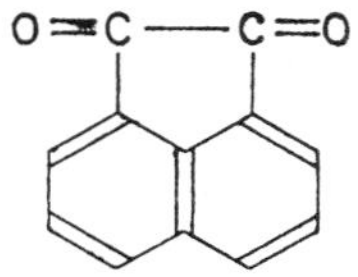

Camphorquinone
or 2,3-Bornanedione

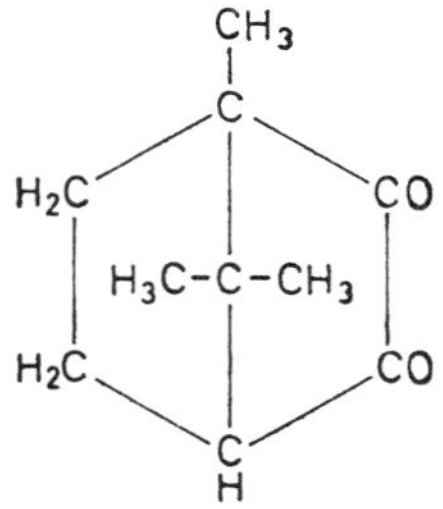

317.2—Radicals formed by loss of hydrogen from a quinone are named by changing the ending " -quinone " to " -quinonyl ", the quinone grouping having priority over the point of attachment for lowest locants.

Example:

7-(1,4-Benzoquinonyl)-2-naphthoic acid

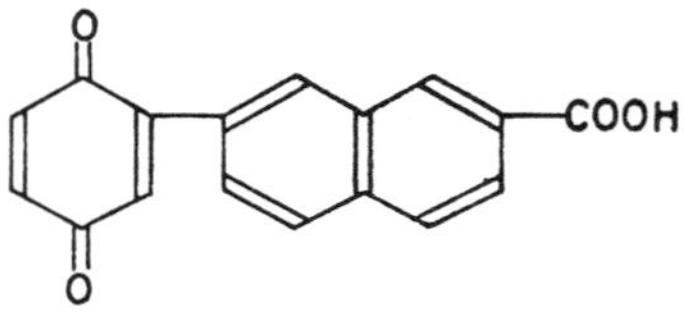

Rule C–318

318.1—In the presence of a group having priority over carbonyl for citation as principal group, or when all the carbonyl groups cannot be included in the ketone functional class name, the presence of carbonyl-oxygen atoms is indicated by the prefixes " oxo- ", " dioxo- ", *etc.*, or by use of the name of a carbonyl-containing radical. Such radicals are: (*a*) acyl radicals, for example, acetyl, valeryl, hexanoyl, benzoyl (see also Rule C–318.2); (*b*) oxo-substituted radicals; (*c*) certain radicals denoted by trivial names.

Examples:

Trivial names of radicals (see *c* above)

Acetonyl	CH_3—CO—CH_2—
Acetonylidene	CH_3—CO—CH=
Phenacyl	C_6H_5—CO—CH_2—
Phenacylidene	C_6H_5—CO—CH=

Complete names

(*a*) 1,3,5-Tribenzoylbenzene

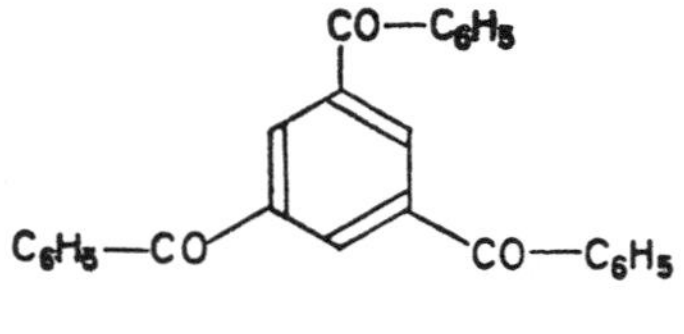

(b) 2-(3-Oxobutyl)-
       1-naphthoic acid

(c) 2-Acetonyl-
       1,4-naphthoquinone

318.2—When two cyclic systems are directly linked by a carbonyl group and another group is also present that has priority for citation as principal group, the compound is named by one of the following methods: (a) The cyclic component having the larger number of groups with such priority is named as parent and the other is named as part of an acyl group substituted in it. (b) If the ring systems are identical and have the same number of groups with such priority the nomenclature for assemblies of identical units (Subsection C–0.7) is used with the name " carbonyl " to denote the bivalent >CO radical. (c) If different cyclic components have the same number of groups with such priority, the senior ring system is named as parent and the other is named as part of an acyl group substituted in it.

Examples:

(a) o-(2-Furoyl)benzoic acid

(b) 4,4'-Carbonyldibenzoic
                acid

(c) 5-(o-Carboxybenzoyl)-
       2-furoic acid

C–3.2. KETENES

Rule C–321

321.1—The compound $CH_2{=}C{=}O$ is named ketene. Derivatives of ketene are named by substitutive nomenclature.

Examples:

Phenylketene $C_6H_5{-}CH{=}CO$
Dibutylketene $(C_4H_9)_2C{=}CO$

Propionylketene $CH_3—CH_2—CO—CH=CO$
or 1-Pentene-1,3-dione (see Rule C–321.3)
p-Phenylenebisketene $OC=CH—C_6H_4—CH=CO$ (*p*-)

Note: " Bis " is used to avoid ambiguity, since diketene is sometimes used for dimeric ketene.

321.2—Ketenes in which the carbon atom of the CH_2 group of ketene has been incorporated into a ring system are named by use of the prefix " carbonyl- ".

Example:

9-Carbonylfluorene

Note: In such cases the use of " carbonyl " for $=CO$ is analogous to use of " methylene " for $=CH_2$.

321.3—An acyl derivative of ketene may be named as a polyketone.

Example:

1-Pentene-1,3-dione $CH_3—CH_2—CO—CH=CO$
or Propionylketene (see Rule C–321.1)

C–3.3. ACETALS AND ACYLALS

ACETALS

Rule C–331

331.1—Compounds containing the group $\begin{matrix} \diagdown \\ \diagup \end{matrix}C\begin{matrix} \diagup OR^1 \\ \diagdown OR^2 \end{matrix}$ are termed acetals*

and are named substitutively as dialkoxy (*etc.*) compounds; alternatively, the name of the corresponding aldehyde or ketone is followed by that of the hydrocarbon radical(s) (if necessary with the prefix " di- ", " bis ", *etc.*), which in turn is followed by the word " acetal "*.

Examples:

1,1-Diethoxypropane $CH_3—CH_2—CH(O—CH_2—CH_3)_2$
Propionaldehyde diethyl acetal

* The name " ketal " is abandoned.

1,1-Dimethoxycyclohexane
or   Cyclohexanone dimethyl acetal

1-Ethoxy-1-methoxycyclohexane
or   Cyclohexanone ethyl methyl acetal

4,4-Diethoxyvaleric acid
or   Levulinic acid diethyl acetal

$$(CH_3-CH_2-O)_2C-CH_2-CH_2-COOH$$
$$\underset{CH_3}{|}$$

331.2—A cyclic acetal in which the two acetal-oxygen atoms form part of a ring may be named: (*a*) as a heterocyclic compound; (*b*) by use of the prefix " methylenedioxy- " for the group $H_2C \big\langle \genfrac{}{}{0pt}{}{O-}{O-}$ as a substituent in the remainder of the molecule, derivatives of the methylenedioxy group being named as substitution products thereof (see also Rule C–205.2); or (*c*) particularly when stereochemistry is implied in the name of the corresponding alcohol or ketone, as the alkylene or alkylidene derivative of that alcohol or as the alkylene or alkylidene acetal of that ketone (see Rule **C–331.1**).

Examples:

(*a*) 1,3-Dioxolane

(*a*) 1,4-Dioxaspiro[4.5]decane

(*a*) Naphtho[2,3-*d*]-1,3-dioxole
or   (*b*) 2,3-Methylenedioxynaphthalene

(*a*) 2,2-Dichloro-1,3-benzo[*d*]-
       dioxole-5-sulfonic acid

or (*b*) 3,4-(Dichloromethylenedioxy)-
        benzenesulfonic acid

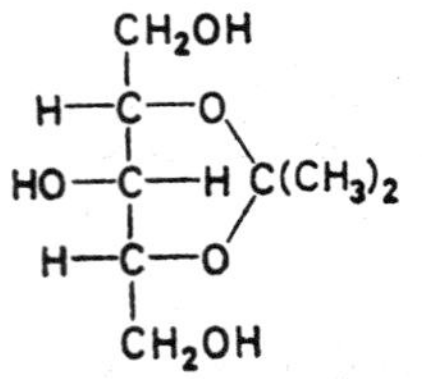

(c) 2,4-O-Isopropylidenexylitol

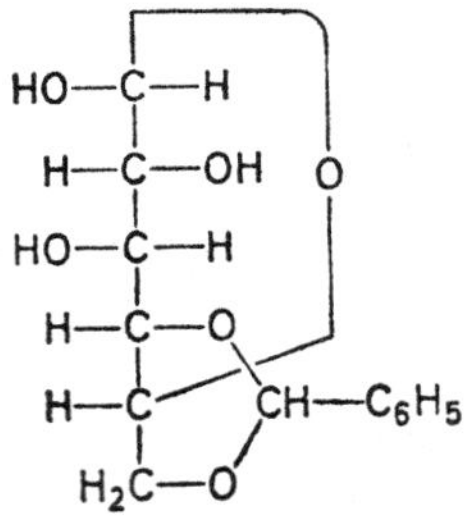

(c) 4,6-O-Benzylidene-β-D-
glucopyranose

(c) 5β-Androstan-3-one ethylene acetal

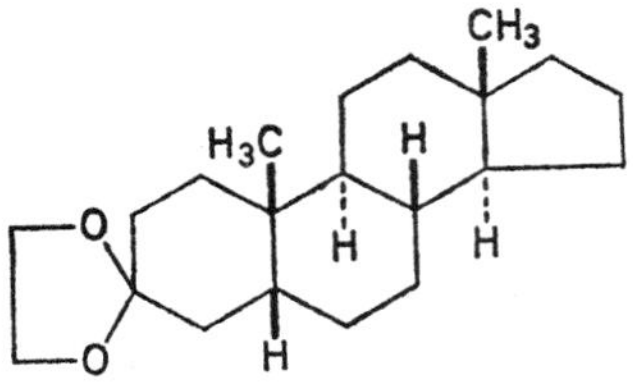

331.3—The following are examples of trivial names that are retained:

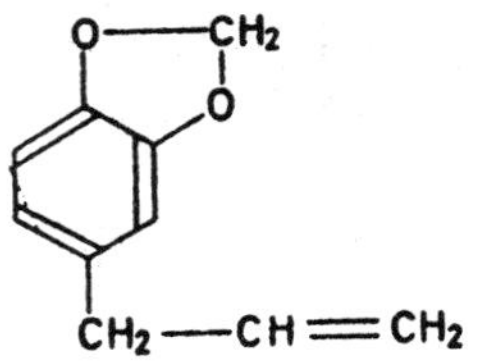

Safrole

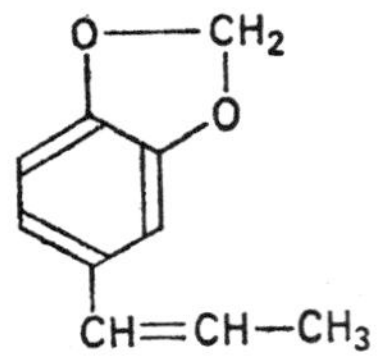

Isosafrole

331.4—Compounds having the general formulae $R^1CH(OH)OR^2$ or $R^1R^2C(OH)OR^3$ are named substitutively as alkoxy (etc.) derivatives of the hydroxy compounds.*

Example:

1-Ethoxy-1-butanol

$$CH_3—CH_2—CH_2—CH{\overset{OH}{\underset{O—CH_2—CH_3}{<}}}$$

---

* The term " hemiacetal " is used only in the generic sense and in certain specialized fields, such as carbohydrate chemistry. The name " hemiketal " is abandoned.

ACYLALS

Rule C–332

332.1—Compounds $R^1CH(OCOR^2)_2$, $R^1R^2C(OCOR^3)_2$, *etc.*, are called generically " acylals ". Specific compounds are named as esters.

Examples:

Ethylidene dipropionate $CH_3—CH(O—CO—CH_2—CH_3)_2$

Benzylidene acetate benzoate

$$C_6H_5—CH\begin{cases} O—CO—CH_3 \\ O—CO—C_6H_5 \end{cases}$$

ACYLOINS (α-HYDROXY KETONES)

Rule C–333

333.1—α-Hydroxy ketones $RCH(OH)—COR$ in which R is an alkyl, aryl, or heterocyclic radical are called generically " acyloins ". They are named by changing the ending " -ic acid " or " -oic acid " of the trivial name of the corresponding acid $RCOOH$ to " -oin ".

Examples:

Acetoin $CH_3—CH(OH)—CO—CH_3$

Benzoin $C_6H_5—CH(OH)—CO—C_6H_5$

2,2′-Furoin

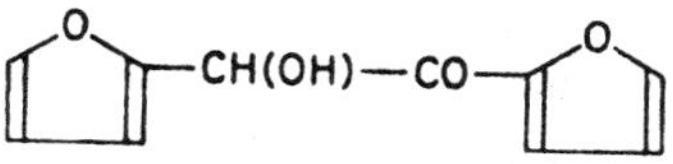

C-4. CARBOXYLIC ACIDS AND THEIR DERIVATIVES

C-4.0. SIMPLE CARBOXYLIC ACIDS

General principles [see Rules C–13.11 (*a*) and C–15.11 (*b*)] require that when carboxyl is the principal group, the numbering of an aliphatic acid is always arranged to give the lowest numbers to this group irrespective of the numbering of the parent compound. The same applies to derivatives of carboxylic acids.

Example:

4-Hexenoic acid

$$\overset{6}{C}H_3\overset{5}{-}CH\overset{4}{=}CH\overset{3}{-}CH_2\overset{2}{-}CH_2\overset{1}{-}COOH$$

or 3-Pentene-1-carboxylic acid

$$\overset{5}{C}H_3\overset{4}{-}CH\overset{3}{=}CH\overset{2}{-}CH_2\overset{1}{-}CH_2-COOH$$

Rule C–401

401.1—Carboxyl groups COOH replacing CH_3 at the end of the main chain of an acyclic hydrocarbon are denoted by adding " -oic acid " or " -dioic acid " to the name of this hydrocarbon, with elision of terminal " e " (if present) before " oic ".

Examples:

Heptanoic acid

$$CH_3-[CH_2]_5-COOH$$

Heptanedioic acid

$$HOOC-[CH_2]_5-COOH$$

4-Pentyl-2,4-pentadienoic acid

$$\overset{5}{C}H_2\overset{4}{=}C\overset{3}{-}CH\overset{2}{=}CH\overset{1}{-}COOH$$
$$CH_3-CH_2-CH_2-CH_2-CH_2$$

3-Vinyl-4-hexynoic acid

$$CH_2=CH$$
$$\overset{6}{C}H_3\overset{5}{-}C\overset{4}{\equiv}C\overset{3}{-}CH\overset{2}{-}CH_2\overset{1}{-}COOH$$

2-Hexen-4-ynedioic acid

$$HOOC\overset{6}{-}C\overset{5}{\equiv}C\overset{4}{-}CH\overset{3}{=}CH\overset{2}{-}COOH$$

$$HOOC\overset{6}{-}\overset{5}{C}H_2\overset{}{-}\overset{4}{C}H\overset{3}{-}CH_2\overset{2}{-}CH\overset{1}{-}COOH$$
$$CH_3-CH_2 \qquad CH_2-CH_2-CH_3$$

4-Ethyl-2-propylhexanedioic acid

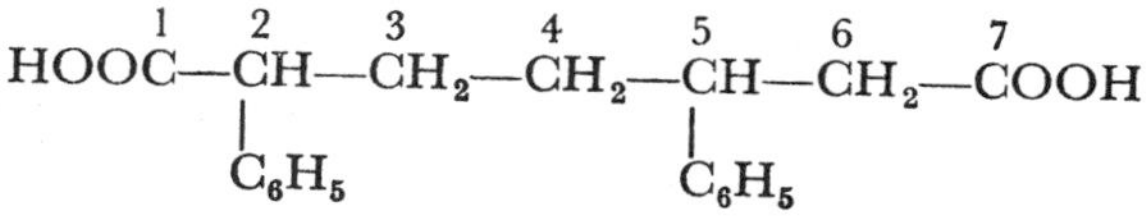

$$\overset{1}{\text{HOOC}}-\overset{2}{\text{CH}}-\overset{3}{\text{CH}_2}-\overset{4}{\text{CH}_2}-\overset{5}{\text{CH}}-\overset{6}{\text{CH}_2}-\overset{7}{\text{COOH}}$$

$$\underset{\text{C}_6\text{H}_5}{|}\qquad\qquad\underset{\text{C}_6\text{H}_5}{|}$$

2,5-Diphenylheptanedioic acid

401.2—Alternatively, the name may be formed by substitutive use of the suffix " -carboxylic acid ". In the aliphatic series the numbering of the chain does not then include the carbon atom of the carboxyl group. For simple aliphatic acids the nomenclature of Rule **C–401.1** is preferred (see Rule **C–13.21**).

Examples:

1-Hexanecarboxylic acid
  (Heptanoic acid preferred)

$$\text{CH}_3—[\text{CH}_2]_5—\text{COOH}$$

1-Pyrrolecarboxylic acid

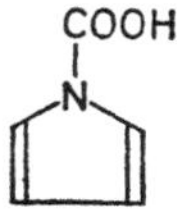

Cyclohexanecarboxylic acid

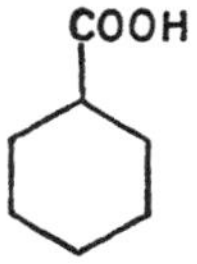

3-Biphenylcarboxylic acid

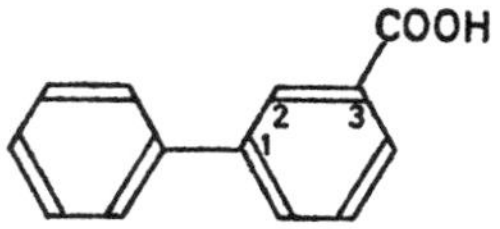

3-(3-Phenylpropyl)-1-
  cyclobutanecarboxylic acid

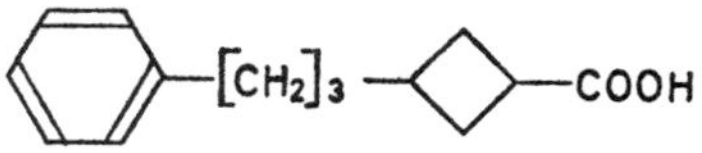

1,3,5-Naphthalenetricarboxylic acid

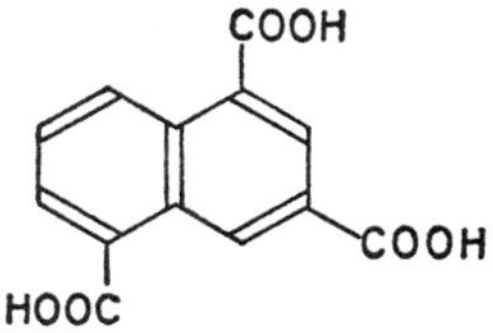

401.3—Carboxyl groups are designated by the prefix " carboxy- " when attached to a group named as a substituent or when another group is also present that has priority for citation as principal group.

Examples:

3-Carboxy-1-methylpyridinium
  chloride

3-(Carboxymethyl)heptanedioic acid

$$\underset{7}{HOOC}-\underset{6}{CH_2}-\underset{5}{CH_2}-\underset{4}{CH_2}-\underset{3}{CH}-\underset{2}{CH_2}-\underset{1}{COOH}$$

with $\underset{3}{\overset{CH_2-COOH}{|}}$ at position 3

401.4—When the carboxyl group is separated from a ring by a chain of carbon atoms the compound is named (*a*) as a derivative of the acyclic system or (*b*) by conjunctive nomenclature. However, the use of method (*b*) for benzene derivatives is restricted as described in Rule C–53.1.

Examples:

$$\underset{\underset{6}{\varepsilon}}{CH_2}-\underset{\underset{5}{\delta}}{CH_2}-\underset{\underset{4}{\gamma}}{CH_2}-\underset{\underset{3}{\beta}}{CH_2}-\underset{\underset{2}{\alpha}}{CH_2}-\underset{1}{COOH}$$

(*a*) 6-(2-Naphthyl)hexanoic acid
or (*b*) 2-Naphthalenehexanoic acid

$$\underset{6}{CH_3}-\underset{5}{\underset{\delta}{CH}}-\underset{4}{\underset{\gamma}{CH_2}}-\underset{3}{\underset{\beta}{CH}}-\underset{2}{\underset{\alpha}{CH_2}}-\underset{1}{COOH}$$

with CH_3 on the β carbon

(*a*) 3-Methyl-5-(1-naphthyl)hexanoic acid
or   (*b*) β,δ-Dimethyl-1-naphthalenevaleric acid

1,3,6-Naphthalenetriacetic acid

6-Carboxy-1-(carboxymethyl)-2-naphthalenepropionic acid

Rule C-402

402.1—If an unbranched chain is linked directly to more than two carboxyl groups, these carboxyl groups are named from the hydrocarbon by substitutive use of the suffix " -tricarboxylic acid ", *etc.* (compare Rule C-401.2). The principal chain should be linked directly to as many carboxyl groups as possible. The carboxyl groups which are not directly linked to the principal chain are expressed by " carboxyalkyl- " prefixes.

Examples:

$$\begin{array}{ccccccc} & COOH & & COOH & COOH & \\ 6 & 5| & 4 & 3| & 2| & 1 \\ CH_3 & CH & CH_2 & CH & CH & CH_3 \end{array}$$

2,3,5-Hexanetricarboxylic acid

$$\begin{array}{cccc} & COOH & COOH & \\ 4 & 3| & 2| & 1 \\ CH_3 & CH & CH & CH_2 & COOH \end{array}$$

1,2,3-Butanetricarboxylic acid

$$\begin{array}{c} (1)\quad(2)\quad(3) \\ CH_2{-}CH_2{-}CH_2{-}COOH \\ 2| \qquad 1 \end{array}$$

$$\begin{array}{ccccccc} 7 & 6 & 5 & 4 & 3 & 2| & 1 \\ CH_3 & CH & CH & CH_2 & CH_2 & CH & CH{-}COOH \\ & | & | & & & & | \\ & HOOC & COOH & & & & COOH \end{array}$$

2-(3-Carboxypropyl)-1,1,5,6-heptanetetracarboxylic acid

In the last example, the principal chain is not the longest, but is the one directly linked to the greatest number of carboxyl groups.

Rule C-403

403.1—When the acid is named according to Rule C-401.1, the name of a univalent or bivalent acyl radical formed by removal of hydroxyl from all the carboxyl groups is derived from the name of the corresponding acid by changing the ending " -oic " to " -oyl ".

Examples:

Heptanoyl	$CH_3{-}CH_2{-}CH_2{-}CH_2{-}CH_2{-}CH_2{-}CO{-}$
Heptanedioyl	$-CO{-}CH_2{-}CH_2{-}CH_2{-}CH_2{-}CH_2{-}CO-$

$$\begin{array}{cccccc} & 6 & 5 & 4 & 3 & 2 & 1 \end{array}$$
2-Hexen-4-ynedioyl　$-CO-C{\equiv}C-CH{=}CH-CO-$

403.2—If an acid name is formed with the ending " -carboxylic ", the name of the radical obtained by removal of hydroxyl from all carboxyl groups is formed by changing the ending " -carboxylic " to " -carbonyl ".

TABLE VI.   Trivial names of some acids and their radicals

Names of acids		Acyl radicals	
Systematic	Trivial	Trivial name	Formula

(a) Saturated aliphatic monocarboxylic acids (limiting list)

Names of acids		Acyl radicals	
Methanoic†	Formic	Formyl	HCO—
Ethanoic†	Acetic	Acetyl	CH_3—CO—
Propanoic†	Propionic	Propionyl	CH_3—CH_2—CO—
Butanoic†	Butyric	Butyryl	CH_3—$[CH_2]_2$—CO—
2-Methylpropanoic†	Isobutyric*	Isobutyryl*	$(CH_3)_2$CH—CO—
Pentanoic†	Valeric	Valeryl	CH_3—$[CH_2]_3$—CO—
3-Methylbutanoic†	Isovaleric*	Isovaleryl*	$(CH_3)_2$CH—CH_2—CO—
2,2-Dimethylpropanoic	Pivalic*	Pivaloyl*	$(CH_3)_3$C—CO—
Dodecanoic	Lauric*	Lauroyl*	CH_3—$[CH_2]_{10}$—CO—
Tetradecanoic	Myristic*	Myristoyl*	CH_3—$[CH_2]_{12}$—CO—
Hexadecanoic	Palmitic*	Palmitoyl*	CH_3—$[CH_2]_{14}$—CO—
Octadecanoic	Stearic*	Stearoyl*	CH_3—$[CH_2]_{16}$—CO—

Note: The names caproic, caprylic, and capric acid (for hexanoic, octanoic, and decanoic acid, respectively) are abandoned.

(b) Saturated aliphatic dicarboxylic acids (limiting list)

Names of acids		Acyl radicals	
Ethanedioic†	Oxalic	Oxalyl	—CO—CO—
Propanedioic†	Malonic	Malonyl	—CO—CH_2—CO—
Butanedioic†	Succinic	Succinyl	—CO—$[CH_2]_2$—CO—
Pentanedioic†	Glutaric	Glutaryl	—CO—$[CH_2]_3$—CO—
Hexanedioic†	Adipic	Adipoyl	—CO—$[CH_2]_4$—CO—
Heptanedioic	Pimelic*	Pimeloyl*	—CO—$[CH_2]_5$—CO—
Octanedioic	Suberic*	Suberoyl*	—CO—$[CH_2]_6$—CO—
Nonanedioic	Azelaic*	Azelaoyl*	—CO—$[CH_2]_7$—CO—
Decanedioic	Sebacic*	Sebacoyl*	—CO—$[CH_2]_8$—CO—

(c) Unsaturated aliphatic acids (non-limiting list)

Names of acids		Acyl radicals	
Propenoic†	Acrylic	Acryloyl	CH_2=CH—CO—
Propynoic†	Propiolic	Propioloyl	CH≡C—CO—
2-Methylpropenoic†	Methacrylic	Methacryloyl	CH_2=C(CH_3)—CO—
trans-2-Butenoic†	Crotonic	Crotonoyl	CH_3—CH=CH—CO—
cis-2-Butenoic†	Isocrotonic	Isocrotonoyl	
cis-9-Octadecenoic†	Oleic	Oleoyl	CH—$[CH_2]_7$—CH_3 ‖ CH—$[CH_2]_7$—CO—
trans-9-Octadecenoic†	Elaidic	Elaidoyl	
cis-Butenedioic†	Maleic	Maleoyl	CH—CO— ‖ CH—CO—
trans-Butenedioic†	Fumaric	Fumaroyl	—CO—CH ‖ CH—CO—
cis-Methylbutenedioic	Citraconic*	Citraconoyl*	CH—CO— ‖ CH_3—C—CO—
trans-Methyl-butenedioic	Mesaconic*	Mesaconoyl*	—CO—CH ‖ CH_3—C—CO—

* Systematic names are recommended for derivatives formed by substitution on a carbon atom.
† The trivial name is normally preferred.

TABLE VI (contd.). Trivial names of some acids and their radicals *

Names of Acids		Acyl radicals	
Systematic†	Trivial	Trivial name	Formula

(d) Carbocyclic carboxylic acids (non-limiting list)

Systematic†	Trivial	Trivial name	Formula
1,2,2-Trimethyl-1,3-cyclopentanedicarboxylic acid	Camphoric	Camphoroyl	
Benzenecarboxylic	Benzoic	Benzoyl	C_6H_5—CO—
1,2-Benzenedicarboxylic	Phthalic	Phthaloyl	
1,3-Benzenedicarboxylic	Isophthalic	Isophthaloyl	
1,4-Benzenedicarboxylic	Terephthalic	Terephthaloyl	
Naphthalenecarboxylic	Naphthoic	Naphthoyl (2- shown)	
Methylbenzenecarboxylic	Toluic	Toluoyl (o- shown)	
2-Phenylpropanoic	Hydratropic	Hydratropoyl	C_6H_5—CH(CH_3)—CO—
2-Phenylpropenoic	Atropic	Atropoyl	C_6H_5—C(=CH_2)—CO—
3-Phenylpropenoic (trans-)	Cinnamic	Cinnamoyl	C_6H_5—CH=CH—CO—

(e) Heterocyclic carboxylic acids (non-limiting list)

Systematic†	Trivial	Trivial name	Formula
Furancarboxylic	Furoic	Furoyl (3- shown)	
Thiophenecarboxylic	Thenoic	Thenoyl (2- shown)	
3-Pyridinecarboxylic	Nicotinic	Nicotinoyl	
4-Pyridinecarboxylic	Isonicotinic	Isonicotinoyl	

† The trivial name is normally preferred.

*For trivial names of carboxylic acids (and fatty acids) that are constituents of lipids see Nomenclature of Lipids, *IUPAC Information Bulletin* Appendix No. 67, December 1977.

Examples:

2,3,5-Hexanetricarbonyl

$$\overset{CO-}{\underset{6}{CH_3}}\overset{CO-}{\underset{5}{CH}}\overset{}{\underset{4}{CH_2}}\overset{CO-}{\underset{3}{CH}}\overset{CO-}{\underset{2}{CH}}\overset{}{\underset{1}{CH_3}}$$

Cyclohexanecarbonyl
Cyclohexylcarbonyl

Note: In radicofunctional nomenclature, names of the type cyclo-hexanecarbonyl are used, as in:

Cyclohexanecarbonyl chloride

When, however, the acyl group is a substituent, the compound radical-prefix form of type cyclohexylcarbonyl is used, as in:

2-(Cyclohexylcarbonyl)-
1-naphthoic acid

4-(Cyclohexylcarbonyl)-
2-morpholinecarboxylic acid

These principles apply to all acyl radicals whose names end in " car-bonyl ", " sulfonyl ", " sulfinyl ", etc. (cf. Rules C-543.3, C-631.1, C-641.7, C-641.8, and C-641.9).

Rule C-404

404.1—Examples of trivial names for acids and acyl radicals are given in Table VI.   When a trivial name is given to an acyclic monoacid or diacid, the number 1 is always given to the carbon atom of a carboxyl group in the acid, or to a carbon atom with a free valence in the radical RCO—.

Rule C-405

405.1—When hydroxyl has been removed from one or more, but not from all, the carboxyl groups present in an acid, the name of the radical is formed by denoting the carboxyl groups by prefixes " carboxy- " and the carbonyl group by the suffix " -oyl " or " -carbonyl ".

Examples:

4-Carboxybutanoyl

$$HOOC\overset{}{\underset{4}{-CH_2}}\overset{}{\underset{3}{-CH_2}}\overset{}{\underset{2}{-CH_2}}\overset{}{\underset{1}{-CO-}}$$

$$\overset{}{\underset{6}{-CO}}\overset{CH_3}{\underset{5}{-CH}}\overset{}{\underset{4}{-CH_2}}\overset{COOH}{\underset{3}{-CH}}\overset{COOH}{\underset{2}{-CH}}\overset{}{\underset{1}{-CO-}}$$

2,3-Dicarboxy-5-methylhexanedioyl

405.2—The following names of radicals are retained (see also Rule C-416.3 and, for amino-acyl radicals, Rule C-421):

Oxalo	HOOC—CO— (univalent radical)*		
Oxalyl	—CO—CO— (bivalent radical)*		
Methoxalyl	CH_3OOC—CO—	Ethoxalyl	C_2H_5OOC—CO—

* Compare Rule C-416.3.

C–4.1. HYDROXY, ALKOXY, AND OXO ACIDS

Rule C–411

411.1—Examples of trivial names for hydroxy and alkoxy carboxylic acids and for the corresponding acyl radicals are given in Table VII. The numbering of aliphatic hydroxy, alkoxy, and oxo acids is made so as to give the number 1 to a carboxyl group and the lowest possible locants to hydroxy, alkoxy, oxo and other groups [see Rule **C–15.11** (*d*)].

Rule C–415

415.1 (Partly alternative to Rules **C–416.1** and **C–416.2**)—The names of carboxylic acids containing an aldehydic and/or one or more ketonic groups in the principal chain or parent ring system are generally derived from the names of the corresponding simple carboxylic acids by adding prefixes " oxo- ", " dioxo- ", *etc.* (see Rule **C–316.2**).

Examples:

5-Oxovaleric acid
(see also Rule **C–416.1**) $OHC-CH_2-CH_2-CH_2-COOH$

3-Oxovaleric acid $CH_3-CH_2-CO-CH_2-COOH$

3,5-Dioxovaleric acid $OHC-CH_2-CO-CH_2-COOH$

2-Oxo-1-cyclohexanecarboxylic acid

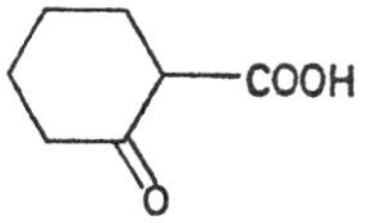

415.2—The prefixes " oxo- ", " dioxo- ", *etc.*, are used also in conjunctive names to indicate substitution.

Examples:

β-Oxocyclohexanepropionic acid

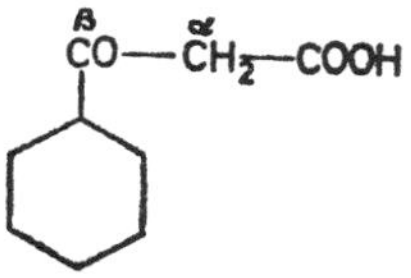

γ-Oxo-6-chrysenebutyric acid

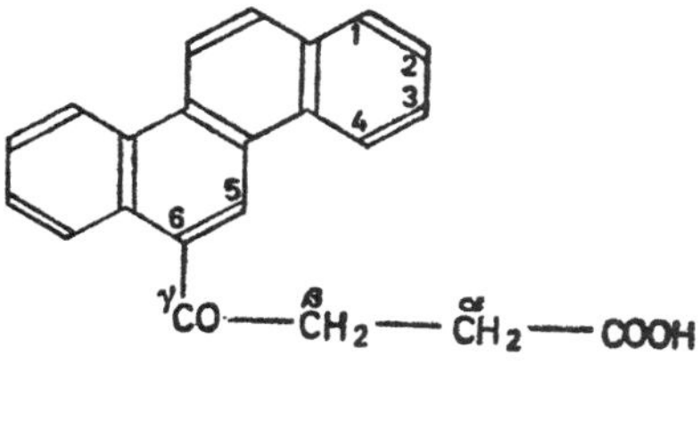

β-Oxo-3-pyridinepropionic acid

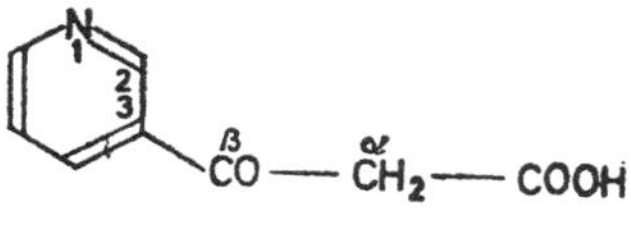

189

TABLE VII.   Trivial names of hydroxy and alkoxy carboxylic acids (non-limiting list)

Names of acids		Acyl radicals	
Systematic	Trivial†	Trivial name	Formula
Hydroxyethanoic	Glycolic	Glycoloyl	$HO—CH_2—CO—$
2-Hydroxypropanoic	Lactic	Lactoyl	$CH_3—CH(OH)—CO—$
2,3-Dihydroxy-propanoic	Glyceric	Glyceroyl	$CH_2OH—CH(OH)—CO—$
Hydroxy-propanedioic	Tartronic	Tartronoyl	$—CO—CH(OH)—CO—$
Hydroxybutanedioic	Malic	Maloyl	$—CO—CH_2—CH(OH)—CO—$
2,3-Dihydroxybutane-dioic	Tartaric	Tartaroyl	$—CO—CH(OH)—CH(OH)—CO—$
3-Hydroxy-2-phenylpropanoic	Tropic	Tropoyl	$C_6H_5—\overset{\displaystyle CH_2OH}{\underset{\displaystyle \vert}{CH}}—CO—$
2-Hydroxy-2,2-diphenylethanoic	Benzilic	Benziloyl	$(C_6H_5)_2C(OH)—CO—$
o-Hydroxybenzoic	Salicylic	Salicyloyl	
Methoxybenzoic	Anisic	Anisoyl (p- shown)	
4-Hydroxy-3-methoxybenzoic	Vanillic	Vanilloyl	
3,4-Dimethoxy-benzoic	Veratric	Veratroyl	
3,4-Methylenedioxy-benzoic	Piperonylic	Piperonyloyl	
3,4-Dihydroxybenzoic	Proto-catechuic	Proto-catechuoyl	
3,4,5-Trihydroxy-benzoic	Gallic	Galloyl	

† Normally preferred.

190

415.3—When a carbonyl group is at the point of attachment of a side chain in a carboxylic acid, the name of the side chain is the name of the acyl radical RCO— (including formyl HCO—).

Examples:

Formylsuccinic acid

$$\underset{\displaystyle HOOC—CH_2—\overset{\displaystyle \overset{CHO}{|}}{CH}—COOH}{}$$

3-(2-Naphthoyl)butyric acid

415.4—When a carbonyl group is present in a side chain but not at either end thereof, an " oxoalkyl " prefix is used.

Example:

p-(2-Oxobutyl)benzoic acid $CH_3—CH_2—CO—CH_2$ —COOH

415.5—When ring systems are united by a carbonyl group only and one of these ring systems carries a carboxyl group, the compound is named as an acid containing an acyl substituent (see Rules **C–403.2** and **C–404**).

Examples:

3-Cyclohexylcarbonyl-1-cyclo-pentanecarboxylic acid

p-Benzoylbenzoic acid

Rule C–416

416.1 (Alternative to part of Rule **C–415.1**)—As exceptions to Rule **C–415.1**, the following acyl radical names may be used even if the carbonyl group is included in the main chain of the parent acid or acyl radical: formyl, benzoyl (and substituted benzoyl groups such as toluoyl, xyloyl), 1- and 2-naphthoyl.

Examples:

4-Formylbutyric acid (see also Rule **C–415.1**)

$$OHC—\overset{4}{C}H_2—\overset{3}{C}H_2—\overset{2}{C}H_2—\overset{1}{C}OOH$$

6-Chloro-6-formylhexanoic acid

$$\overset{6}{OHC}-\overset{5}{CHCl}-\overset{4}{CH_2}-\overset{3}{CH_2}-\overset{2}{CH_2}-\overset{1}{CH_2}-COOH$$

3-Benzoylpropionic acid $C_6H_5-CO-\overset{3}{CH_2}-\overset{2}{CH_2}-\overset{1}{COOH}$

p-(3-Formylpropyl)benzoic acid   OHC—CH₂—CH₂—CH₂— —COOH

416.2—When a dicarboxylic acid has a trivial name, the change of one of the carboxyl groups into an aldehyde group may be denoted by changing the syllables " -ic acid " into " -aldehydic acid ".

Examples:

Malonaldehydic acid $OHC-CH_2-COOH$

Glutaraldehydic acid $OHC-CH_2-CH_2-CH_2-COOH$

Phthalaldehydic acid   Isophthalaldehydic acid

Terephthalaldehydic acid

416.3—The following are examples of trivial or semitrivial names for oxo carboxylic acids and for the corresponding radicals (see also Rule **C–405.2**):

Name of the acid	Acyl radicals	
	Name	*Formula*
Glyoxylic	Glyoxyloyl	$OHC-CO-$
Pyruvic	Pyruvoyl	$CH_3-CO-CO-$
Acetoacetic	Acetoacetyl	$CH_3-CO-CH_2-CO-$
Mesoxalic	Mesoxalyl	$-CO-CO-CO-$
	Mesoxalo	$HOOC-CO-CO-$
Oxalacetic	Oxalacetyl	$-CO-CH_2-CO-CO-$
	Oxalaceto	$HOOC-CO-CH_2-CO-$

C–4.2. AMINO ACIDS**

Rule C–421

421.1—Table VIII gives examples of trivial names for α-amino carboxylic acids. The name of the radical obtained by removal of hydroxyl from the carboxyl group is derived from the trivial name of the amino acid by replacing the ending " -ine " with " -yl ".

TABLE VIII.   Some α-amino acids and their radicals

Parent acid	Radical name	Formula of the radical
Alanine	Alanyl	CH_3—$CH(NH_2)$—CO—
β-Alanine	β-Alanyl	H_2N—CH_2—CH_2—CO—

Arginine	Arginyl	$HN{=}C$—NH—$[CH_2]_3$—CH—CO—
		     $\mid$ $\mid$
		     NH_2 NH_2

Cystathionine	Cystathionyl	$S{<}$ CH_2—$CH(NH_2)$—CO— CH_2—CH_2—$CH(NH_2)$—CO—
Cystine	Cystyl	S—CH_2—$CH(NH_2)$—CO— $\mid$ S—CH_2—$CH(NH_2)$—CO—
Glycine	Glycyl	H_2N—CH_2—CO—
Histidine	Histidyl	$HC{=}C$—CH_2—$CH(NH_2)$—CO— $HN\diagdown$ N $\underset{H}{C}$
Homoserine	Homoseryl	HO—CH_2—CH_2—$CH(NH_2)$—CO—
Isoleucine	Isoleucyl	C_2H_5—$CH(CH_3)$—$CH(NH_2)$—CO—
Lanthionine	Lanthionyl	$S{<}$ CH_2—$CH(NH_2)$—CO— CH_2—$CH(NH_2)$—CO—
Leucine	Leucyl	$(CH_3)_2CH$—CH_2—$CH(NH_2)$—CO—
Lysine	Lysyl	H_2N—$[CH_2]_4$—$CH(NH_2)$—CO—
Methionine	Methionyl	CH_3—S—CH_2—CH_2—$CH(NH_2)$—CO—
Norleucine*	Norleucyl*	CH_3—$[CH_2]_3$—$CH(NH_2)$—CO—
Norvaline*	Norvalyl*	CH_3—CH_2—CH_2—$CH(NH_2)$—CO—
Ornithine	Ornithyl	H_2N—$[CH_2]_3$—$CH(NH_2)$—CO—
Proline	Prolyl	H_2C——CH—CO— H_2C NH $\underset{H_2}{C}$
Sarcosine	Sarcosyl	CH_3—NH—CH_2—CO—
Serine	Seryl	HO—CH_2—$CH(NH_2)$—CO—
Threonine	Threonyl	CH_3—$CH(OH)$—$CH(NH_2)$—CO—
Thyronine	Thyronyl	HO—⟨ ⟩—O—⟨ ⟩—CH_2—$CH(NH_2)$—CO—
Tyrosine	Tyrosyl	HO—⟨ ⟩—CH_2—$CH(NH_2)$—CO—
Valine	Valyl	$(CH_3)_2CH$—$CH(NH_2)$—CO—

* Use of " nor " in these names to denote a " normal " (unbranched) chain conflicts with more recent usage (see Rules **C–42** and **C–43**).

**See also Nomenclature of α-Amino Acids, *IUPAC Information Bulletin* Appendix No. 46, September 1975 or *Biochemistry* (1975) 14, pp. 449-462.

421.2—The following names are also recommended, the radical names being exceptions to the preceding rule:

Cysteine	Cysteinyl	$HS—CH_2—CH(NH_2)—CO—$
Homocysteine	Homocysteinyl	$HS—CH_2—CH_2—CH(NH_2)—CO—$

Tryptophan   Tryptophyl

421.3—(*a*) 2-Aminobutanedioic acid is named aspartic acid. (*b*) 2-Aminopentanedioic acid is named glutamic acid. (*c*) The derived univalent acyl radicals are named aspartyl and glutamyl, while the corresponding bivalent acyl radicals are named aspartoyl and glutamoyl (compare Rule C–403.1 and C–405.2). (*d*) Monoamides of these acids in which the carboxyl group numbered 1 is intact are respectively named asparagine and glutamine, and univalent acyl radicals corresponding to these monoacids are named asparaginyl and glutaminyl (see Table IX).

TABLE IX.   Some amino dicarboxylic acids and their amides and radicals

Name	Formula
Aspartic acid	$\overset{4}{H}OOC—\overset{3}{C}H_2—\overset{2}{C}H(NH_2)—\overset{1}{C}OOH$
α-Aspartyl	$HOOC—CH_2—CH(NH_2)—CO—$
β-Aspartyl	$—CO—CH_2—CH(NH_2)—COOH$
Aspartoyl	$—CO—CH_2—CH(NH_2)—CO—$
Asparagine	$\overset{4}{H}_2N—CO—\overset{3}{C}H_2—\overset{2}{C}H(NH_2)—\overset{1}{C}OOH$
Asparaginyl	$H_2N—CO—CH_2—CH(NH_2)—CO—$
Glutamic acid	$\overset{5}{H}OOC—\overset{4}{C}H_2—\overset{3}{C}H_2—\overset{2}{C}H(NH_2)—\overset{1}{C}OOH$
α-Glutamyl	$HOOC—CH_2—CH_2—CH(NH_2)—CO–·$
γ-Glutamyl	$—CO—CH_2—CH_2—CH(NH_2)—COOH$
Glutamoyl	$—CO—CH_2—CH_2—CH(NH_2)—CO—$
Glutamine	$\overset{5}{H}_2N—CO—\overset{4}{C}H_2—\overset{3}{C}H_2—\overset{2}{C}H(NH_2)—\overset{1}{C}OOH$
Glutaminyl	$H_2N—CO—CH_2—CH_2—CH(NH_2)—CO—$

421.4—The following trivial names are also retained:

Anthranilic acid

Hippuric acid $C_6H_5—CO—NH—CH_2—COOH$

C-4.3. AMIC ACIDS

Rule C-431

431.1—When a dicarboxylic acid has a trivial name and when one of its carboxyl groups is replaced by a carboxamide group —C(=O)—NH$_2$, the resulting amic acid is named by replacing the suffix " -ic " of the trivial name of the dicarboxylic acid by the suffix " -amic ".

Examples:

Succinamic acid $\qquad$ H$_2$N—CO—CH$_2$—CH$_2$—COOH

3-Bromophthalamic acid

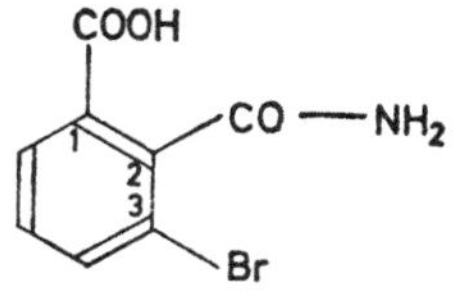

The following names are also recommended:

$\qquad$ Carbamic acid $\qquad$ H$_2$N—COOH

$\qquad$ Oxamic acid $\qquad$ H$_2$N—CO—COOH

431.2—Acyl radicals derived from the above amic acids are named by replacing the " -amic acid " ending by " -amoyl ".

Examples:

$\qquad$ Carbamoyl (from carbamic acid) $\qquad$ H$_2$N—CO—

$\qquad$ Oxamoyl (from oxamic acid) $\qquad$ H$_2$N—CO—CO—

$\qquad$ Succinamoyl $\qquad$ H$_2$N—CO—CH$_2$—CH$_2$—CO—
$\qquad$ (from succinamic acid)

431.3—When the corresponding dicarboxylic acid has a trivial or semi-trivial name, *N*-phenyl derivatives of -amic acids are named by changing the ending " -amic acid " to " -anilic acid ".

Examples:

Succinanilic acid $\qquad$ C$_6$H$_5$—NH—CO—CH$_2$—CH$_2$—COOH

2'-Methylsuccinanilic acid

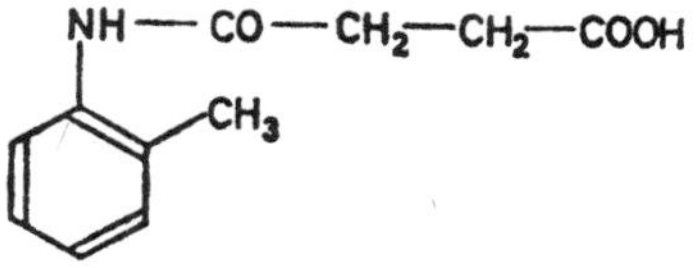

$\qquad$ Carbanilic acid $\qquad$ C$_6$H$_5$—NH—COOH
or Phenylcarbamic acid

431.4—Monoamides of dicarboxylic acids without trivial names and the mono- or di-amides of tricarboxylic acids, *etc.*, are designated as carbamoyl derivatives of the acid named on the basis of the remaining carboxyl groups.

(This Rule is partly alternative to Rule C-431.1.)

Examples:

$$\overset{6}{H_2N-CO-CH_2}-\overset{5}{CH_2}-\overset{4}{CH_2}-\overset{3}{CH_2}-\overset{2}{CH_2}-\overset{1}{COOH}$$

6-Carbamoylhexanoic acid

$$C_6H_5-NH-CO-\overset{6}{CH_2}-\overset{5}{CH_2}-\overset{4}{CH_2}-\overset{3}{CH_2}-\overset{2}{CH_2}-\overset{1}{COOH}$$

6-(Phenylcarbamoyl)hexanoic acid

$$HOOC-CH_2-\underset{}{\overset{C_6H_5}{CH}}-\underset{CH_3}{\overset{CO-NH_2}{C}}-COOH$$

2-Carbamoyl-2-methyl-3-phenylglutaric acid

C–4.4. PEROXY ACIDS

Rule C-441

441.1—Acids containing the group —C(=O)—OOH are called peroxy acids. In individual names the prefix " peroxy- " is used before the trivial or -oic acid name of the acid or before carboxylic (as the case may be). For diacids, " monoperoxy- " or " diperoxy- " is used as appropriate. Names of derivatives are formed in the usual way but with *O-* or *OO-* prefixes where necessary.

Examples:

Peroxypropionic acid $CH_3-CH_2-\underset{O}{\overset{\parallel}{C}}-OOH$

Peroxyhexanoic acid $CH_3-CH_2-CH_2-CH_2-CH_2-\underset{O}{\overset{\parallel}{C}}-OOH$

Cyclohexaneperoxy-
carboxylic acid

Diperoxy-
phthalic acid

OO-Ethyl hydrogen
monoperoxyphthalate

Exceptions: The following well-established names are retained:

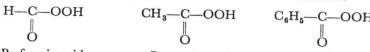

H—C—OOH	CH$_3$—C—OOH	C$_6$H$_5$—C—OOH
Performic acid	Peracetic acid	Perbenzoic acid

C–4.5. IMIDIC, HYDRAZONIC, AND HYDROXAMIC ACIDS

Rule C–451

451.1—The name of an acid in which the carbonyl-oxygen atom of a carboxyl group has been replaced by =NH, =N—NH$_2$, or =N—OH is formed by modifying the ending " -oic " or " -carboxylic " of the systematic name of the carboxylic acid, or the ending " -ic " of the trivial name, as shown in Table X. A letter " o " is added for euphony between " h " and a preceding consonant.

Examples:

Butyrimidic acid	CH$_3$—CH$_2$—CH$_2$—C(=NH)—OH
Heptanimidic acid	CH$_3$—[CH$_2$]$_5$—C(=NH)—OH
Heptanediimidic acid	HO—(HN=)C—[CH$_2$]$_5$—C(=NH)—OH

2-Pyrrolecarboximidic acid

Heptanehydrazonic acid   CH$_3$—[CH$_2$]$_5$—C(=N—NH$_2$)—OH

Cyclohexanecarbohydrazonic acid

Benzohydroximic acid   C$_6$H$_5$—C(=N—OH)—OH

451.2—Radicals RC(=X)—, where =X is =NH, =N—NH$_2$, or =NOH, are formed from the names of the corresponding acids RC(=X)—OH by changing " -ic " of the endings given in Table X to " -oyl ".

Examples:

Formimidoyl	CH(=NH)—
Hexanimidoyl	CH$_3$—[CH$_2$]$_4$—C(=NH)—
Cyclohexanecarbohydroximoyl	C$_6$H$_{11}$—C(=NOH)—
Acetohydrazonoyl	CH$_3$—C(=N—NH$_2$)—
Succinimidoyl	—C(=NH)—CH$_2$—CH$_2$—C(=NH)—
p-Formimidoylbenzoic acid	HN=CH—C$_6$H$_4$—COOH-*p*

TABLE X.   Modified endings for some
nitrogen-containing acids

Group	Ending
Rule C–451.1	
$-(C){\Large\langle}\genfrac{}{}{0pt}{}{NH}{OH}$	-imidic
$-C{\Large\langle}\genfrac{}{}{0pt}{}{NH}{OH}$	-carboximidic
$-(C){\Large\langle}\genfrac{}{}{0pt}{}{N-NH_2}{OH}$	-hydrazonic
$-C{\Large\langle}\genfrac{}{}{0pt}{}{N-NH_2}{OH}$	-carbohydrazonic
$-(C){\Large\langle}\genfrac{}{}{0pt}{}{N-OH}{OH}$	-hydroximic
$-C{\Large\langle}\genfrac{}{}{0pt}{}{N-OH}{OH}$	-carbohydroximic
Rule C–451.3	
$-(C){\Large\langle}\genfrac{}{}{0pt}{}{NH-OH}{O}$	-hydroxamic*
$-C{\Large\langle}\genfrac{}{}{0pt}{}{NH-OH}{O}$	-carbohydroxamic*

* Such compounds may also be named as N-hydroxy-
amides (see Rule C–841.3).

451.3—The name of an acid in which the hydroxyl group of the carboxyl group has been changed to —NH—OH is formed by changing the
ending " -oic acid " of the systematic name of the carboxylic acid, or " -ic
acid " of the trivial name, to " -ohydroxamic acid ", or " -carboxylic
acid " to " -carbohydroxamic acid ". (However, see also Rule C–841.3.)

Examples:

Acetohydroxamic acid CH_3—$C(=O)$—NH—OH

2*H*-Pyran-3-carbohydroxamic acid

451.4—The hypothetical compound $HN=C(OH)_2$ is named carbonimidic acid, and the bivalent radical $>C=NH$ is named carbonimidoyl.

Example:

4,4′-Carbonimidoyldibenzoic acid

C–4.6. SALTS AND ESTERS

Rule C–461

461.1—Neutral salts of carboxylic, amic, imidic, carboximidic, hydroxamic, carbohydroxamic, *etc.*, acids are named by citing the cation(s) and then the anion (see Rule C–84.1) or, when the carboxyl groups of the acid are not all named as affixes, by use of a periphrase, such as " (metal) salt of (the acid) ". When different acidic residues are present in one structure, prefixes are formed by changing the anion ending " -ate " to " -ato- " or " -ide " to " -ido- " (see Rule C–86.1); the prefix " carboxylato- " may be used to denote the ionic group —$\overset{-}{C}OO^-$.

Examples:

Sodium heptanoate	CH_3—$[CH_2]_5$—COONa
Potassium acetate	CH_3—COOK
Ammonium cyclohexanecarboxylate	C_6H_{11}—$COONH_4$
Disodium succinate	CH_2—COONa | CH_2—COONa
Calcium diacetate	$(CH_3$—$COO)_2Ca$
Potassium sodium succinate	CH_2—COONa | CH_2—COOK

Disodium 2-carboxylatocyclohexaneacetate

Sodium salt of methionine $CH_3S—CH_2—CH_2—CH(NH_2)—COONa$

Sodium hexanimidate $CH_3—[CH_2]_4—C(=NH)ONa$

Potassium cyclohexane-
carbohydroxamate

2Na⁺

Disodium 3-(carboxylatomethyl)-2-naphthalenepropionate

Rule C–462

462.1—Acid salts of carboxylic acids, *etc.*, are named in the same way as neutral salts; the word " hydrogen " (or " dihydrogen ", *etc.*, as appropriate) is inserted between the name of the cation and (*i*) the name of the anion or (*ii*) the word " salt "*. When different acidic residues are present in one structure, the prefix " carboxylato- ", *etc.*, may be used to denote the ionic group —COO⁻ (see Rule C-86.1 and C-461.1).

Examples:

Potassium hydrogen heptanedioate $HOOC—[CH_2]_5—COOK$

Sodium dihydrogen citrate

Na⁺ 2H⁺

Na⁺H⁺

Sodium hydrogen 3-(carboxylatomethyl)-2-naphthalenepropionate
or Monosodium salt of 3-(carboxymethyl)-2-naphthalenepropionic acid

$^-O(HON=)C—[CH_2]_6—C(=NOH)O^-$ Na⁺ H⁺
Sodium hydrogen octanedihydroximate

Rule C–463

463.1—Neutral esters of carboxylic acids, *etc.*, are named in the same way as their neutral salts except that (*a*) the name of the alkyl or aryl,

---

* I.U.P.A.C. *Nomenclature of Inorganic Chemistry* 1970 (Butterworths Scientific Publications, London, 1971) recommends (Rule **6.2**) joining hydrogen to the following name of the anion for acid salts, thus: potassium hydrogenheptanedioate. Previous practice is, however, followed in the present rules.

etc., radical replaces the name of the cation and (*b*) a periphrase such as " (alkyl or aryl) ester " replaces " (metal) salt ".

Examples:

Ethyl acetate	$CH_3-COOC_2H_5$
Diethyl malonate	$CH_2(COOC_2H_5)_2$
Ethyl methyl malonate	$C_2H_5OOC-CH_2-COOCH_3$

Ethylene diacetate

$$CH_2-OOC-CH_3$$
$$|$$
$$CH_2-OOC-CH_3$$

Ethyl benzohydrazonate

$$C_6H_5-\underset{\underset{N-NH_2}{\|}}{C}-OC_2H_5$$

Ethyl chloroformate $Cl-CO-OC_2H_5$

Ethyl cyclohexanecarboxylate

    $\langle\hexagon\rangle-COOC_2H_5$

$CH_3-COO-\langle\hexagon\rangle-OOC-CH_2Cl$

p-Phenylene acetate chloroacetate

$$CH_3S-CH_2-CH_2-CH(NH_2)-COO-CH_2-C_6H_5$$
Benzyl ester of methionine or Methionine benzyl ester

463.2—Acid esters of carboxylic acids, *etc.*, and their salts, are named analogously to neutral esters, but the components are cited in the order: cation, alkyl or aryl, *etc.*, radical, hydrogen, anion. Locants are added if necessary.

Examples:

$$\underset{}{\overset{COOC_2H_5}{\underset{COOH}{\bigcirc}}}$$

1-Ethyl hydrogen 3-chlorophthalate
(Note that the numbering of the acid is retained.)

Sodium ethyl succinate

$$CH_2-COOC_2H_5$$
$$|$$
$$CH_2-COONa$$

Potassium 1-ethyl hydrogen citrate

$$CH_2-COOC_2H_5$$
$$|$$
$$C(OH)-COO^- \quad K^+ \; H^+$$
$$|$$
$$CH_2-COO^-$$

$$CH_3O-\underset{\underset{HN}{\|}}{C}--\underset{\underset{NH}{\|}}{C}-OH$$

Methyl hydrogen 2,7-naphthalenedicarboximidate

463.3—Ester groups in compounds R^1—CO—OR^2 are named (*a*) by use of a prefix " alkoxycarbonyl- " or " aryloxycarbonyl- ", *etc.*, for the group —CO—OR^2 when the radical R^1 contains a substituent with priority for citation as principal group, or (*b*) by use of a prefix " acyloxy- " for the group R^1—CO—O— when the radical R^2 contains a substituent with priority for citation as principal group.

Examples:

(*a*) Methyl 3-methoxycarbonyl-2-naphthalenepropionate

$$C_2H_5OOC-CH_2-CH_2-\overset{+}{N}(CH_3)_3\ Br^-$$

(*a*) [2-(Ethoxycarbonyl)ethyl]trimethylammonium bromide

$$C_6H_5-CO-O-CH_2-CH_2-COOH$$

(*b*) 3-Benzoyloxypropionic acid

(*b*) *p*-(Cyclohexylcarbonyloxy)benzoic acid

Exception:
The following abbreviated form is retained:

      Acetoxy CH_3—CO—O—

Rule C–464

464.1—Compounds $R^1C(OR^2)_3$, *etc.*, are named as R^2 esters of the hypothetical ortho acids $R^1C(OH)_3$.

Example:
    Trimethyl orthoacetate CH_3—$C(OCH_3)_3$

Note: Tetramethyl orthocarbonate $C(OCH_3)_4$

C-4.7. LACTONES, LACTIDES, LACTAMS, AND LACTIMS

Rule C-471

471.1—Compounds that may be considered to be derived by intramolecular loss of water from a hydroxyl and a carboxyl group of a hydroxy acid, leading to formation of a cyclic ester, are called " lactones ".

Rule C-472 (Alternative to Rule C-473)

472.1—When the corresponding hydroxy acid from which water may be considered to have been eliminated has a trivial name, the lactone is designated by substituting " -olactone " for " -ic acid " in the name of the hydroxy acid. Any necessary locants are added, that for a carbonyl group being as low as possible. If the locants for both the hydroxyl and the carbonyl group require to be cited, that for the carbonyl group is placed first.

Examples:

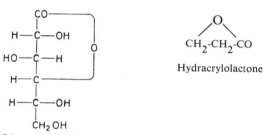

D-Glucono-1,4-lactone

Hydracrylolactone

472.2—Lactones formed from aliphatic acids are named by adding " -olide " to the name of the (non-hydroxylated) hydrocarbon with the same number of carbon atoms, with elision of terminal " e ", if present, before a vowel. Locants are added to define the position of ring closure.

Examples:

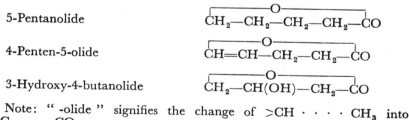

5-Pentanolide

4-Penten-5-olide

3-Hydroxy-4-butanolide

Note: " -olide " signifies the change of $>$CH · · · · CH$_3$ into $>$C . . . . CO with —O— below.

For naming certain lactones of hydroxy acids in the steroid series see the Rules for Nomenclature of Steroids, *Pure and Applied Chemistry*, 1972, Vol. 31, Nos. 1-2 (Rules 2S-3 and 2S-4).

472.3—Structures in which one or more (but not all) rings of an aggregate are lactone rings are named by placing " -carbolactone " (denoting the —O—CO— bridge) after the names of the structures which remain when each —O—CO— is replaced by two hydrogen atoms. Multiplying affixes "bis- ", " tris- ", *etc.*, and/or locants are used as necessary (see Rule C–472.1). The locant for —CO— is cited before that for the ester —O—.

Example:

1,10:9,8-Phenanthrenebiscarbolactone

Note: " carbolactone " signifies the change of >CH . . . . CH< into

$$\begin{matrix} O & - - - - - & CO \\ | & & | \\ >C & . . . . & C< \end{matrix}$$, an additional carbon atom being incorporated.

472.4—The following common names, derived from trivial names of non-hydroxylated acids, are permitted and may be used in naming their substitution products, but the principle of their formation should not be extended to other lactones:

γ-Butyrolactone
$$\overset{\displaystyle\overbrace{\hspace{3em}O\hspace{3em}}}{CH_2-CH_2-CH_2-CO}$$

γ-Valerolactone
$$\overset{\displaystyle\overbrace{\hspace{3em}O\hspace{3em}}}{CH_3-CH-CH_2-CH_2-CO}$$

δ-Valerolactone
$$\overset{\displaystyle\overbrace{\hspace{3em}O\hspace{3em}}}{CH_2-CH_2-CH_2-CH_2-CO}$$

Rule C–473 (Alternative to C–472)

473.1—Names based on the rules of Section B for heterocycles may be used for all lactones, and also trivial names which are considered to pertain to the heterocyclic series.

Examples:

Tetrahydro-2-furanone
or Dihydro-2(3H)-furanone

Tetrahydro-2-pyrone
(See Note at Rule C-314.1)

1-Oxaacenaphthen-2-one
or Naphtho[1,8-*bc*]furan-2-one

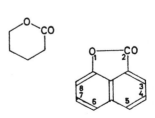

Coumarin

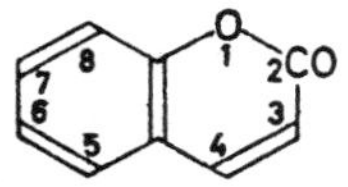

Isocoumarin

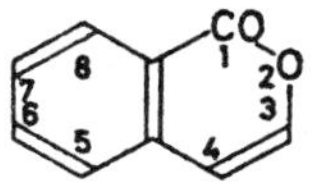

2(3H)-Benzofuranone

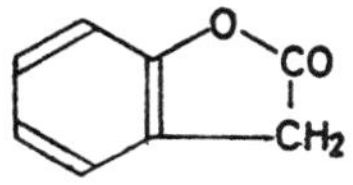

Phthalide

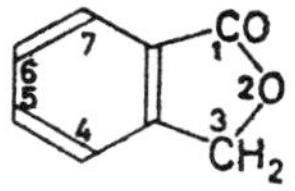

Rule C–474

474.1—Lactides are intermolecular cyclic esters formed by self-esterification from two or more molecules of a hydroxy acid. They are named as heterocycles by the rules of Section B unless the hydroxy acid has a trivial name. In the latter case the syllables " -ic acid " in that trivial name are replaced by " -ide " and a prefix " di- " or " tri- ", *etc.*, is added to denote the number of molecules involved.

Examples:

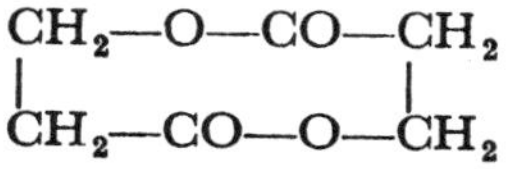

$$CH_2—O—CO—CH_2$$
$$CH_2—CO—O—CH_2$$

1,5-Dioxacyclooctane-2,6-dione

Dilactide
(preferred to
Lactide)

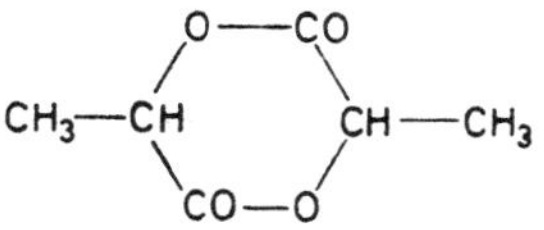

$$CH_3—CH \qquad CH—CH_3$$

Trisalicylide

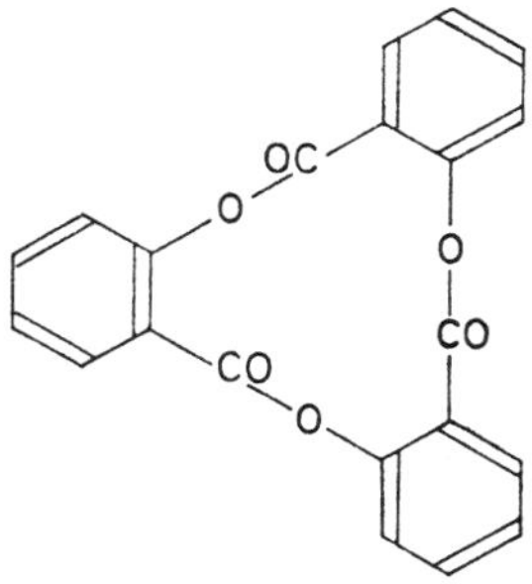

Rule C–475

475.1—Compounds containing a group —CO—NH— or —C(OH)=N— as part of a ring are preferably named as heterocyclic compounds in accordance with the rules of Section B, but may also be named as described in Rule C–472.2 but with " -lactam " or " -lactim ", respectively, in place of " -olide ".

Examples:

<table>
<tr><td>2-Pyrrolidinone
or 4-Butanelactam</td><td></td></tr>
</table>

2-Pyrrolidinone
or 4-Butanelactam

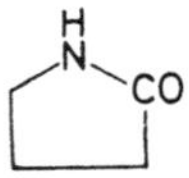

2-Hydroxy-1-pyrroline
or 4-Butanelactim

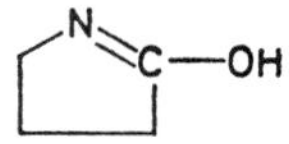

Exception: The name ε-caprolactam may be used for the compound

$$NH \underline{\hspace{4cm}} CO$$
$$CH_2—CH_2—CH_2—CH_2—CH_2$$

in place of 6-hexanelactam or hexahydro-2-azepinone.

C–4.8.   ACYL HALIDES

Rule C–481

481.1—Acyl halides, that is, compounds in which the hydroxyl group of a carboxyl group is replaced by halogen, are named by placing the name of the corresponding halide after that of the acyl radical.

Examples:

Acetyl chloride	$CH_3—COCl$
Hexanoyl bromide	$CH_3—CH_2—CH_2—CH_2—CH_2—COBr$
Benzoyl iodide	$C_6H_5—COI$
p-Anisoyl chloride	$p\text{-}CH_3O—C_6H_4—COCl$
Malonyl dibromide	$CH_2(COBr)_2$

$$ClOC—CH_2—CH_2—CH_2—CH_2—CH_2—COCl$$
Heptanedioyl dichloride

Terephthaloyl dichloride

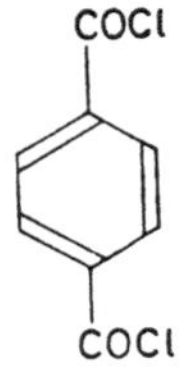

Cyclohexanecarbonyl chloride

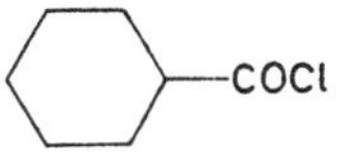

Acetoacetyl chloride $CH_3-CO-CH_2-COCl$

Aminoacetyl chloride H_2N-CH_2-COCl
(not glycine chloride)

Salicyloyl chloride

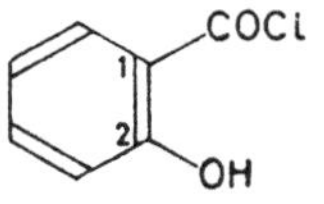

1,3,5-Pentanetricarbonyl $ClOC-[CH_2]_2-CH-[CH_2]_2-COCl$
  trichloride $|$
                                               $COCl$

481.2—When another group is present that has priority over acyl halide for citation as principal group, or when the acyl halide group is attached to a side chain, the prefix " fluoroformyl- ", " chloroformyl- ", " bromo-formyl- ", or " iodoformyl- " is used.

Examples:

Methyl 2-chloroformyl-1-cyclohexane-
  carboxylate

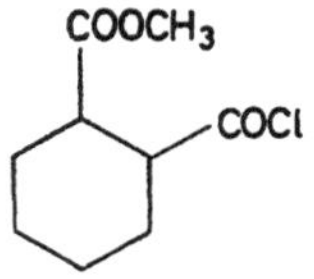

Ethyl o-chloroformylbenzoate

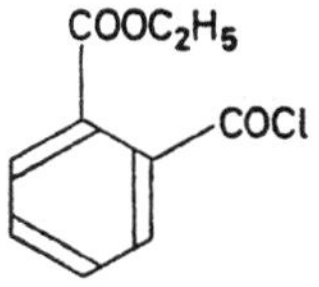

(Chloroformyl)acetic acid $ClOC-CH_2-COOH$

$$ClOC-CH_2 \qquad CH_2-COCl$$
$$| \qquad\qquad\qquad |$$
$$CH_3OOC-CH-CH_2-CH-COOCH_3$$

Dimethyl 2,4-bis[(chloroformyl)methyl]pentanedioate

C–4.9.   ACID ANHYDRIDES

Rule C–491

491.1—Symmetrical anhydrides of monocarboxylic acids, when unsubstituted, are named by replacing the word " acid " by " anhydride ".

Examples:

Acetic anhydride	$(CH_3—CO)_2O$
Butyric anhydride	$(CH_3—CH_2—CH_2—CO)_2O$
Hexanoic anhydride	$(CH_3—CH_2—CH_2—CH_2—CH_2—CO)_2O$
Cyclohexanecarboxylic anhydride	$(C_6H_{11}—CO)_2O$

Note: In some languages the word " acid " is included before " anhydride ".

491.2—Anhydrides of substituted monocarboxylic acids, if symmetrically substituted, are named by prefixing " bis- " to the name of the acid and replacing the word " acid " by " anhydride ". The " bis " may, however, be omitted (*cf.* di-, p. 81)

Examples:

Bis(chloroacetic) anhydride	$(ClCH_2—CO)_2O$
Bis(2,4-dibromobenzoic) anhydride	$(2,4-Br_2C_6H_3—CO)_2O$
Bis(6-aminohexanoic) anhydride	$(H_2N—[CH_2]_5—CO)_2O$

491.3—Mixed anhydrides (anhydrides of different monobasic acids) are named by giving in alphabetical order the first (*i.e.*, the specific) part of the names of the two acids, followed by "anhydride".

Examples:

Acetic propionic anhydride	$CH_3—CO—O—CO—CH_2—CH_3$
Benzoic butyric anhydride	$C_6H_5—CO—O—CO—CH_2—CH_2—CH_3$
Acetic chloroacetic anhydride	$CH_3—CO—O—CO—CH_2Cl$
Chloroacetic *p*-toluene-sulfonic anhydride	$CH_2Cl—CO—O—SO_2—C_6H_4—CH_3\text{-}p$

491.4—Cyclic anhydrides of polycarboxylic acids, although possessing a heterocyclic structure, are preferably named as acid anhydrides, as in Rule C–491.1 but with any necessary locants.

Examples:

Succinic anhydride

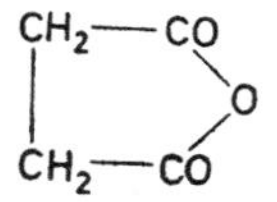

$$CH_2\text{---}CO$$
$$CH_2\text{---}CO$$
O

Phthalic anhydride

Image structure

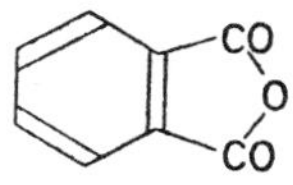

1,2-Cyclohexanedicarboxylic anhydride

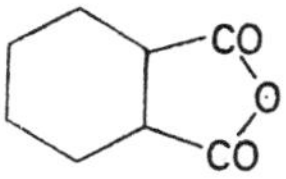

Image structure

1,8:4,5-Naphthalenetetracarboxylic dianhydride

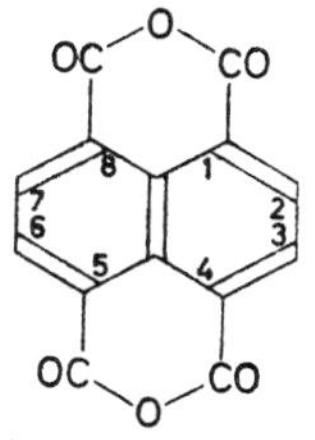

Image structure

1,2,3,4-Cyclohexanetetracarboxylic acid
   1,2-anhydride

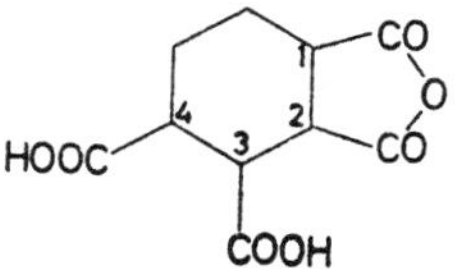

Image structure

C-5. COMPOUNDS CONTAINING BIVALENT SULFUR

C-5.0. INTRODUCTION

501—Various methods, in part conflicting, have been used for naming sulfur compounds. To eliminate all confusion a completely new system would be required. However, in accord with the present general policy of the Commission, the following rules present those parts of the existing methods that appear most useful, and their applications are rigidly prescribed.

502—A key role has long been played by the syllables " thio ", and this practice is retained. In nearly all its applications "thio" denotes replacement of oxygen by sulfur; the distinction from " thia " which denotes replacement of carbon in a ring or chain (see Rules **B–1.1** and **C–61.1**) should be noted.* Usually, " thio " (denoting replacement of oxygen) is placed in front of the name of an affix that denotes an oxygen-containing group or an oxygen atom; for example, " ol " denotes OH, so " thiol " denotes SH; " one " denotes (C)=O, so " thione " denotes (C)=S; " oic acid " denotes $(C){<}^{O}_{OH}$, so " thioic acid " denotes $(C){<}^{S}_{OH} \rightleftharpoons (C){<}^{SH}_{O}$ and " dithioic acid " denotes $(C){<}^{S}_{SH}$. However, there are some types of name where this procedure is not followed. For instance, (1) by long custom, where an acid has a trivial name, " thio " is placed in front of that name, so that, for example, thioacetic acid is the name for the acid $CH_3{-}C{<}^{S}_{OH} \rightleftharpoons CH_3{-}C{<}^{SH}_{O}$. (2) For sulfur derivatives of carboxylic acids, names ending in " -carbothioic acid " $({-}C{<}^{S}_{OH} \rightleftharpoons {-}C{<}^{SH}_{O})$ and " -carbodithioic acid " are used. (3) Particular difficulty arises with prefixes for the groups HS— and RS—, largely owing to differing uses of " hydroxy "and " oxy " in various languages; of the various possibilities

---

* Nevertheless, " thia " has been used in trivial names of cyclic compounds to denote replacement of oxygen, for example in 2*H*-thiapyran, but this usage is not recommended.

210

it has seemed best to retain " mercapto- " as prefix for HS—, and " thio- " for the bivalent atom —S—. Occasionally this last use of " thio " may require parentheses for distinction from its use for replacement of oxygen in a trivially named acid, as in:

p-Ethyl(thiobenzoic acid)	p-C_2H_5—C_6H_4—COSH
p-(Ethylthio)benzoic acid	p-C_2H_5—S—C_6H_4—COOH
p-(Ethylthio)thiobenzoic acid	p-C_2H_5—S—C_6H_4—COSH

C–5.1. THIOLS AND RELATED COMPOUNDS

THIOLS

Rule C–511

511.1—Compounds containing —SH as the principal group directly attached to carbon are named " thiols ".* In substitutive nomenclature their names are formed by adding " -thiol " as a suffix to the name of the parent compound. When —SH is not the principal group, the prefix " mercapto- " is placed before the name of the parent compound to denote an unsubstituted —SH group. Conjunctive names are formed in accordance with the rules of Subsection C–0.5.

Note: " Thiol " is not to be confused with " thiole " which denotes a five-membered ring embodying one sulfur atom (see Rule **B–1**).

Examples:

Ethanethiol	CH_3—CH_2—SH
Benzenethiol	C_6H_5—SH

Phenylmethanethiol or α-Toluenethiol†	

p-Xylene-α,α′-dithiol†	

Cyclopentanethiol	

---

* The class name " mercaptan " is abandoned. The root is retained only in the prefix " mercapto- " for an unsubstituted —SH group.
† For methyl-substituted benzene derivatives α, α′, α″, *etc.*, are used as locants for substituents in the methyl groups. Note that the locant for the CH_2 group of benzyl is α (see Rule **A–13.3**).

2-Quinolinethiol

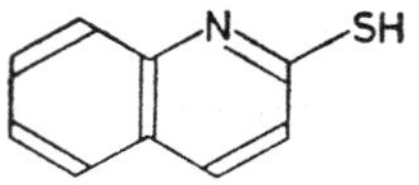

1,4-Butanedithiol   HS—CH$_2$—CH$_2$—CH$_2$—CH$_2$—SH

p-Mercaptobenzoic acid   HS—⟨benzene ring⟩—COOH

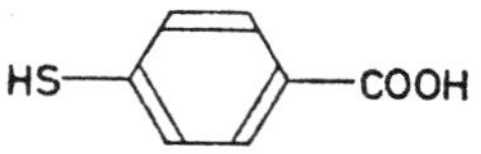

4-Mercapto-2-pyridine-
    carboxylic acid

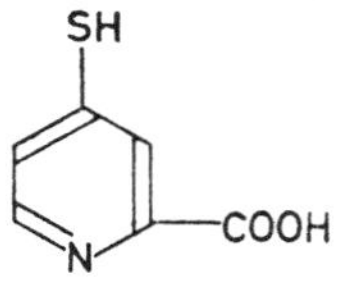

2,3-Dimercaptosuccinic acid   HS—CH—COOH
                                 |
                                 HS—CH—COOH

1-Anthracenemethanethiol
or 1-Anthrylmethanethiol

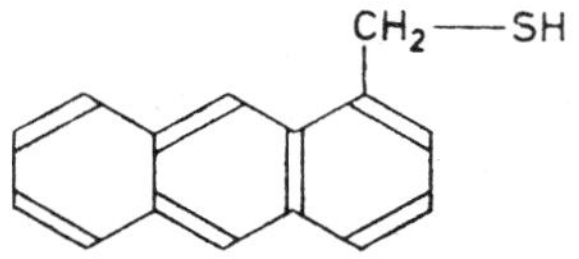

511.2—The use of " thio- " prefixed to the trivial name of a phenol, indicating replacement of the hydroxylic oxygen by sulfur, may be continued in simple instances, but the nomenclature of Rule **C–511.1** is preferred.

Examples:

Thiophenol

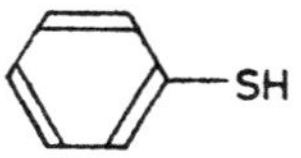

2-Bromo(thio-*p*-cresol)

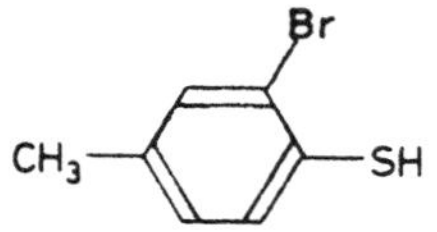

511.3—Salts of thiols are named analogously to salts of hydroxy compounds (see Rule **C–206**), with " thiolate " in place of " olate ", or " sulfide " in place of " oxide ".

Example:

Sodium ethanethiolate   CH$_3$—CH$_2$—SNa
or Sodium ethyl sulfide

511.4—When " mercapto- " is used as a prefix in the name of a thiol the salts of the latter are named by use of the prefix " sulfido- " for —S⁻. Example:

$$NaS-C_6H_4-SO_2-ONa$$

Disodium p-sulfidobenzenesulfonate

511.5—When radicofunctional nomenclature is used for a compound RSH, the name of the radical R is followed by the functional class name hydrosulfide.

Example:

Ethyl hydrosulfide $\quad\quad$ CH_3-CH_2-SH

HYDROPOLYSULFIDES

Rule C–513

513.1—Compounds $R-S_2H$, $R-S_3H$ . . . $R-S_nH$ are designated by radicofunctional nomenclature, generically as " hydropolysulfides ", specifically by name of the radical R followed by " hydrodisulfide ", " hydrotrisulfide " . . . . " hydropolysulfide ". But see also Rule C–641.3 and Table XII.

Examples:

Ethyl hydrodisulfide $\quad\quad$ $C_2H_5-S_2H$

Phenyl hydropolysulfide $\quad\quad$ $C_6H_5-S_nH$

513.2—When the presence of an unbranched chain of sulfur atoms is to be stressed, compounds $R-S-SH$, $R-S-S-SH$ . . . . $R[-S-]_nH$ are named as disulfanes, trisulfanes . . . . polysulfanes substituted by a radical R.

Examples:

Ethyldisulfane $\quad\quad$ C_2H_5-S-SH

Phenylpolysulfane $\quad\quad$ $C_6H_5[-S-]_nH$

SULFIDES

Rule C–514

514.1—Compounds R^1-S-R^2 have the generic name " sulfides "* Specifically they are named analogously to ethers, in substitutive nomenclature with " alkylthio- " or " arylthio- ", *etc.*, in place of " alkyloxy- ", *etc.*, in radicofunctional nomenclature with " sulfide " in place of " ether "

* The term " thioether " is not recommended.

213

or " oxide ", and for assemblies of identical units with " thio- " in place of " oxy- ".

Examples:

Diethyl sulfide $CH_3—CH_2—S—CH_2—CH_3$

Methyl propyl sulfide $CH_3—S—CH_2—CH_2—CH_3$
or 1-(Methylthio)propane

Diphenyl sulfide $C_6H_5—S—C_6H_5$

Phenyl 4-piperidyl sulfide
or 4-(Phenylthio)piperidine

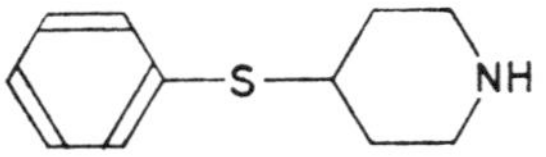

Hexyl 4-piperidyl sulfide
or 4-(Hexylthio)piperidine

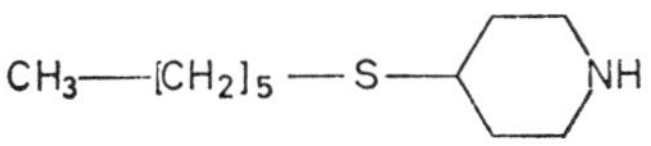

$CH_3—[CH_2]_5—S—$

4,4'-Thiodibenzoic acid

HOOC—⬡—S—⬡—COOH

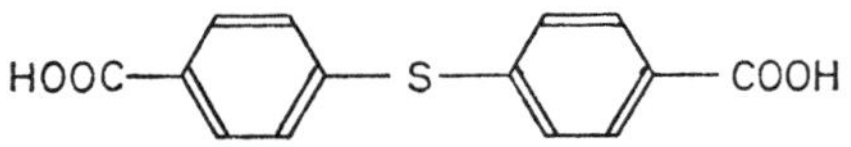

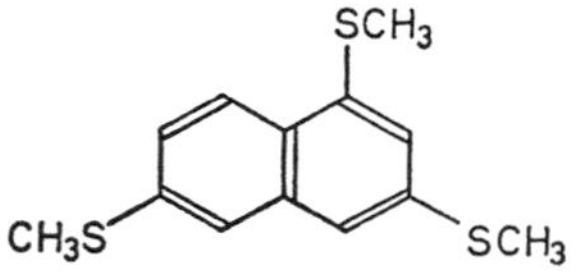

1,3,6 - Tris (methylthio) naphthalene

$$CH_3—[CH_2]_4—CH_2—S—CH=CH—[CH_2]_3—CH_3$$
1-Hexenyl hexyl sulfide or 1-(Hexylthio)-1-hexene

1,2-Bis(methylthio)ethane $CH_3—S—CH_2—CH_2—S—CH_3$

Bis[(ethylthio)methyl] sulfide $C_2H_5—S—CH_2—S—CH_2—S—C_2H_5$

514.2—When nomenclature according to Rule C–**514.1** becomes complex, as, for example, when several sulfur atoms are present in an acyclic compound, replacement nomenclature can be used (see Subsection C–0.6).

Example:

$$\overset{11}{CH_3}—S—CH_2—\overset{9}{S}—CH—CH_2—\overset{6}{S}—CH_2—\overset{4}{S}—CH_2—\overset{2}{S}—CH_3$$
$$|$$
$$CH_3$$

8-Methyl-2,4,6,9,11-pentathiadodecane

514.3—Cyclic sulfides are named by the following methods: (*a*) by prefixing " thia- " to the name of the parent carbocyclic compound; (*b*) by using an extension of the Hantzsch–Widman method (Rule **B–1**); or (*c*) by trivial or semitrivial names for cyclic sulfur compounds set out in Rule **B–2.11**.

Examples:

(*a*) Thiacyclooctane

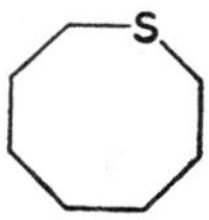

(*a*) 2,3-Dihydro-1,4-dithianaphthalene
or (*a*) 1,4-Dithia-1,2,3,4-tetrahydronaphthalene
or (*b*) 2,3-Dihydro-1,4-benzodithiin

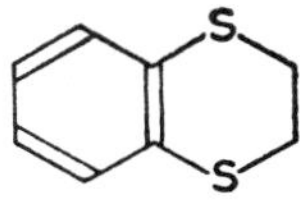

(*a*) 1-Thiaspiro[4.5]decane

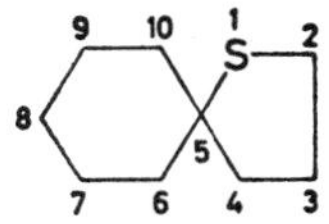

(*c*) Thiophene

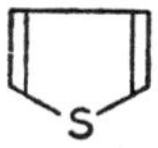

514.4—A sulfur atom directly attached to two carbon atoms already forming part of a ring, or to two adjacent carbon atoms in a chain, may be indicated (*a*) by the prefix " epithio " or (*b*) by means of Rule C–514.3 (*a*).

Examples:

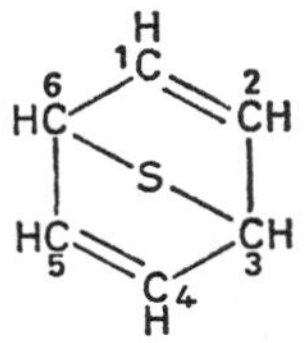

(*a*) 3,6-Epithio-1,4-   or   (*b*) 7-Thiabicyclo[2.2.1]-
     cyclohexadiene   hepta-2,5-diene

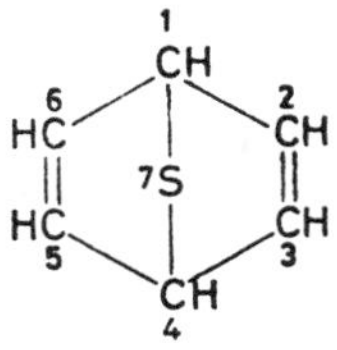

(*a*) 4,7-Dihydro-4,7-epithioindole

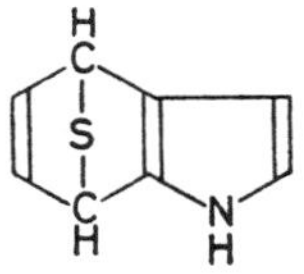

(*a*) 2,3-Epithiohexane CH_3—CH—CH—CH_2—CH_2—CH_3
                                  S

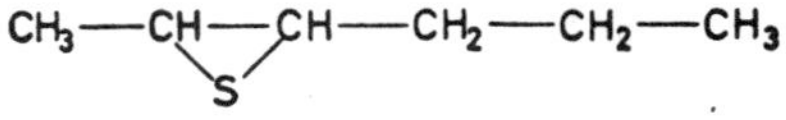

Note, however:

2-Ethyltetrahydro-5-methylthiophene
  (not 2,5-Epithioheptane) CH_3—⬠—CH_2—CH_3

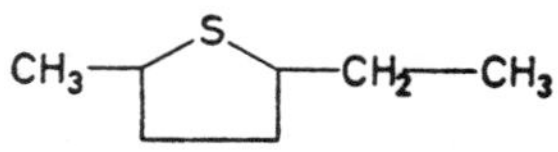

POLYSULFIDES

Rule C–515

515.1—Disulfides, trisulfides, *etc.*, and polysulfides are named analogously to sulfides, except that " di- ", " tri- ", or " poly-sulfide " replaces " sulfide ", and that " di- ", " tri- ", or " poly-thio- " replaces " thio- ".

Examples:

Diethyl disulfide	CH_3—CH_2—S_2—CH_2—CH_3
Diphenyl trisulfide	C_6H_5—S_3—C_6H_5
1-(Methyltrithio)propane or Methyl propyl trisulfide	CH_3—S_3—CH_2—CH_2—CH_3

1,4-Bis(ethyldithio)cyclohexane

tert-Butyl 2-naphthyl disulfide

2-(*o*-Nitrophenyldithio)benzothiazole

515.2—When a compound contains a straight chain of bivalent sulfur atoms and it is desired to stress that fact, the compound may be named substitutively as a derivative of " disulfane " (HS—SH), " trisulfane " (HS—S—SH), " tetrasulfane " (HS—S—S—SH), . . . or " polysulfane " (H—[S]$_n$—H). All substituents (including those which in substitutive nomenclature would be principal groups) are named as prefixes. Radicals RS—S—, RS—S—S—, . . . R—[S]$_n$— are named by adding " -yl " to the name of the corresponding sulfane, with elision of terminal " e ".

Examples:

Methyltetrasulfane	CH_3S—S—S—SH
Ethylphenyldisulfane	CH_3—CH_2—S—S—C_6H_5
Diphenyltrisulfane	C_6H_5—S—S—S—C_6H_5
Diethylpolysulfane	CH_3—CH_2—[S]$_n$—CH_2—CH_3
Sodium phenyltetrasulfanide	C_6H_5—S—S—S—SNa
Phenylsulfodisulfane	C_6H_5—S—S—SO_3H

1-Hydroxy-3-
    methyltrisulfane $CH_3-S-S-S-OH$

Sodium 1-oxido-2-
    phenyldisulfane $C_6H_5-S-S-ONa$

1-Ethoxy-3-hexyltrisulfane $CH_3-[CH_2]_5-S-S-S-O-CH_2-CH_3$

1-Diethylamino-3-
    phenyltrisulfane $C_6H_5-S-S-S-N(C_2H_5)_2$

Phenyldisulfanylformic acid $C_6H_5-S-S-COOH$
or 1-Carboxy-2-phenyldisulfane

Sodium 1-phenyl-3-
    sulfonatotrisulfane $C_6H_5-S-S-S-SO_3Na$

p-(Phenyldisulfanyl)
    benzoic acid C_6H_5-S-S-⬡$-COOH$

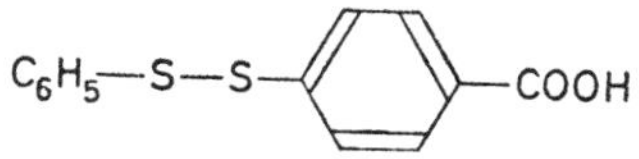

515.3—The bivalent radicals $-S-S-$, $-S-S-S-$ . . . $-[S]_n-$ are named " dithio- ", " trithio- " . . . " polythio- " or " disulfanediyl- ", " trisulfanediyl " . . . " polysulfanediyl ", respectively, in the naming of assemblies of identical units (see Rule C–72.1).

Example:

3,3′-Trithiodipropionic acid
or 3,3′-Trisulfanediyldi-
    propionic acid $S\big\langle{}^{S-CH_2-CH_2-COOH}_{S-CH_2-CH_2-COOH}$

515.4—The prefixes " epidithio- " and " epitrithio- " are used to indicate bridges consisting of two or three sulfur atoms, respectively, across rings.

Example:

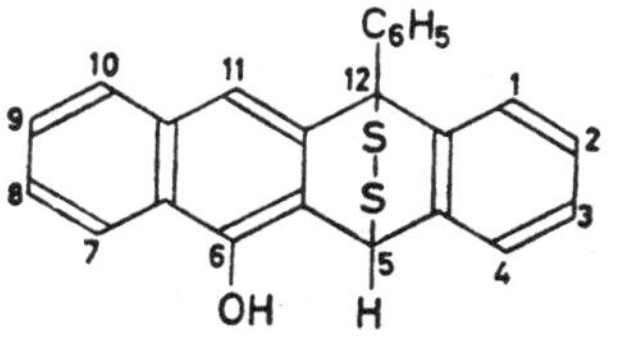

5,12-Dihydro-12-phenyl-5,12-epidithionaphthacen-6-ol

C–5.2. SULFENIC ACIDS AND THEIR DERIVATIVES

Rule C–521

521.1—Sulfenic acids RS—OH and their derivatives are named according to the same principles as sulfinic and sulfonic acids and their derivatives, as described in Rule C–641 and Subsection C–6.4, with the exception of the " thio " derivatives R—S—SH (hydrodisulfides; see Rule C–513) and R¹—S—S—R² (disulfides; see Rules C–515.1, C–515.2, and C–515.3).

C–5.3. THIOALDEHYDES (MONOMERIC), THIOKETONES, AND THIOACETALS

THIOALDEHYDES

Rule C–531

531.1 Where —(C)HS is the principal group the monomeric compound RCHS is named by adding " -thial " to the name of the hydrocarbon RCH$_3$.

Example:

Hexanethial

$$\overset{6}{CH_3}-\overset{5}{CH_2}-\overset{4}{CH_2}-\overset{3}{CH_2}-\overset{2}{CH_2}-\overset{1}{CHS}$$

531.2—Alternatively, the —CHS group is named " -carbothialdehyde ", this term being used as suffix, the principles of Rule C-303.1 being applied.

Examples:

Cyclohexanecarbothialdehyde

1,2,4-Butanetricarbo-thialdehyde

$$\overset{4}{SHC}-\overset{3}{CH_2}-\overset{2}{CH_2}-\overset{}{\underset{|}{CH}}-\overset{1}{CH_2}-CHS$$

with CHS on carbon 2.

531.3—Where —CHS is not the principal group, the prefix " thioformyl- " is used.

Examples:

7-Thioformylheptanoic acid

$$\overset{7}{SHC}-\overset{}{CH_2}-[CH_2]_4-\overset{2}{CH_2}-\overset{1}{COOH}$$

p-Thioformylbenzoic acid

531.4—The use of " thio " and " aldehyde " in trivial names, as in " thiobenzaldehyde " (C$_6$H$_5$—CHS) is retained for simple instances, but the nomenclature of Rules C–531.1 to C–531.3 is in general preferred.

THIOKETONES

Rule C–532

532.1—The suffix " thione " is used to signify the presence of ═S at a non-terminal carbon atom.

Examples:

4-Heptanethione

$$CH_3-CH_2-CH_2-\underset{\underset{S}{\|}}{C}-CH_2-CH_2-CH_3$$

2,4-Pentanedithione

$$CH_3\!-\!\underset{\underset{S}{\|}}{C}\!-\!CH_2\!-\!\underset{\underset{S}{\|}}{C}\!-\!CH_3$$

Cyclohexanethione

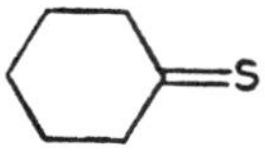

1-(Tetrahydro-2-furyl)-1-butanethione

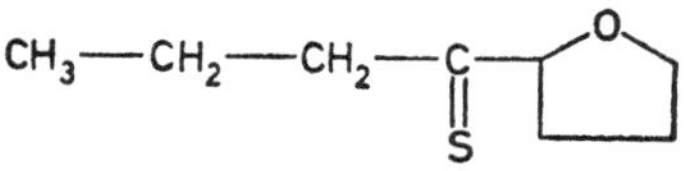

1,3-Dithiolane-2-thione
or 1,3-Dithiacyclopentane-2-thione

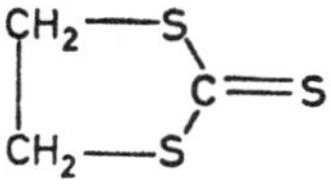

532.2—Names of thiones formed by placing the syllables " thio- " before the name of the corresponding ketone are retained for thio analogues of ketones having accepted non-systematic names, but the nomenclature of the preceding Rule (C–532.1) is in general preferred.

Examples:

Thioacetone

$$CH_3\!-\!\underset{\underset{S}{\|}}{C}\!-\!CH_3$$

Thiobenzophenone

$$C_6H_5\!-\!\underset{\underset{S}{\|}}{C}\!-\!C_6H_5$$

532.3—Where a prefix is needed for naming =S in a thioketone, the prefix " thioxo- " is used*.

Examples:

p-(3-Thioxobutyl)benzoic acid

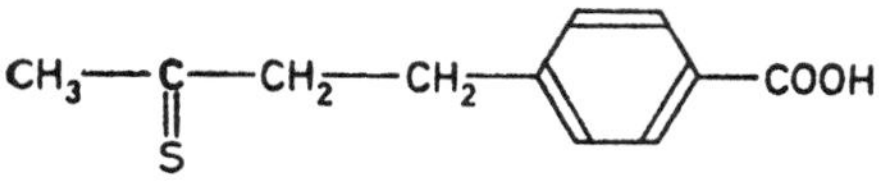

1,9,10-Trithioxo-1,2,9,10-tetrahydroanthracene
or   1,9,10(2H)-Anthracenetrithione

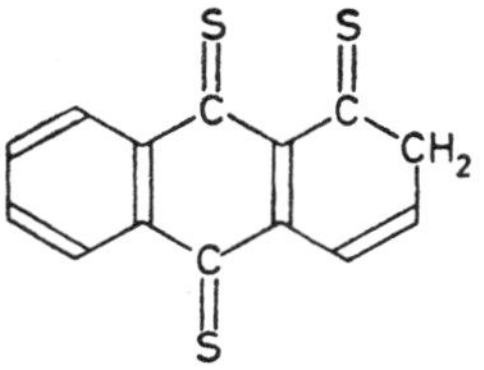

* " Thiono- " has been used in some languages in place of " thioxo ", but is not recommended.

I.U.P.A.C A, B & (C) —8

532.4—In radicofunctional nomenclature, compounds R^1—CS—R^2 are named by use of the functional class name " thioketone ".

Example:

2-Naphthyl phenyl thioketone

THIOACETALS

Rule C–533

533.1—Sulfur analogues of acetals, namely, those containing the group

$$\begin{array}{c}\diagdown\\ \diagup\end{array}\!C\!\begin{array}{c}\diagup SR^1\\ \diagdown SR^2\end{array}\quad\text{or}\quad\begin{array}{c}\diagdown\\ \diagup\end{array}\!C\!\begin{array}{c}\diagup OR^1\\ \diagdown SR^2\end{array}$$

are named substitutively by the use of the prefixes " alkylthio- " and " alkyloxy- ", " arylthio- " and " aryloxy- ", *etc.* Such compounds are known generically as " dithioacetals " and " monothioacetals "*, respectively, but these names are not used for systematic nomenclature except in certain specialized fields such as that of the carbohydrates.

Examples:

$$CH_3—CH_2—CH_2—CH\begin{array}{c}\diagup S—CH_2—CH_3\\ \diagdown O—CH_2—CH_3\end{array}$$

1-Ethoxy-1-(ethylthio)butane

$$CH_3—CH_2—CH_2—CH_2—CH\begin{array}{c}\diagup S—CH_2—CH_3\\ \diagdown S—CH_2—CH_3\end{array}$$

1,1-Bis(ethylthio)pentane

$$\begin{array}{c}CH_3\\ CH_3\end{array}\!\!\begin{array}{c}\diagdown\\ \diagup\end{array}\!C\!\begin{array}{c}\diagup S—CH_2—CH_3\\ \diagdown S—CH_2—CH_2—C_6H_5\end{array}$$

2-(Ethylthio)-2-(phenethylthio)propane

533.2—Monothiohemiacetals $\begin{array}{c}\diagdown\\ \diagup\end{array}\!C\!\begin{array}{c}\diagup OH\\ \diagdown SR\end{array}$ are named as alcohols

* The terms " mercaptal " and " mercaptole " are not recommended.

substituted by a group —SR; monothiohemiacetals $\begin{array}{c} \diagup C \diagdown \end{array}\begin{array}{c} OR \\ SH \end{array}$ are named as

thiols substituted by a group —OR.

Examples:

1-(Ethylthio)-1-propanol $CH_3—CH_2—CH\diagup^{S—CH_2—CH_3}_{\diagdown OH}$

1-Ethoxy-1-propanethiol $CH_3—CH_2—CH\diagup^{SH}_{\diagdown O—CH_2—CH_3}$

C–5.4. THIOCARBOXYLIC ACIDS AND THIOCARBONIC ACID DERIVATIVES

Rule C–541

541.1—Replacement of oxygen of a carboxyl group by sulfur is indicated in the name of the acid by a change of " -oic " to " -thioic " or " -dithioic ", of " -carboxylic " to " -carbothioic " or " -carbodithioic ", or of " carboxy- " to " thiocarboxy- " or " dithiocarboxy- "; or, when a group having priority for citation as principal group is also present, the groups HO—SC— and HS—CO—, are named by the prefixes " hydroxy(thiocarbonyl)- " and " mercaptocarbonyl- ", respectively. The generic name " thiocarboxylic acid " is retained. When it is not desired to indicate whether the acid hydrogen atom of a monothio acid should be attached to oxygen or to sulfur, this is indicated by $C\left\{{O \atop S}\right\}H$[*] the two forms may be differentiated by use of the words " S-acid " for the form —C(=O)—SH and " O-acid " for the form —C(=S)—OH.

Examples:

Hexanethioic acid $CH_3—CH_2—CH_2—CH_2—CH_2—C\left\{{O \atop S}\right\}H$

Hexanedithioic acid $CH_3—CH_2—CH_2—CH_2—CH_2—CS—SH$

Hexanebis(thioic acid) $H\left\{{O \atop S}\right\}C—CH_2—CH_2—CH_2—CH_2—C\left\{{O \atop S}\right\}H$

Hexanebis(dithioic acid)

         $HS—CS—CH_2—CH_2—CH_2—CH_2—CS—SH$

*The name of the prefix form is "thiocarboxy".

3-(Thiocarboxy)propionic acid

$$H\left\{\begin{matrix}O\\S\end{matrix}\right\}C-CH_2-CH_2-COOH$$

Hexanethioic S-acid $CH_3-CH_2-CH_2-CH_2-CH_2-CO-SH$

1-Piperidinecarbodithioic acid   N—CS—SH

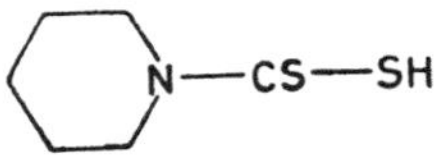

p-(Dithiocarboxy)benzoic acid $HS-CS-C_6H_4-COOH\text{-}p$

p-[Hydroxy(thiocarbonyl)]benzoic acid $HO-SC-C_6H_4-COOH\text{-}p$

p-(Mercaptocarbonyl)benzoic acid $HS-CO-C_6H_4-COOH\text{-}p$

541.2—In simple instances the names of sulfur-containing acids, obtained by prefixing " thio- " or " dithio- " to the trivial or semitrivial name of the analogous carboxylic acid, are retained.

Examples:

Dithio-2-naphthoic acid

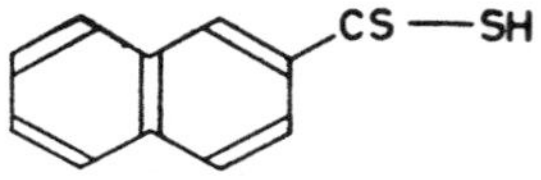

CS —SH

Thioacetic O-acid $CH_3-CS-OH$

Thioacetic S-acid $CH_3-CO-SH$

Rule C–542

542.1—The names of salts of thiocarboxylic acids are formed from the name of the acid by replacing " -ic acid " by " -ate " and stating first the name of the cation.

Examples:

Potassium hexanethioate $CH_3-CH_2-CH_2-CH_2-CH_2-C\left\{\begin{matrix}O\\S\end{matrix}\right\}K$

Sodium 2-naphthalenecarbodithioate
or Sodium dithio-2-naphthoate

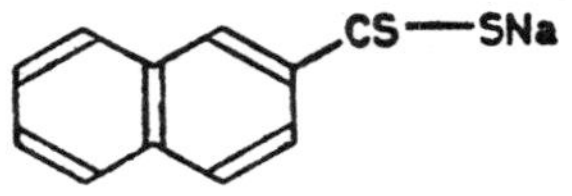

CS —SNa

Potassium 1-piperidinecarbodithioate   N—CS—SK

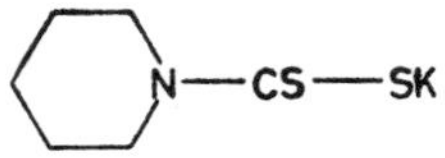

Rule C–543

543.1—Esters of thiocarboxylic acids are named analogously to the salts of these acids (see preceding rule). An *S*- or *O*-prefix is used when necessary.

Examples:

S-Ethyl hexanethioate $\qquad$ $CH_3—[CH_2]_4—CO—SC_2H_5$

O-Ethyl hexanethioate $\qquad$ $CH_3—[CH_2]_4—CS—OC_2H_5$

Ethyl 1-piperidinecarbodithioate

543.2—When a group having priority for citation as principal group is also present, ester groups RO—CS—, RS—CO—, and RS—CS— are named by use of prefixes " alkoxythiocarbonyl- ", " (alkylthio)carbonyl- ", and " (alkylthio)thiocarbonyl- ", respectively, " aryloxythiocarbonyl- ", etc.

Examples:

p-[Methoxy(thiocarbonyl)]dithiobenzoic acid

$$CH_3S—CO—[CH_2]_4—CSSH$$

5-[(Methylthio)carbonyl]pentanedithioic acid

$$CH_3S—CS—[CH_2]_4—COOH$$

5-[(Methylthio)thiocarbonyl]valeric acid

543.3—For monothio acids whose names end in " -thioic acid " or " -carbothioic acid " (see Rule C–541.1), radicals formed by loss of the hydroxyl group are named by changing these endings to " -thioyl " and " -carbothioyl ", respectively. For naming of amides and similar derivatives (for example, hydrazides) of such acids these endings are changed to " -thioamide " and " -carbothioamide ", respectively.

Examples:

Hexanethioyl chloride $\qquad$ $CH_3—[CH_2]_4—C(=S)Cl$

Hexanebis(thioyl chloride) $\qquad$ $Cl(S=)C—[CH_2]_4—C(=S)Cl$

2-Thiophenecarbothioyl chloride	
Hexanethioamide	$CH_3—[CH_2]_4—CS—NH_2$
2-Thiazolecarbothioamide	
Cyclopentanecarbothio-hydrazide	
Cyclohexanecarbothioyl or Cyclohexyl(thiocarbonyl)	

Note: Names of the type cyclohexanecarbothioyl are used in radico-functional nomenclature, and those of the type cyclohexyl(thiocarbonyl) for substituents, paralleling the usages for cyclohexanecarbonyl and cyclo-hexylcarbonyl set out in the Note to Rule C–403.2. However, see also Rule C–545.1.

543.4—For monothio acids that are named in accordance with Rule C–541.2, the names of derivatives in which the hydroxyl group has been replaced are formed by adding the prefix " thio- " to those of the sulfur-free analogues.

Examples:

Thioacetyl chloride	$CH_3—C(=S)Cl$
Thiobenzamide	$C_6H_5—CS—NH_2$

543.5—Anhydrides of thio acids are named in the same way as their oxy-gen analogues (see Subsection C–4.9), with the use of " thio- " prefixes as illustrated in Table XI; the sulfur linkage in Acyl—S—Acyl is denoted by " thioanhydride ".

Examples:

Bis(thiobenzoic) anhydride	$(C_6H_5—CS)_2O$
Di(cyclohexanecarbothioic) thioanhydride	
Benzoic thioacetic anhydride	$CH_3—CS—O—CO—C_6H_5$
Acetic benzoic thioanhydride	$CH_3—CO—S—CO—C_6H_5$

TABLE XI. Examples of mixed anhydrides of thio acids, arranged in order of decreasing priority of the anhydride grouping for citation as principal group

Mixed anhydride*	Formula
Acetic propionic anhydride	CH_3—CO—O—CO—CH_2—CH_3
Acetic propionic thioanhydride	CH_3—CO—S—CO—CH_2—CH_3
Acetic thiopropionic anhydride	CH_3—CO—O—CS—CH_2—CH_3
Propionic thioacetic anhydride	CH_3—CS—O—CO—CH_2—CH_3
Acetic thiopropionic thioanhydride	CH_3—CO—S—CS—CH_2—CH_3
Propionic thioacetic thioanhydride	CH_3—CS—S—CO—CH_2—CH_3
Thioacetic thiopropionic anhydride	CH_3—CS—O—CS—CH_2—CH_3
Thioacetic thiopropionic thioanhydride	CH_3—CS—S—CS—CH_2—CH_3

* In some languages the word "acid" is included before "anhydride".

Rule C–544

544.1—Derivatives of the acids $HO—C{S \atop O}H$, $HS—C{S \atop O}H$, and $S{=}C(SH)_2$ are named as derivatives of thiocarbonic, dithiocarbonic, and trithiocarbonic acid, respectively. Prefixes O- and S- are used, where necessary, to denote attachment of a radical or metal to oxygen and sulfur, respectively.

Examples:

S-Methyl hydrogen thiocarbonate	$HO—CO—SCH_3$
Sodium S-methyl dithiocarbonate	$CH_3S—C{S \atop O}Na$
Dimethyl trithiocarbonate	$CH_3S—CS—SCH_3$
Potassium O-ethyl dithiocarbonate†	$[C_2H_5O—CS—S]K$

Rule C–545

545.1—The name "thiocarbonyl" is used for the bivalent radical —CS— in nomenclature of assemblies of identical units and for designation of compounds by radicofunctional nomenclature.

† The xanthate nomenclature for this type of compound is not recommended.

Examples:

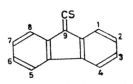

4,4′-Thiocarbonyldibenzoic acid

9-Thiocarbonylfluorene

CSF$_2$

CSCl$_2$

Thiocarbonyl difluoride
(see Rule C–108.2)

or   Thiocarbonyl dichloride
       Thiophosgene

Rule C–546

546.1—When the above nomenclature is difficult or inconvenient, thio derivatives of carbonic acid may be named by replacement nomenclature (see Subsection C–0.6).

Example:

$$\overset{9}{C}H_3—\overset{8}{C}H_2—\overset{7}{O}—\overset{6}{C}S—\overset{5}{S}—\overset{4}{C}O—\overset{3}{O}—\overset{2}{C}H_2—\overset{1}{C}H_3$$

6-Thioxo-3,7-dioxa-5-thia-4-nonanone

Rule C–547

547.1—" Thio- " prefixes are used when —CS— replaces —CO— in urea, semicarbazide, carbonohydrazide, or carbamic acid, and their derivatives, whose nomenclature is described in Rules C–974, C–975, C–984, and C–431.1.

Examples:

4,4-Dimethylthiosemicarbazide

$(CH_3)_2\overset{4}{N}—\overset{3}{C}S—\overset{2}{N}H—\overset{1}{N}H_2$

O-Ethyl thiocarbamate

$NH_2—CS—O—C_2H_5$

Rule C–548

548.1—Compounds containing a —CS— group as a ring member are named as heterocyclic compounds by the Rules of Section B and the suffix " -thione ".

Examples:

4-Thiazoline-2-thione

2-Thiazolidinethione

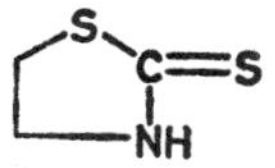

Exceptions:

The following are examples of trivial and semitrivial names retained:

Thiohydantoin (2- shown)

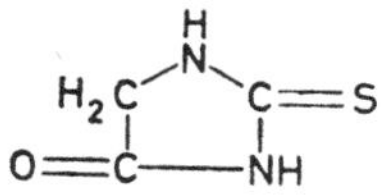

Dithiohydantoin

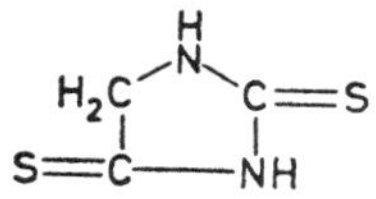

Rhodanine

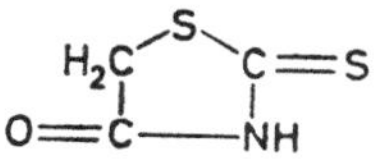

C–5.5. SULFONIUM COMPOUNDS

Rule C–551

551.1—Sulfonium compounds $R^1R^2R^3S^+X^-$ are named by citing the radical names corresponding to $R^1R^2R^3$ followed by " -sulfonium " and the name of the anion X^-.

Example:

Benzylethylmethylsulfonium chloride

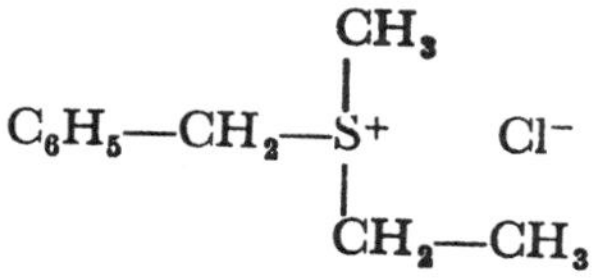

$$C_6H_5-CH_2-\overset{\displaystyle CH_3}{\underset{\displaystyle CH_2-CH_3}{S^+}} \quad Cl^-$$

551.2—When heterocyclic nomenclature according to the principles of Rule **B–1** is applied to compounds containing sulfonium sulfur, " -ium " is added to the name of the ring system with elision of terminal " e " if present; when the principles of Rule **B–6** are applied, replacement of CH by sulfonium sulfur is denoted by the prefix " thionia- ". The name of the anion is finally added.

Examples:

1-Ethylthiazolium bromide

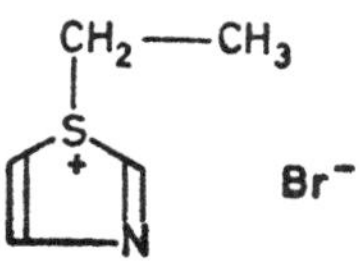

1-Methyl-1,4-dithianium monochloride

1-Thioniabicyclo[2.2.1]heptane chloride

3aH-1,6-Dithia-6a-thioniapentalene chloride

C–6. SULFUR HALIDES, SULFOXIDES, SULFONES, AND SULFUR ACIDS AND THEIR DERIVATIVES

C–6.1. INTRODUCTION

611—Among thio acids of sulfur, that is, those with empirical formulae $R—S_2H$, $R—S_2OH$, $R—S_3H$, $R—S_2O_2H$, $R—S_3OH$, and $R—S_4H$, and their derivatives, it sometimes occurs that well-characterized compounds exist for which the structure is not known with certainty. The present I.U.P.A.C. rules for the nomenclature of organic chemistry refer only to compounds of known structure. The names described, including those for thio acids of sulfur, refer always to the structures displayed in formulae or described in words. If a compound is shown to have a structure different from that specified by the name usually assigned to it, then that name should be changed to one specifying that new structure. For instance, in a compound $C_6H_{13}—S_3—OCH_3$, where the bonding is not to be closely specified, the compound will be called *O*-methyl hexanedithiosulfonate (Rule **C–641.1**); if it be later shown to have the structure $CH_3O—S—S—S—C_6H_{13}$ its name should be 1-hexyl-3-methoxytrisulfane (Rule **C–515.2**).

612—In some formulae in Subsection C–6, oxygen or sulfur atoms, or NH or NR groups, are shown as united to sulfur without bonds or in parentheses. This is a purely formal device designed to avoid specifying a particular electronic structure of the bond when this structure is open to dispute.

Examples:

Ethyl ethanesulfonohydrazonimidate	$\overset{NH}{\underset{N—NH_2}{C_2H_5—S—OC_2H_5}}$
Sodium benzenethiosulfonate	$\left.\overset{S}{\underset{O}{C_6H_5—SO}}\right\}Na$ (*cf.* Table XII)
S-Ethyl ethanethiosulfonimidate	$C_2H_5S(O)(NH)—SC_2H_5$

C–6.2. ORGANOSULFUR HALIDES

Rule C–621

621.1—Compounds in which sulfur is directly linked only to an organic radical and to halogen atoms are named by attaching the name of the radical in front of the word " sulfur " and then stating, as a separate word (or words), the number and name(s) of the halide.

Examples:

Phenylsulfur trichloride	$C_6H_5—SCl_3$

1-Naphthylsulfur chloride difluoride

$$SClF_2$$

Perfluoroethylsulfur pentafluoride $CF_3—CF_2—SF_5$

Methylsulfur monochloride $CH_3—SCl$
(or Methanesulfenyl chloride; see Rules **C–641.1** and **C–641.7** and Table XII)

621.2—When in the compounds described in the preceding **Rule** the organic radical contains also a substituent to be named as **principal group**, a composite prefix is formed for the halogenosulfur group by **attaching the number and substitutive name(s) of the halogen atoms** (see Rule **C–102**) in front of the syllable " thio ".

Examples:

p-(Chlorothio)benzoic acid $ClS—\langle\rangle—COOH$

(Trichlorothio)acetic acid $Cl_3S—CH_2—COOH$

Octafluoro-2-(pentafluorothio)-1-butanol

$$CF_3—CF_2—CF—CF_2—OH$$
$$SF_5$$

Methyl *p*-(chlorodifluorothio)benzoate $F_2ClS—\langle\rangle—COOCH_3$

C–6.3. SULFOXIDES AND SULFONES

Rule C–631

631.1—Compounds $R^1—SO—R^2$ and $R^1—SO_2—R^2$ are named " sulfoxides " and " sulfones ", respectively. Specific names are formed by placing the names of the radicals R^1 and R^2 (in alphabetical order) before "sulfoxide" or "sulfone". Alternatively compounds $R^1—SO—R^2$ and $R^1—SO_2—R^2$ may be named from the radical R^1 followed by "sulfinyl-" or "-sulfonyl-" and the name of the parent compound corresponding to R^2, where R^2 is senior to R^1.

Note: In the latter type of nomenclature the groups $R^1—SO_2—$and $R^1—SO—$ are named as acyl radicals substituted into a molecule R^2H, and thus names of the type phenylsulfonyl, phenylsulfinyl, *etc.*, are used. However, in radicofunctional nomenclature names of the type benzene-sulfonyl, benzenesulfinyl, *etc.*, are used (see Rule **C-641.7**; and compare Rule **C-543.3**). This difference parallels that for cyclohexanecarbonyl and cyclohexylcarbonyl (see Note to Rule **C-403.2**) and it applies also to derived radicals such as benzenesulfonimidoyl or phenylsulfonimidoyl, *etc.* (Rule **C-642.4**).

Examples:

Diphenyl sulfoxide C_6H_5—SO—C_6H_5

Butyl methyl sulfoxide
or 1-(Methylsulfinyl)butane CH_3—SO —CH_2—CH_2—CH_2—CH_3

p-Bis(ethylsulfinyl)benzene

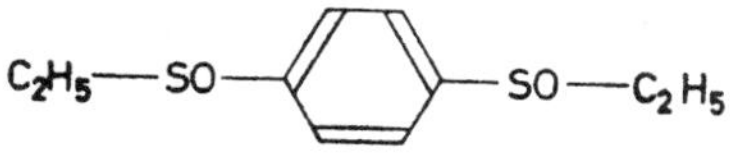

Diethyl sulfone C_2H_5—SO_2—C_2H_5

7-(Phenylsulfonyl)quinoline
or Phenyl 7-quinolyl sulfone

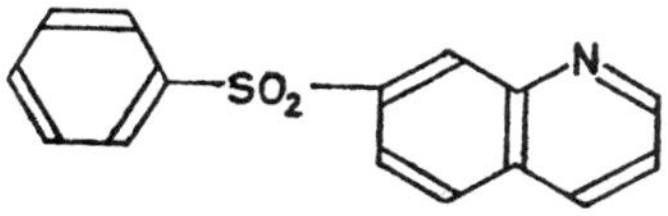

2,2-Bis(ethylsulfonyl)propane

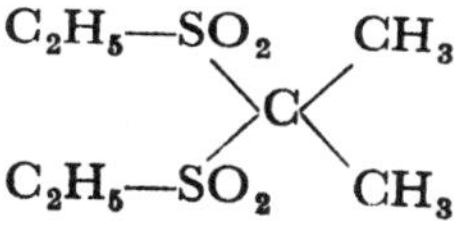

631.2—The prefixes "sulfinyl- " and "sulfonyl- " are used for the bivalent groups >SO and >SO_2, respectively, in conjunction with the nomenclature of assemblies of identical units (see Subsection C-0.7), when substituents are also present that have priority over these groups for citation as principal group.

Examples:

2,2'-Sulfinyldiethanol $OS(CH_2$—$CH_2OH)_2$

2,4'-Sulfonyldibenzoic acid

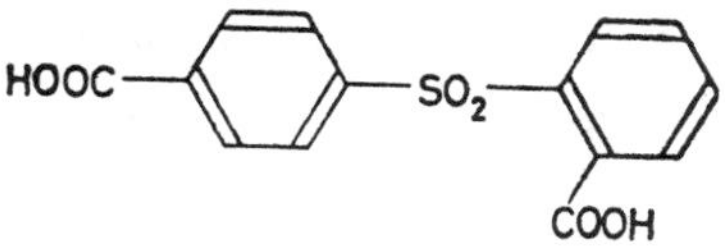

631.3—When an $>$SO or $>$SO$_2$ group is incorporated in a ring the compound is named as an oxide.

Examples:

Thiophene 1,1-dioxide

Thianthrene 5-oxide

Rule C–632

632.1—Compounds having two or more contiguous $>$SO or $>$SO$_2$ groups are named as disulfoxides, trisulfoxides . . . disulfones, trisulfones, *etc.* However, compounds having a group—SR contiguous to a group $\geq$SO or $>$SO$_2$ are named as derivatives of thiosulfur acids (see Rule **C—641**).

Example:

Diphenyl disulfone C_6H_5—SO_2—SO_2—C_6H_5

Rule C–633

633.1—The following parent names are used for compounds derived by replacement of hydrogen by an organic radical:

Sulfimide $H_2S(NH)$ (hypothetical)

Sulfoximide $H_2S(O)(NH)$ (hypothetical)

Examples:

$(C_2H_5)_2S(N—SO_2—C_6H_5)$

S,S-Diethyl-*N*-phenylsulfonylsulfimide

$(CH_3)_2S(O)(N—SO_2—C_6H_4—CH_3\text{-}p)$

S,S-Dimethyl-*N*-*p*-tolylsulfonylsulfoximide

633.2—Compounds RHC$=$SO and RHC$=$SO$_2$ are named as thioaldehyde oxides and dioxides, respectively (see Rule C–531), and compounds R^{1}R^{2}C$=$SO and R^{1}R^{2}C$=$SO$_2$ as thioketone oxides and dioxides, respectively (see Rule C–532).

Examples:

Thiobenzaldehyde oxide C_6H_5—CH$=$SO

3-Heptanethione dioxide C_4H_9—C—C_2H_5
                                              $\overset{\|}{SO_2}$

C-6.4. SULFUR ACIDS AND THEIR DERIVATIVES CONTAINING SULFUR DIRECTLY LINKED TO AN ORGANIC RADICAL

Rule C-641

641.1—Organic oxy acids of sulfur in which an organic portion is linked directly to sulfur are named according to substitutive nomenclature by adding to the name of the parent compound the appropriate suffix given in Table XII.

Examples:

2,4-Toluenedisulfonic acid

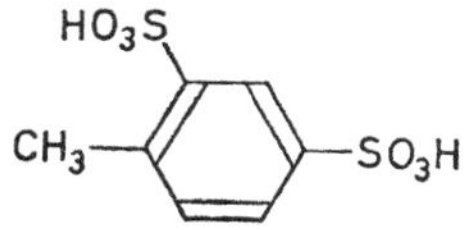

HO_3S

CH_3— —SO_3H

1-Piperidinesulfonic acid (cf. Rule C-661.1)

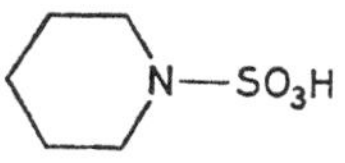

N—SO_3H

2-Butanesulfinic acid

CH_3—CH_2
$\diagdown$CH—SO_2H
$CH_3$$\diagup$

Benzenesulfenic acid

C_6H_5—SOH

Exceptions:

Sulfanilic acid

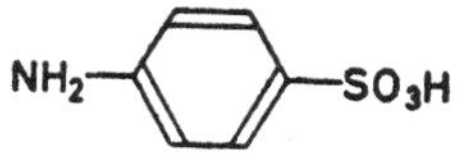

NH_2— —SO_3H

Naphthionic acid

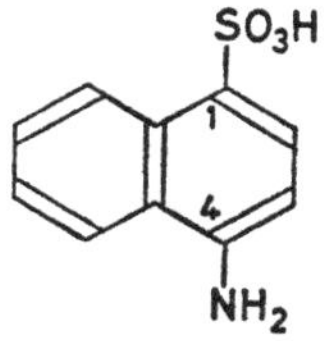

SO_3H

NH_2

Taurine (cf. Rule C-641.7)

NH_2—CH_2—CH_2—SO_3H

641.2—When another group is also present that has priority for citation as principal group, organic oxy acids of sulfur are named by adding to the name of the parent compound the appropriate prefix given in Table XII.

TABLE XII. Suffixes and prefixes for oxy acids of sulfur, arranged in order of decreasing priority for citation as principal group

Substituent (cf. Rule 612)	Suffix for acid	Prefix for group
$\begin{matrix} O \\ -SO \\ O \end{matrix}$ H	-sulfonic	Sulfo-
$-S_O^O$ H	-sulfinic	Sulfino-
—S—OH	-sulfenic	Sulfeno-

Examples:

4,6-Disulfino-1-naphthoic acid

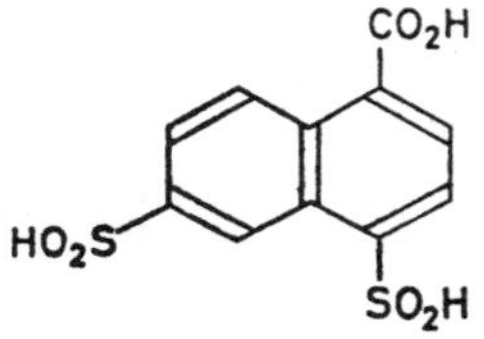

p-Sulfobenzoic acid

641.3—Organic acids derived from an oxy acid named according to Rule C–641.1 and in which one, two, or three atoms of oxygen of the acid group have been replaced by sulfur are named by placing " thio- ", " dithio- ", or " trithio- " immediately before the appropriate affix for the oxygen analogue given in Table XII.

Examples:

Benzenedithiosulfonic acid

$C_6H_5-\begin{matrix} S \\ S \ S \\ O \end{matrix} H$

Ethanethiosulfinic acid

$$C_2H_5—S\!\!\begin{array}{c}S\\O\end{array}\!\!\Big)H$$

p-Thiosulfobenzoic acid

$$H\left\{\begin{array}{c}O\\S\,S\\O\end{array}\right.\!\!—\!\!\langle\ \rangle\!\!—COOH$$

Note: A compound R—S_nH is named as a hydropolysulfide if the structure of the sulfur-containing group is unknown (see Rule C–513.1) and as a polysulfane derivative if the presence of an unbranched chain of sulfur atoms is to be stressed (see Rule C–515.2). In the latter case, the sulfane nomenclature may also be applied to compounds having structures such as R—S_n—OH.

641.4—The two forms of a monothiosulfinic, monothiosulfonic, or dithiosulfonic acid in which hydrogen is attached to oxygen or sulfur may be differentiated when required by use of S-acid for the form —S(O)—SH and O-acid for the form —S(S)—OH.

Examples:

Benzenethiosulfinic S-acid $C_6H_5—S(O)—SH$

Benzenethiosulfinic O-acid $C_6H_5—S(S)—OH$

641.5—For the nomenclature of esters and salts of sulfur acids the ending " -ic acid " is changed to " -ate " (see Rule C–84.1), the resulting word being preceded by the name of the cation or the radical.

Examples:

Sodium benzenethiosulfonate

$$C_6H_5—S\!\!\begin{array}{c}S\\O\end{array}\!\!\Big)Na$$

S-Methyl benzenethiosulfonate

$$C_6H_5—\underset{O}{\overset{O}{S}}—SCH_3$$

Ethyl ethanesulfinate

$$C_2H_5—\underset{O}{S}—OC_2H_5$$

Methyl p-toluenesulfonate
(not methyl tosate or tosylate)

$$p\text{-}CH_3\text{—}C_6H_4\text{—}SO_2\text{—}OCH_3$$

Phenyl methanesulfonate
(not mesate or mesylate)

$$CH_3\text{—}SO_2\text{—}OC_6H_5$$

641.6—When a group having priority for citation as principal group is also present, esters of sulfur acids are designated by composite prefixes formed from the following group names: RO—, alkoxy, *etc.*; RS—, alkylthio, *etc.*; —S(O)—, sulfinyl; —S(O)$_2$—, sulfonyl; —S(S)—, thiosulfinyl; —S(S)(O)—, thiosulfonyl.

Examples:

p-(Methoxysulfonyl)benzoic acid

Ethyl 4-phenoxy(thiosulfinyl)-1-naphthoate

Ethyl 4-(phenylthio)sulfinyl-1-naphthoate

641.7—Radicals derived from an organic oxy-sulfur acid by loss of a hydroxyl group are named by changing the syllables " -ic acid " to " -yl ". (For the difference between names of the types benzenesulfonyl and phenylsulfonyl, *etc.*, see Note to Rule C–631.1 and compare Rules C–403.2, C–543.3, C–641.8, and C–641.9).

Examples:

Trichloromethanesulfenyl chloride $CCl_3\text{—}SCl$

Benzenesulfinyl azide $C_6H_5\text{—}SO\text{—}N_3$

Ethanesulfonyl bromide $CH_3—CH_2—SO_2Br$

4-Chloro-1,3-benzenedisulfonyl dichloride

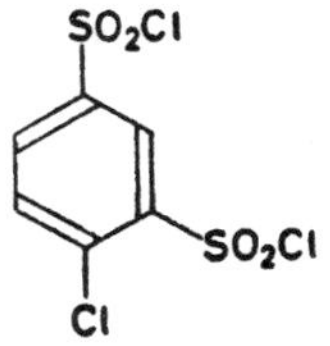

Exceptions:

Tosyl
(*p*- only)

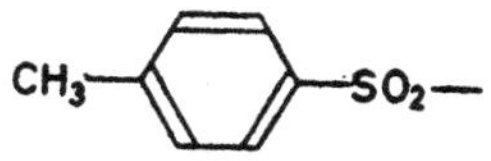

Mesyl $CH_3—SO_2—$

Tauryl (from Taurine; see Rule C–641.1) $NH_2—CH_2—CH_2—SO_2—$

641.8—(*a*) Amides derived by substituting NH_2 for OH in an organic sulfur acid are named by changing the syllables " -ic acid " to " -amide ", or as acyl derivatives (see Rule C–641.7) of the amine (compare Rule C–822; see also Note to Rule C–631.1).

Examples:

Ethanesulfenamide $C_2H_5—S—NH_2$

N-Methylbenzenesulfinamide $C_6H_5—SO—NH—CH_3$

p-Toluenesulfonamide

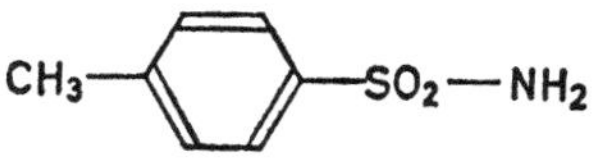

Benzenethiosulfonamide

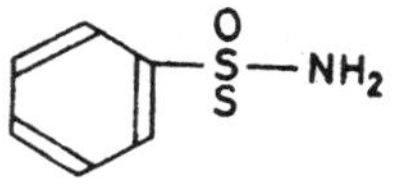

1-Methylsulfonylpiperidine

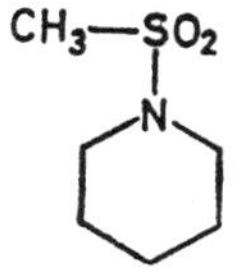

Exception:

N-Phenyl-substituted amides may be named with " anilide " in place of *N*-phenylamide, locants for substituents in the aniline residue being primed, as in:

or   2′,3-Dichloro-2-naphthalene-
sulfonanilide
3-Chloro-*N*-(*o*-chlorophenyl)-2-
naphthalenesulfonamide

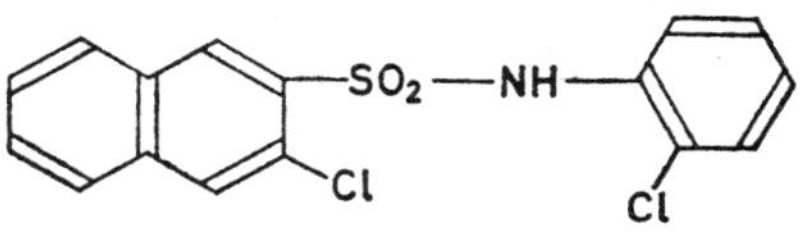

(b) When a group having priority over the amide group for citation as principal group is also present, the groups —SO$_2$—NH$_2$, —SO—NH$_2$, and —S—NH$_2$ are designated by the prefixes "sulfamoyl-", "sulfinamoyl-" and "sulfenamoyl-", respectively. Substituents on the nitrogen atom are cited as N-prefixes, and replacement of oxygen by sulfur is denoted by thio prefixes as described in Rule C–641.3 (see Note to Rule C–631.1).

Examples:

p-Sulfamoylbenzoic acid   p-NH$_2$—SO$_2$—C$_6$H$_4$—COOH

p-(N-Methylsulfinamoyl)benzamide

p-CH$_3$—NH—SO—C$_6$H$_4$—CO—NH$_2$

p-[N-Methyl(thiosulfamoyl)]benzoic acid

$$\text{p-CH}_3\text{—NH—}\overset{\text{O}}{\underset{\text{S}}{\text{S}}}\text{—C}_6\text{H}_4\text{—COOH}$$

(c) A radical formed from the amide of a sulfur acid by loss of one hydrogen atom from the NH$_2$ group is designated by changing the ending "-amide" to "-amido-", the complex radical name being used as a prefix.

Examples:

2-Benzenesulfonamidoimidazole-5-carboxylic acid

o-Benzenethiosulfonamidobenzoic acid

2-Sulfanilamidothiazole-5-carboxylic acid

641.9—Other nitrogen derivatives of organic sulfur acids are named as described in Subsections C–8 and C–9 for similar derivatives of carbon acids, but with "-carbox-" replaced by "-sulfon-", "-sulfin-", or "-sulfen-" before a vowel, and "-carbo-" replaced by "-sulfono-", "-sulfino-", or "-sulfeno-" before a consonant, as appropriate.

Examples:

Benzenesulfonohydrazide   C$_6$H$_5$—SO$_2$—NH—NH$_2$

Benzenesulfinamidine   C$_6$H$_5$—S(=NH)—NH$_2$

Benzenesulfonimidamide C_6H_5—S(O)(NH)—NH_2

Benzenesulfonodiimidamide C_6H_5—S(NH)$_2$—NH_2

Note: Benzenesulfonamidine is not to be used.

641.10—As exceptions, because of precedent, hydroxy-derivatives of benzene and hydroxy or amino derivatives of naphthalene (but not similar derivatives of other cyclic compounds), when substituted also by —SO_3H groups, may be named phenolsulfonic, naphtholsulfonic, and naphthylaminesulfonic acids, respectively; the hydroxyl or amino group(s) have priority for lowest available numbers. Derivatives of these acids may be named and numbered similarly. Nevertheless, the systematic names are preferred.

Examples:

Phenol-4-sulfonic acid
or *p*-Hydroxybenzenesulfonic acid (preferred)

Resorcinol-5-sulfonamide
or 3,5-Dihydroxybenzenesulfonamide (preferred)

Sodium 1-naphthol-5-sulfonate
or Sodium 5-hydroxy-1-naphthalenesulfonate (preferred)

1-Naphthylamine-4,8-disulfonic acid
or 4-Amino-1,5-naphthalenedisulfonic acid (preferred)

Rule C–642

642.1—If an oxygen atom of a sulfonic or sulfinic acid group is replaced by a bivalent nitrogenous group, the ending sulfonic acid or sulfinic acid of the name of the acid is modified as follows (compare Rule **C–451.1** and **C–451.2**), derivatives being named as customary for other acids:

Group	*Modified ending*
—S(OH)=NH	-sulfinimidic acid
—S(O)(OH)=NH	-sulfonimidic acid
—S(OH)=N—NH$_2$	-sulfinohydrazonic acid
—S(O)(OH)=N—NH$_2$	-sulfonohydrazonic acid
—S(OH)=NOH	-sulfinohydroximic acid
—S(O)(OH)=NOH	-sulfonohydroximic acid

Examples:

Benzenesulfinimidic acid	C_6H_5—S(OH)=NH
Ethyl ethanesulfonohydroximate	C_2H_5—S(O)(OC$_2$H$_5$)=NOH

642.2—(*a*) If both sulfonyl-oxygen atoms are replaced by the same nitrogenous group, the syllable "di" is placed immediately before the name prescribed for that group in the preceding rule. (*b*) If the two nitrogenous groups are different, the ending for these groups is "hydrazonimidic", "-hydroximimidic", or "-hydrazonohydroximic", as appropriate. *(c)* A letter "o" is inserted to separate two consonants on such endings (see Rule **C—641.9**).

Examples:

2-Naphthalenesulfonodiimidic acid

$$C_2H_5—\overset{\displaystyle NH}{\underset{\displaystyle N—NH_2}{S}}—OC_2H_5$$

Ethyl ethanesulfonohydrazonimidate

642.3—If sulfur replaces oxygen in the acids named in Rules **C–642.1** and **C–642.2**, a prefix "thio-" or "dithio-" is placed before the syllable "sulf", an *O*- or *S*- prefix being used if necessary.

Examples:

Ethanethiosulfinimidic acid	C_2H_5—S(NH)—SH
S-Ethyl ethanethiosulfonimidate	C_2H_5—S(O)(NH)—SC$_2$H$_5$

642.4—Radicals derived by loss of a hydroxyl group from an acid named in the three preceding rules are named by changing the ending " -ic acid " of the name of the acid to " -oyl ".

Examples:

N-Acetylbenzenesulfino-
hydrazonoyl chloride

$$C_6H_5—S(Cl)=N—NH—CO—CH_3$$

p-(Phenylsulfonimidoyl)benzoic acid

(For the change of benzenesulfonimidoyl to phenylsulfonimidoyl see the Note to Rule C–631.1.)

Rule C–643

643.1—(a) Compounds containing the grouping $—SO_2—O—SO_2—$ or $—SO—O—SO—$ are named sulfonic and sulfinic anhydrides, analogously to carboxylic anhydrides (Subsection C–4.9), and those containing the grouping $—SO_2—O—SO—$ are named sulfinic sulfonic anhydrides, analogously to mixed carboxylic anhydrides (Rule C–491.3 and C–543.5).

Examples:

Benzenesulfonic anhydride	$C_6H_5—SO_2—O—SO_2—C_6H_5$
Benzenesulfonic ethanesulfinic anhydride	$C_6H_5—SO_2—O—SO—C_2H_5$
Ethanesulfinic ethanesulfonic anhydride	$C_2H_5—SO_2—O—SO—C_2H_5$

1,2-Benzenedisulfonic anhydride

(b) When the oxygen atom linking the two acid residues of an acid anhydride is replaced by sulfur, the anhydride is named according to the principles of Rule C–491 but with the word "anhydride" replaced by "thioanhydride".

Examples:

Ethanesulfonic thioanhydride	$C_2H_5—SO_2—S—SO_2—C_2H_5$
Benzenesulfenic benzene-sulfinic thioanhydride	$C_6H_5—S—S—SO—C_6H_5$
Benzoic ethanesulfonic thioanhydride	$C_6H_5—CO—S—SO_2—C_2H_5$

643.2—Cyclic anhydrides formed by loss of water between a sulfur acid and a carboxylic acid group, and the analogous imides, are named by replacing "acid" in the name of the corresponding acid by "cyclic anhydride" or "imide", respectively; alternatively, they are named as cyclic compounds by the rules of Section B, with use of " oxide " and a locant to denote oxygen attached to sulfur.

Examples:

o-Sulfobenzoic cyclic anhydride
or 2,1-Benzoxathiol-3-one 1,1-dioxide

3-Sulfopropionic imide
or 3-Isothiazolidinone 1,1-dioxide

C–6.5. SULFUR ACIDS AND THEIR DERIVATIVES IN WHICH SULFUR IS LINKED ONLY THROUGH OXYGEN TO THE ORGANIC RADICAL

Rule C–651

651.1—Neutral esters of sulfur acids (and their derivatives) listed in the table pertaining to Rule **5.214** of the I.U.P.A.C. *Nomenclature of Inorganic Chemistry*, 1970* (p. 32), are designated by the name of the organic radical(s) followed by that of the inorganic anion.   Prefixes *O-* and *S-* are used where necessary, to denote attachment to oxygen and sulfur, respectively.

Examples:

Dimethyl sulfate	$(CH_3O)_2SO_2$
Ethyl methyl sulfate	$CH_3O—SO_2—O—CH_2—CH_3$
S-Ethyl *O*-methyl thiosulfate	$CH_3O—SO_2—S—CH_2—CH_3$
Ethyl chlorosulfite	$Cl—SO—O—CH_2—CH_3$

651.2—Acid esters of sulfur acids and their salts are designated in the same way as the corresponding derivatives of carboxylic acids (see Rule C–**463.2**).

Examples:

Methyl hydrogen sulfate	$CH_3O—SO_2—OH$
Sodium *S*-ethyl thiosulfite	$CH_3—CH_2—S—SO—ONa$

* I.U.P.A.C. *Nomenclature of Inorganic Chemistry*, 1970 (Butterworths Scientific Publications, London, 1971).

C–6.6. SULFUR ACIDS AND THEIR DERIVATIVES IN WHICH SULFUR IS LINKED ONLY THROUGH NITROGEN OR THROUGH NITROGEN AND OXYGEN TO THE ORGANIC RADICAL

Rule C–661

661.1—*N*-Substituted amides of sulfur acids listed in the table pertaining to Rule **5.214** of the I.U.P.A.C. *Nomenclature of Inorganic Chemistry*, 1970* (p. 32), are designated as *N*-derivatives of the sulfur amides, these amides being named as described in the inorganic Rule **5.34** (p. 35 of those Rules).

Examples:

or Phenylamidosulfuric acid
or Phenylsulfamidic acid C_6H_5—NH—SO_3H

  (but see also Rule C–**661.2**)

  Sulfuric ethylamide CH_3—NH—SO_2—NH—CH_2—CH_3
   methylamide

or *N*-Ethyl-*N'*-methylsulfuric diamide

  (but see also Rule C-661.3)

661.2 (Alternative to part of Rule C–**661.1**)—Compounds R—NH—SO_3H and R^1R^2N—SO_3H may be named as *N*-substituted sulfamic acids, or by means of a prefix " sulfoamino- " to denote the group HO_3S—NH—.

Examples:

  N-Phenylsulfamic acid C_6H_5—NH—SO_3H
   (see also Rule C–**661.1**)

  p-Sulfoaminobenzoic acid p-HO_3S—NH—C_6H_4—COOH

661.3 (Alternative to part of Rule C–**661.1**)—(*a*) Compounds $S(NR^1R^2)_2$, $SO(NR^1R^2)_2$, and $SO_2(NR^1R^2)_2$, where R^1 may be H, may be named as assemblies of identical units (see Subsection C–0.7), by use of the prefix " thio- " for >S (Rule C–**514.1**) or " sulfinyl- " for >SO or " sulfonyl- " for >SO_2 (Rule C–**631.2**). (*b*) Compounds $SO_2(NR^1R^2)_2$ may be named as substitution products of sulfamide, $SO_2(NH_2)_2$. (*c*) If R^1 or R^2 is phenyl, " aniline " replaces " phenylamine ", and " anilide " replaces " phenylamide " in the resulting names. Prefixes *N* and *N'* are used for clarity when required.

Examples:

N,N'-Thiobis(diethylamine)	$(CH_3$—$CH_2)_2N$—S—$N(CH_2$—$CH_3)_2$
N,N'-Sulfinyldianiline	C_6H_5—NH—SO—NH—C_6H_5
N-Ethyl-*N'*-methylsulfamide	CH_3—CH_2—NH—SO_2—NH—CH_3

* I.U.P.A.C. *Nomenclature of Inorganic Chemistry*, 1970 (Butterworths Scientific Publications, London, 1971).

661.4—Names of esters, salts, and other derivatives of amides of sulfur acids are derived by altering the ending of the names given in the preceding rule in the ways described in Rule **C–651**.

Example:

Methyl phenylamidosulfate

or Methyl phenylsulfamidate $C_6H_5—NH—SO_2—OCH_3$

or Methyl phenylsulfamate

661.5—Compounds containing, formally, a group $—N{=}SO$ or $—N{=}SO_2$ are named as sulfinylamines or sulfonylamines, respectively.

Examples:

N-Sulfinylaniline	$C_6H_5—N{=}SO$
N-Sulfonylmethylamine	$CH_3—N{=}SO_2$

661.6—The following trivial names for parent compounds are retained:

$S<\begin{matrix}C({=}S)—NH_2\\C({=}S)—NH_2\end{matrix}$

Thiuram monosulfide

$\begin{matrix}S—C({=}S)—NH_2\\|\\S—C({=}S)—NH_2\end{matrix}$

Thiuram disulfide

Examples:

Tetramethylthiuram monosulfide $S<\begin{matrix}C({=}S)—N(CH_3)_2\\C({=}S)—N(CH_3)_2\end{matrix}$

N,*N*-Diethyl-*N'*,*N'*-bis(*p*-nitrophenyl)thiuram disulfide

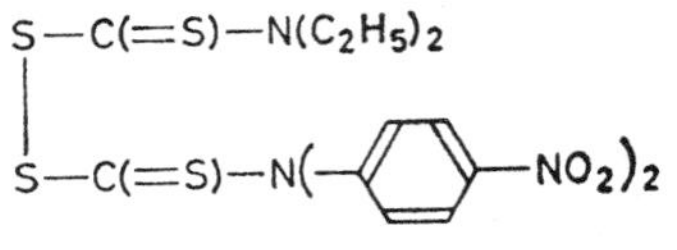

Rule C–662

662.1—If a sulfur-containing group is attached by oxygen or nitrogen to a compound that contains also another substituent having priority over the sulfur-containing group for citation as principal group, then the sulfur-containing group is named by a composite prefix built from prefixes named in other parts of these rules and arranged in the order in which the components occur in the compound.

Examples:

$$HO—SO_2—O—CH_2—CH_2—COOH$$
3-(Hydroxysulfonyloxy)propionic acid

$$CH_3O—SO—O—CH_2—CH_2—COOH$$
3-(Methoxysulfinyloxy)propionic acid

$$Cl—SO_2—O—CH_2—CH_2—COOH$$
3-(Chlorosulfonyloxy)propionic acid

$$NH_2—SO—O—CH_2—CH_2—COOH$$
3-(Sulfinamoyloxy)propionic acid

$$CH_3O—SO_2—NH—CH_2—CH_2—COOH$$
3-(Methoxysulfonylamino)propionic acid

C–6.7.   SULTONES AND SULTAMS

Rule C–671

671.1—Compounds containing a group —SO_2—O— as part of a ring are named as described in Rule C–472.3 for lactones but with the ending " -sultone " in place of " -carbolactone ". The —SO_2— group has priority over the —O— atom for lowest available number.

Examples:

1,8-Naphthalenesultone

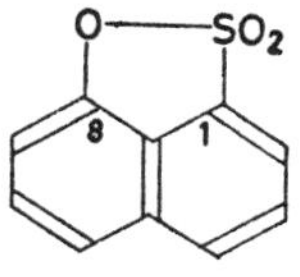

1,4-Butanesultone

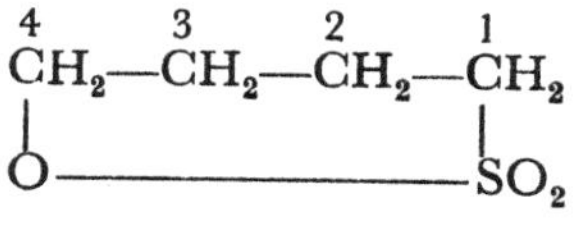

671.2—Compounds containing a group —SO_2—N< as part of a ring are named by adding "-sultam" to the name of the hydrocarbon with the same number of carbon atoms. The —SO_2— group has priority over the —NH— group for lowest available number.

Example:

7-Chloro-1,8-naphthalenesultam

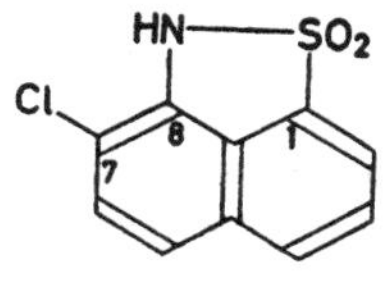

671.3—Sultones and sultams may alternatively be named as heterocyclic compounds by the rules of Section B, the oxygen attached to sulfur being denoted as " dioxide ", with a locant.

Example:

1,2-Oxathiane 2,2-dioxide

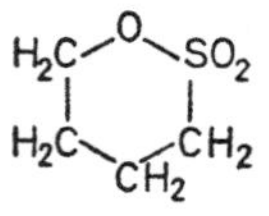

$$H_2C-O-SO_2$$
$$\quad|\qquad\quad|$$
$$H_2C\qquad CH_2$$
$$\quad\backslash CH_2/$$

C–7. COMPOUNDS CONTAINING SELENIUM OR TELLURIUM

LINKED TO AN ORGANIC RADICAL

Rule C–701

701.1—Organic compounds of selenium are named as far as possible analogously to the corresponding sulfur compounds. The prefixes and suffixes used, with examples of their application, are shown in Table XIII. Structures not listed in the Table are named by placing the syllables "seleno-" before the name of the corresponding oxygen compound.

Example:

Selenourea $\qquad Se{=}C(NH_2)_2$

Examples of use of prefixes and suffixes

(1) Ethaneselenol $\qquad CH_3{-}CH_2{-}SeH$

(2) 3-(Hydroseleno)propionic acid $\qquad HSe{-}CH_2{-}CH_2{-}COOH$

(3) Diethyl selenide $\qquad CH_3{-}CH_2{-}Se{-}CH_2{-}CH_3$

(4) (Ethylseleno)acetic acid $\qquad CH_3{-}CH_2{-}Se{-}CH_2{-}COOH$

(5) Selenacyclopentane
or Tetrahydroselenole

$$\begin{matrix} CH_2{-}CH_2 \\ | \qquad\quad \backslash \\ \qquad\qquad Se \\ | \qquad\quad / \\ CH_2{-}CH_2 \end{matrix}$$

(6) 1,4-Episeleno-1,4-dihydro-naphthalene

(7) Hexaneselenal

(8) 1-Pentanecarboselen-aldehyde

$$CH_3{-}[CH_2]_4{-}CH{=}Se$$

(9) *p*-Selenoformylbenzoic acid

(10) 3-Selenoxohexane
or 3-Hexaneselone

$$CH_3{-}CH_2{-}CH_2{-}\underset{\underset{Se}{\|}}{C}{-}CH_2{-}CH_3$$

(11) Selenobenzoic acid

$$C_6H_5-C{\overset{Se}{\underset{O}{\big<}}}H$$

(12) Cyclohexanecarboseleno-
thioic acid

$$\text{(cyclohexane)}-C{\overset{Se}{\underset{S}{\big<}}}H$$

(13) Ethyl phenyl selenoxide

(14) (Ethylseleninyl)benzene

$$C_6H_5-\underset{\underset{O}{\|}}{Se}-CH_2-CH_3$$

TABLE XIII.   Prefixes and suffixes for use with compounds
containing selenium

Group	Prefix	Suffix or functional-class name	See example
—SeH	Hydroseleno	-selenol	(1) (2)
—Se—	Seleno	selenide	(3) (4)
—Se—	Selena		(5)
—Se— bridge	Episeleno		(6)
—Se—Se— bridge	Epidiseleno		
$-(C){\overset{Se}{\underset{H}{\big<}}}$		-selenal	(7)
$-C{\overset{Se}{\underset{H}{\big<}}}$	Selenoformyl	-carboselenaldehyde	(8) (9)
>(C)=Se	Selenoxo	-selone	(10)
$-C{\overset{Se}{\underset{O}{\big<}}}H$		-carboselenoic acid	(11)
$-C{\overset{Se}{\underset{S}{\big<}}}H$		-carboselenothioic acid	(12)
>SeO	Seleninyl	selenoxide	(13) (14)
>SeO₂	Selenonyl	selenone	
—SeO₃H	Selenono	-selenonic acid	
—SeO₂H	Selenino	-seleninic acid	
—SeOH	Seleneno	-selenenic acid	

701.2—Insofar as their structures are known and conform to those of the selenium derivatives, organic compounds of tellurium are named in the same way as those of selenium, with " tellur-" in place of " selen-".

Example:

Diethyl telluride $(CH_3-CH_2)_2Te$

Note: The nomenclature of tellurium compounds will be dealt with in greater detail in Section D of these Rules.

C-8.   GROUPS CONTAINING ONE NITROGEN ATOM

C-8.1.   AMINES

GENERAL

Rule C-811

811.1—Bases in which nitrogen forms part of a ring are named as heterocyclic compounds in accordance with Section B of these Rules (see p. 53). The termination " -ine ", " -ole ", or " -ete " of such names is not considered to be a suffix in the sense of Rule C-10.3 and so may be followed by endings to denote a substituent of the heterocycle, as, for example, in 2-quinolinol (I) or 4-thiazolecarboxylic acid (II). However, the endings denoting quaternary atoms (see Rule C-816) are considered as suffixes, and substituents must then be named by prefixes, as, for example, in 2-carboxypyridinium chloride (III). In the same way as " -ium " the ending " -amine " (see below) is considered as a suffix.

I   II   III

811.2—The generic name " amine " is applied to compounds NH_2R, NHR^1R^2, and $NR^1R^2R^3$, which are called primary, secondary, and tertiary amines, respectively. In a wider sense, compounds containing nitrogen in a ring and owing their basicity to this atom may also be referred to as " amines ".

811.3—An —NH_2 group, when not the principal group, is named by the prefix " amino- ".

Examples:

2-Aminoethyl- NH_2—CH_2—CH_2—

p-Aminobenzoic acid

Note: In Beilstein's *Handbuch der organischen Chemie* use of the prefix " amino- " is permitted as an alternative, when —NH_2 is the principal group, but this method is not recommended here; it leads to names such as 1-aminobutane, 2-aminopentane, and 1-aminoanthracene.

811.4—Radicals RNH—, R$_2$N—, and R^{1}R^{2}N— are named as substituted amino groups by changing "ine" of the amine to "ino".

Examples:

Methylamino-	CH$_3$—NH—
Diphenylamino-	(C$_6$H$_5$)$_2$N—

N-2-Naphthyl- *N*-propylamino-

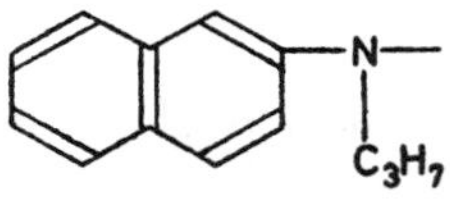

The following are examples of trivial names which are retained:

Anilino-	C$_6$H$_5$—NH—
Anisidino- (*o-*, *m-*, *p-*)	CH$_3$O—C$_6$H$_4$—NH—
Phenetidino- (*o-*, *m-*, *p-*)	C$_2$H$_5$O—C$_6$H$_4$—NH—
Toluidino- (*o-*, *m-*, *p-*)	CH$_3$—C$_6$H$_4$—NH—

Xylidino- (2,3- shown)

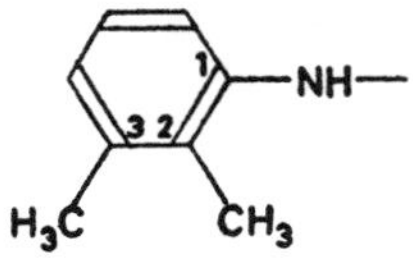

PRIMARY AMINES

Rule C–812

812.1—Primary monoamines, RNH$_2$, are named by adding the suffix " -amine " to (*a*) the name of the radical R or (*b*) the name of the parent compound RH.   Method (*a*) is generally preferred for derivatives of simple parent compounds, and method (*b*) for complex cyclic compounds.

Examples:

(*a*) Ethylamine   CH$_3$—CH$_2$—NH$_2$

(*a*) 1-Ethylbutylamine   CH$_3$—CH$_2$—CH$_2$—CH—CH$_2$—CH$_3$
                             |
                             NH$_2$

(*a*) Cyclohexylamine

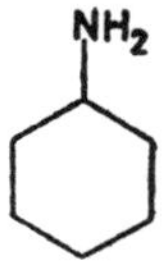

(a) 2-Naphthylamine

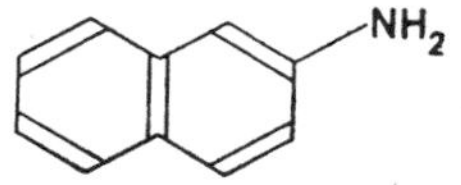

(b) 2-Benzofuranamine

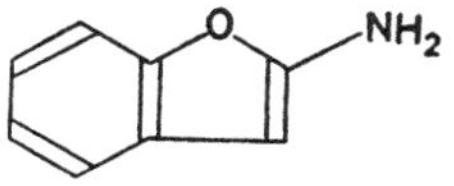

The following are examples of trivial names which are retained:

Aniline $C_6H_5-NH_2$

Anisidine (o-, m-, p-) $CH_3O-C_6H_4-NH_2$

Phenetidine (o-, m-, p-) $C_2H_5O-C_6H_4-NH_2$

Toluidine (o-, m-, p-) $CH_3-C_6H_4-NH_2$

Xylidine (2,3- shown)

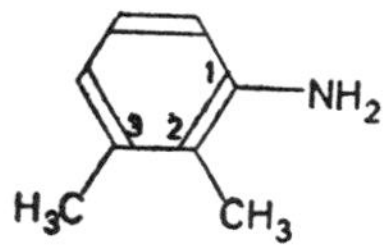

812.2—Primary amines, RNH_2, in which R is a heterocyclic nucleus itself containing nitrogen, are named by adding (a) the suffix " -amine " to the name of the radical, (b) the suffix " -amine " to the name of the parent compound with elision of terminal " e " (if present), or (c) the prefix " amino- " to the name of the parent compound.

Examples:

   (a) 4-Quinolylamine
or (b) 4-Quinolinamine
or (c) 4-Aminoquinoline

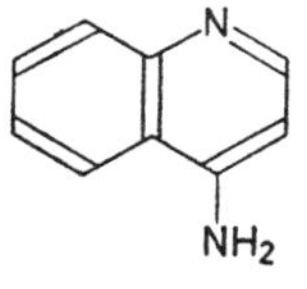

   (a) 1,3,5-Triazin-2-ylamine
or (b) 1,3,5-Triazin-2-amine
or (c) 2-Amino-1,3,5-triazine

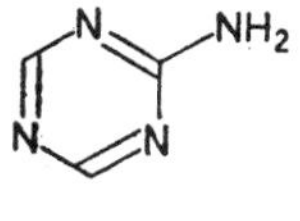

The following is an example of trivial names and numbering which are retained:

Adenine
(6-Aminopurine)

812.3—Primary monoamines, RNH_2, in which R is an alkyl group substituted terminally by a cyclic group are named (*a*) by adding the suffix " -amine " to the name of the radical, (*b*) for complex compounds by conjunctive nomenclature, or (*c*) as substituted cyclic compounds.

Examples:

(*a*) Benzylamine

$$C_6H_5—CH_2—NH_2$$

(*a*) 2-(Fluoren-2-yl)ethylamine
or (*b*) Fluorene-2-ethylamine

(*a*) (Imidazol-4-ylmethyl)amine
or (*b*) Imidazole-4-methylamine
or (*c*) 4-(Aminomethyl)imidazole

Rule C–813

813.1—Primary diamines and polyamines, where all the amino groups are attached to an aliphatic chain or directly to a cyclic nucleus, are named by adding the suffix " -diamine ", " -triamine ", *etc.*, to (*a*) the name of the parent compound or (*b*) the name of the multivalent radical.

Examples:

(*a*) 1,4-Butanediamine
or (*b*) Tetramethylenediamine

$$NH_2—CH_2—CH_2—CH_2—CH_2—NH_2$$

(*a*) 1,4-Pentanediamine
or (*b*) 1-Methyltetramethylenediamine

$$NH_2—\overset{\underset{|}{CH_3}}{CH}—CH_2—CH_2—CH_2—NH_2$$

(*a*) 1,2,5-Pentanetriamine

$$NH_2—CH_2—\overset{\underset{|}{NH_2}}{CH}—CH_2—CH_2—CH_2—NH_2$$

(a) 1,3-Cyclopentanediamine

or (b) 1,3-Cyclopentylenediamine

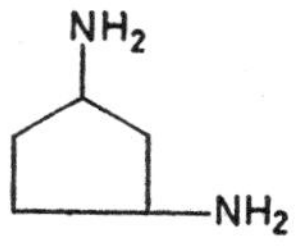

(a) 1,4-Naphthalenediamine

or (b) 1,4-Naphthylenediamine

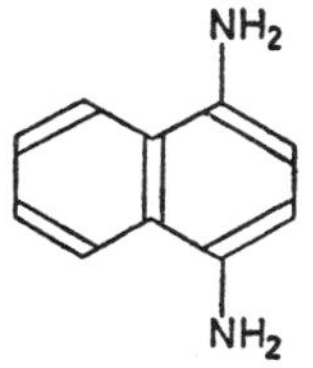

The following is an example of trivial names which are retained:

Benzidine

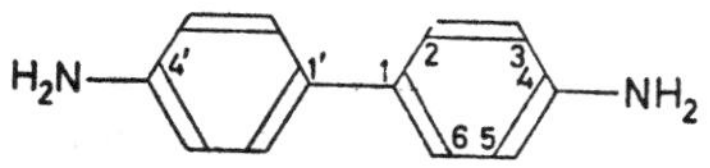

813.2—Primary diamines and polyamines in which the amino groups are attached to a heterocyclic nucleus itself containing nitrogen are named by adding (a) the suffix " -diamine ", " -triamine ", etc., to the name of the parent compound, or (b) the prefix " diamino- ", " triamino- ", etc., to the name of the parent compound.

Examples:

(a) 2,4-Pyridinediamine

or (b) 2,4-Diaminopyridine

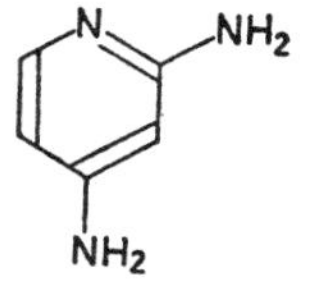

(a) Carbazole-2,7-diamine

or (b) 2,7-Diaminocarbazole

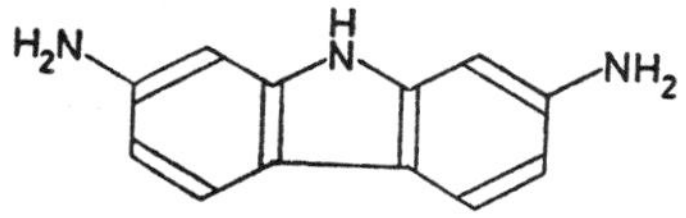

813.3—Primary diamines and polyamines where all the amino groups are attached to identical side chains linked to a cyclic nucleus are named (a) by conjunctive nomenclature or (b) as aminoalkyl derivatives of the cyclic parent.

Example:

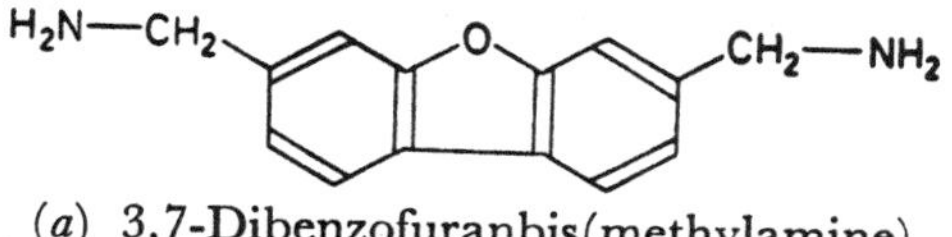

(a) 3,7-Dibenzofuranbis(methylamine)

or (b) 3,7-Bis(aminomethyl)dibenzofuran

813.4—When primary diamines and polyamines contain amino groups in different side chains linked to a cyclic nucleus, then (*a*) the senior side chain or side chains are named by conjunctive nomenclature and the remainder as prefixes, or (*b*) all the side chains are named by means of prefixes.

Examples:

(*a*) 7-(Aminomethyl)-3-dibenzofuranethylamine

or (*b*) 3-(2-Aminoethyl)-7-(aminomethyl)dibenzofuran

(*a*) 7-(Aminomethyl)-2,3-dibenzofuranbis(ethylamine)

or (*b*) 2,3-Bis(2-aminoethyl)-7-(aminomethyl)dibenzofuran

SECONDARY AND TERTIARY AMINES

Rule C–814

814.1—Symmetrical secondary and tertiary amines are named by adding to the name of the radical a prefix " di- " or " tri- ", respectively, and the suffix " -amine ".

Examples:

Diphenylamine $NH(C_6H_5)_2$

Triethylamine $N(C_2H_5)_3$

Di-2-quinolylamine

814.2—In names of symmetrically substituted derivatives of symmetrical secondary and tertiary amines the locants of the substituents are distinguished by primes, or the names of the complete substituted radicals are enclosed in parentheses.

Example:

2,2′-Dichlorodiethylamine $NH(CH_2—CH_2Cl)_2$
or Bis(2-chloroethyl)amine

814.3—Unsymmetrically substituted derivatives of symmetrical secondary and tertiary amines are named (*a*) in the same way as symmetrically substituted derivatives, the locants being distinguished by primes, or (*b*) in the same way as unsymmetrical secondary and tertiary amines (see Rule C–814.4).

Examples:

      (*a*) 1,2'-Dichlorodiethylamine
or (*b*) 1-Chloro-*N*-(2-chloroethyl)ethylamine
or   (1-Chloroethyl)(2-chloroethyl)amine

$$HN \begin{cases} CH_2—CH_2Cl \\ CHCl—CH_3 \end{cases}$$

      (*a*) 2,2,2,2'-Tetrachlorodiethylamine
or (*b*) 2,2,2-Trichloro-*N*-(2-chloroethyl)-
        ethylamine
or   (2-Chloroethyl)(2,2,2-trichloroethyl)-
        amine

$$HN \begin{cases} CH_2—CCl_3 \\ CH_2—CH_2Cl \end{cases}$$

      (*a*) 1,2',3"-Trichlorotripropylamine
or (*b*) 1-Chloro-*N*-(2-chloropropyl)-*N*-
        (3-chloropropyl)propylamine
or   (1-Chloropropyl)(2-chloropropyl)-
        (3-chloropropyl)amine

$$N \begin{cases} CH_2—CH_2—CH_2Cl \\ CH_2—CHCl—CH_3 \\ CHCl—CH_2—CH_3 \end{cases}$$

814.4—Unsymmetrical secondary and tertiary amines, NHR^1R^2, $NR^1R^2R^3$, and $NR^1_2R^2$, if not too complex, are named as *N*-substitution products of a primary amine. The most senior of the radicals R is chosen for the parent primary amine. (cf. Rules C-13.1 and C-14.1)

Examples:

$$CH_3—CH_2—CH_2—CH_2—CH_2—N(CH_3)_2$$

N,*N*-Dimethylpentylamine

$$CH_3—CH_2—CH_2—CH_2—N—CH_2—CH_3$$
$$\underset{CH_3}{|}$$

N-Ethyl-*N*-methylbutylamine

N,*N*-Dimethylcyclohexylamine

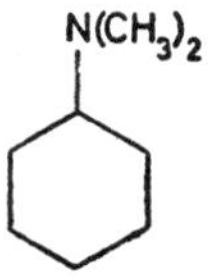

N-Ethyl-*N*-propylaniline

$$C_6H_5—N—CH_2—CH_2—CH_3$$
$$\underset{CH_2—CH_3}{|}$$

N-Phenyl-2-naphthylamine

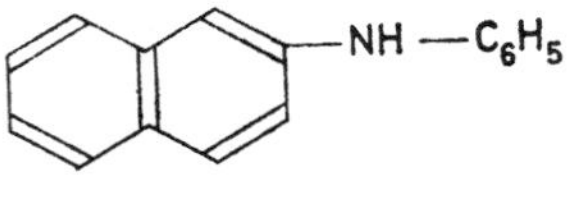

N,*N*-Diethyl-2-furanamine
or *N*,*N*-Diethyl-2-furylamine

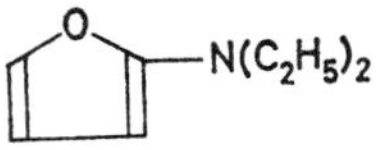

814.5—More complex secondary and tertiary amines in which the amino group is attached directly to a cyclic structure are named (*a*) by the method of Rule C–814.4 or (*b*) as substituted derivatives of the cyclic parent compound.

Examples:

(*a*) *N*,*N*-Dimethylbenz[*a*]anthracen-6-amine

or (*b*) 6-Dimethylaminobenz[*a*]anthracene

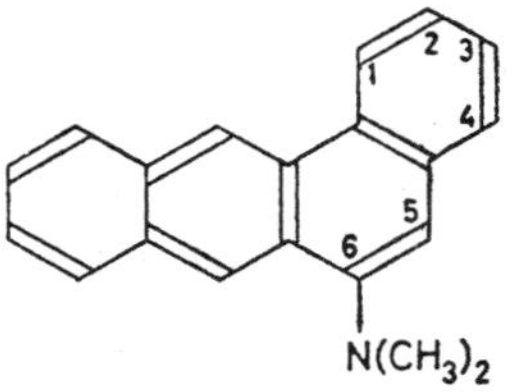

(*a*) *N*-Ethyl-*N*-propyl-1-acridinamine

or (*b*) 1-(Ethylpropylamino)acridine

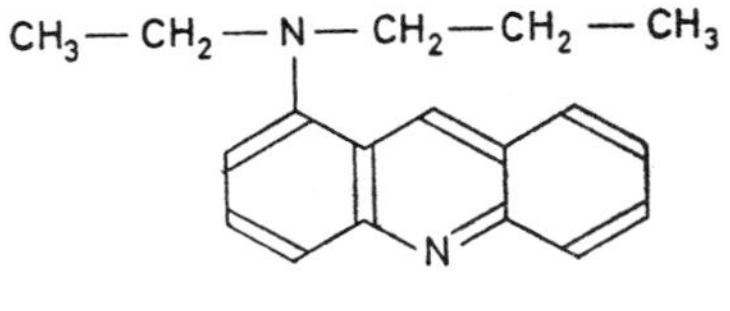

(*a*) *N*-Phenyl-4-quinolinamine

or (*b*) 4-Anilinoquinoline

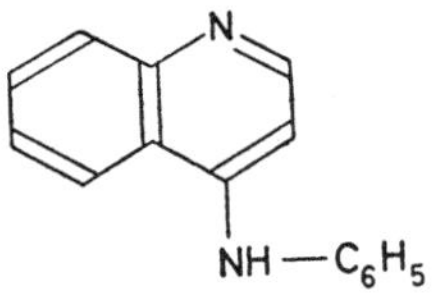

814.6—Complex linear polyamines are best designated by replacement nomenclature (see Subsection C–0.6).

Example:

$$\underset{6}{NH}-\underset{5}{CH_2}-\underset{4}{CH_2}-\underset{3}{N(CH_3)}-\underset{2}{CH_2}-\underset{1}{COOH}$$

$$\underset{7}{CH_2}-\underset{8}{CH_2}-\underset{9}{N(CH_3)}-\underset{10}{CH_2}-\underset{11}{CH_2}-\underset{12}{COOH}$$

3,9-Dimethyl-3,6,9-triazadodecanedioic acid

Rule C–815

815.1—In the nomenclature of assemblies of identical units a bivalent —NH— group linked to two identical radicals R, or a tervalent ≥N atom linked to three identical radicals R, is denoted by the prefix " imino- " or " nitrilo- ", respectively, when the radicals R contain also a group having priority over amine for citation as principal group (see Subsection C–0.7).

Examples:

2,2′-Iminodi-4-quinolineacetic acid

4,4′-Iminodibenzoic acid

3,3′-(2,6-Pyridinediyldiimino)dipropionic acid

3,3′,3″-Nitrilotripropionic acid

Exception:

Ethylenediaminetetraacetic acid

Note: By established custom the above name may be used in place of the systematic name Ethylenedinitrilotetraacetic acid. Extension of the former nomenclature to analogous cases should be restricted to cases where intelligibility is thereby clearly improved; a border line example is:

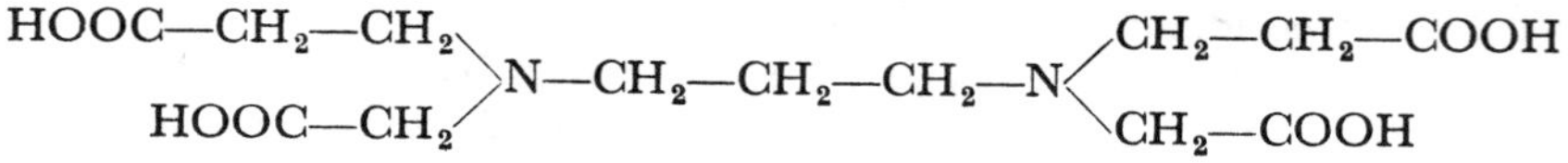

$$\begin{array}{c}
HOOC-CH_2-CH_2 \\
HOOC-CH_2
\end{array} \!\! N-CH_2-CH_2-CH_2-N \!\! \begin{array}{c}
CH_2-CH_2-COOH \\
CH_2-COOH
\end{array}$$

Trimethylenediamine-N,N'-diacetic-N,N'-dipropionic acid

Although these are conjunctive names they do not conform to Rule C-0.5.

815.2—A group —NH— forming a bridge between two carbon atoms of a ring may be named by a prefix " epimino- " or " imino- " (see Rule B–15.1).

Example:

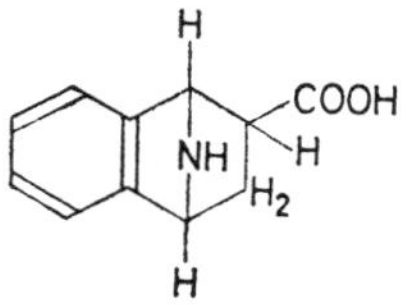

1,2,3,4-Tetrahydro-1,4-epimino-2-naphthoic acid
or 1,4-Epimino-1,2,3,4-tetrahydro-2-naphthoic acid
or 1, 2, 3, 4- Tetrahydro-1,4-iminonaphthalene-2-carboxylic acid

815.3—A compound containing a group $>$C$=$NH may be named (*a*) from the corresponding $>$CH$_2$ compound by means of a suffix " -imine " or, if a group having priority for citation as principal group is also present, by a prefix " imino- ", or (*b*) by citing the name of the bivalent radical $R^1R^2C=$ as prefix to " amine ". Compounds $R^1R^2C=NR^3$ have the class name " azomethines ". When the nitrogen atom is substituted, this class of compound has the generic name " Schiff's bases ".

Examples:

(*a*) 5-Imino-2-pyrrolidinone

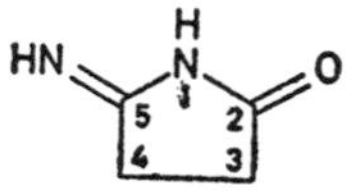

or (*a*) 1-Hexanimine $CH_3-CH_2-CH_2-CH_2-CH_2-CH=NH$
    (*b*) Hexylideneamine

(*b*) *N*-Ethylidene- $CH_3-CH=N-CH_3$
        methylamine

(*b*) 2-Chloro-*N*-(3-chloropropyl-
        idene)ethylamine $ClCH_2-CH_2-CH=N-CH_2-CH_2Cl$

(b) 4-Chloro-*N*-(4-chlorobenz-
ylidene)aniline

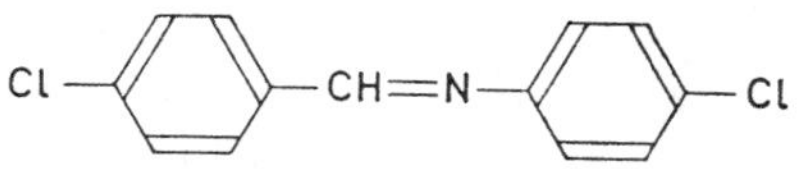

$$Cl - \langle \ \rangle - CH = N - \langle \ \rangle - Cl$$

Exception: Quinones from which one or more atoms of quinonoid oxygen have been replaced by =NH or =NR are named by following the name of the quinone with the word " imine ", " diimine ", *etc.*; substituents on nitrogen are named as prefixes.

Examples:

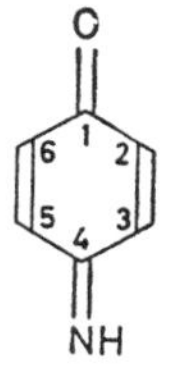

p-Benzoquinone monoimine
or 4-Imino-2,5-cyclohexadien-1-one

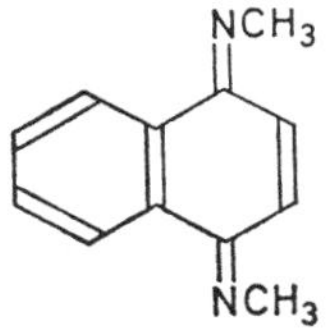

N,N'-Dimethyl-1,4-naphtho-
quinone diimine

AMMONIUM COMPOUNDS

Rule C–816

Salts and hydroxides containing quadricovalent nitrogen, $R_4N^+X^-$ (where the R's may be the same or different), are named by one of the following methods:

816.1—The compound is named as a substituted ammonium salt or hydroxide; the names of the substituting radicals precede the word " ammonium " and then the name of the anion is added.

Examples:

Benzyltrimethylammonium
hydroxide $\qquad$ $[C_6H_5-CH_2-N(CH_3)_3]OH$

Tetramethylammonium iodide $\qquad$ $[N(CH_3)_4]I$

$$CH_3-CH_2-CH_2-CH_2-CH_2-CH_2-\overset{+}{\underset{|}{\overset{|}{N}}}-CH_3 \qquad Cl^-$$

where the N bears CH_2-CH_3 above and CH_2-CH_3 below

Diethylhexylmethylammonium chloride

Exceptions:

The following names are retained for the unsubstituted compounds:

Choline (chloride, *etc.*) $\qquad$ $HO-CH_2-CH_2-\overset{+}{N}(CH_3)_3$ (Cl$^-$, *etc.*)

Betaine* $\qquad$ $^-OOC-CH_2-\overset{+}{N}(CH_3)_3$

Betaine (hydrochloride, *etc.*) $HOOC-CH_2-\overset{+}{N}(CH_3)_3$ (Cl$^-$, *etc.*)

* " Betaine " is also used as a class name for such zwitterionic compounds.

816.2—When the compound can be considered as derived from a base whose name does not end in " -amine ", its quaternary nature is denoted by adding " -ium " to the name of that base, with elision of terminal " e " if present; substituent groups are cited as prefixes, and the name of the anion is added at the end.

Examples:

Anilinium chloride $[C_6H_5—NH_3]Cl$
(preferred to Aniline hydrochloride)

Anilinium hydrogen sulfate $[C_6H_5—NH_3]HSO_4$
(preferred to Aniline hydrogen sulfate)

Dianilinium hexachloroplatinate $[C_6H_5—NH_3]_2[PtCl_6]$
(preferred to Dianiline hexachloroplatinate)

3-Methylthiazolium bromide

1-Methylaziridinium p-nitrobenzoate

816.3—In complex cases, the prefixes " amino- " and " imino- " may be changed to " ammonio- " and " iminio- " and are followed by the name of the molecule representing the most complex group attached to this nitrogen atom and preceded by the names of the other radicals attached to this nitrogen atom; finally the name of the anion is added (see also Rules C–85 and C–87).

Examples:

1-Trimethylammonioacridine chloride

or, by Rule **C–816.1**, 1-Acridinyltrimethylammonium chloride

10-Methyl-1-trimethylammonioacridinium dichloride

816.4—When the above rules cannot be applied or lead to inconvenient names, recourse may be had to two traditional methods of naming salts of organic bases, namely: (*a*) the unaltered name of the base is followed by the name of the anion; and (*b*), for salts of hydrohalogen acids only, the unaltered name of the base is followed by hydrofluoride, hydrochloride, hydrobromide, or hydriodide, as the case may be. These methods are retained for use also in simpler cases but the procedure of Rule **C–816.2** is preferred.

Examples, where the structure is indefinite and Rule **C–816.2** thus cannot be applied:

$$2\left(\underset{\text{NH}}{\overset{\text{S}}{\bigcirc}}-\text{N(CH}_3)_2\right)\text{H}_2\text{SO}_4$$

Bis(2-dimethylaminothiazolidine) sulfate

$$\text{NH}_2\text{—}\overset{\text{C}_2\text{H}_5}{\bigcirc}\text{—NH}_2 \quad \text{HCl}$$

2-Ethyl-*p*-phenylenediamine monohydrochloride

816.5—Complexes formed from bases and phenols are named by citing the name of the base followed by that of the phenol in its anion form.

Examples:

Aniline picrate	$C_6H_5NH_2,C_6H_3N_3O_7$
Indole styphnate	$C_8H_7N,C_6H_3N_3O_8$

C–8.2. AMIDES AND IMIDES

Rule C–821

821.1—Compounds containing one, two, or three acyl groups attached to nitrogen bear the generic name "amide". When only one acyl group is attached to a nitrogen atom, the generic name "primary amide" may be used; when two acyl groups are so attached, the generic name "secondary amide" may be used; and when three acyl groups are so attached, the generic name "tertiary amide" may be used. Amides derived from carbon acids may be termed "carboxamides", those from sulfonic acids "sulfonamides", *etc.*; *N*-substituted primary and secondary amides may also be described as acylamines or, more specifically as monoacylamines and diacylamines, respectively.

Note: As the nomenclature of amides of sulfur acids is described (though less extensively) in Rule **C–641.8**, the present Subsection is illustrated mostly for carbon acids.

MONOACYLAMINES

Rule C–822

822.1—Names of primary amides in which the NH_2 group is unsubstituted are derived from the systematic names of the corresponding acids by replacing " -oic acid " or " -ic acid " by " -amide " or " -carboxylic acid " by " -carboxamide ". When trivial names of the acids are used which end in " -ic acid " this ending is replaced by " -amide ".

(cf. Rule C-971).

Examples:

Hexanamide	$CH_3—CH_2—CH_2—CH_2—CH_2—CO—NH_2$
Acetamide	$CH_3—CO—NH_2$
Benzenesulfonamide	$C_6H_5—SO_2—NH_2$

Imidazole-2-carboxamide

Succinamide	$H_2N—CO—CH_2—CH_2—CO—NH_2$
Maleamide	$cis\text{-}H_2N—CO—CH=CH—CO—NH_2$
Malamide	$H_2N—CO—CH_2—CH(OH)—CO—NH_2$
Decanediamide	$NH_2—CO—(CH_2)_8—CO—NH_2$

The following abbreviated name is retained:
Oxamide $NH_2—CO—CO—NH_2$

822.2—For amides of amino acids having trivial names ending in " -ine ", " -amide " or " -diamide " is added after the name of the acid, with elision of terminal " e " (for monoamides).*

Examples:

Alaninamide	$CH_3—CH(NH_2)—CO—NH_2$
Cystinediamide	$[NH_2—CO—CH(NH_2)—CH_2—S—]_2$

Note: For monoamides of dicarboxylic acids see Rules **C–421.3** and **C–431.1**.

Rule C–823

823.1—(a) Names of radicals $RCO—NH—$, $R^1CO—NR^2—$, $R—SO_2—NH—$, etc., formed by loss of a hydrogen atom from the NH_2 or NHR group of an amide, are derived from the systematic or semi-trivial name of the amide by changing the ending " -amide " to " -amido- ", such radical names being used when the compound contains also a group having priority for citation as principal group. (b) Alternatively the radicals are named as acylamino radicals.

*See also Nomenclature of α-Amino Acids, *IUPAC Information Bulletin* Appendix No. 46, September 1975 or *Biochemistry* (1975) *14*, pp. 449-462.

Examples:

(a) 4-Acetamido-l-naphthoic acid

(b) 4-Acetylamino-l-naphthoic acid

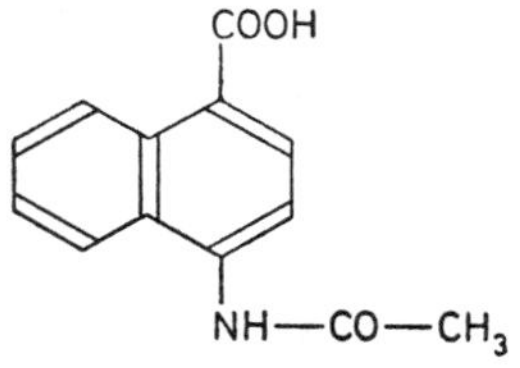

(a) 7-Benzenesulfonamido-
quinolinium chloride

(b) 7-Phenylsulfonylamino-
quinolinium chloride

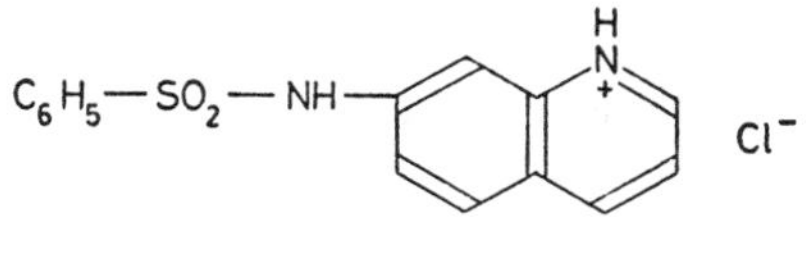

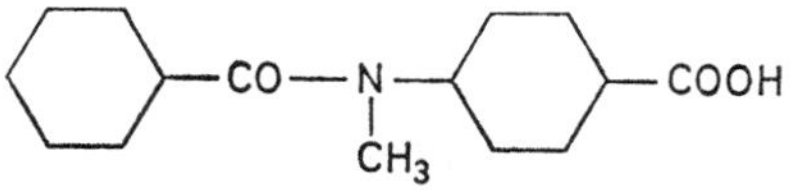

(a) 4-(N-Methylcyclohexanecarboxamido)-
1-cyclohexanecarboxylic acid

(b) 4-(N-Methylcyclohexylcarbonylamino)-
1-cyclohexanecarboxylic acid

823.2—Despite Rule **C-823.1**, radicals $RCH(NH_2)$—CO—NH— derived from amino acids that have trivial names are named only as acylamino radicals, the acyl radical names for the groups $RCH(NH_2)$—CO— being used (see Rule **C-421**).

Example:

p-Glycylaminobenzoic acid

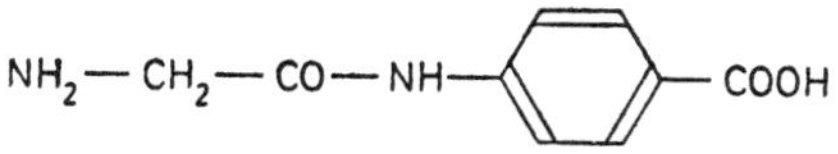

Rule C-824

N-Substituted primary amides R^1—CO—NHR^2 and R^1—CO—NR^2R^3 (also analogous sulfonamides, *etc.*) are named by one of the following methods:*

824.1—The compound is named as an *N*-substituted amide, the substituents R^2 and R^3 being cited as prefixes.

Note: This method is particularly suitable when the group R^1 in the acyl residue R^1CO— is more complex than the group R^2, or R^2 and R^3, in the amine residue R^2NH— or R^2R^3N—.

* These compounds have been called, respectively, secondary and tertiary amides, but this usage is not recommended.

Examples:

N-Methylbenzamide C_6H_5—CO—NH—CH_3

N,N-Diethyl-2-furamide

N-Ethyl-N-methyl-8-quinolinecarboxamide

824.2—Alternatively, the acyl group is named as an N-substituent of the base.

Note: This method is particularly suitable when the group R^1 in the acyl residue R^1CO— is less complex than the group R^2, or R^2 and R^3, in the amine residue R^2NH— or R^2R^3N—, and for use with derivatives of nitrogen-containing heterocyclic compounds.

Examples:

1-Acetyl-1,2,3,4-tetrahydroquinoline

1-Acetylaminoacridine (but see also Rule C–824.3)

3-(Acetylmethylamino)dibenzofuran (but see also Rule C–824.4)

C–824.4

3-Formylthiazolidine
(see Rule C–304.3)
or 3-Thiazolidinecarbaldehyde
(see Rule C–304.1)

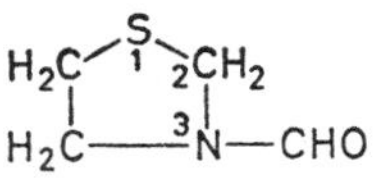

824.3—As a further alternative, for mono-*N*-substituted primary amides R¹—CO—NHR² the group R¹—CO—NH— is treated as a substituent into the compound HR² and is named by changing the ending " -amide " to " amido- " or " -carboxamide " to " carboxamido- ".

Examples:

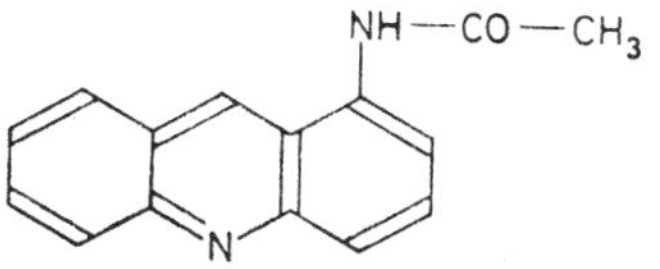

1-Acetamidoacridine (but see also Rule C–824.2)

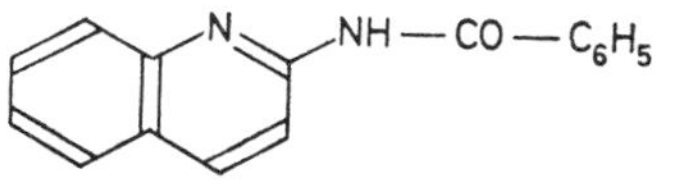

2-Benzamidoquinoline

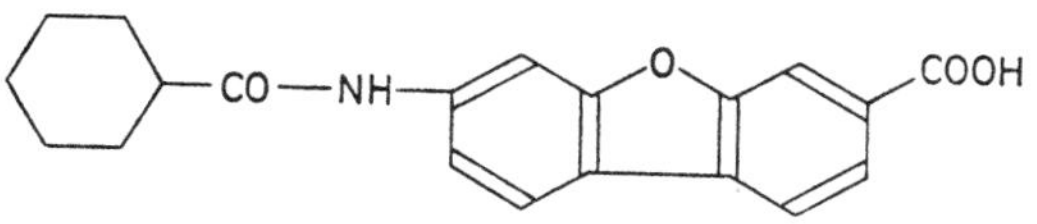

7-Cyclohexanecarboxamido-3-dibenzofurancarboxylic acid

824.4—For di-*N*-substituted primary amides R¹—CO—NR²R³ the procedure of Rule C–824.3 is applied, the less complex of the groups R² and R³ being named as substituent of the amido or carboxamido group with locant *N*.

Example:

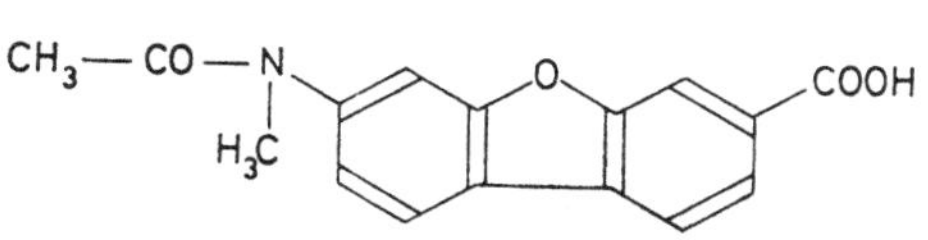

7-(*N*-Methylacetamido)-3-dibenzofurancarboxylic acid

Note: The procedures of Rules C–824.3 and C–824.4 are particularly suitable when the group R¹ in the acyl residue R¹CO— is less complex than the group R², or R² and R³, in the amine residue R²NH— or R²R³N—; they are not applicable to derivatives of *N*-acyl nitrogen-containing heterocyclic systems.

Rule C–825

825.1—The ending " -anilide " is retained for N-phenyl-substituted amido groups, being used in the same way as the ending "-amide " (see Rules C–822 and C–824). Locants for substituents in the amine residue are primed. Alkyl substituents on the nitrogen atom are named as prefixes with the locant N-.

Examples:

Acetanilide $CH_3—CO—NH—C_6H_5$

2′,3′,4-Triethylbenzanilide

N-Methylbenzanilide $C_6H_5—CO—N(CH_3)—C_6H_5$

Hexananilide $CH_3—[CH_2]_4—CO—NH—C_6H_5$

DI- AND TRIACYLAMINES

Rule C–826

826.1—Symmetrical compounds $HN(CO—R)_2$ and $N(CO—R)_3$ are named as diacylamines and triacylamines, respectively.

Examples:

Di-2-thenoylamine

Tris(cyclohexylcarbonyl)amine $(C_6H_{11}—CO)_3N$

The following names are retained and may be used for naming of derivatives:

Diacetamide $NH(CO—CH_3)_2$

Triacetamide $N(CO—CH_3)_3$

Dibenzamide $NH(CO—C_6H_5)_2$

Tribenzamide $N(CO—C_6H_5)_3$

Examples:

N-Methyldibenzamide $CH_3—N(CO—C_6H_5)_2$

N-Phenyldiacetamide $C_6H_5—N(CO—CH_3)_2$
(but see also Rule C–826.3)

266

826.2—Unsymmetrical compounds R^1CO—NH—COR^2 and R^1CO—$N(COR^2)$—COR^3 are named as N-acyl derivatives of the most complex amide.

Examples:

N-Acetylbenzamide C_6H_5—CO—NH—CO—CH_3

$$C_6H_5-CH{=}CH-CO-N-CO-C_6H_{11}$$
$$|$$
$$CO-CH_3$$

N-Acetyl-N-cyclohexylcarbonylcinnamamide

826.3—Compounds $R^1N(COR^2)_2$ and $R^1N(COR^2)$—COR^3 are named (*a*) as diacyl derivatives of the amine R^1NH_2 or (*b*) as N-substituted secondary amides. Method (*a*) includes use of a prefix "diacylamino-", which is specifically required when a group having priority for citation as principal group is also present.

Examples:

(*a*) N,N-Diacetylaniline C_6H_5—$N(CO$—$CH_3)_2$
or (*b*) N-Phenyldiacetamide
     (see Rule C–826.1)

(*a*) N-Acetyl-N-benzoyl-2-naphthylamine
or (*b*) N-Acetyl-N-2-naphthylbenzamide

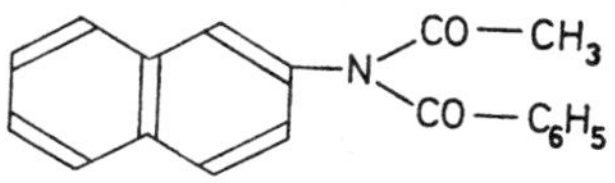

(*a*)   9-(Diacetylamino)phenanthrene
or (*b*) N-9-Phenanthryldiacetamide
     (see Rule C–826.1)

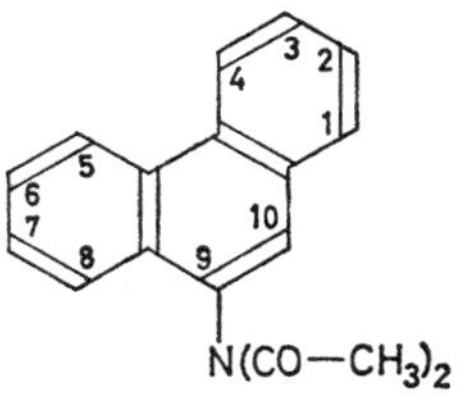

(*a*)   2-(Dipropionylamino)-7-
     quinolinecarboxylic acid

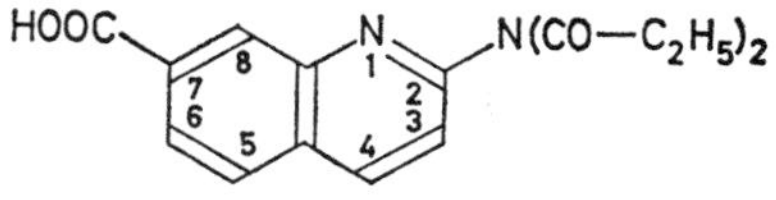

(*a*)   2-(Dibenzoylamino)-5-
     imidazolecarboxylic acid

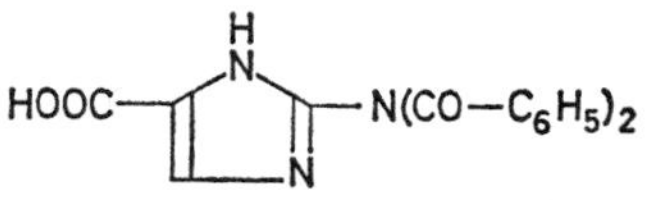

(*a*)   2-(Acetylbenzoylamino)-
     6-quinoxalinecarboxylic acid

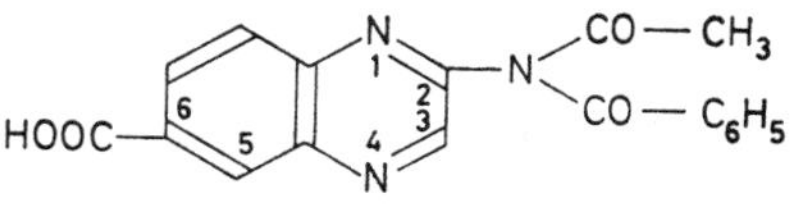

IMIDES

Rule C–827

827.1—Imides of dicarboxylic acids are named by replacing the ending "-carboxylic acid" by "-carboximide", or, for trivial names, "-ic acid" by "-imide". Such compounds bearing a substituent on the nitrogen atom are named as *N*-substituted imides.

Examples:

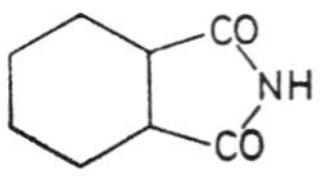

1,2-Cyclohexanedicarboximide

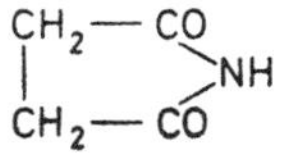

Succinimide

N-Phenylphthalimide

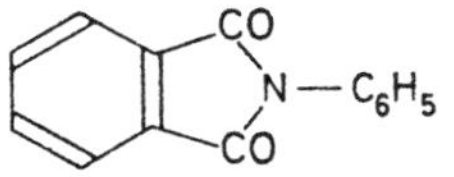

827.2—Names of radicals derived from imides by removal of the hydrogen atom attached to the imide-nitrogen atom are formed from the names of the corresponding imides by changing the ending "-imide" to "-imido-".

Examples:

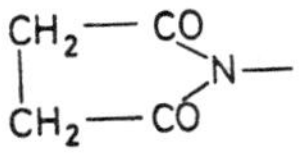

Succinimido

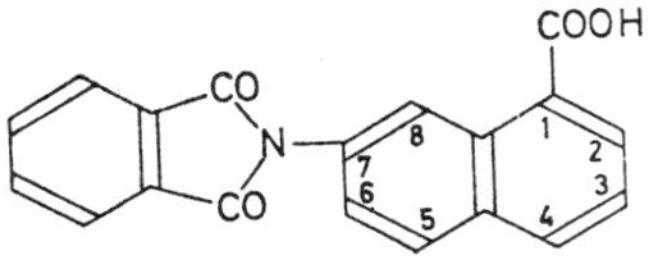

7-Phthalimido-1-naphthoic acid

827.3—When the nomenclature of Rules C–827.1 and C–827.2 is inconvenient, imides may be named as heterocyclic compounds.

Examples:

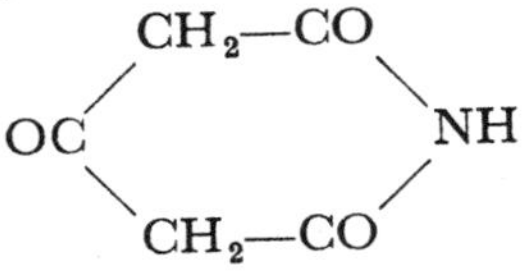

2,4,6-Piperidinetrione

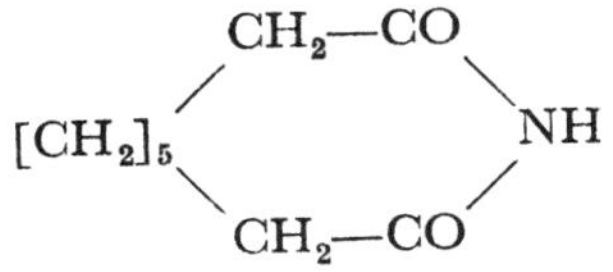

Perhydroazecine-2,10-dione or
Hexahydro-2,10(1*H*,3*H*)-azecinedione

C-8.3 NITRILES, ISOCYANIDES, AND THEIR DERIVATIVES

Rule C–831

831.1—Compounds containing the group —CN are termed generically "nitriles" (substitutive nomenclature) or "cyanides" (radicofunctional nomenclature).

Rule C–832

By substitutive nomenclature, nitriles are named by one of the following methods:

832.1—Compounds RCN, in which $\equiv$N replaces H_3 at the end of the main chain of an acyclic hydrocarbon are denoted by adding "-nitrile" or "-dinitrile" to the name of this hydrocarbon.

Examples:

Hexanenitrile

$$\overset{6}{C}H_3-\overset{5}{C}H_2-\overset{4}{C}H_2-\overset{3}{C}H_2-\overset{2}{C}H_2-\overset{1}{C}N$$

Hexanedinitrile

$$NC-CH_2-CH_2-CH_2-CH_2-CN$$

Note: "Nitrile", here and in Rule C–832.3, denotes the triply bound nitrogen atom, $\equiv$N, and not the carbon atom attached to it. Numbering begins with that carbon atom.

832.2—Compounds RCN, when considered as derived from acids R—COOH whose systematic names end in "-carboxylic acid", are named by changing this ending to "-carbonitrile".

Note: "Carbonitrile" denotes the group —C$\equiv$N, including the carbon atom contained therein. That carbon atom is excluded from the numbering of a chain to which that group is attached.

Examples:

Cyclohexanecarbonitrile C_6H_{11}—CN

$$NC-\overset{6}{C}H_2-\overset{5}{C}H_2-\overset{4}{C}H_2-\overset{3}{C}H(CN)-\overset{2}{C}H_2-\overset{1}{C}H_2-CN$$
1,3,6-Hexanetricarbonitrile

2-Thiazolecarbonitrile

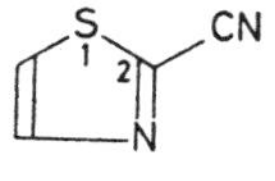

832.3—Names of compounds RCN, when considered as derived from acids R—COOH having trivial names, are formed by changing the syllables "-oic acid" to "-onitrile", or, if the name of the acid does not end in "-oic acid", then by changing "-ic acid" to "-onitrile".

Examples:

Benzonitrile C_6H_5—CN

Propiononitrile CH_3—CH_2—CN
   (but see Rules C–832.1 and C–832.4)

Oxalonitrile   NC—CN

832.4—By the radicofunctional procedure, compounds RCN are named by stating the name of the radical R, followed by the name " cyanide " for the group —CN.

Examples:

Ethyl cyanide CH_3—CH_2—CN
   (but see Rules C–832.1 and C–832.3)

Benzoyl cyanide C_6H_5—CO—CN

832.5—When the compound contains also a group that has priority over —CN for citation as principal group, the —CN group is named by the prefix " cyano- ".

Examples:

5-Cyano-2-furoic acid

2,4-Dicyanobenzamide

Rule C–833

833.1—Compounds containing a group X listed in the first column of Table XIV are named by methods analogous to those described for cyanides in Rules C–832.4 and C–832.5; the functional class names given in the second column of the Table are used in place of " cyanide ", or the prefixes given in the third column of the Table in place of "cyano- ". The functional class names are also the generic names of the respective classes of compound.

Note: " Cyanide ", like " carbonitrile " but unlike " nitrile ", denotes the group —CN including its carbon atom. Its carbon atom is excluded from the numbering of a chain to which the group is attached.

TABLE XIV. Cyanide and related groups in order of decreasing priority for citation as functional class name

Group X in RX	Functional class ending and generic name of class	Prefix
—CN	cyanide	cyano-
—NC	isocyanide*	isocyano-
—OCN	cyanate	cyanato-
—NCO	isocyanate	isocyanato-
—ONC	fulminate	—
—SCN	thiocyanate	thiocyanato-
—NCS	isothiocyanate	isothiocyanato-
—SeCN	selenocyanate	selenocyanato-
—NCSe	isoselenocyanate	isoselenocyanato-

* Not isonitrile or carbylamine.

Examples:

Phenyl isocyanide C_6H_5—NC

p-Isocyanobenzoic acid 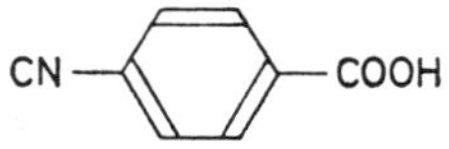 CN—⟨⟩—COOH

$$\overset{4}{CH_3}—\overset{3}{CH_2}—\overset{2}{CH_2}—\overset{1}{CH_2}—CN$$

Butyl cyanide

Thiocyanatoacetic acid NCS—CH_2—COOH

Ethyl selenocyanate CH_3—CH_2—SeCN

Cyclohexyl isocyanate C_6H_{11}—NCO

p-Tolyl isothiocyanate CH_3—⟨⟩—NCS

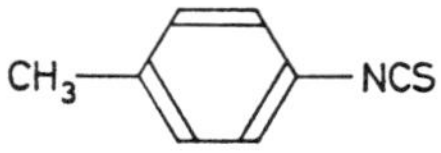

Rule C–834

834.1—Compounds RC≡NO have the generic name " nitrile oxides ". In specific cases, " oxide " is added as a separate word after the name of the compound RCN, formed as a nitrile (not as a cyanide) (see Rules **C–832.1** and **C–832.3**).

Examples:

Benzonitrile oxide C_6H_5—C≡NO

Hexanenitrile oxide CH_3—CH_2—CH_2—CH_2—CH_2—C≡NO

C-8.4. HYDROXYLAMINES AND RELATED COMPOUNDS

Rule C–841. Hydroxylamines and their derivatives

841.1—Compounds RNH—OH are named by prefixing the name of the radical R to " hydroxylamine " or, when another substituent is present having priority for citation as principal group, by attaching the prefix " hydroxyamino- " to the name of the parent compound RH.

Examples:

N-Phenylhydroxylamine C_6H_5—NH—OH

p-(Hydroxyamino)phenol

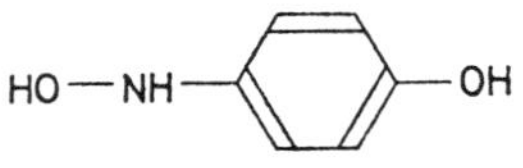

HO—NH—⟨benzene ring⟩—OH

841.2—Compounds R^1NH—OR^2 are named as (*a*) alkoxyamino derivatives of the compound R^1H, (*b*) as *N,O*-substituted hydroxylamines, (*c*) as alkoxyamines (even if R^1 is hydrogen), or (*d*), if the group R^2 has a substituent with priority for citation as principal group, by use of the prefix " aminooxy- ".

Examples:

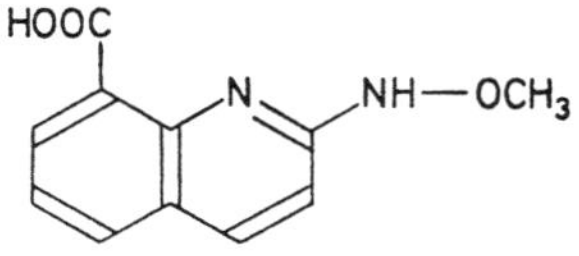

(*a*) 2-(Methoxyamino)-8-quinolinecarboxylic acid
(*b*) *O*-Phenylhydroxylamine (preferred) ⎫
(*c*) Phenoxyamine   ⎬ C_6H_5O—NH$_2$
(*d*) Ethyl (aminooxy)acetate NH$_2$—O—CH$_2$—CO—O—CH$_2$—CH$_3$

841.3—Acyl derivatives RCO—NH—OH and NH$_2$—O—COR are named as *N*-hydroxy derivatives of amides and as *O*-acylhydroxylamines, respectively. Alternatively the former may be named as hydroxamic acids (see Rule C–451.3).

Examples:

N-Hydroxyacetamide CH_3—CO—NH—OH
(see also Rule **C–451.3**)

O-Acetylhydroxylamine   NH$_2$—O—CO—CH$_3$

841.4—Further substitution on the nitrogen or oxygen atoms of hydroxylamine is denoted by prefixes for the substituents, with *O*- and/or *N*-locants. Alternatively, compounds $R^1R^2N—OR^3$ and $R^1—NH—OR^2$ may be named as derivatives of the alkoxyamine or of the hydroxamic acid (see Rule **C–451.3**).

Examples:

N,N-Dimethylhydroxylamine $(CH_3)_2N—OH$

O-Ethyl-*N*-phenylhydroxylamine $C_6H_5—NH—OC_2H_5$
or *N*-Ethoxyaniline

p-(*N*-Hydroxy-*N*-methylamino)phenol

O-Acetyl-*N*-methylhydroxylamine $CH_3—NH—O—CO—CH_3$

Note, however:

N-Hydroxy-*N*-methylacetamide
(see Rule **C–841.3**) $\Big\}$ $CH_3—CO—N(CH_3)—OH$

or *N*-Methylacetohydroxamic acid
(see Rule **C–451.3**)

Rule C–842.   Oximes and their derivatives

842.1—Compounds $RHC\!=\!N—OH$ or $R^1R^2C\!=\!N—OH$, termed generically " oximes ", are named (*a*) by placing the word " oxime " after the name of the aldehyde RCHO or the ketone R^1R^2CO, respectively, or (*b*), if the compound contains also a group having priority over a carbonyl group for citation as principal group, by use of the prefix " hydroxyimino- " attached to the name of the corresponding compound RCH_3 or $CH_2R^1R^2$.

Examples:

(*a*) Cinnamaldehyde oxime $C_6H_5—CH\!=\!CH—CH\!=\!N—OH$

(*a*) Benzophenone oxime $(C_6H_5)_2C\!=\!N—OH$

(*a*) Cyclohexanone oxime

(*a*) 2-Hexanone oxime $CH_3—CH_2—CH_2—CH_2—C—CH_3$
$$\underset{\displaystyle N—OH}{\overset{\displaystyle \|}{}}$$

(*b*)　4-(Hydroxyimino)-1-methyl-2,5-cyclohexadiene-1-carboxylic acid

(*a*)　Benzil dioxime

$$C_6H_5-C-C-C_6H_5$$
$$HO-N \quad N-OH$$

842.2—Compounds containing the group $=N-OR$ are named (*a*) by a prefix of the type " alkyloxyimino- ", " aryloxyimino- ", *etc.*, or (*b*) as oxime *O*-ethers, or (*c*) as *O*-substituted oximes.

Examples:

(*a*)　4-(Ethoxyimino)-3,4-dihydro-3,3-dimethyl-1-naphthalenesulfonic acid

(*b*)　Benzil dioxime *O,O′*-dimethyl ether

or　(*c*)　Benzil bis(*O*-methyloxime)

$$C_6H_5-C-C-C_6H_5$$
$$CH_3O-N \quad N-OCH_3$$

842.3—Compounds containing the grouping $>C=N(O)R$ are named by adding the word " *N*-oxide " after the name of the alkylideneamino (*etc.*) compound.　The class name " nitrone " is retained (see also Rule C–843.1).

Example:

N-(α-Methylbenzylidene)methylamine
N-oxide

$$C_6H_5-C-CH_3$$
$$ON-CH_3$$

Rule C–843.　Amine oxides

843.1—Compounds R_3NO (where the R's may be the same or different) are named by adding the word " oxide " (or, for cyclic bases, " oxide " and a locant) after the name of the base R_3N.

Examples:

Trimethylamine oxide　　　　$(CH_3)_3NO$

Pyridine 1-oxide
or　Pyridine *N*-oxide

C–8.5. NITROSO AND NITRO COMPOUNDS

Rule C–851

851.1—Compounds containing a —NO group are named only by means of the prefix " nitroso- ".

Examples:

Nitrosobenzene C_6H_5—NO

1-Methyl-1-nitrosourea $\overset{3}{N}H_2$—$\overset{2}{C}O$—$\overset{1}{N}(CH_3)$—NO

Rule C–852

852.1—Compounds containing a —NO_2 group are named only by means of a prefix " nitro- ".

Examples:

Nitromethane CH_3—NO_2

1-Nitronaphthalene

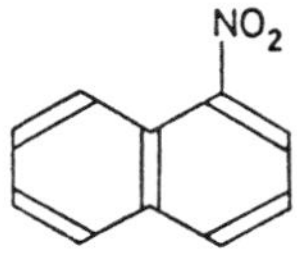

o-Nitrobenzoic acid

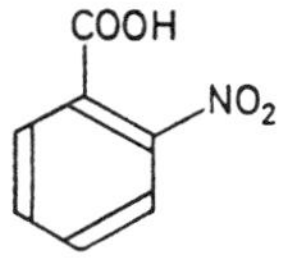

852.2—Compounds containing the group $X{=}N(O)OH$ are named by adding the prefix " *aci*-nitro- " to the name of the parent compound XH_2.

Examples:

aci-Nitromethane $CH_2{=}N(O)OH$

2-*aci*-Nitro-1-cyclohexanone

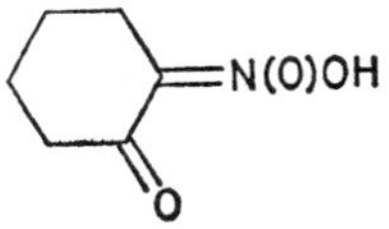

Sodium salt of 2-*aci*-nitropropane $(CH_3)_2C{=}N(O)ONa$

C–8.6. AMINE RADICAL IONS

Rule C–861

861.1—Ions that can be considered to be derived by adding a proton to an amino radical $R^1R^2N\cdot$ are named by adding the suffix " -yl " to the name of the ammonium cation (see Rule C–83.3).

Example:

$$(p\text{-}CH_3\text{---}C_6H_4)_3\overset{+}{N}\cdot \qquad ClO_4^-$$

Tri-p-tolylammoniumyl perchlorate

C–9.   GROUPS CONTAINING MORE THAN
ONE NITROGEN ATOM

In the rules below, the term "parent molecule" denotes a molecule R^1H, R^2H, XH_2, *etc.*, from which an azo compound, $R^1N=NR^2$, or $R^1N=N—X—N=NR^2$, *etc.*, (or the analogous azoxy-, hydrazo-, *etc.*, compounds) can be considered to be derived. "Unsubstituted parent molecule" denotes those parent molecules from which all substituents have been removed.

C–9.1.   AZO AND AZOXY COMPOUNDS

Except for certain simple cases where simplification of names is achieved, azo compounds may be named by either of two methods. In an older method (Rule C–911) the components R linked by the azo groups are named as molecules RH, each cited with its own substituents, and the resultant partial names are joined together by the designation "azo" (for the group —N=N—) and the locants of this group.

In the *Chemical Abstracts* method (Rule C–912) two principles apply: (*a*) one component is chosen as parent which is regarded as substituted by RN=N—, this group R being named as a radical; or (*b*) the nomenclature is that for assemblies of identical units (see Subsection C–0.7), with "azo" as a bivalent linking group.

The older method was designed to display the separate components in a complete, often complex structure. The *Chemical Abstracts* method gives prominence to the principal group.

Rule C–911.   Azo compounds (Alternative to Rule C–912)

911.1—Monoazo compounds RN=NR, in which the azo group —N=N— links radicals derived from parent molecules that, when unsubstituted, are identical, are named by adding the prefix "azo-" to the name of the unsubstituted parent molecule. Substituents are denoted in the usual way by prefixes and suffixes. The azo group has priority for lowest available numbers; one set of locants is distinguished by primes.

Examples:

Azomethane	$CH_3—N=N—CH_3$
Azobenzene	$C_6H_5—N=N—C_6H_5$

1,2'-Azonaphthalene

Azobenzene-4-sulfonic acid

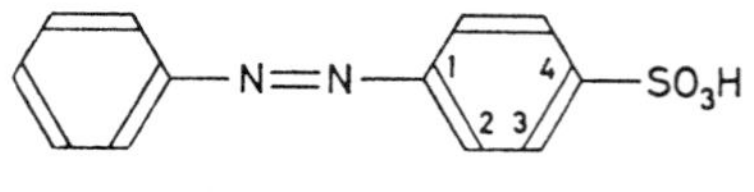

4'-Amino-2-chloroazobenzene-
4-sulfonic acid

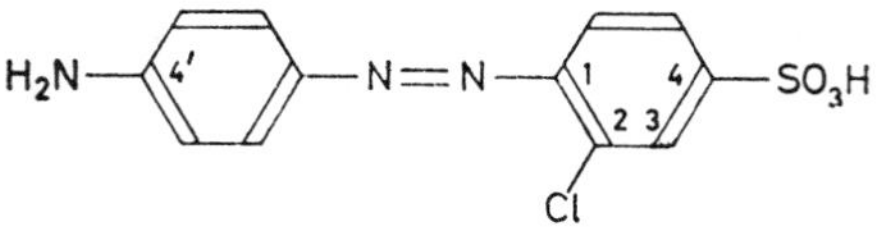

2',4-Dichloroazobenzene-2,4'-
disulfonic acid

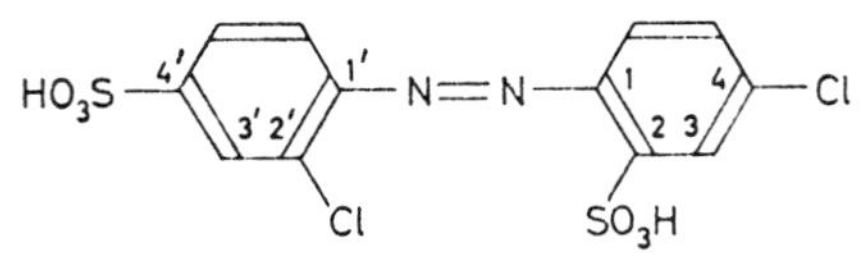

911.2—Monoazo compounds $R^1N{=}NR^2$, in which the azo group links radicals derived from parent molecules that, when unsubstituted, are different, are named by placing " azo " between the complete names of the (substituted) parent molecules; when locants are needed to indicate the position of the azo group they are placed between the affix " azo " and the names of the molecules to which each respectively refers. Preference is given to the more complex parent molecule for citation as the first component. The azo group has priority for lowest permissible numbers. All substituents in the component named first are denoted by prefixes unless their presence is indicated in the trivial or semi-trivial name of that component. Locants for substituents in the first-named component are not primed; the others are primed.

Examples:

Naphthalene-2-azobenzene

$$2\text{-}C_{10}H_7\text{—}N{=}N\text{—}C_6H_5$$

1-Carboxynaphthalene-2-azobenzene

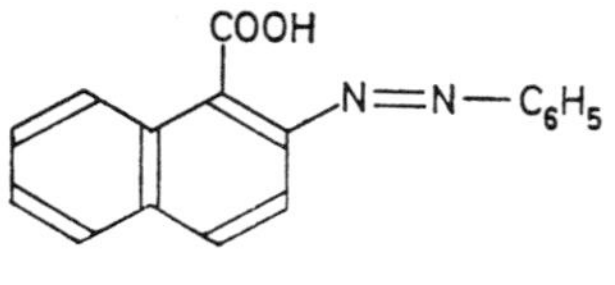

2-Aminonaphthalene-1-azo-
(4'-chloro-2'-methylbenzene)

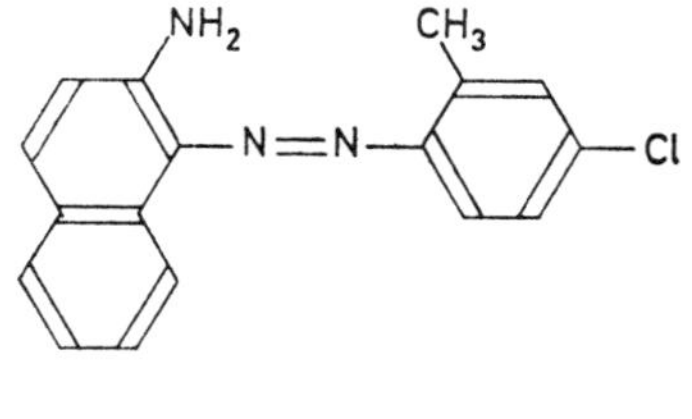

278

3-Methyl-1-(*p*-nitrophenyl)-5-pyrazolone-4-azo-3'-
(4'-hydroxybenzamide)

2-Sulfoanthraquinone-1-azo-2'-naphthalene-1'-sulfonic acid

911.3—Bisazo compounds and more complex analogues are similarly named by citing the components serially, starting with the more complex terminal component; or, when the molecule consists of a central component from which identical $RN{=}N{-}$ groups radiate, by citing these radiating groups as substituents of the central component.

Examples:

Anthracene-2-azo-2'-naphthalene-7'-azobenzene

[4-(*p*-Methoxyphenyl)-2,2'-disulfostilbene]-4'-azo-1-benzene-4-azo-2-
[(7-dimethylamino-1-hydroxy)-3-naphthalenesulfonic acid]

Bis(benzeneazo)-2,7-(1,8-naphthalenediol)

Bis(5-carbamoyl-2-hydroxybenzene-1-azo)-6,6'-(5,5'-dihydroxy-7,7'-disulfo-2,2'-dinaphthylamine)

Rule C–912. Azo compounds: Chemical Abstracts method (Alternative to Rule C–911)

912.1—As described in Rule C–911, but only in absence of a group that can be cited as a suffix, monoazo compounds RN=NR, in which the azo group —N=N— links radicals derived from parent molecules that, when unsubstituted, are identical, are named by adding the prefix " azo- " to the name of the unsubstituted parent molecule. Substituents are denoted in the usual way by prefixes and the shorter set (or, if of equal length, that with higher numbers) is distinguished by primes. The azo group has priority for lowest available numbers.

Example:

3,4'-Dichloroazobenzene

912.2—When each group R in the azo compound RN=NR is (a) identical when unsubstituted and (b) carries the same number of the same group that is to be cited as suffix, then the azo compound is named as an assembly of identical units (see Subsection C–0.7). The name of the unsubstituted parent compound is preceded by a prefix " azodi- ", and this is preceded by prefixes for the other substituents; the group to be cited as suffix is placed last. The suffix has a locant as low as possible and, if a choice remains, lower locants are next given to the azo links; the higher azo locant is primed, and so are the locants of any other substituents in the ring whose azo locant is primed.

Example:

3',5-Dichloro-2,4'-azodibenzenesulfonic acid

912.3—For other monoazo compounds RN=NR, derived from parent molecules RH that, when unsubstituted, are identical, the molecule RH carrying the larger number of groups to be cited as suffix is chosen as parent and named as substituted by all the remaining groups; the group R in the substituting RN=N— group is named as a radical.

Examples:

p-Phenylazobenzenesulfonic acid $C_6H_5—N=N—C_6H_4—SO_3H$-*p*

p-(2,4-Dihydroxyphenylazo)benzenesulfonic acid

4′-Hydroxy-3′- [(3,4,5-trihydroxyphenyl)azo]butyranilide
(For numbering see Rule **C–825.1**)

912.4—In the name of a monoazo compound $R^1N=NR^2$, where the azo group links groups which, when unsubstituted, are different, a parent molecule R^1H is treated as substituted by a radical $R^2N=N—$, that parent R^1H being chosen which has the larger number of the same group for citation as suffix; or, if the number in each is equal, the more complex group is chosen for the parent molecule R^1H.

Examples:

p-(2-Hydroxy-1-naphthylazo)benzenesulfonic acid

4-Hydroxy-3-[3-methyl-1-(*p*-nitrophenyl)-5-oxo-
2-pyrazolin-4-ylazo]benzamide

1-(1-Sulfo-2-naphthylazo)-2-anthraquinonesulfonic acid

912.5—Bisazo compounds and more complex analogues are named as follows: (*a*) as substituted derivatives of a parent chosen according to the principles of Rule **C–912.4**; or (*b*), when the terminal units are derived from identical molecules and when these units carry the greater number of the groups having priority for citation as suffix, then the whole may be named as an assembly of identical units (see Rule **C–912.2**).

Examples:

(*a*) 2-(7-Phenylazo-2-naphthylazo)anthracene

(*a*) 4-[*p*-(7-Dimethylamino-1-hydroxy-3-sulfo-2-naphthylazo)-
phenylazo]-4'-(*p*-methoxyphenyl)-2,2'-stilbenedisulfonic acid

(*a*) 2,7-Bis(phenylazo)-1,8-naphthalenediol

I.U.P.A.C. A;B & (C)-8

(b) 6,6'-Iminobis[2-(5-carbamoyl-2-hydroxyphenylazo)-1-
naphthol-3-sulfonic acid]

(b) 4,4'-(1,8-Dihydroxy-2,7-naphthylenebisazo)dibenzenesulfonic acid

Rule C–913.   Azoxy compounds

913.1—Azoxy compounds, R—N₂O—R and R¹—N₂O—R², where the
position of the azoxy-oxygen atom is unknown or immaterial, are named in
accordance with the principles of Rule **C–911** or **C–912**, with " azo "
replaced by " azoxy ".

Examples:

Azoxybenzene $C_6H_5—N_2O—C_6H_5$

1-Carboxynaphthalene-2-azoxybenzene
(compare Rule C–911.2)
or 2-Phenylazoxy-1-naphthoic acid
(compare Rule C–912.4)

2,2',4-Trichloroazoxybenzene

913.2—When it is desired to express the position of the azoxy-oxygen atom
in an unsymmetrical compound, a prefix *NNO-* or *ONN-* is used.   (a) When
both of the groups attached to the azoxy radical are cited in the name of the
compound, the prefix *NNO-* specifies that the second of these two groups is
attached directly to NO; and the prefix *ONN-* specifies that the first of
these two groups is attached directly to NO.   (b) When only one parent
compound is cited in the name, the prefixes *ONN-* and *NNO-* specify that

the group carrying respectively primed and unprimed substituents is attached directly to NO. (*c*) To distinguish it from the *NNO*- or *ONN*-compound a substance in which the position of the azoxy-oxygen atom is unknown may be designated by a prefix *NON*-.

Examples:

1-Carboxynaphthalene-2-*NNO*-azoxy-
    benzene (compare Rule **C–911.2**)

or 2-(Phenyl-*ONN*-azoxy)-1-naphthoic
    acid (compare Rule **C–912.4**)

1-Carboxynaphthalene-2-*ONN*-
    azoxybenzene
    (compare Rule **C–911.2**)

or 2-(Phenyl-*NNO*-azoxy)-
    1-naphthoic acid
    (compare Rule **C–912.4**)

2,2′,4-Trichloro-*NNO*-azoxybenzene

2,2′,4-Trichloro-*ONN*-azoxybenzene

2,2′,4-Trichloro-*NON*-azoxybenzene

C–9.2. HYDRAZINES AND THEIR DERIVATIVES

Rule C–921

921.1—Compounds derived by replacement of hydrogen in hydrazine by groups other than acyl (for which see Rule **C—921.5**) are named (*a*) as substitution products of hydrazine or (*b*), if a group having priority for citation as principal group is also present, by use of the prefix " hydrazino- ". The nitrogen atoms are denoted by locants *N* and *N′* or 1 and 2. For method (*a*) numerical locants are the lowest possible, or primes as few as possible; for method (*b*) the nitrogen atom at the point of attachment is unprimed or has locant 1.

I.U.P.A.C. A, B & (C)-8*

Examples:

Phenylhydrazine $\qquad$ NH_2—NH—C_6H_5

N-Methyl-N'-phenylhydrazine $\qquad$ C_6H_5—$\overset{2}{NH}$—$\overset{1}{NH}$—CH_3
or 1-Methyl-2-phenylhydrazine

N,N-Diethyl-N',N'-dimethylhydrazine $\quad$ $(CH_3)_2\overset{2}{N}$—$\overset{1}{N}(C_2H_5)_2$
or 1,1-Diethyl-2,2-dimethylhydrazine

p-Hydrazinophenol

NH_2—NH—⟨benzene ring⟩—OH

p-(N'-Methylhydrazino)benzoic acid
or p-2-Methylhydrazinobenzoic acid $\quad$ CH_3—$\overset{2}{NH}$—$\overset{1}{NH}$—⟨benzene ring⟩—COOH

921.2—Alternatively, for compounds R^1NH—NHR^2 in which the radicals R^1 and R^2 are derived from parent molecules that, when unsubstituted, are identical, names may be formed by methods analogous to those used for azo compounds (Rules C–911 and C–912) but with " hydrazo" in place of " azo ". The names so obtained are available for indication of further substitution on the nitrogen atoms provided that no ambiguity is caused thereby; when ambiguity would be caused thereby, the compounds are named according to Rule C–921.1.

Examples:

Cl—⟨benzene ring⟩—NH—NH—⟨benzene ring⟩—Cl

4,4'-Dichlorohydrazobenzene

⟨naphthalene⟩—NH—NH—⟨naphthalene⟩
SO₃H $\qquad$ SO₃H

1,2'-Hydrazonaphthalene-4',5-disulfonic acid
(compare Rule C–911.1)
or 3,5'-Hydrazodi-(1-naphthalenesulfonic acid)
(compare Rule C–912.2)

Cl—⟨benzene ring⟩—NH—N—⟨benzene ring⟩—Cl
$\qquad$ |
$\qquad$ CH₃

4,4'-Dichloro-N-methylhydrazobenzene

But

N'-Methyl-1,2'-hydrazonaphthalene-4',5-disulfonic acid (compare Rule **C–911**)

or

3-[1-Methyl-2-(5-sulfo-1-naphthyl)hydrazino]-1-naphthalenesulfonic acid (compare Rule **C–912**)

921.3—When one nitrogen atom of a hydrazine group forms part of a ring the compound is named as an amino derivative of the heterocycle.

Example:

1-Aminopiperazine

921.4—A group —NH—NH— attached to a single carbon atom is denoted by a prefix " hydrazi- " when another group having priority for citation as principal group is also present.   See, however, Rule **B–1**.

Example:

Hydraziacetic acid
or   Diaziridine-3-carboxylic acid

921.5—Compounds formed by replacement of hydrogen of a hydrazine group by an acyl group, and their further substitution products, are named by one of the following methods: (*a*) If a substituent having priority for citation as principal group is also present, the compound is named as an acyl-hydrazino compound (compare Rule **C–921.1**); the nitrogen atom attached to the parent carrying the principal group has the locant *N* (unprimed) or 1.   (*b*) If there is no substituent having priority for citation as principal group, the ending " ic " acid " or " -oic acid " of the name of the acid is changed to " -ohydrazide "; or if the ending of the name of the acid is " -carboxylic acid ", this is changed to " -carbohydrazide "; the nitrogen atom attached to the acyl group then has the locant *N* or 1'.

Examples:

$$CH_3—CO$$

p-(*N'*-Acetyl-*N'*-ethylhydrazino)benzoic acid
or *p*-(2-Acetyl-2-ethylhydrazino)benzoic acid

p-(*N*-Acetyl-*N'*-ethylhydrazino)benzoic acid
or *p*-(1-Acetyl-2-ethylhydrazino)benzoic acid

$$\overset{3}{CH_3}—\overset{2}{CH_2}—\overset{1}{CO}—\overset{1'}{NH}—\overset{2'}{NH}—CH_2—CH_3$$

N'-Ethylpropionohydrazide
or 2'-Ethylpropionohydrazide

921.6—Salts of hydrazines are named as derivatives of hydrazinium(1+) or hydrazinium(2+), according to whether one or both nitrogen atoms bear a charge (see Rule **3.17** of I.U.P.A.C. *Nomenclature of Inorganic Chemistry*, 1970*). If only one nitrogen atom bears a charge and it is known which, then that atom has the locant *N* (unprimed) or 1.

Examples:

N,N-Dimethylhydrazinium chloride
or 1,1-Dimethylhydrazinium chloride

$$NH_2—\overset{+}{N}H(CH_3)_2\ Cl^-$$

N,N'-Dimethylhydrazinium sulfate
or 1,2-Dimethylhydrazinium sulfate

$$CH_3—\overset{+}{N}H_2—\overset{+}{N}H_2—CH_3$$
$$SO_4^{2-}$$

* I.U.P.A.C. *Nomenclature of Inorganic Chemistry*, 1970 (Butterworths Scientific Publications, London, 1971).

921.7—Acyclic polyhydrazines may be named by replacement nomenclature (see Subsection C–0.6).

Example:

$$\underset{9}{CH_2}-\underset{8}{NH}-\underset{7}{NH}-\underset{6}{CH_2}-\underset{5}{CH_2}-\underset{4}{NH}-\underset{3}{NH}-\underset{2}{CH_2}-\underset{1}{CH_2}-OH$$

$$\underset{10}{CH_2}-\underset{11}{O}-\underset{12}{CH_2}-\underset{13}{CH_2}-\underset{14}{NH}-\underset{15}{NH}-\underset{16}{CH_2}-\underset{17}{CH_2}-OH$$

11-Oxa-3,4,7,8,14,15-hexaazaheptadecane-1,17-diol

Rule C-922

922.1—Compounds $RCH{=}N{-}NH_2$ and $R^1R^2C{=}N{-}NH_2$ are named (*a*) by placing the word " hydrazone " after the name of the corresponding aldehyde or ketone, or (*b*) by means of a prefix " hydrazono- " before the name of the compound CH_3R or $CH_2R^1R^2$ if this compound has also a substituent to be cited as principal group. Generically these compounds are called " hydrazones ".

Examples:

(*a*) Acetaldehyde hydrazone $CH_3{-}CH{=}N{-}NH_2$

(*b*) 2-Hydrazono-1-cyclohexanecarboxylic acid

922.2—Compounds $X{=}N{-}NHR$ or $X{=}N{-}NR^1R^2$ are named by one of the following methods: (*a*) as substituted hydrazones if no other group is present having priority for citation as principal group; (*b*) as hydrazino derivatives of the compound RH, substituted by a bivalent radical $=X$, if the group R has a substituent having priority for citation as principal group; or (*c*) as substituted hydrazono derivatives of the compound XH_2 if the group X has a substituent having priority for citation as principal group.

Examples:

(*a*) Butanone dimethylhydrazone $\begin{array}{c}CH_3{-}CH_2\\ \\CH_3\end{array}\!\!\!\!\!\!>C{=}N{-}N(CH_3)_2$

(*b*) *p*-Isopropylidenehydrazino-
benzoic acid

$$(CH_3)_2C=N-NH-\!\!\!\bigcirc\!\!\!-COOH$$

(*c*) 4-Dimethylhydrazono-1-cyclo-
hexanecarboxylic acid

$$(CH_3)_2N-N=\!\!\!\bigcirc\!\!\!\langle^{H}_{COOH}$$

Rule C-923

923.1—Compounds X=N—N=X are named by one of the following methods: (*a*) by adding the word " azine " after the name of the corresponding aldehyde or ketone when no other group is present having priority for citation as principal group; or (*b*) by use of the prefix " azino- " for the radical =N—N= and the nomenclature for assemblies of identical units (see Subsection C-0.7) when the group X contains a substituent having priority for citation as principal group.

Examples:

Acetone azine

$$(CH_3)_2C=N-N=C(CH_3)_2$$

2,2′-Azinodi-1-cyclohexane-
carboxylic acid

$$\bigcirc\!\!\!=\!\!N-\!\!\!-\!\!N\!\!=\!\!\!\bigcirc$$
$$\text{COOH HOOC}$$

923.2—Compounds X=N—N=Y (where X is different from Y) are named as hydrazones NH$_2$—N=Y substituted by an alkylidene, cycloalkylidene, or bivalent heterocyclic radical X=. The group Y is that one of the pair X, Y that carries the substituent with priority for citation as suffix, or, if that is not decisive, is the senior.

Example:

$$(CH_3)_2C=N-N=\!\!\!\bigcirc\!\!\!\langle^{H}_{COOH}$$

4-Isopropylidenehydrazono-2,5-cyclohexadiene-1-carboxylic acid

C–9.3. DIAZONIUM AND RELATED GROUPS

Rule C-931

931.1—Compounds RN$_2^+$X$^-$ are named by adding the suffix " -diazonium " to the name of the parent substance RH, the whole being followed by the name of the ion X$^-$.

Examples:

Benzenediazonium chloride $\qquad$ C_6H_5—N_2^+ Cl^-

8-Hydroxy-2-naphthalenediazonium tetrafluoroborate

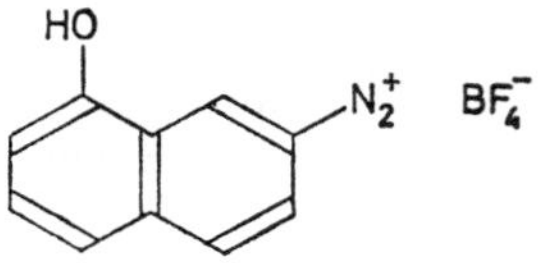

931.2—Compounds RN=NX are named by adding, after the name of the parent compound RH, the syllables " diazo " joined to the designation of the atom or group X.

Note: These compounds are distinguished from azo compounds by the fact that the group X is not joined to the nitrogen atom by a link from a carbon atom (except for the cyanide).

Examples:

Benzenediazohydroxide $\qquad$ C_6H_5—N=N—OH
Benzenediazocyanide $\qquad$ C_6H_5—N=N—CN
Sodium benzenediazosulfonate $\qquad$ C_6H_5—N=N—SO_3Na

931.3—Compounds RN=N—OM, where M is a metal, are named metal diazoates.

Example:

Sodium benzenediazoate $\qquad$ C_6H_5—N=N—ONa

931.4—Compounds containing a group N_2 attached by one atom to carbon are named by adding a prefix " diazo- " for this substituent to the name of the parent compound.

Examples:

Diazomethane $\qquad$ CH_2N_2
Ethyl diazoacetate $\qquad$ N_2CH—CO—OC_2H_5
ω-Diazoacetophenone $\qquad$ N_2CH—CO—C_6H_5

931.5—Compounds in which a group —N=N— forms part of a ring are named as (a) cyclic compounds (see Section B) or (b), if the group creates a spiro ring system, by use of the prefix " -azi ".

Examples:

(a) Ethyl 3*H*-diazirine-3-carboxylate

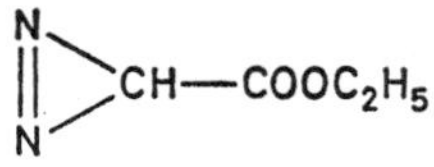

290

(b) 3-Azi-5β-androstan-17-one

C–9.4. GROUPS CONTAINING THREE OR MORE CONTIGUOUS NITROGEN ATOMS

Rule C–941

941.1—Compounds R—N_3 are named (a), in radicofunctional nomenclature, by placing the word " azide " after the name of the radical R or (b), in substitutive nomenclature, by adding the prefix " azido- " to the name of the compound RH.

Examples:

(a) Phenyl azide (radicofunctional nomenclature) C_6H_5—N_3

or (b) Azidobenzene (substitutive nomenclature)

(a) Benzoyl azide C_6H_5—CO—N_3

(b) 1-Azido-2-naphthalenesulfonic acid

Rule C–942

942.1—Compounds derived by substitution from NH_2—NH—NH_2, NH_2—N=NH, NH_2—NH—NH—NH_2, NH_2—NH—N=NH, NH=N—N=NH, or NH_2—NH—NH—NH—NH_2, etc., are named as substitution products of triazane, triazene, tetrazane, 1-tetrazene, 1,3-tetrazadiene, pentazane, etc. The chain of nitrogen atoms is numbered consecutively from one end to the other. Double bonds are designated, and assigned lowest available numbers, as in the aliphatic series (Section A), and they have priority over substituents for lowest numbers. (For naming triazanes, etc., by the replacement method for open chains, see Subsection C–0.6).

Examples:

1-Methyltriazane	$\overset{3}{NH_2}$—$\overset{2}{NH}$—$\overset{1}{NH}$—CH_3
1-Methyl-3-propyltriazane	$\overset{3}{C_3H_7}$—NH—$\overset{2}{NH}$—$\overset{1}{NH}$—CH_3
3-Methyltriazene	$\overset{3}{CH_3}$—$\overset{2}{NH}$—$\overset{1}{N}$=NH
3-Methyl-1-tetrazene	$\overset{4}{NH_2}$—$\overset{3}{N}(CH_3)$—$\overset{2}{N}$=$\overset{1}{N}H$

942.2—Alternatively, compounds containing the same radical R at each end of an —N=N—NH— group may be named by attaching the prefix " diazoamino- " to the name of the compound RH, the whole preceded by locants if necessary.

Examples:

Diazoaminobenzene C_6H_5—NH—N=N—C_6H_5

2,2′-Diazoaminonaphthalene

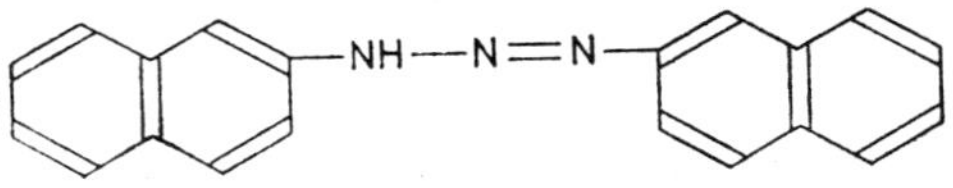

N-Methyldiazoaminobenzene C_6H_5—N—N=N—C_6H_5
                                                                    |
                                                                    CH_3

942.3—When the compounds described in Rule C–942.1 are converted into radicals by loss of a terminal hydrogen atom the terminal " -ne " of their names is changed to " -no "; the point of attachment of the radical then has the locant 1.

Examples:

p-2-Triazenobenzoic acid

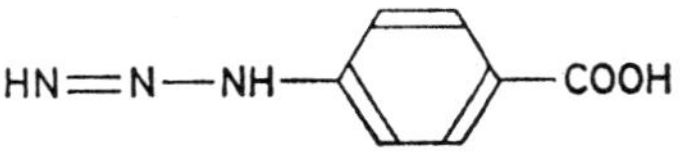

HN=N—NH—〈 〉—COOH

Ethyl tetrazanoacetate H_2N—NH—NH—NH—CH_2—CO—OC_2H_5

C–9.5. COMPOUNDS CONTAINING A N=C—N OR N=C=N GROUP

Rule C–951. Amidines

951.1—Compounds RC(=NH)—NH_2 are termed generically " amidines " or, if the name of the corresponding acid ends in carboxylic acid, may be termed " carboxamidines ".

951.2—Names of amidines are derived from the names of the corresponding acids by replacing " -oic acid " or " -ic acid " by " -amidine " or " -carboxylic acid " by " -carboxamidine ".   (For sulfinamidines see Rule C–641.9).

Examples:

Hexanamidine C_5H_{11}—C(=NH)—NH_2
Cyclohexanecarboxamidine C_6H_{11}—C(=NH)—NH_2
Acetamidine CH_3—C(=NH)—NH_2

951.3—N-Substituted amidines are named by prefixing the name of the appropriate radical to the name of the unsubstituted amidine, with N^1 or N^2 as locant, N^1 referring to the NH_2 and N^2 to the $=NH$ group. If the position of the double bond is not known the locants N and N' are used.

Examples:

N^2-Methyl-N^1,N^1-diphenylbenzamidine

$$C_6H_5—C(=N—CH_3)—N(C_6H_5)_2$$

N^1-Ethyl-N^2-methylbenzamidine $C_6H_5—C(=N—CH_3)—NH—C_2H_5$

N^2-Ethyl-N^1-methylbenzamidine $C_6H_5—C(=N—C_2H_5)—NH—CH_3$

N-Ethyl-N'-methylbenzamidine
$$C_6H_5—C\begin{Bmatrix} N—C_2H_5 \\ N—CH_3 \end{Bmatrix}H$$

951.4—The systematic prefix name for the radical $—C(=NH)—NH_2$ is " carbamimidoyl- " but, because of precedent, the name " amidino- " is retained.*

Example:

p-Amidinobenzoic acid $NH_2—C(=NH)—$⟨benzene ring⟩$—COOH$

Note: The isomeric radicals $HN=CH—NH—$ and $H_2N—CH=N—$ are " iminomethylamino- ", and " aminomethyleneamino- ", respectively.

Example:
Iminomethylaminoacetic acid $HN=CH—NH—CH_2—COOH$

951.5—The compound $[H_2N(HN=)C—S—]_2$ and its substitution products have the generic name " formamidine disulfides ". Individual compounds are named as derivatives of assemblies of identical units (see Subsection C–0.7); locants are assigned as in the example.

Example:

N^1-Ethyl-N^3-methyl-α,α'-dithio-bisformamidine

$$\overset{\alpha}{S}—\overset{2}{C}(=NH)—NH—\overset{1}{CH_2}—CH_3$$
$$\underset{\alpha'}{S}—\underset{4}{C}(=NH)—\underset{3}{NH}—CH_3$$

* The name " guanyl- " is not recommended.

Rule C–952.   Amide oximes

952.1—Names of compounds $RC(NH_2)\!=\!NOH$ are derived from the names of the corresponding acids by changing the ending " -ic acid " or " -oic acid " to " -amide oxime ", or " -carboxylic acid " to " -carboxamide oxime ". *O*- and *N*-Substituents are identified by *O* and *N* locants, respectively.

Examples:

N-Methylacetamide oxime $CH_3\!-\!C(NH\!-\!CH_3)\!=\!NOH$

Imidazole-2-carboxamide
O-ethyloxime

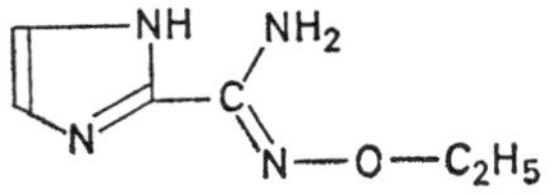

Rule C–953.   Amidrazones

953.1—Compounds $RC(NH_2)\!=\!N\!-\!NH_2$ or $RC(\!=\!NH)\!-\!NH\!-\!NH_2$ are termed generically " amidrazones " or, if the name of the corresponding acid ends in " -carboxylic acid ", are termed " carboxamidrazones ".

953.2—Names of individual compounds $RC(NH_2)\!=\!N\!-\!NH_2$ or $RC(\!=\!NH)\!-\!NH\!-\!NH_2$ are derived from the systematic names of the corresponding acids by replacing " -oic acid " by " -amide hydrazone " or " -ohydrazide imide ", respectively, by replacing " -carboxylic acid " by " -carboxamide hydrazone " or " -carbohydrazide imide ", respectively, or by replacing " -ic acid " by " -amide hydrazone " or " -ohydrazide imide ", respectively. If the position of the double bond is unknown the compounds are named similarly but with the ending " -amidrazone ".

Examples:

Benzamide hydrazone $C_6H_5\!-\!C(NH_2)\!=\!N\!-\!NH_2$
Benzohydrazide imide $C_6H_5\!-\!C(\!=\!NH)\!-\!NH\!-\!NH_2$

Benzamidrazone: the name for a compound which may have one or other of the above two structures.

953.3—For *N*-substituted amidrazones, where the position of the double bond is known, the substituents receive locants laid down in preceding rules for amides, imides, hydrazides, and hydrazones. When the position of the double bond is unknown, primes are assigned in accordance with the system $N''\!-\!CR\!-\!N'\!-\!N$.

Examples:

$$C_2H_5\!-\!N\!=\!C\!-\!N(CH_3)\!-\!NH\!-\!C_6H_5$$
$$\vert$$
$$C_6H_5$$

N^1-Methyl-N^2-phenylbenzohydrazide ethylimide

N-Phenylbenzamidrazone is the name used when it is uncertain whether a compound has the structure $HN\!=\!C(C_6H_5)\!-\!NH\!-\!NH\!-\!C_6H_5$ or one of its double-bond isomers.

Rule C-954. Hydrazidines

954.1—Compounds $RC(NH-NH_2)=N-NH_2$ are termed generically " hydrazidines " or, if the name of the corresponding acid ends in " -carboxylic acid ", are termed " -carbohydrazidines ".

954.2—Names of individual compounds $RC(NH-NH_2)=N-NH_2$ are derived from the name of the corresponding acids by changing the ending " -ic acid " or " -oic acid " to " -ohydrazide hydrazone ", or " -carboxylic acid " to " -carbohydrazide hydrazone ". The nitrogen atoms are numbered as shown in the example.

Examples:

N^2, N^4-Dimethyl-2-thenohydrazide hydrazone

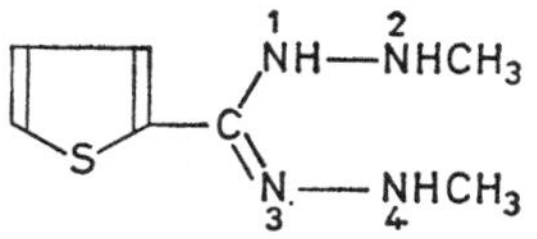

N^2, N^4-Dibenzylidene-4-thiazole-carbohydrazide hydrazone

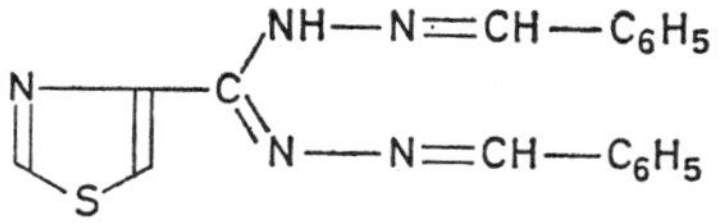

Rule C-955. Formazans

955.1—The compound $\overset{5}{N}H_2-\overset{4}{N}=\overset{3}{C}H-\overset{2}{N}=\overset{1}{N}H$ is named " formazan " and its derivatives are named as substitution products with numerical locants as indicated. If the positions of the double bonds are unknown, primes are used in accordance with the system $N'''-N''-\overset{3}{C}-N'-N$.

Examples:

1,3-Diphenylformazan $\overset{5}{N}H_2-\overset{4}{N}=\overset{3}{C}(C_6H_5)-\overset{2}{N}=\overset{1}{N}C_6H_5$

3,5-Diphenylformazan $C_6H_5-NH-N=C(C_6H_5)-N=NH$

(If there is uncertainty whether the first or the second of the above structures applies, the name is $N,3$-diphenylformazan.)

1,5-Diphenylformazan $C_6H_5-NH-N=CH-N=N-C_6H_5$
(preferred to Formazyl)

Rule C-956. Carbodiimides

956.1—The hypothetical compound $NH=C=NH$ is named " carbodiimide ". Its derivatives are named as substitution products thereof.

Example:

Dicyclohexylcarbodiimide $C_6H_{11}-N=C=N-C_6H_{11}$

C–9.6.   COMPOUNDS CONTAINING A N—C=N GROUP
$$\underset{N}{|}$$

Rule C–961

961.1—The compound $H_2\overset{3}{N}$—$\overset{2}{C}$(=NH)—$\overset{1}{N}H_2$ is named " guanidine " and numbered as shown.

961.2—Derivatives of guanidine are named (*a*) as substitution products of guanidine or (*b*), if the substituent carries a group to be cited as suffix, by use of the prefix " guanidino- " for the radical NH_2—C(=NH)—NH—. Locants 1 and 3 are used for substituents of an NH_2 group, and 2 for a substituent of the =NH group. If the position of the double bond is not proved, primes are used (where there is a choice, as few primes as possible). The group $(NH_2)_2$C=N— is named " diaminomethyleneamino- ".

Examples:

$$(CH_3)_2\overset{3}{N}—\overset{2}{C}(=\overset{}{N}—C_6H_5)—\overset{1}{N}(CH_3)_2$$

Tetramethyl-2-phenylguanidine

1,1,2-Trimethylguanidine $\overset{3}{N}H_2$—$\overset{2}{C}$(=$\overset{}{N}$—CH_3)—$\overset{1}{N}$(CH_3)_2

(When the position of the double bond is uncertain, the last compound is named *N,N,N'*-trimethylguanidine.)

1,1,3-Trimethylguanidine CH_3—$\overset{3}{N}H$—$\overset{2}{C}$(=$\overset{}{N}H$)—$\overset{1}{N}$(CH_3)_2

p-(1,3-Dimethyl-3-phenylguanidino)benzoic acid

$$(NH_2)_2C=N—CH_2—COOH$$

(Diaminomethyleneamino)acetic acid

961.3—Salts formed from guanidine and its derivatives are named **as** guanidinium salts. Locants for substituents are assigned as in Rule **C–961.2.**

Example:

$$[(C_2H_5)_2N—C(=\overset{+}{N}H_2)—NH—C_2H_5]\ Cl^-$$
1,1,3-Triethylguanidinium chloride

Rule C–962

962.1—The following names for condensed products from guanidine are retained and may be used for derivatives. Substituents have numerical locants as indicated.

Biguanide

$$\overset{5}{NH_2}—C—\overset{3}{NH}—C—\overset{1}{NH_2}$$
$$\underset{4}{\overset{\|}{NH}}\qquad\underset{2}{\overset{\|}{NH}}$$

Triguanide

$$\overset{7}{NH_2}—C—\overset{5}{NH}—C—\overset{3}{NH}—C—\overset{1}{NH_2}$$
$$\underset{6}{\overset{\|}{NH}}\qquad\underset{4}{\overset{\|}{NH}}\qquad\underset{2}{\overset{\|}{NH}}$$

etc.

Example:

2-Ethyl-1,1-diphenylbiguanide

$$NH_2—C—NH—C—N(C_6H_5)_2$$
$$\overset{\|}{NH}\qquad\overset{\|}{N}—C_2H_5$$

C–9.7. COMPOUNDS CONTAINING A N—CO—N OR RELATED GROUP

UREA AND ITS DERIVATIVES

Rule C–971

971.1—The compound $\overset{3}{NH_2}—\overset{2}{CO}—\overset{1}{NH_2}$ is named " urea " and numbered as shown.

971.2—Derivatives of urea formed by replacement of hydrogen are named (*a*) as substitution products of urea or (*b*), if the substituent carries also a group to be cited as principal group, by use of the prefix " ureido- " for the group $NH_2—CO—NH—$. If necessary, locants N and N' or 1 and 3, respectively, are added; the point of attachment of a ureido radical is denoted N or 1.

Examples:

 (*a*) *N*,*N*′-Dimethylurea $CH_3—NH—CO—NH—CH_3$
or 1,3-Dimethylurea

 (*a*) *N*,*N*-Diethylurea $NH_2—CO—N(C_2H_5)_2$
or 1,1-Diethylurea

 (*a*) *N*-Acetyl-*N*-phenylurea $NH_2—CO—N(C_6H_5)—CO—CH_3$
or 1-Acetyl-1-phenylurea

(*a*) Isopropylideneurea $NH_2—CO—N=C(CH_3)_2$

(*b*)   2-(*N'*-Methylureido)-1-naphthoic acid
or   2-(3-Methylureido)-1-naphthoic acid

The following trivial names are retained for urea derivatives (numbered as shown) and may be associated with prefixes denoting substituents, derivatives being named in the usual way:

Allophanic acid

$$\overset{4}{N}H_2—\overset{3}{C}O—\overset{2}{N}H—\overset{1}{C}OOH$$

Hydantoic acid

$$\overset{5}{N}H_2—\overset{4}{C}O—\overset{3}{N}H—\overset{2}{C}H_2—\overset{1}{C}OOH$$

Allophanoyl

$$NH_2—CO—NH—CO—$$

971.3—The bivalent radical —NH—CO—NH— is named " ureylene " and is used in nomenclature for assemblies of identical units (see Subsection C–0.7) when the terminal nitrogen atoms are attached to radicals carrying the same group to be cited as a suffix, but it is not used for compounds in which the —NH—CO—NH— group forms part of a ring (for example, not for barbituric acid).

Example:

4,4'-Ureylenedi-1-naphthalenesulfonic acid

Rule C–972

972.1—The compound $\overset{3}{H}N=\overset{2}{C}(OH)—\overset{1}{N}H_2$ is named " isourea " and this name is used as a basis for naming derivatives, together with the locants 1, 2, and 3 as shown, when no other group is present that has priority for citation as principal group. When the position of the double bond is unknown, *N*, *N'*, and *O* are used as locants.

Examples:

2-Ethyl-1,1-diphenylisourea

$$HN=\overset{\overset{\displaystyle O—C_2H_5}{|}}{C}—N(C_6H_5)_2$$

2-Ethyl-1-phenylisourea

$$\underset{\displaystyle HN\!=\!\overset{\textstyle |}{C}\!-\!NH\!-\!C_6H_5}{\overset{\textstyle O\!-\!C_2H_5}{}}$$

2-Ethyl-3-phenylisourea

$$\underset{\displaystyle C_6H_5\!-\!N\!=\!\overset{\textstyle |}{C}\!-\!NH_2}{\overset{\textstyle O\!-\!C_2H_5}{}}$$

If the position of the double bond is uncertain, so that choice cannot be made between the last two structures, then the name is: *O*-Ethyl-*N*-phenylisourea.

972.2—The radical $NH\!=\!C(OH)\!-\!NH\!-$ is named " 1-isoureido- ", and $NH_2\!-\!C(OH)\!=\!N\!-$ is named " 3-isoureido- ". " Isoureido- ", without a numerical prefix, denotes a radical in which the position of the double bond is not proved. These names are used as a basis for designation of derivatives in which another group is to be cited as principal group.

Example:

4-(2-Ethyl-1,1-dimethyl-3-isoureido)-1-naphthol

Rule C–973

973.1—Quaternary derivatives of urea (isourea) are named as " uronium " salts.

Example:

2-(2-Naphthyl)uronium picrate

Rule C–974

974.1—Compounds formed from urea, isourea, or their derivatives by replacement of oxygen by sulfur, selenium, or tellurium are named by attaching a prefix " thio- ", " seleno- ", or " telluro- ", respectively, directly before urea, ureido-, ureylene-, or uronium. Locants are used in the same way as for urea and its derivatives; when necessary, a locant *S*, *Se*, or *Te* is used to assign substituents to these atoms.

Examples:

Thiourea $\qquad$ NH_2—CS—NH_2

N-Methylselenourea $\qquad$ NH_2—CSe—NH—CH_3
or   1-Methyl-2-selenourea

1-Methyl-2-propylisothiourea
$$\overset{2}{S}\text{—}C_3H_7$$
$$HN=\overset{\mid}{\underset{3}{C}}\text{—}\underset{1}{NH}\text{—}CH_3$$

3-Methyl-2-propylisothiourea
$$\overset{2}{S}\text{—}C_3H_7$$
$$CH_3\text{—}\underset{3}{N}=\overset{\mid}{\underset{1}{C}}\text{—}NH_2$$

When a choice between the two preceding structures cannot be made, the name is: *N*-Methyl-*S*-propylisothiourea.

$$\left[\begin{array}{ccc} \overset{S\text{—}C_6H_5}{\mid} & & \overset{S\text{—}C_6H_5}{\mid} \\ H_2N=C\text{—}N(C_2H_5)_2 & \longleftrightarrow & H_2N\text{—}C=N(C_2H_5)_2 \\ & \begin{array}{c} S\text{—}C_6H_5 \\ \| \\ H_2N\text{—}C\text{—}N(C_2H_5)_2 \end{array} & \end{array}\right]^+$$

For the above mesomeric cation the name is:

N,N-Diethyl-S-phenylthiouronium

Rule C-975

975.1—The following names for condensed products (numbered as shown) are retained and may be used with prefixes and numerical locants for assignments of substituents.

Biuret
$$\underset{5}{NH_2}\text{—}\underset{4}{CO}\text{—}\underset{3}{NH}\text{—}\underset{2}{CO}\text{—}\underset{1}{NH_2}$$

Thiobiuret
$$\underset{5}{NH_2}\text{—}\underset{4}{CO}\text{—}\underset{3}{NH}\text{—}\underset{2}{CS}\text{—}\underset{1}{NH_2}$$

Dithiobiuret
$$\underset{5}{NH_2}\text{—}\underset{4}{CS}\text{—}\underset{3}{NH}\text{—}\underset{2}{CS}\text{—}\underset{1}{NH_2}$$

Triuret
$$\underset{7}{NH_2}\text{—}\underset{6}{CO}\text{—}\underset{5}{NH}\text{—}\underset{4}{CO}\text{—}\underset{3}{NH}\text{—}\underset{2}{CO}\text{—}\underset{1}{NH_2}$$

4-Thiotriuret
$$\underset{7}{NH_2}\text{—}\underset{6}{CO}\text{—}\underset{5}{NH}\text{—}\underset{4}{CS}\text{—}\underset{3}{NH}\text{—}\underset{2}{CO}\text{—}\underset{1}{NH_2}$$

etc.

Example:

1-Methyl-2-thiobiuret $\qquad$ NH_2—CO—NH—CS—NH—CH_3

C-9.8. COMPOUNDS CONTAINING A N—CO—N—N OR MORE COMPLEX GROUP

Rule C-981

981.1—The following compounds are named and numbered as shown:

Semicarbazide

$$\overset{4}{N}H_2\overset{3}{-}CO\overset{2}{-}NH\overset{1}{-}NH_2$$

Carbonohydrazide (preferred to Carbohydrazide or Carbazide)

$$\overset{5}{N}H_2\overset{4}{-}NH\overset{3}{-}CO\overset{2}{-}NH\overset{1}{-}NH_2$$

Carbazone*

$$\overset{5}{N}H{=}\overset{4}{N}\overset{3}{-}CO\overset{2}{-}NH\overset{1}{-}NH_2$$

Carbodiazone*

$$\overset{5}{N}H{=}\overset{4}{N}\overset{3}{-}CO\overset{2}{-}N{=}\overset{1}{N}H$$

981.2—Radicals formed by removal of one hydrogen atom from position 1 of the compounds listed in Rule C–981.1 are named by changing the ending " -ide " to " -ido ", or " -one " to " -ono "; these names are used when a substituent is also present that has priority for citation as principal group.

Examples:

Semicarbazido

$$\overset{4}{N}H_2\overset{3}{-}CO\overset{2}{-}NH\overset{1}{-}NH-$$

Carbazono

$$\overset{5}{N}H{=}\overset{4}{N}\overset{3}{-}CO\overset{2}{-}NH\overset{1}{-}NH-$$

$$NH_2-CO-NH-NH-\left\langle\bigcirc\right\rangle-CO-O-C_2H_5$$

Ethyl 4-semicarbazido-1-cyclohexanecarboxylate

Rule C-982

982.1—Derivatives of the types NH_2—CO—NH—N=CHR and NH_2—CO—NH—N=CR^1R^2 are named as follows: (a) by adding the word " semicarbazone " after the name of the aldehyde RCHO or ketone COR^1R^2; (b), if another substituent having priority for citation as principal group is also present, by use of the prefix " semicarbazono- " for the radical NH_2—CO—NH—N=; or (c) as a semicarbazide derivative with the prefix denoting the bivalent radical =CHR or =CR^1R^2 (alkylidene, etc.). Other substituents on nitrogen are indicated by prefixes and numerical locants.

Examples:

$$(C_6H_5)_2N-CO-NH-N{=}C(C_6H_5)_2$$

(a) Benzophenone 4,4-diphenylsemicarbazone

---

* Only derivatives are known.

301

$$NH_2—CO—NH—N{=}\underset{}{\bigcirc}—COOH$$

(*b*) 4-Semicarbazono-1-cyclohexanecarboxylic acid

$$CH_3—NH—CO—N(CH_3)—N{=}C(CH_3)_2$$

(*c*) 1-Isopropylidene-2,4-dimethylsemicarbazide

982.2—Derivatives of semicarbazide, carbonohydrazide, carbazone, and carbodiazone are named as substitution products of these compounds except as provided by Rules **C-981.2** and **C-982.1**.

Examples:

$$C_6H_5—NH—NH—CO—NH—NH—C_6H_5$$
1,5-Diphenylcarbonohydrazide

$$\overset{5}{NH}—\overset{4}{NH}—\overset{3}{CO}—\overset{2}{NH}—\overset{1}{N}{=}CH—CH_3$$

1-Ethylidene-5-(2-naphthyl)carbonohydrazide

Rule C-983

983.1—The names "isosemicarbazide" and "isocarbonohydrazide" are used to cover compounds $\overset{4}{N}H{=}\overset{3}{C}(OH)—\overset{2}{N}H—\overset{1}{N}H_2$ and $\overset{1}{N}H_2—\overset{2}{N}H—\overset{3}{C}(OH){=}\overset{4}{N}—\overset{5}{N}H_2$, respectively, and their tautomeric forms. Radicals derived by loss of one hydrogen atom from position 1 are termed "isosemicarbazido-" and "isocarbonohydrazido-", respectively.

Example:

$$NH_2—N{=}C(OCH_3)—NH—CH_3 \text{ and its tautomers}$$

3,4-Dimethylisosemicarbazide

Rule C-984

984.1—The name "carbazic acid" is retained for the compound $\overset{3}{N}H_2—\overset{2}{N}H—\overset{1}{C}OOH$, with the numbering shown. Derivatives are named in the usual way.

Examples:

Ethyl carbazate	$NH_2—NH—COOC_2H_5$
Carbazoyl	$NH_2—NH—CO—$
3-Ethylcarbazic acid	$C_2H_5—NH—NH—COOH$
Note, however, 2-Carboxyhydrazino	$HOOC—NH—NH—$

Rule C-985

985.1—(*a*) The following sulfur analogues of the preceding compounds are named and numbered as shown:

Thiosemicarbazide
$$\overset{4}{N}H_2-\overset{3}{C}S-\overset{2}{N}H-\overset{1}{N}H_2$$

Thiocarbonohydrazide
$$\overset{5}{N}H_2-\overset{4}{N}H-\overset{3}{C}S-\overset{2}{N}H-\overset{1}{N}H_2$$

Thiocarbazone*
$$\overset{5}{N}H=\overset{4}{N}-\overset{3}{C}S-\overset{2}{N}H-\overset{1}{N}H_2$$

Thiocarbodiazone*
$$\overset{5}{N}H=\overset{4}{N}-\overset{3}{C}S-\overset{2}{N}=\overset{1}{N}H$$

(*b*) The derived radicals are named by changing the final " e " to " o "

(*c*) Selenium analogues are named similarly, by use of " seleno- " in place of " thio- ".

(*d*) Numbering of these compounds and radicals, and formation of names of derivatives, are carried out as indicated for the oxygen compounds.

Examples:

1-Ethyl-3-thiocarbonohydrazide
$$NH_2-NH-CS-NH-NH-C_2H_5$$

Acetophenone selenosemicarbazone
$$NH_2-CSe-NH-N=C(CH_3)-C_6H_5$$

* Only derivatives are known.

LIST OF RADICAL NAMES

This list is compiled from Sections A, B, and C of the Rules. It includes, besides organic radicals, substituents such as halogens, oxo, amino, nitro, whose names are laid down in the Rules of these Sections.

Composite radicals will usually be found under methane, methyl, acetyl, benzene, phenyl, or cyclohexane (for example, methylthio, acetylimino, cyclohexane-carbonyl), but some other important examples as well as exceptions to the systematic rules are also listed.

The Rules cited in the third column are the principal references but are not to be considered as necessarily an exhaustive list.

Except at the point(s) of attachment, decimal points are used for brevity, in place of lines to denote single bonds, and colons to denote double bonds.

Radical name	Formula	Based on rule no.
Aceanthrenyl	$C_{16}H_{11}$—	A–23.1, A–24.2
Aceanthrylenyl	$C_{16}H_9$—	A–21.1, A–24.2
Acenaphthenyl	$C_{12}H_9$—	A–23.1, A–24.2
Acenaphthenylene	—$C_{12}H_8$—	A–21.3, A–24.4
Acenaphthenylidene	$C_{12}H_8$=	A–23.1, A–24.3
Acenaphthylenyl	$C_{12}H_7$—	A–21.1, A–24.2
Acephenanthrenyl	$C_{16}H_{11}$—	A–23.1, A–24.2
Acephenanthrylenyl	$C_{16}H_9$—	A–21.1, A–24.2
Acetamido	$CH_3 \cdot CO \cdot NH$—	C–823.1
Acetimidoyl	$CH_3 \cdot C(:NH)$—	C–451.2
Acetoacetyl	$CH_3 \cdot CO \cdot CH_2 \cdot CO$—	C–416.3
Acetohydrazonoyl	$CH_3 \cdot C(:N \cdot NH_2)$—	C–451.2
Acetohydroximoyl	$CH_3 \cdot C(:N \cdot OH)$—	C–451.2
Acetonyl	$CH_3 \cdot CO \cdot CH_2$—	C–318.1
Acetonylidene	$CH_3 \cdot CO \cdot CH$=	C–318.1
Acetoxy	$CH_3 \cdot CO \cdot O$—	C–463.3
Acetyl (*preferred to ethanoyl*)	$CH_3 \cdot CO$—	C–404.1
Acetylamino	$CH_3 \cdot CO \cdot NH$—	C–823.1
Acetylhydrazino (*N'- or 2- shown*)	$CH_3 \cdot CO \cdot NH \cdot NH$—	C–921.5
Acetylimino	$CH_3 \cdot CO \cdot N$=	C–815.3
Acridinyl	$NC_{13}H_8$—	B–2.11, B–5.11
Acryloyl (*preferred to propenoyl*)	$CH_2:CH \cdot CO$—	C–404.1
Adipoyl (*preferred to hexanedioyl*)	—$CO \cdot [CH_2]_4 \cdot CO$—	C–404.1
Alanyl	$CH_3 \cdot CH(NH_2) \cdot CO$—	C–421.1
β-Alanyl	$H_2N \cdot CH_2 \cdot CH_2 \cdot CO$—	C–421.1
Allophanoyl	$NH_2 \cdot CO \cdot NH \cdot CO$—	C–971.2
Allyl (*preferred to 2-propenyl*)	$CH_2:CH \cdot CH_2$—	A–3.5
Allylidene	$CH_2:CH \cdot CH$=	A–3.5, A–4.1
Allyloxy	$CH_2:CH \cdot CH_2 \cdot O$—	C–205.1
Amidino (*replacing guanyl*)	$H_2N \cdot C(:NH)$—	C–951.4
Amino	NH_2—	C–811.3, C–812.2
Aminomethyleneamino	$H_2N \cdot CH:N$—	C–951.4
Aminooxy	$H_2N \cdot O$—	C–841.2
Ammonio	^+H_3N—	C–82.1, C–85, C–87.1, C–816.3
amyl, see Pentyl		
Anilino	$C_6H_5 \cdot NH$—	C–811.4
Anisidino (*o-, m-, or p-*)	$CH_3O \cdot C_6H_4 \cdot NH$—	C–811.4
Anisoyl (*o-, m-, or p-*) (*preferred to methoxybenzoyl*)	$CH_3O \cdot C_6H_4 \cdot CO$—	C–411.1

Radical name	Formula	Based on rule no.
Anthraniloyl	o-$NH_2 \cdot C_6H_4 \cdot CO-$	C–404.1, C–421.4
Anthryl	$C_{14}H_9-$	A–24.2
Anthrylene	$-C_{14}H_8-$	A–24.4
Arginyl	$NH_2 \cdot C(:NH) \cdot NH \cdot [CH_2]_3 \cdot CH(NH_2) \cdot CO-$	
		C–421.1
Asparaginyl	$NH_2 \cdot CO \cdot CH_2 \cdot CH(NH_2) \cdot CO-$	C–421.3
Aspartoyl	$-CO \cdot CH_2 \cdot CH(NH_2) \cdot CO-$	C–421.3
α-Aspartyl	$HO_2C \cdot CH_2 \cdot CH(NH_2) \cdot CO-$	C–421.3
β-Aspartyl	$HO_2C \cdot CH(NH_2) \cdot CH_2 \cdot CO-$	C–421.3
Atropoyl (preferred to	$C_6H_5 \cdot C(:CH_2) \cdot CO-$	C–404.1
2-phenylpropenoyl)		
Azabicyclo[2.2.1]heptyl	$NC_6H_{10}-$	B–4.1, B–5.21
Azelaoyl (unsubstituted only)	$-CO \cdot [CH_2]_7 \cdot CO-$	C–404.1
Azi	$-N:N-$ (to a single atom)	C–931.5
Azido	N_3-	C–10.1, C–941.1
Azino	$=N \cdot N=$	C–923.1
Azo	$-N:N-$	C–911, C–912
Azoxy	$-N(O) \cdot N-$	C–913.1
Azulenyl	$C_{10}H_7-$	A–21.1, A–24.2
Benzamido	$C_6H_5 \cdot CO \cdot NH-$	C–823.1
Benzeneazo	$C_6H_5 \cdot N:N-$	C–911.2, C–911.3
Benzeneazoxy	$C_6H_5 \cdot N_2O-$	C–913.1
benzenecarbonyl, see Benzoyl		
1,2-benzenedicarbonyl, see Phthaloyl		
1,3-benzenedicarbonyl, see Isophthaloyl		
1,4-benzenedicarbonyl, see Terephthaloyl		
Benzenesulfinyl	$C_6H_5 \cdot SO-$	C–631.1
Benzenesulfonamido	$C_6H_5 \cdot SO_2 \cdot NH-$	C–631.1, C–641.7, C–823.1
Benzenesulfonyl	$C_6H_5 \cdot SO_2-$	C–631.1
Benzenesulfonylamino	$C_6H_5 \cdot SO_2 \cdot NH-$	C–823·1
Benzenetriyl	$C_6H_3{\Big\langle}$	A–24.4
Benzhydryl (alternative to	$(C_6H_5)_2CH-$	A–13.3
Diphenylmethyl)		
Benzhydrylidene (alternative to	$(C_6H_5)_2C=$	A–13.4
Diphenylmethylene)		
Benzidino	p-$NH_2 \cdot C_6H_4 \cdot C_6H_4 \cdot NH-$	C–811.4, C–813.1
Benziloyl (preferred to 2-hydroxy-	$(C_6H_5)_2C(OH) \cdot CO-$	C–411.1
2,2-diphenylethanoyl)		
Benzimidazolyl	$N_2C_7H_5-$	B–3, B–5.11
Benzimidoyl	$C_6H_5 \cdot C(:NH)-$	C–451.2
Benzofuranyl	OC_8H_5-	B–3, B–5.11
Benzopyranyl	OC_9H_7-	B–3, B–5.11
Benzoquinonyl (1,2- or 1,4-)	$(O:)_2C_6H_3-$	C–317.2
Benzo[b]thienyl (replacing	SC_8H_5	B–2.11, B–5.11
thianaphthenyl)		
Benzoxazinyl	ONC_8H_6-	B–3, B–5.11
Benzoxazolyl	ONC_7H_4-	B–3, B–5.11
Benzoyl (preferred to benzenecarbonyl)	$C_6H_5 \cdot CO-$	C–404.1
Benzoylamino	$C_6H_5 \cdot CO \cdot NH-$	C–823.1
Benzoylhydrazino (N'- or 2-shown)	$C_6H_5 \cdot CO \cdot NH \cdot NH-$	C–921.5
Benzoylimino	$C_6H_5 \cdot CO \cdot N=$	C–815.3
Benzoyloxy	$C_6H_5 \cdot CO \cdot O-$	C–463.3
Benzyl	$C_6H_5 \cdot CH_2-$	A–13.3
Benzylidene	$C_6H_5 \cdot CH=$	A–13.4
Benzylidyne	$C_6H_5 \cdot C\equiv$	A–13.4

Radical name	Formula	Based on rule no.
Benzyloxy	$C_6H_5 \cdot CH_2 \cdot O-$	C–205.1
Benzyloxycarbonyl	$C_6H_5 \cdot CH_2 \cdot O \cdot CO-$	C–463.3
Benzylthio	$C_6H_5 \cdot CH_2 \cdot S-$	C–514.1
Bicyclo[2.2.1]hept-5-en-2-yl	C_7H_9-	A–31.4
Bi(cyclohexan)yl	$C_6H_{11} \cdot C_6H_{10}-$	A–52.1, C–71.2
Bi(cyclohexyl)yl	$C_6H_{11} \cdot C_6H_{10}-$	A–52.1, C–71.2
Binaphthalenyl	$C_{10}H_7 \cdot C_{10}H_6-$	A–52.1, C–71.2
Binaphthylyl	$C_{10}H_7 \cdot C_{10}H_6-$	A–52.4, C–71.2
Biphenylenyl	$C_{12}H_7-$	A–21.1, A–24.2
Biphenylyl	$C_6H_5 \cdot C_6H_4-$	A–52.4, C–71.2
Bornenyl	$C_{10}H_{15}-$	A–72.1, A–75.2
Bornyl (*replacing camphyl and bornylyl*)	$C_{10}H_{17}-$	A–72.1, A–75.2
bornylyl, see Bornyl		
Bromo	$Br-$	C–10.1, C–102
Bromoformyl	$Br \cdot CO-$	C–481.2
Bromonio	$+HBr-$	C–82.1
Butadienyl ((1,3- *shown*)	$CH_2 : CH \cdot CH : CH-$	A–3.5
butanedioyl, see Succinyl		
Butanediylidene	$=CH \cdot CH_2 \cdot CH_2 \cdot CH=$	A–4.4
Butanediylidyne	$\equiv C \cdot CH_2 \cdot CH_2 \cdot C\equiv$	A–4.4
1,2,3-Butanetricarbonyl	$CH_3 \cdot CH{-}CH \cdot CH \cdot CO-$	C–403.2
	$\qquad\quad -CO \quad CO-$	
butanoyl, see Butyryl		
1-Butanyliden-4-ylidyne	$\equiv C \cdot CH_2 \cdot CH_2 \cdot CH=$	A–4.4
cis-*butenedioyl, see* Maleoyl		
trans-*butenedioyl, see* Fumaroyl		
butenoyl, see Crotonoyl *and* Isocrotonoyl		
1-Butenyl	$CH_3 \cdot CH_2 \cdot CH : CH-$	A–3.5
2-Butenyl (*replacing crotyl*)	$CH_3 \cdot CH : CH \cdot CH_2-$	A–3.5
2-Butenylene	$-CH_2 \cdot CH : CH \cdot CH_2-$	A–4.3
Butenylidene (2- *shown*)	$CH_3 \cdot CH : CH \cdot CH=$	A–4.1
Butenylidyne (2- *shown*)	$CH_3 \cdot CH : CH \cdot C\equiv$	A–4.1
Butoxy	$CH_3 \cdot [CH_2]_2 \cdot CH_2 \cdot O-$	C–205.1
sec-Butoxy (*unsubstituted only*)	$C_2H_5 \cdot CH(CH_3) \cdot O-$	C–205.1
tert-Butoxy (*unsubstituted only*)	$(CH_3)_3C \cdot O-$	C–205.1
Butyl	$CH_3 \cdot [CH_2]_2 \cdot CH_2-$	A–1.2
sec-Butyl (*unsubstituted only*)	$C_2H_5 \cdot CH(CH_3)-$	A–2.25
tert-Butyl (*unsubstituted only*)	$(CH_3)_3C-$	A–2.25
Butylidene	$CH_3 \cdot CH_2 \cdot CH_2 \cdot CH=$	A–4.1
sec-Butylidene (*unsubstituted only*)	$C_2H_5 \cdot C(CH_3)=$	A–4.1
Butylidyne	$CH_3 \cdot [CH_2]_2 \cdot C\equiv$	A–4.1
Butyryl (*preferred to butanoyl*)	$CH_3 \cdot CH_2 \cdot CH_2 \cdot CO-$	C–404.1
Camphoroyl (*preferred to 1,2,2-trimethyl-1,3-cyclopentanedicarbonyl*)	$C_{10}H_{14}O_2-$	C–404.1
camphyl, see Bornyl		
Carbamoyl	$NH_2 \cdot CO-$	C–431.2
carbazido, see Carbonohydrazido		
Carbazolyl	$NC_{12}H_8-$	B–2.11, B–5.11
Carbazono	$HN : N \cdot CO \cdot NH \cdot NH-$	C–981.1, C–981.2
Carbazoyl	$NH_2 \cdot NH \cdot CO-$	C–984.1
Carbodiazono	$HN = N \cdot CO \cdot N = N-$	C–981.1, C–981.2
Carbolinyl (α-, β-, γ-)	$N_2C_{11}H_7-$	B–2.11, B–5.11
Carbonimidoyl	$-C(:NH)-$	C–451.4

Radical name	Formula	Based on rule no.
Carbonohydrazido (*preferred to* carbohydrazido *or* carbazido)	$H_2N \cdot NH \cdot CO \cdot NH \cdot NH-$	C–981.1, C–981.2
Carbonyl	$-CO-$, $OC<$	C–72.1, C–108.2, C–403.2
Carbonyldioxy	$-O \cdot CO \cdot O-$	C–205.2
Carboxy	HO_2C-	C–10.3, C–401.3
Carboxylato	$^-OOC-$	C–86.1
Carenyl	$C_{10}H_{15}-$	A–72.1, A–75.2
Caryl	$C_{10}H_{17}-$	A–72.1, A–75.2
Chloro	$Cl-$	C–10.1, C–102
chlorocarbonyl, see Chloroformyl		
Chloroformyl (*replacing* chlorocarbonyl)	$Cl \cdot OC-$	C–10.3, C–481.2
Chloronio	$^+HCl-$	C–82.1
Chlorosyl	$OCl-$	C–10.1, C–106.2
Chlorothio	$ClS-$	C–621.2
Chloryl	O_2Cl-	C–10.1, C–106.2
Cholanthrenyl	$C_{20}H_{13}-$	A–23.1, A–24.2
Chromanyl	OC_9H_9-	B–2.12, B–5.11
Chromenyl	OC_9H_7-	B–2.11, B–5.11
Chrysenyl	$C_{18}H_{11}-$	A–21.1, A–24.2
Cinnamoyl (*preferred to* 3-phenylpropenoyl)	$C_6H_5 \cdot CH:CH \cdot CO-$	C–404.1
Cinnamyl	$C_6H_5 \cdot CH:CH \cdot CH_2-$	A–13.3
Cinnamylidene	$C_6H_5 \cdot CH:CH \cdot CH=$	A–13.4
Cinnolinyl	$N_2C_8H_5-$	B–2.11, B–5.11
Citraconoyl (*unsubstituted only*)	$HC \cdot CO-$ $\|$ $CH_3 \cdot C \cdot CO-$	C–404.1
Coronenyl	$C_{24}H_{11}-$	A–21.1, A–24.2
Crotonoyl (*preferred to* trans-2-butenoyl)	$CH_3 \cdot CH:CH \cdot CO-$ (*trans*)	C–404.1
crotyl, see 2-Butenyl		
Cumenyl (o-, m-, or p-)	$(CH_3)_2CH \cdot C_6H_4-$	A–13.1
Cyanato	$NCO-$	C–833.1
Cyano	$N\equiv C-$	C–10.3, C–832.5
Cyclobutyl	C_4H_7-	A–11.2
Cycloheptyl	$C_7H_{13}-$	A–11.2
Cyclohexadienyl (2,4- *shown*)	$CH_2 \cdot CH:CH \cdot CH:CH \cdot CH-$	A–11.4
Cyclohexadienylene	$-C_6H_6-$	A–11.6
Cyclohexadienylidene (2,4- *shown*)	$CH:CH \cdot CH:CH \cdot CH_2 \cdot C=$	A–11.5
Cyclohexanecarbohydrazonoyl	$C_6H_{11} \cdot C(:N \cdot NH_2)-$	C–451.2
Cyclohexanecarbohydroximoyl	$C_6H_{11} \cdot C(:N \cdot OH)-$	C–451.2
Cyclohexanecarbonyl	$C_6H_{11} \cdot CO-$	C–403.2
Cyclohexanecarbothioyl	$C_6H_{11} \cdot CS-$	C–543.3
Cyclohexanecarboxamido	$C_6H_{11} \cdot CO \cdot NH-$	C–824.3
Cyclohexanecarboximidoyl	$C_6H_{11} \cdot C(:NH)-$	C–451.2
Cyclohexenyl	C_6H_9-	A–11.4
Cyclohexenylene	$-C_6H_8-$	A–11.6
2-Cyclohexenylidene	$CH_2 \cdot CH_2 \cdot CH_2 \cdot CH:CH \cdot C=$	A–11.5
Cyclohexyl	$C_6H_{11}-$	A–11.2
Cyclohexylcarbonyl	$C_6H_{11} \cdot CO-$	C–403.2

Radical name	Formula	Based on rule no.
Cyclohexylene	$-C_6H_{10}-$	A–11.6
Cyclohexylidene	$CH_2 \cdot CH_2 \cdot CH_2 \cdot CH_2 \cdot CH_2 \cdot C=$	A–11.5
Cyclohexylthiocarbonyl	$C_6H_{11} \cdot CS-$	C–543.3
Cyclopentadienyl	C_5H_5-	A–11.4
Cyclopentadienylidene	$CH:CH \cdot CH:CH \cdot C=$	A–11.5
Cyclopentanespirocyclobutyl	$C_8H_{13}-$	A–11.2, A–42.1
Cyclopenta[a]phenanthryl		A–21. A–23.1, A–24.2
1,2-Cyclopentenophenanthryl	$\}$ $C_{17}H_{11}$	A–21, A–23.5, A–24.2
Cyclopentenyl	C_5H_7-	A–11.4
Cyclopentenylidene (2- *shown*)	$CH_2 \cdot CH_2 \cdot CH:CH \cdot C=$	A–11.5
Cyclopentyl	C_5H_9-	A–11.2
Cyclopentylene	$-C_5H_8-$	A–11.6
Cyclopentylidene	$CH_2 \cdot CH_2 \cdot CH_2 \cdot CH_2 \cdot C=$	A–11.5
Cyclopropyl	C_3H_5-	A–11.2
Cystathionyl	$S{\Large<}^{CH_2 \cdot CH(NH_2) \cdot CO-}_{CH_2 \cdot CH_2 \cdot CH(NH_2) \cdot CO-}$	C–421.1
Cysteinyl	$HS \cdot CH_2 \cdot CH(NH_2) \cdot CO-$	C–421.2
Cystyl	$-CO \cdot CH(NH_2) \cdot CH_2 \cdot S \cdot S \cdot CH_2 \cdot CH(NH_2) \cdot CO-$	C–421.1
Decanedioyl	$-CO \cdot [CH_2]_8 \cdot CO-$	C–403.1
Decanoyl	$CH_3 \cdot [CH_2]_8 \cdot CO-$	C–403.1
Decyl	$CH_3 \cdot [CH_2]_8 \cdot CH_2-$	A–1.2
Diacetoxyiodo	$(CH_3 \cdot CO \cdot O)_2 I-$	C–10.1, C–106.3
Diacetylamino	$(CH_3 \cdot CO)_2 N-$	C–826.3
Diaminomethyleneamino	$(NH_2)_2 C:N-$	C–961.2
Diazaanthryl	$N_2 C_{12}H_7-$	B–4.1, B–5.21
Diazo	$=N_2$	C–10.1, C–931.4
Diazoamino	$-N:N \cdot NH-$	C–942.2
Dibenzoylamino	$(C_6H_5 \cdot CO)_2 N-$	C–826.3
Dichloroiodo	$Cl_2 I-$	C–10.1, C–106.3
Diethylamino	$(C_2H_5)_2 N-$	C–811.4
3,4-*dihydroxybenzoyl, see* Protocatechuoyl		
2,3-*dihydroxybutanedioyl, see* Tartaroyl		
Dihydroxyiodo	$(HO)_2 I-$	C–10.1, C–106.3
2,3-*dihydroxypropanoyl, see* Glyceroyl		
3,4-*dimethoxybenzoyl, see* Veratroyl		
3,4-Dimethoxyphenethyl	$3,4-(CH_3O)_2 C_6H_3 \cdot CH_2 \cdot CH-$	A–13.3
3,4-Dimethoxyphenylacetyl	$3,4-(CH_3O)_2 C_6H_3 \cdot CH_2 \cdot CO-$	C–401.4
Dimethylamino	$(CH_3)_2 N-$	C–811.4
Dimethylbenzoyl	$(CH_3)_2 C_6H_3 \cdot CO-$	C–404.1
Dioxacyclohexyl	$O_2 C_4H_7-$	B–4.1, B–5.21
Dioxy	$-O \cdot O-$	C–218.2
Diphenylamino	$(C_6H_5)_2 N-$	C–811.4
Diphenylmethyl (*alternative to* Benzhydryl)	$(C_6H_5)_2 CH-$	A–13.3
Diphenylmethylene (*alternative to* Benzyhydrylidene)	$(C_6H_5)_2 C=$	A–13.4
Dithianaphthyl	$S_2 C_6H_7-$	B–4.1, B–5.21
Dithio	$-S_2-$	C–515.1

Radical name	Formula	Based on rule no.
Dithiocarboxy	$HSSC-$	C–541.1
Dithiosulfo	HOS_3-	C–641.3
Docosyl	$CH_3 \cdot [CH_2]_{20} \cdot CH_2-$	A–1.2
Dodecanoyl	$CH_3 \cdot [CH_2]_{10} \cdot CO-$	C–403.1
Dodecyl	$CH_3 \cdot [CH_2]_{10} \cdot CH_2-$	A–1.2
Dotriacontyl	$CH_3 \cdot [CH_2]_{30} \cdot CH_2-$	A–1.2

Elaidoyl (*preferred to* trans-9-*octadecenoyl*)	$CH_3 \cdot [CH_2]_7 \cdot CH:CH \cdot [CH_2]_7 \cdot CO-$	C–404.1
Epidioxy (*as a bridge*)	$-O \cdot O-$	C–218.2
Epidiseleno (*as a bridge*)	$-Se_2-$	C–701.1
Epidithio (*as a bridge*)	$-S_2-$	C–515.4
Epimino (*as a bridge*)	$-NH-$	C–815.2
Episeleno (*as a bridge*)	$-Se-$	C–701.1
Epithio (*as a bridge*)	$-S-$	C–514.4
Epoxy (*as bridge*)	$-O-$	C–212.2
ethanedioyl, see Oxalyl		
Ethanediylidene	$=CH \cdot CH=$	A–4.4
Ethanesulfonamido	$C_2H_5 \cdot SO_2 \cdot NH-$	C–641.7, C–641.8, C–823.1
ethanoyl, see Acetyl		
ethenyl, see Vinyl		
Ethoxalyl	$C_2H_5OOC \cdot CO-$	C–405.2
Ethoxy	$C_2H_5 \cdot O-$	C–205.1
Ethoxycarbonyl	$C_2H_5 \cdot O \cdot OC-$	C–463.3
Ethyl	$CH_3 \cdot CH_2-$	A–1.2
Ethylamino	$C_2H_5 \cdot NH-$	C–811.4
Ethylene	$-CH_2 \cdot CH_2-$	A–4.2
Ethylenedioxy	$-O \cdot CH_2 \cdot CH_2 \cdot O-$	C–72.2, C–205.2, C–212.3
Ethylidene	$CH_3 \cdot CH=$	A–4.1
Ethylidyne	$CH_3 \cdot C\equiv$	A–4.1
Ethylsulfonylamino	$C_2H_5 \cdot SO_2 \cdot NH-$	C–641.7, C–641.8, C–823.1
Ethylthio	$CH_3 \cdot CH_2 \cdot S-$	C–514.1
Ethynyl	$HC:C-$	A–3.5
Ethynylene	$-C:C-$	A–4.3

Fluoranthenyl	$C_{16}H_9-$	A–21.1, A–24.2
Fluorenyl	$C_{13}H_9-$	A–21.1, A–24.2
Fluorenylidene	$C_{13}H_8=$	A–21.1, A–24.3
Fluoro	$F-$	C–10.1, C–102.1
Fluoroformyl	$F \cdot CO-$	C–481.2
Formamido	$OCH \cdot NH-$	C–823.1
Formimidoyl	$CH(:NH)-$	C–451.2
Formyl (*preferred to methanoyl*)	$OHC-$	C–10.3, C–304.2, C–404.1
Formylamino	$H \cdot CO \cdot NH-$	C–823.1
Formylimino	$H \cdot CO \cdot N=$	C–815.1
Formyloxy	$H \cdot CO \cdot O-$	C–463.3
Fumaroyl (*preferred to* trans-*butenedioyl*)	$-CO \cdot CH:CH \cdot CO-$ (*trans*)	C–404.1
furancarbonyl, see Furoyl		
Furazanyl	N_2OC_2H-	B–2.11, B–5.11
Furfuryl (2- *only*) (*preferred to* 2-*furylmethyl*)	$OC_4H_3 \cdot CH_2-$	B–5.11
Furfurylidene (2- *only*)	$O \cdot CH:CH \cdot CH:C \cdot CH=$	B–5.11
Furoyl (3- *shown*) (*preferred to furancarbonyl*)	$CH:CH \cdot O \cdot CH:C \cdot CO-$	C–404.1

Radical name	Formula	Based on rule no.
Furyl	OC$_4$H$_3$—	B–2.11, B–5.11
3-Furylmethyl (*contrast* Furfuryl)	CH:CH·O·CH:C·CH$_2$—	C–12.7
Galloyl (*preferred to* 3,4,5-*trihydroxybenzoyl*)	3,4,5-(HO)$_3$C$_6$H$_2$·CO—	C–411.1
Geranyl	C$_{10}$H$_{17}$—	A–75.1
Glutaminyl	NH$_2$·CO·CH$_2$·CH$_2$·CH(NH$_2$)·CO—	C–421.3
Glutamoyl	—CO·CH$_2$·CH$_2$·CH(NH$_2$)·CO—	C–421.3
α-Glutamyl	HOOC·[CH$_2$]$_2$·CH(NH$_2$)·CO—	C–421.3
γ-Glutamyl	HOOC·CH(NH$_2$)·[CH$_2$]$_2$·CO—	C–421.3
Glutaryl (*preferred to pentanedioyl*)	—CO·[CH$_2$]$_3$·CO—	C–404.1
Glyceroyl (*preferred to* 2,3-*dihydroxypropanoyl*)	HO·CH$_2$·CH(OH)·CO—	C–411.1
Glycoloyl (*preferred to* *hydroxyethanoyl*)	HO·CH$_2$·CO—	C–411.1
Glycyl	NH$_2$·CH$_2$·CO—	C–421.1
Glycylamino	NH$_2$·CH$_2$·CO·NH—	C–823.2
Glyoxyloyl	OHC·CO—	C–416.3
Guanidino	NH$_2$·C(:NH)·NH—	C–961.2
Guanyl, see Amidino		
Hectyl	CH$_3$·[CH$_2$]$_{98}$·CH$_2$—	A–1.2
Henicosyl	CH$_3$·[CH$_2$]$_{19}$·CH$_2$—	A–1.2
Hentriacontyl	CH$_3$·[CH$_2$]$_{29}$·CH$_2$—	A–1.2
Heptacenyl	C$_{30}$H$_{17}$—	A–21.1, A–24.2
Heptacontyl	CH$_3$·[CH$_2$]$_{68}$·CH$_2$—	A–1.2
Heptacosyl	CH$_3$·[CH$_2$]$_{25}$·CH$_2$—	A–1.2
Heptadecanoyl	CH$_3$·[CH$_2$]$_{15}$·CO—	C–403.1
Heptadecyl	CH$_3$·[CH$_2$]$_{15}$·CH$_2$—	A–1.2
Heptalenyl	C$_{12}$H$_9$—	A–21.1, A–24.2
Heptanamido	CH$_3$·[CH$_2$]$_5$·CO·NH—	C–823.1
Heptanedioyl	—CO·[CH$_2$]$_5$·CO—	C–403.1
Heptanoyl	CH$_3$·[CH$_2$]$_5$·CO—	C–403.1
Heptaphenyl (from heptaphene)	C$_{30}$H$_{17}$—	A–21.1, A–24.2
Heptyl	CH$_3$·[CH$_2$]$_5$·CH$_2$—	A–1.2
Hexacenyl	C$_{26}$H$_{15}$—	A–21.2, A–24.2
Hexacontyl	CH$_3$·[CH$_2$]$_{58}$·CH$_2$—	A–1.2
Hexacosyl	CH$_3$·[CH$_2$]$_{24}$·CH$_2$—	A–1.2
Hexadecanoyl	CH$_3$·[CH$_2$]$_{14}$·CO—	C–403.1
Hexadecyl	CH$_3$·[CH$_2$]$_{14}$·CH$_2$—	A–1.2
Hexamethylene	—CH$_2$·[CH$_2$]$_4$·CH$_2$—	A–4.2
Hexanamido	C$_5$H$_{11}$·CO·NH—	C–823.1
hexanedioyl, see Adipoyl		
Hexanimidoyl	C$_5$H$_{11}$·C(:NH)—	C–451.2
Hexanohydrazonoyl	C$_5$H$_{11}$·C(:N·NH$_2$)—	C–451.2
Hexanohydroximoyl	C$_5$H$_{11}$·C(:N·OH)—	C–451.2
Hexanoyl	CH$_3$·[CH$_2$]$_4$·CO—	C–403.1
Hexanoylamino	C$_5$H$_{11}$·CO·NH—	C–823.1
Hexapheneyl (from hexaphene)	C$_{26}$H$_{15}$—	A–21.1, A–24.2
Hexyl	CH$_3$·[CH$_2$]$_4$·CH$_2$—	A–1.2
Hexylidene	CH$_3$·[CH$_2$]$_4$·CH=	A–4.1
Hexylidyne	CH$_3$·[CH$_2$]$_4$·C≡	A–4.1
Hexyloxy	CH$_3$·[CH$_2$]$_5$·O—	C–205.1
Hippuroyl	C$_6$H$_5$·CO·NH·CH$_2$·CO—	C–421.1, C–421.4
Histidyl	N$_2$C$_3$H$_3$·CH$_2$·CH(NH$_2$)·CO—	C–421.1

Radical name	Formula	Based on rule no.
Homocysteinyl	$HS \cdot CH_2 \cdot CH_2 \cdot CH(NH_2) \cdot CO—$	C–421.2
Homoseryl	$HO \cdot CH_2 \cdot CH_2 \cdot CH(NH_2) \cdot CO—$	C–421.1
Hydantoyl	$NH_2 \cdot CO \cdot NH \cdot CH_2 \cdot CO—$	C–971.2
Hydratropoyl (*preferred to*	$C_6H_5 \cdot CH(CH_3) \cdot CO—$	C–404.1
2-phenylpropanoyl)		
Hydrazi	$—NH \cdot NH—$ (to a single atom)	C–921.4
Hydrazino	$NH_2 \cdot NH—$	C–921.1
Hydrazo	$—NH \cdot NH—$ (to different atoms)	C–921.2
Hydrazono	$NH_2 \cdot N=$	C–922.1
Hydroperoxy	$HO \cdot O—$	C–218.1
Hydroseleno	$HSe—$	C–701.1
Hydroxy	$HO—$	C–10.3, C–201.2
Hydroxyamino	$HO \cdot NH—$	C–841.1
m-Hydroxybenzoyl	$m\text{-}HO \cdot C_6H_4 \cdot CO—$	C–411.1
o-hydroxybenzoyl, see Salicyloyl		
p-Hydroxybenzoyl	$p\text{-}HO \cdot C_6H_4 \cdot CO—$	C–411.1
2-hydroxybenzyl, see Salicyl		
2-hydroxybenzylidene, see		
Salicylidene		
hydroxybutanedioyl, see Maloyl		
2-hydroxy-2,2-diphenylethanoyl, see		
Benziloyl		
hydroxyethanoyl, see Glycoloyl		
Hydroxyimino	$HO \cdot N=$	C–842.1
4-hydroxy-3-methoxybenzoyl, see		
Vanilloyl		
3-hydroxy-2-phenylpropanoyl, see		
Tropoyl		
hydroxypropanedioyl, see Tartronoyl		
2-hydroxypropanoyl, see Lactoyl		
Icosyl	$CH_3 \cdot [CH_2]_{18} \cdot CH_2—$	A-1.2
Imidazolidinyl	$N_2C_3H_7—$	B-2.12, B-5.11
Imidazolinyl	$N_2C_3H_5—$	B-2.12, B-5.11
Imidazolyl	$N_2C_3H_3—$	B-2.11, B-5.11
Imino	$—HN—, HN<$	C–815.1, C–815.3
Iminomethylamino	$HN=CH—NH—$	C–951.4
Indacenyl	$C_{12}H_7—$	A-21.1, A-24.2
Indanyl	$C_9H_9—$	A-23.1, A-24.2
Indazolyl	$N_2C_7H_5—$	B-2.11, B-5.11
Indenyl	$C_9H_7—$	A-21.1, A-24.2
Indolinyl	$NC_8H_8—$	B-2.12, B-5.11
Indolinylidene (3- *shown*)	$CH_2 \cdot NH \cdot C_6H_4 \cdot C=$	B-5.12
Indolizinyl	$NC_8H_6—$	B-2.11, B-5.11
Indolyl	$NC_8H_6—$	B-2.11, B-5.11
Iodo	$I—$	C–10.1, C–102.1
Iodoformyl	$I \cdot CO—$	C–481.2
Iodonio	$+HI—$	C–82.1
iodoso, see Iodosyl		
Iodosyl, *replacing iodoso*	$OI—$	C–10.1, C–106.1
iodoxy, see Iodyl		
Iodyl, *replacing iodoxy*	$O_2I—$	C–10.1, C–106.1
Isobenzofuranyl	$OC_8H_5—$	B-2.11, B-5.11
Isobutoxy (*unsubstituted only*)	$(CH_3)_2CH \cdot CH_2 \cdot O—$	C–205.1
Isobutyl (*unsubstituted only*)	$(CH_3)_2CH \cdot CH_2—$	A-2.25
Isobutylidene (*unsubstituted only*)	$(CH_3)_2CH \cdot CH=$	A-4.1
Isobutylidyne (*unsubstituted only*)	$(CH_3)_2CH \cdot C\equiv$	A-4.1

Radical name	Formula	Based on rule no.
Isobutyryl (*preferred to 2-methyl-propanoyl*) (*unsubstituted only*)	$(CH_3)_2CH \cdot CO$—	C–404.1
Isocarbonohydrazido	$H_2N \cdot N : C(OH) \cdot NH \cdot NH$— (and tautomers)	C–983.1
Isochromanyl	OC_7H_9	B–2.12, B–5.11
Isocoumarinyl	$O_2C_7H_5$—	B–5.11, C–473.1
Isocrotonoyl (*preferred to cis-2-butenoyl*)	$CH_3 \cdot CH : CH \cdot CO$— (*cis*)	C–404.1
Isocyanato	OCN—	C–833.1
Isocyano	CN—	C–833.1
Isohexyl (*unsubstituted only*)	$(CH_3)_2CH \cdot [CH_2]_2 \cdot CH_2$—	A–2.25
Isohexylidene (*unsubstituted only*)	$(CH_3)_2CH \cdot [CH_2]_2 \cdot CH=$	A–4.1
Isohexylidyne (*unsubstituted only*)	$(CH_3)_2CH \cdot [CH_2]_2 \cdot C\equiv$	A–4.1
Isoindolinyl	NC_8H_8—	B–2.12, B–5.11
Isoindolyl	NC_8H_6—	B–2.11, B–5.11
Isoleucyl	$C_2H_5 \cdot CH(CH_3) \cdot CH(NH_2) \cdot CO$—	C–421.1
Isonicotinoyl (*preferred to 4-pyridinecarbonyl*)	$NC_5H_4 \cdot CO$— (4-)	C–404.1
Isoxazolyl	ONC_3H_2—	B–2.11, B–5.11
Isopentyl (*unsubstituted only*)	$(CH_3)_2CH \cdot CH_2 \cdot CH_2$—	A–2.25
Isopentylidene (*unsubstituted only*)	$(CH_3)_2CH \cdot CH_2 \cdot CH=$	A–4.1
Isopentylidyne (*unsubstituted only*)	$(CH_3)_2CH \cdot CH_2 \cdot C\equiv$	A–4.1
Isopentyloxy (*unsubstituted only*)	$(CH_3)_2CH \cdot CH_2 \cdot CH_2 \cdot O$—	C–205.1
Isophthaloyl (*preferred to 1,3-benzenedicarbonyl*)	—$CO \cdot C_6H_4 \cdot CO$— (*m*-)	C–404.1
Isopropenyl (*replacing 1-methyl-vinyl*) (*unsubstituted only*)	$CH_2 : C(CH_3)$—	A–3.5
Isopropoxy (*unsubstituted only*)	$(CH_3)_2CH \cdot O$—	C–205.1
Isopropyl (*unsubstituted only*)	$(CH_3)_2CH$—	A–2.25
p-Isopropylbenzoyl	p-$(CH_3)_2CH \cdot C_6H_4 \cdot CO$—	C–404.1
Isopropylbenzyl	$(CH_3)_2CH \cdot C_6H_4 \cdot CH_2$—	A–13.3
Isopropylidene	$(CH_3)_2C=$	A–4.1
Isoquinolyl	NC_9H_6—	B–2.11, B–5.11
Isoselenocyanato	$SeCN$—	C–833.1
Isosemicarbazido	$H_2N \cdot NH \cdot C(OH) : N$— (and tautomers)	C–983.1
Isothiazolyl	SNC_3H_2—	B–2.11, B–5.11
Isothiocyanato	SCN—	C–833.1
Isothioureido	$HN : C(SH) \cdot NH$—, $H_2N \cdot C(SH) : N$—	C–972.2, C–974.1
Isoureido	$HN : C(OH) \cdot NH$—, $H_2N \cdot C(OH) : N$—	C–972.2
Isovaleryl (*preferred to 3-methyl-butanoyl*) (*unsubstituted only*)	$(CH_3)_2CH \cdot CH_2 \cdot CO$—	C–404.1
Isoviolanthrenyl	$C_{34}H_{19}$—	A–23.1, A–24.2
Lactoyl (*preferred to 2-hydroxypropanoyl*)	$CH_3 \cdot CH(OH) \cdot CO$—	C–411.1
Lanthionyl	—$CO \cdot CH(NH_2) \cdot CH_2 \cdot S \cdot CH_2 \cdot CH(NH_2) \cdot CO$—	C–421.1
Lauroyl (*unsubstituted only*)	$CH_3 \cdot [CH_2]_{10} \cdot CO$—	C–404.1
Leucyl	$(CH_3)_2CH \cdot CH_2 \cdot CH(NH_2) \cdot CO$—	C–421.1
Linalyl	$C_{10}H_{17}$—	A–75.1
Lysyl	$NH_2 \cdot [CH_2]_4 \cdot CH(NH_2) \cdot CO$—	C–421.1
Maleoyl (*preferred to cis-butenedioyl*)	—$CO \cdot CH : CH \cdot CO$—	C–404.1
Malonyl (*preferred to propanedioyl*)	—$CO \cdot CH_2 \cdot CO$—	C–404.1
Maloyl (*preferred to hydroxybutanedioyl*)	—$CO \cdot CH(OH) \cdot CH_2 \cdot CO$—	C–411.1
Menthenyl	$C_{10}H_{17}$—	A–73.1, A–75.2

Radical name	Formula	Based on rule no.
Menthyl	$C_{10}H_{19}$—	A–72.1, A–75.2
Mercapto	HS—	C–10.3, C–511.1
Mesaconoyl (*unsubstituted only*)	—CO·CH$\parallel$CH$_3$·C·CO—	C–404.1
Mesityl	2,4,6-$(CH_3)_3C_6H_2$—	A–13.1
Mesoxalo	HOOC·CO·CO—	C–416.3
Mesoxalyl	—CO·CO·CO—	C–416.3
Mesyl	CH_3·SO$_2$—	C–641.7
Methacryloyl (*preferred to* 2-*methylpropenoyl*)	CH_2:C(CH$_3$)·CO—	C–404.1
Methaneazo	CH_3·N:N—	C–911.2
Methaneazoxy	CH_3·N$_2$O—	C–913.1
Methanesulfinamido	CH_3·SO·NH—	C–641.8
Methanesulfinyl	CH_3·SO—	C–631.1
Methanesulfonamido	CH_3·SO$_2$·NH—	C–641.8
methanesulfonyl, see Mesyl		
methanoyl, see Formyl		
Methionyl	CH_3·S·CH$_2$·CH$_2$·CH(NH$_2$)·CO—	C–421.1
Methoxalyl	CH_3OOC·CO—	C–405.2
Methoxy	CH_3O—	C–205.1
methoxybenzoyl, see Anisoyl		
Methoxycarbonyl	CH_3O·OC—	C–463.3
Methoxyimino	CH_3·O·N=	C–842.2
Methoxyphenyl	CH_3O·C$_6$H$_4$—	A–13.1
Methoxysulfinyl	CH_3·O·SO—	C–641.6
Methoxysulfonyl	CH_3·O·SO$_2$—	C–641.6
Methoxy(thiosulfonyl)	CH_3·O·S$_2$O—	C–641.6
Methyl	CH_3—	A–1.2
α-*methylacryloyl, see* Methacryloyl		
Methylallyl	CH_2:C(CH$_3$)·CH$_2$—	A–3.5
Methylamino	CH_3·NH—	C–811.4
Methylazo	CH_3·N:N—	C–912.3
Methylazoxy	CH_3·N$_2$O—	C–913.1
methylbenzenecarbonoyl, see Toluoyl		
α-Methylbenzyl	C_6H_5·CH(CH$_3$)—	A–13.3
Methylbenzyl	CH_3·C$_6$H$_4$·CH$_2$—	A–13.3
3-*methylbutanoyl, see* Isovaleryl		
cis-Methylbutenedioyl	HC·CO—$\parallel$CH$_3$·C·CO—	C–404.1
trans-Methylbutenedioyl	—CO·CH$\parallel$CH$_3$·C·CO—	C–40ł.1
Methyldithio	CH_3·S$_2$—	C–515.1
Methylene	—CH$_2$—, H$_2$C<	A–4.1
Methylenedioxy	—O·CH$_2$·O—	C–72.2, C–205.2, C–212.3
3,4-*methylenedioxybenzoyl, see* Piperonyloyl		
5-Methylhexyl	$(CH_3)_2$CH·CH$_2$·CH$_2$·CH$_2$·CH$_2$—	A–2.25
Methylidyne	HC≡	A–4.1
1-Methylpentyl	CH_3·CH$_2$·CH$_2$·CH$_2$·CH(CH$_3$)—	A–2.25
2-Methylpentyl	CH_3·CH$_2$·CH$_2$·CH(CH$_3$)·CH$_2$—	A–2.25
2-*methylpropenoyl, see* Methacryloyl		
Methylsulfinimidoyl	CH_3·S(:NH)—	C–642

314

Radical name	Formula	Based on rule no.
Methylsulfinohydrazonoyl	$CH_3 \cdot S(:N \cdot NH_2)$—	C–642.2
Methylsulfinohydroximoyl	$CH_3 \cdot S(:N \cdot OH)$—	C–642.2
Methylsulfinyl	$CH_3 \cdot SO$—	C–631.1
Methylsulfinylamino	$CH_3 \cdot SO \cdot NH$—	C–641.7, C–641.8, C–823.1
Methylsulfonimidoyl	$CH_3 \cdot S(O)(NH)$—	C–642
Methylsulfonohydrazonoyl	$CH_3 \cdot S(O)(N \cdot NH_2)$—	C–642
Methylsulfonohydroxamoyl	$CH_3 \cdot S(O)(N \cdot OH)$—	C–642
methylsulfonyl, see Mesyl		
Methylthio	CH_3S—	C–514.1
(Methylthio)sulfonyl	$CH_3 \cdot S \cdot SO_2$—	C–641.6
1-*methylvinvl, see* Isopropenyl		
Morpholino (4-*position only*)	$CH_2 \cdot CH_2 \cdot O \cdot CH_2 \cdot CH_2 \cdot N$—	B–2.12, B–5.11
Morpholinyl (3- *shown*)	$NH \cdot CH_2 \cdot CH_2 \cdot O \cdot CH_2 \cdot CH$—	B–2.12, B–5.11
Myristoyl (*unsubstituted only*)	$CH_3 \cdot [CH_2]_{12} \cdot CO$—	C–404.1
Naphthacenyl	$C_{18}H_{11}$—	A–21.1, A–24.2
Naphthaleneazo	$C_{10}H_7 \cdot N:N$—	C–911.2
naphthalenecarbonyl, see Naphthoyl		
Naphthalenetetrayl	$> C_{10}H_4 <$	A–24.4
Naphtho[2,3-*b*]thienyl (*replacing thiophanthrenyl*)	$SC_{12}H_7$—	B–2.11, B–5.11
Naphthoyl (*preferred to naphthalenecarbonyl*)	$C_{10}H_7 \cdot CO$—	C–404.1
Naphthoyloxy	$C_{10}H_7 \cdot CO \cdot O$—	C–463.3
Naphthyl	$C_{10}H_7$—	A–24.2
Naphthylazo	$C_{10}H_7 \cdot N:N$—	C–912.3
Naphthylene	$-C_{10}H_6$—	A–24.4
Naphthylenebisazo	$-N:N \cdot C_{10}H_6 \cdot N:N$—	C–912.5
Naphthylmethylene	$C_{10}H_7 \cdot CH=$	A–4.1
Naphthylmethylidyne	$C_{10}H_7 \cdot C\equiv$	A–4.1
Naphthyloxy	$C_{10}H_7 \cdot O$—	C–205.1
Naphthyridinyl	$N_2C_8H_5$—	B–2.11, B–5.11
Neopentyl (*unsubstituted only*)	$(CH_3)_3C \cdot CH_2$—	A–2.25
Neryl	$C_{10}H_{17}$—	A–75.1
Nicotinoyl (*preferred to 3-pyridinecarbonyl*)	$NC_5H_4 \cdot CO$— (3-)	C–404.1
Nitrilo	$N\equiv$	C–72.1, C–815.1
Nitro	O_2N—	C–10.1, C–852.1
aci-Nitro	$HO \cdot (O:)N=$	C–10.1, C–852.2
Nitroso	ON—	C–10.1, C–851.1
Nonacontyl	$CH_3 \cdot [CH_2]_{88} \cdot CH_2$—	A–1.2
Nonacosyl	$CH_3 \cdot [CH_2]_{27} \cdot CH_2$—	A–1.2
Nonadecyl	$CH_3 \cdot [CH_2]_{17} \cdot CH_2$—	A–1.2
Nonanedioyl	$-CO \cdot [CH_2]_7 \cdot CO$—	C–403.1
Nonanoyl	$CH_3 \cdot [CH_2]_7 \cdot CO$—	C–403.1
Nonyl	$CH_3 \cdot [CH_2]_7 \cdot CH_2$—	A–1.2
Norbornyl (*replacing norcamphyl and norbornylyl*)	C_7H_{11}—	A–75.2
norbornylyl, see Norbornyl		
norcamphyl, see Norbornyl		
Norcaryl	C_7H_{11}—	A–75.2
Norleucyl	$CH_3 \cdot [CH_2]_3 \cdot CH(NH_2) \cdot CO$—	C–421.1
Norpinanyl	C_7H_{11}—	A–75.2
Norvalyl	$CH_3 \cdot CH_2 \cdot CH_2 \cdot CH(NH_2) \cdot CO$—	C–421.1

Radical name	Formula	Based on rule no.
Octacontyl	$CH_3 \cdot [CH_2]_{78} \cdot CH—$	A–1.2
Octacosyl	$CH_3 \cdot [CH_2]_{26} \cdot CH_2—$	A–1.2
Octadecanoyl	$CH_3 \cdot [CH_2]_{16} \cdot CO—$	C–403.1
cis-9-*octadecenoyl, see* Oleoyl		
Octadecyl	$CH_3 \cdot [CH_2]_{16} \cdot CH_2—$	A–1.2
Octanedioyl	$—CO \cdot [CH_2]_6 \cdot CO—$	C–403.1
Octanoyl	$CH_3 \cdot [CH_2]_6 \cdot CO—$	C–403.1
Octyl	$CH_3 \cdot [CH_2]_6 \cdot CH_2—$	A–1.2
Oleoyl (*preferred to* cis-9-*octadecenoyl*	$CH_3 \cdot [CH_2]_7 \cdot CH:CH \cdot [CH_2]_7 \cdot CO—$	C–404.1
Ornithyl	$NH_2 \cdot [CH_2]_3 \cdot CH(NH_2) \cdot CO—$	C–421.1
Ovalenyl	$C_{32}H_{13}—$	A–21.1, A–24.2
Oxalaceto	$HOOC \cdot CO \cdot CH_2 \cdot CO—$	C–416.3
Oxalacetyl	$—CO \cdot CH_2 \cdot CO \cdot CO—$	C–416.3
Oxalo	$HOOC \cdot CO—$	C–405.2
Oxalyl (*preferred to ethanedioyl*)	$—CO \cdot CO—$	C–404.1, C–405.2
Oxamoyl	$NH_2 \cdot CO \cdot CO—$	C–431.2
Oxapyrenyl	$OC_{15}H_9—$	B–4.1, B–5.21
Oxazinyl	$ONC_4H_4—$	B–1, B–5.11
Oxazolidinyl	$ONC_3H_6—$	B–1, B–5.11
Oxazolinyl	$ONC_3H_4—$	B–1, B–5.11
Oxazolyl	$ONC_3H_2—$	B–1, B–5.11
Oxido	$^-O—$ (ion)	C–86.2
Oxo	$O=$	C–10.3, C–316
Oxonio	$^+H_2O—$	C–82.1, C–85, C–87.1
Oxy	$—O—$	C–72.2, C–212.1
Palmitoyl (*unsubstituted only*)	$CH_3 \cdot [CH_2]_{14} \cdot CO—$	C–404.1
Pentacenyl	$C_{22}H_{13}—$	A–21.1, A–24.2
Pentacontyl	$CH_3 \cdot [CH_2]_{48} \cdot CH_2—$	A–1.2
Pentacosyl	$CH_3 \cdot [CH_2]_{23} \cdot CH_2—$	A–1.2
Pentadecanoyl	$CH_3 \cdot [CH_2]_{13} \cdot CO—$	C–403.1
Pentadecyl	$CH_3 \cdot [CH_2]_{13} \cdot CH_2—$	A–1.2
Pentafluorothio	$F_5S—$	C–621.2
Pentalenyl	$C_8H_5—$	A–21.1, A–24.2
Pentamethylene	$—CH_2 \cdot [CH_2]_3 \cdot CH_2—$	A–4.2
pentanedioyl, see Glutaryl		
pentanoyl, see Valeryl		
Pentapheneyl (from pentaphene)	$C_{22}H_{13}—$	A–21.1, A–24.2
Pentazolyl	$N:N \cdot N:N \cdot N—$	B–1, B–5.11
Pentenyl (2- *shown*)	$CH_3 \cdot CH_2 \cdot CH:CH \cdot CH_2—$	A–3.5
2-Penten-4-ynyl	$CH\equiv C \cdot CH=CH \cdot CH_2—$	A–3.5
Pentyl (*replacing amyl*)	$CH_3 \cdot [CH_2]_3 \cdot CH_2—$	A–1.2
tert-Pentyl (*unsubstituted only*)	$C_2H_5 \cdot C(CH_3)_2—$	A–2.25
Pentylidene	$CH_3 \cdot [CH_2]_3 \cdot CH=$	A–4.1
Pentylidyne	$CH_3 \cdot [CH_2]_3 \cdot C\equiv$	A–4.1
Pentyloxy	$CH_3 \cdot [CH_2]_3 \cdot CH_2 \cdot O—$	C–205.1
Perchloryl	$O_3Cl—$	C–10.1, C–106.2
Perimidinyl	$N_2C_{11}H_7—$	B–2.11, B–5.11
Perylenyl	$C_{20}H_{11}—$	A–21.1, A–24.2
Phenacyl	$C_6H_5 \cdot CO \cdot CH_2—$	C–318.1
Phenacylidene	$C_6H_5 \cdot CO \cdot CH=$	C–318.1
Phenalenyl	$C_{13}H_9—$	A–21.1, A–24.2
Phenanthridinyl	$NC_{13}H_8—$	B–2.11, B–5.11
Phenanthrolinyl	$H_2C_{12}H_7—$	B–2.11, B–5.11

Radical name	Formula	Based on rule no.
Phenanthryl	$C_{14}H_9—$	A–24.2
Phenanthrylene	$—C_{14}H_8—$	A–24.4
Phenarsazinyl	$AsNC_{12}H_7—$	B–2.11, B–5.11
Phenazinyl	$N_2C_{12}H_7—$	B–2.11, B–5.11
Phenethyl	$C_6H_5 \cdot CH_2 \cdot CH_2—$	A–13.3
Phenetidino (*o-, m-, or p-*)	$C_2H_5O \cdot C_6H_4 \cdot NH—$	C–811.4
Phenothiazinyl	$SNC_{12}H_8—$	B–2.11, B–5.11
Phenoxathiinyl	$OSC_{12}H_7—$	B–2.11, B–5.11
Phenoxazinyl	$ONC_{12}H_8—$	B–2.11, B–5.11
Phenoxy	$C_6H_5 \cdot O—$	C–205.1
Phenyl	$C_6H_5—$	A–13.1
Phenylacetyl	$C_6H_5 \cdot CH_2 \cdot CO—$	C–12.7, C–404.1
Phenylazo	$C_6H_5 \cdot N:N—$	C–912.3
Phenylazoxy	$C_6H_5 \cdot N_2O—$	C–913.1
Phenylcarbamoyl	$C_6H_5 \cdot NH \cdot CO—$	C–431.2
Phenylene	$—C_6H_4—$	A–11.6
Phenylenebisazo	$—N:N \cdot C_6H_4 \cdot N:N—$	C–912.5
Phenylimino	$C_6H_5 \cdot N=$	C–815.3
2-*phenylpropanoyl, see* Hydratropoyl		
3-*phenylpropenoyl, see* Cinnamoyl		
3-Phenylpropyl	$C_6H_5 \cdot CH_2 \cdot CH_2 \cdot CH_2—$	A–61.2
Phenylsulfamoyl	$C_6H_5 \cdot NH \cdot SO_2—$	C–661.2
Phenylsulfinyl	$C_6H_5 \cdot SO—$	C–631.1
Phenylsulfonyl	$C_6H_5 \cdot SO_2—$	C–631.1
Phenylsulfonylamino	$C_6H_5 \cdot SO_2 \cdot NH—$ C–641.7, C–641.8, C–823.1	
Phenylthio	$C_6H_5 \cdot S—$	C–514.1
3-Phenylureido	$C_6H_5 \cdot NH \cdot CO \cdot NH—$	C–971.2
Phthalamoyl	$NH_2 \cdot CO \cdot C_6H_4 \cdot CO—$ (*o-*)	C–431.1, C–431.2
Phthalazinyl	$N_2C_8H_5—$	B–2.11, B–5.11
Phthalidyl	$C_6H_4 \cdot CO \cdot O \cdot CH—$ $\vert \underline{\qquad\qquad} \vert$	B–5.11, C–473.1
Phthalidylidene	$C_6H_4 \cdot CO \cdot O \cdot C=$ $\vert \underline{\qquad\qquad} \vert$	B–5.12, C–473.1
Phthalimido	$CO \cdot C_6H_4 \cdot CO \cdot N—$ $\vert \underline{\qquad\qquad} \vert$	C–827.2
Phthaloyl (*preferred to* 1,2-*benzenedicarbonyl*)	$—CO \cdot C_6H_4 \cdot CO—$ (*o-*)	C–404.1
Phytyl [*for (E)-(7R,11R)-3,7,11,15-tetramethyl-2-hexadecenyl*]	$C_{20}H_{39}—$	A–75.1
Picenyl	$C_{22}H_{13}—$	A–21.1, A–24.2
Picryl	$2,4,6-(NO_2)_3C_6H_2—$	C–202.2
Pimeloyl (*unsubstituted only*)	$—CO \cdot [CH_2]_5 \cdot CO—$	C–404.1
Pinanyl	$C_{10}H_{17}—$	A–72.1, A–75.2
Pinanylene	$—C_{10}H_{16}—$	A–72.1, A–75.2
Pinanylidene	$C_{10}H_{16}=$	A–72.1, A–75.2
Piperazinyl	$N_2C_4H_9—$	B–2.12, B–5.11
Piperidino (1-*position only*)	$C_5H_{10}N—$	B–2.12, B–5.11
Piperidyl	$NC_5H_{10}—$	B–2.12, B–5.11
Piperidylidene	$NC_5H_9=$	B–5.12
Piperonyl	$3,4-CH_2O_2:C_6H_3 \cdot CH_2—$	(*cf.* C–411.1)
Piperonylidene	$3,4-CH_2O_2:C_6H_3 \cdot CH=$	A–13.4 (*cf.* C–411.1)
Piperonyloyl (*preferred to* 3,4-*methylenedioxybenzoyl*)	$3,4-CH_2O_2:C_6H_3 \cdot CO—$	C–411.1
Pivaloyl (*unsubstituted only*)	$(CH_3)_3C \cdot CO—$	C–404.1

317

Radical name	Formula	Based on rule no.
Pleiadenyl	$C_{18}H_{11}-$	A–21.1, A–24.2
Polythio	$-S_n-$	C–515.1
Prolyl	$NH \cdot CH_2 \cdot CH_2 \cdot CH_2 \cdot CH \cdot CO-$	C–421.1
propanedioyl, see Malonyl		
Propane-1,3-diyl-2-ylidene	$-CH_2-C-CH_2-$	A–4.5
Propane-1,2,3-triyl	$-CH_2-CH-CH_2-$	A–4.5
propanoyl, see Propionyl		
Propan-1-yl-3-ylidene	$=CH \cdot CH_2 \cdot CH_2-$	A–4.4
propargyl, see 2-Propynyl		
propenoyl, see Acryloyl		
1-Propenyl	$CH_3 \cdot CH : CH-$	A–3.5
2-propenyl, see Allyl		
Propenylene	$-CH_2 \cdot CH : CH-$	A–4.3
Propioloyl (*preferred to propynoyl*)	$CH : C \cdot CO-$	C–404.1
Propionamido	$CH_3 \cdot CH_2 \cdot CO \cdot NH-$	C–823.1
Propionyl (*preferred to propanoyl*)	$CH_3 \cdot CH_2 \cdot CO-$	C–404.1
Propionylamino	$CH_3 \cdot CH_2 \cdot CO \cdot NH-$	C–823.1
Propionyloxy	$CH_3 \cdot CH_2 \cdot CO \cdot O-$	C–463.3
Propoxy	$CH_3 \cdot CH_2 \cdot CH_2 \cdot O-$	C–205.1
Propyl	$CH_3 \cdot CH_2 \cdot CH_2-$	A–1.2
Propylene	$-CH(CH_3) \cdot CH_2-$	A–4.2
Propylidene	$CH_3 \cdot CH_2 \cdot CH=$	A–4.1
Propylidyne	$CH_3 \cdot CH_2 \cdot C\equiv$	A–4.1
propynoyl, see Propioloyl		
1-Propynyl	$CH_3 \cdot C : C-$	A–3.5
2-Propynyl	$HC : C \cdot CH_2-$	A–3.5
Protocatechuoyl (*preferred to 3,4-dihydroxybenzoyl*)	$3,4-(HO)_2C_6H_3 \cdot CO-$	C–411.1
Pteridinyl	$N_4C_6H_3-$	B–2.11, B–5.11
Purinyl	$N_4C_5H_3-$	B–2.11, B–5.11
Pyranthrenyl	$C_{30}H_{15}-$	A–21.1, A–24.2
Pyranyl	OC_5H_5-	B–2.11, B–5.11
Pyranylidene	$OC_5H_4=$	B–2.11, B–5.12
Pyrazinyl	$N_2C_4H_3-$	B–2.11, B–5.11
Pyrazolidinyl	$N_2C_3H_7-$	B–2.12, B–5.11
Pyrazolinyl	$N_2C_3H_5-$	B–2.12, B–5.11
Pyrazolyl	$N_2C_3H_3-$	B–2.11, B–5.11
Pyrenyl	$C_{16}H_9-$	A–21.1, A–24.2
Pyridazinyl	$N_2C_4H_3-$	B–2.11, B–5.11
2-Pyridinecarbonyl	$NC_5H_4 \cdot CO-$	C–401.2, C–403.2
3-pyridinecarbonyl, see Nicotinoyl		
4-pyridinecarbonyl, see Isonicotinoyl		
Pyridinio	$+NC_5H_5-$	C–82.1, C–85.1
Pyridyl	NC_5H_4-	B–2.11, B–5.11
2-Pyridylcarbonyl	$NC_5H_4 \cdot CO-$	C–403.2
Pyridyloxy	$NC_5H_4 \cdot O-$	C–205.1
Pyrimidinyl	$N_2C_4H_3-$	B–2.11, B–5.11
Pyrrolidinyl	NC_4H_8-	B–2.12, B–5.11
Pyrrolinyl	NC_4H_6-	B–2.12, B–5.11
Pyrrolyl	NC_4H_4-	B–2.11, B–5.11
Pyruvoyl	$CH_3 \cdot CO \cdot CO-$	C–416.3
Quinazolinyl	$N_2C_8H_5-$	B–2.11, B–5.11
Quinolinediyl	$-NC_9H_5-$	B–2.11, B–5.13
Quinolizinyl	NC_9H_8-	B–2.11, B–5.11

Radical name	Formula	Based on rule no.
Quinolyl	NC_9H_6-	B–2.11, B–5.11
Quinoxalinyl	$N_2C_8H_5-$	B–2.11, B–5.11
Quinuclidinyl	$NC_7H_{12}-$	B–2.12, B–5.11
Rubicenyl	$C_{26}H_{13}-$	A–21.1, A–24.2
Salicyl (*preferred to* 2-*hydroxybenzyl*)	$o\text{-}HO \cdot C_6H_4 \cdot CH_2-$	A–13.3, C–201.4
Salicylidene (*preferred to* 2-*hydroxybenzylidene*)	$o\text{-}HO \cdot C_6H_4 \cdot CH=$	A–13.4, C–201.4
Salicyloyl (*preferred to* o-*hydroxybenzoyl*)	$o\text{-}HO \cdot C_6H_4 \cdot CO-$	C–411.1
Sarcosyl	$CH_3 \cdot NH \cdot CH_2 \cdot CO-$	C–421.1
Sebacoyl (*unsubstituted only*)	$-CO \cdot [CH_2]_8 \cdot CO-$	C–404.1
Seleneno	$HOSe-$	C–701.1
Selenino	HO_2Se-	C–701.1
Seleninyl	$OSe{<}$	C–701.1
Seleno	$-Se-$	C–701.1
Selenocyanato	$NC \cdot Se-$	C–833.1
Selenoformyl	$HSeC-$	C–701.1
Selenonio	$^+H_2Se-$ (ion)	C–82.1, C–85, C–87.1
Selenono	HO_3Se-	C–701.1
Selenonyl	$O_2Se{<}$	C–701.1
Selenoureido	$H_2N \cdot CSe \cdot NH-$	C–971.2, C–974.1
Selenoxo	$(C)=Se$	C–701.1
Semicarbazido	$H_2N \cdot CO \cdot NH \cdot NH-$	C–981.2
Semicarbazono	$H_2N \cdot CO \cdot NH \cdot N=$	C–982.1
Seryl	$HO \cdot CH_2 \cdot CH(NH_2) \cdot CO-$	C–421.1
Stearoyl (*unsubstituted only*)	$CH_2 \cdot [CH_2]_{16} \cdot CO-$	C–404.1
Styryl	$C_6H_5 \cdot CH:CH-$	A–13.3
Suberoyl (*unsubstituted only*)	$-CO \cdot [CH_2]_6 \cdot CO-$	C–404.1
Succinamoyl	$NH_2 \cdot CO \cdot CH_2 \cdot CH_2 \cdot CO-$	C–431.2
Succinimido	$\overline{CO \cdot CH_2 \cdot CH_2 \cdot CO} \cdot N-$	C–827.2
Succinimidoyl	$-C(:NH) \cdot CH_2 \cdot CH_2 \cdot C(:NH)-$	C–451.2
Succinyl (*preferred to butanedioyl*)	$-CO \cdot CH_2 \cdot CH_2 \cdot CO-$	C–404.1
Sulfamoyl	$NH_2 \cdot SO_2-$	C–641.8
Sulfanilamido	$p\text{-}NH_2 \cdot C_6H_4 \cdot SO_2 \cdot NH-$	C–641.8
Sulfanilyl	$p\text{-}H_2N \cdot C_6H_4 \cdot SO_2-$	C–641.1, C–641.7
Sulfenamoyl	$NH_2 \cdot S-$	C–641.8
Sulfeno	$HO \cdot S-$	C–641.2
Sulfido	$^-S-$ (ion)	C–86.2, C–511.4
Sulfinamoyl	$NH_2 \cdot SO-$	C–641.8
Sulfino	HO_2S-	C–641.2
Sulfinyl	$-SO-$	C–631.2, C–661.3
Sulfo	$HO \cdot SO_2-$	C–10.3, C–641.2
Sulfoamino	$HO_3S \cdot NH-$	C–661.2
Sulfonato	$^-O_3S-$	C–86.1
Sulfonio	$^+H_2S-$ (ion)	C–82.1, C–85, C–87.1
Sulfonyl	$-SO_2-$	C–631.1, C–631.2, C–661.3
Sulfonyldioxy	$-O \cdot SO_2 \cdot O-$	C–72.2, C–205 2

Radical name	Formula	Based on rule no.
Tartaroyl (*preferred to* 2,3-*dihydroxybutanedioyl*)	$-OC \cdot CH(OH) \cdot CH(OH) \cdot CO-$	C–411.1
Tartronoyl (*preferred to* *hydroxypropanedioyl*)	$-CO \cdot CH(OH) \cdot CO-$	C–411.1
Tauryl	$H_2N \cdot CH_2 \cdot CH_2 \cdot SO_2-$	C–641.7
Telluro	Te replacing O	C–701.2
(Tercyclohexan)yl	$C_6H_{11} \cdot C_6H_{10} \cdot C_6H_{10}-$	A–54.1, C–71.2
Terephthaloyl (*preferred to* 1,4-*benzenedicarbonyl*)	$-CO \cdot C_6H_4 \cdot CO-$ (*p*-)	C–404.1
Terphenylyl	$C_6H_5 \cdot C_6H_4 \cdot C_6H_4-$	A–54.3, C–71.2
(Terthiophen)yl	$SC_4H_3 \cdot SC_4H_2 \cdot SC_4H_2-$	C–71.1, C–71.2
Tetracontyl	$CH_3 \cdot [CH_2]_{38} \cdot CH_2-$	A–1.2
Tetracosyl	$CH_3 \cdot [CH_2]_{22} \cdot CH_2-$	A–1.2
Tetradecanoyl	$CH_3 \cdot [CH_2]_{12} \cdot CO-$	C–403.1
Tetradecyl	$CH_3 \cdot [CH_2]_{12} \cdot CH_2-$	A–1.2
Tetramethylene	$-CH_2 \cdot [CH_2]_2 \cdot CH_2-$	A–4.2
Tetraphenylenyl	$C_{24}H_{15}$	A–21.1, A–24.2
Tetrazolyl	N_4CH-	B–1, B–5.11
Thenoyl (2- *shown*) (*preferred to* *thiophenecarbonyl*)	$S \cdot CH:CH \cdot CH:C \cdot CO-$	C–404.1
Thenyl	$SC_4H_3 \cdot CH_2-$	B–5.11
Thenylidene	$SC_4H_3 \cdot CH=$	B–5.11
Thianthrenyl	$S_2C_{12}H_7-$	B–2.11, B–5.11
Thiazinyl	SNC_4H_4-	B–1, B–5.11
Thiazolidinyl	SNC_3H_6-	B–1, B–5.11
Thiazolinyl	SNC_3H_4-	B–1, B–5.11
Thiazolyl	SNC_3H_2-	B–1, B–5.11
Thienyl	SC_4H_3-	B–2.11, B–5.11
Thio	$-S-$	C–72.1, C–514.1, C–661.3
Thioacetyl	$CH_3 \cdot CS-$	C–541.2
Thiobenzoyl	$C_6H_5 \cdot CS-$	C–541.2
Thiocarbamoyl	$H_2N \cdot CS-$	C–431.2, C–547.1
Thiocarbazono	$HN:N \cdot CS \cdot NH \cdot NH-$ (and tautomers)	C–985.1
Thiocarbodiazono	$HN:N \cdot CS \cdot N:N-$ (and tautomers)	C–985.1
Thiocarbonohydrazido	$H_2N \cdot NH \cdot CS \cdot NH \cdot NH-$	C–985.1
Thiocarbonyl	$-CS-$, $SC{<}$	C–108.2, C–543.2, C–545.1
Thiocarboxy	$HSOC-$	C–541.1
Thiocyanato	$NCS-$	C–833.1
Thioformyl	$SHC-$	C–531.3, C–543.4
thionaphthenyl, see Benzo[*b*]thienyl		
thiophanthrenyl, see Naphtho-[2,3-*b*]thienyl		
thiophenecarbonyl, see Thenoyl		
Thiosemicarbazido	$H_2N \cdot CS \cdot NH \cdot NH-$	C–985.1
Thiosulfino	HOS_2-	C–641.3
Thiosulfo	HO_2S_2-	C–641.3
Thioureido	$H_2N \cdot CS \cdot NH-$	C–971.2, C–974.1
Thioxo	$S=$	C–532.3
Threonyl	$CH_3 \cdot CH(OH) \cdot CH(NH_2) \cdot CO-$	C–421.1
Thujenyl	$C_{10}H_{15}-$	A–72.1, A–75.2
Thujyl	$C_{10}H_{17}-$	A–72.1, A–75.2
Thyronyl	p-(p-$HO \cdot C_6H_4 \cdot O) \cdot C_6H_4 \cdot CH_2 \cdot CH(NH_2) \cdot CO-$	C–421.1
Toluenesulfonyl (*cf.* Tosyl)	$CH_3 \cdot C_6H_4 \cdot SO_2-$	C–631.1, C–641.7
Toluidino (*o*-, *m*-, and *p*-)	$CH_3 \cdot C_6H_4 \cdot NH-$	C–811.4

Radical name	Formula	Based on rule no.
Toluoyl (o-, m-, and p-) (preferred to methylbenzenecarbonyl)	$CH_3 \cdot C_6H_4 \cdot CO—$	C–404.1
Tolyl (o-, m-, and p-)	$CH_3 \cdot C_6H_4—$	A–13.1
Tolylsulfonyl	$CH_3 \cdot C_6H_4 \cdot SO_2—$	C–631.1
Tosyl (p- only)	$p\text{-}CH_3 \cdot C_6H_4 \cdot SO_2—$	C–641.7
Triacontyl	$CH_3 \cdot [CH_2]_{28} \cdot CH_2—$	A–1.2
Triazano	$H_2N \cdot NH \cdot NH—$	C–942.3
Triazaphenanthryl	$N_3C_{11}H_6—$	B–4.1, B–5.21
Triazeno	$NH_2 \cdot N:N—$	C–942.3
Triazinyl	$N_3C_3H_2—$	B–1, B–5.11
Triazolidinyl	$N_3C_2H_6—$	B–1, B–5.11
Triazolyl	$N_3C_2H_2—$	B–1, B–5.11
Trichlorothio	$Cl_3S—$	C–621.2
Tricosyl	$CH_3 \cdot [CH_2]_{21} \cdot CH_2—$	A–1.2
Tridecanoyl	$CH_3 \cdot [CH_2]_{11} \cdot CO—$	C–403.1
Tridecyl	$CH_3 \cdot [CH_2]_{11} \cdot CH_2—$	A–1.2
Trifluorothio	$F_3S—$	C–621.2
3,4,5-trihydroxybenzoyl, see Galloyl		
Trimethylammonio	$(CH_3)_3\overset{+}{N}—$	C–82.1, C–85, C–87.1, C–816.3
Trimethylanilino (all isomers)	$(CH_3)_3C_6H_2 \cdot NH—$	C–811.4
1,2,2-trimethyl-1,3-cyclopentane-dicarbonyl, see Camphoroyl		
Trimethylene	$—CH_2 \cdot CH_2 \cdot CH_2—$	A–4.2
Trimethylenedioxy	$—O \cdot CH_2 \cdot CH_2 \cdot CH_2 \cdot O—$	C–72.2, C–205.2
Trinaphthylenyl	$C_{30}H_{17}—$	A–21.1, A–24.2
Triphenylenyl	$C_{18}H_{11}—$	A–21.1, A–24.2
Triphenylmethyl (cf. Trityl)	$(C_6H_5)_3C—$	C–81.1 (cf. A–13.3)
Trithiadiazaindenyl	$S_3N_2C_4H_7—$	B–4.1, B–5.21
Trithio	$—S_3—$	C–515.1
Trithiosulfo	$HS \cdot S_3—$	C–641.1, C–641.2
Tritriacontyl	$CH_3 \cdot [CH_2]_{31} \cdot CH_2—$	A–1.2
Trityl (cf. Triphenylmethyl)	$(C_6H_5)_3C—$	A–13.3 (cf. C–81.1)
Tropoyl (preferred to 3-hydroxy-2-phenylpropanoyl)	$C_6H_5 \cdot CH(CH_2 \cdot OH) \cdot CO—$	C–411.1
Tryptophyl	$NC_8H_6 \cdot CH_2 \cdot CH(NH_2) \cdot CO—$	C–421.2
Tyrosyl	$p\text{-}HO \cdot C_6H_4 \cdot CH_2 \cdot CH(NH_2) \cdot CO—$	C–421.1
Undecanoyl	$CH_3 \cdot [CH_2]_9 \cdot CO—$	C–403.1
Undecyl	$CH_3 \cdot [CH_2]_9 \cdot CH_2—$	A–1.2
Ureido	$NH_2 \cdot CO \cdot NH—$	C–971.2
Ureylene	$—NH \cdot CO \cdot NH—$	C–72.2, C–971.3
Valeryl	$CH_3 \cdot [CH_2]_3 \cdot CO—$	C–404.1
Valyl	$(CH_3)_2CH \cdot CH(NH_2) \cdot CO—$	C–421.1
Vanilloyl (preferred to 4-hydroxy-3-methoxybenzoyl)	$3,4\text{-}CH_3O \cdot (HO)C_6H_3 \cdot CO—$	C–411.1
Vanillyl	$3,4\text{-}CH_3O \cdot (HO)C_6H_3 \cdot CH_2—$	cf. A–13.3, C–201.4, C–411.1
Vanillylidene	$3,4\text{-}CH_3O \cdot (HO)C_6H_3 \cdot CH=$	cf. A–13.4, C–201.4, C–411.1
Veratroyl (preferred to 3,4-dimethoxybenzoyl)	$3,4\text{-}(CH_3O)_2C_6H_3 \cdot CO—$	C–411.1
Veratryl	$3,4\text{-}(CH_3O)_2C_6H_3 \cdot CH_2—$	cf. A–13.3, C–201.4, C–411.1
Veratrylidene	$3,4\text{-}(CH_3O)_2C_6H_3 \cdot CH=$	cf. A–13.4, C–201.4, C–411.1
Vinyl	$CH_2:CH—$	A–3.5
Vinylene	$—CH:CH—$	A–4.3
Vinylidene	$CH_2:C=$	A–4.1
Violanthrenyl	$C_{34}H_{19}—$	A–23.1, A–24.2

Radical name	Formula	Based on rule no.
Xanthenyl	$OC_{13}H_9—$	B–2.11, B–5.11
Xylidino	$(CH_3)_2C_6H_3 \cdot NH—$	C–811.4
Xylyl	$(CH_3)_2C_6H_3—$	A–13.1

(Provisional Recommendations 1978)

D-0. PREAMBLE

D-0.1 - Scope of Section D

Rules for the general nomenclature of inorganic chemistry and for the nomenclature of specialized areas or groups of inorganic compounds are contained in "IUPAC Nomenclature of Inorganic Chemistry"[*] prepared by Commission on Nomenclature of Inorganic Chemistry[†].

The nomenclature of organic compounds which contain carbon, hydrogen, oxygen, nitrogen, halogen, sulfur, selenium and/or tellurium has been described in the preceding Sections A, B and C. Note, however, that Subsection B-1 also mentions the elements phosphorus, arsenic, antimony, bismuth, silicon, germanium, tin, lead, boron and mercury, therefore permitting their use in the Hantzsch-Widman system and replacement nomenclature (see Subsection C-63).

Section D deals with organic compounds of other elements. As a consequence, it covers fields at the border of organic and inorganic chemistry, and hence compounds that can be named either from an organic or an inorganic point of view.

The main difference between the nomenclature of inorganic chemistry and that of organic chemistry is that the latter is based predominantly on the substitution of hydrogen in parent compounds while the former is generally based on coordination (additive) concepts. Thus, organic nomenclature will often be based on names of hydrides. A good illustration is the name pentaphenylphosphorane, based on the name phosphorane for PH_5. By coordination nomenclature, $P(C_6H_5)_5$ is named pentaphenylphosphorus.

Section D attempts to deal with both organic and inorganic approaches to nomenclature for very similar compounds in many cases, while special nomenclature systems, which may eventually disappear, have been codified in a few areas, it is still the goal of both Commissions to provide the most general rules for future use.

Whether an organic or an inorganic name is more convenient depends upon the class of compounds considered, the nature of the specific compound, the context and, without doubt, upon tradition.

In Section D, the nomenclature of organosilicon compounds is almost exclusively "organic" and that of organometallic compounds predominantly "inorganic". In Subsection D-5 on organophosphorus nomenclature many compounds have been given both organic and inorganic names and it is left to future development to show which will survive.

---

[*] *Nomenclature of Inorganic Chemistry*, 2nd Edition (Definitive Rules 1970). Butterworths, London (1971) but now available from Pergamon Press, Oxford.

[†] IUPAC Commission on the Nomenclature of Inorganic Chemistry: *Members for varying periods from 1959 to 1972:* R.M. ADAMS, H. BASSETT*, J. BÉNARD, L.F. BERTELLO, K.C. BUSCHBECK, J. CHATT, G.H. CHEESMAN, E.J. CRANE*, T. ERDEY-GRÚZ*, W. FEITKNECHT, W.C. FERNELIUS, F. GALLAIS, A.K. HOLLIDAY, Y. JEANNIN, K.A. JENSEN, W. KLEMM, W. KOTOWSKI, J.G. LEIGH, L. MALATESTA, B. MYASOEDOV, A. ÖLANDER, W.H. POWELL, J.E. PRUE*, A.L.G. REES, H. REMY*, C. SCHÄFFER, A. SILVERMAN*, A.A. VLČEK, E. WEISS, K. YAMASAKI.

*Deceased

The Commissions have preferred to codify satisfactory existing practices rather than to elaborate new nomenclature systems . At the same time they have tried to keep the number of permitted names to a minimum, although it is relatively difficult to do so for such diversified topics .

D-0.2 - Conventions

The way in which these Rules are written, the use of multiplying affixes (di, tri, bis, tris, etc.) and the conventions for numbering are the same as in Sections A, B and C, when names are formed according to organic nomenclature, and the same as in the "Nomenclature of Inorganic Chemistry", when names are formed according to inorganic nomenclature rules.

D-0.3 - Terminology

The terminology used in Section D is as follows :

Central atom : central atoms are defined according to the nomenclature of coordination compounds (Subsection D-2) . The extension of the concept of "central atom" to organic substitutive nomenclature may be useful.

Functional class name : a word such as ketone, chloride or alcohol, used in radicofunctional nomenclature as an ending or as a separate word according to the language (see Rules C-21 to C-24) .

Substituent : any atom or group replacing hydrogen of a parent compound .

Characteristic group   * : an atom or group that is incorporated into a parent compound otherwise than by a direct carbon-carbon linkage, but including groups -C≡N and >C=X where X is O, S, Se, Te, NH or substituted NH . It will be noted that the concept of "characteristic group" includes groups such as OH, NH_2, COOH and single atoms such as halogen, =O and ≡N . It does not apply to substituents such as methyl, phenyl, 2-pyridyl, but does include, for example, piperidino and acetyl .

Principal group : the characteristic group chosen for expression as a suffix in a particular name . This is equivalent to the "principal function" of the Liège Rules .

Seniority of characteristic groups : the word "seniority" is used, as in Sections A, B and C, with reference to priority according to a prescribed order given either by a table or a rule .

---

  * This definition has been carried over from Section C (page 82) for the purpose of this tentative version . It will require reconsideration .

Bonding number :   The extension of the principles of substitutive nomenclature to include compounds with heteroatoms of variable valency requires a method for distinguishing between different valency states of a heteroatom in the parent compound. A number of methods have been used in the past, most of which are limited in scope and none of which is entirely satisfactory for general applicability. In this document different valency states of neutral heteroatoms in parent compounds are distinguished by the symbol λ^n, where n is the "bonding number" of the heteroatom.

The "bonding number" of a neutral heteroatom in the corresponding unsubstituted parent compound is defined as the sum of the total number of bonds to adjacent skeletal atoms, if any, *and* the number of hydrogen atoms attached to it, if any.

In cyclic compounds, when cumulative double bonds terminate at a heteroatom of the ring, the symbol δ^c, where c is the number of skeletal cumulative double bonds terminating at the heteroatom, is added to the λ^n symbol for the heteroatom.

D-1. NOMENCLATURE SYSTEMS

D-1.1 - Inorganic nomenclature of binary compounds

Many inorganic compounds consist, or can be regarded as consisting essentially of two parts, an electropositive and an electronegative constituent.

1.11 - Systematic names are formed by indicating the constituents and their proportions, the name of the electropositive constituent being placed before that of the electronegative constituent.

1.12 - Exceptionally if one element is hydrogen the compound may have a **one-word name.**

Examples :

Ammonia	NH_3
Diphosphane	P_2H_4
Germane	GeH_4

1.13 - The name of the electropositive constituent is not modified. The name of the electronegative constituent is modified according to Rules D-1.14, D-1.15 or D-1.16.

1.14 - If the electronegative constituent is monoatomic or homopolyatomic its name is modified to end in "ide".

Examples :

Tin tetrachloride	$SnCl_4$
Dinitrogen oxide	N_2O
Nitrogen dioxide	NO_2

1.15 - In certain exceptional cases heteropolyatomic electronegative groups also have the termination "ide".

Examples :

Hydroxide	HO^-
Cyanide	NC^-
Amide	H_2N^-

1.16 - If the electronegative substituent is heteropolyatomic, its name is modified to end in "ate".

Examples :

Trisodium tetrathiophosphate	$Na_3[PS_4]$
Sodium tetraphenylborate	$Na[B(C_6H_5)_4]$

D-1.2 - Coordination nomenclature

This nomenclature, originally intended for coordination compounds, is spreading to other parts of inorganic and organometallic chemistry because it is often felt that a binary-type name implies salt-like character in the compound . From this point of view, chloro(phenyl)mercury * is preferred to phenylmercury chloride.

In Section D, coordination nomenclature is used consistently to designate heteropolyatomic anions named by inorganic procedures. This method is also applicable to organic compounds and may have advantages because of its simplicity.

1.21 - To name an electronegative heteropolyatomic group or ion, the name of its central atom is generally modified to end in "ate" (Rule D-1.16) . Such a polyatomic group is designated as a coordination entity and the atoms, groups and molecules bound to the central atom are termed ligands.

The ligand names are added, as prefixes, in alphabetical order, to the modified name of the central atom to name the polyatomic group. The details concerning the definition of central atom, the designation of ligands, etc., appear in Subsection D-2.

The termination "ite", to denote a lower oxidation state, is only retained in a few trivial names and should gradually disappear completely . For example, names such as antimonate(III), plumbate(II), etc. , should now be used instead of antimonite, plumbite, etc. . The oxidation state of the central atom may be indicated according to the Stock system . Alternatively, the charge of the entire ion may be given according to the Ewens-Bassett system .

Example :

Hexacyanoferrate(II) ion (Stock system) $[Fe(CN)_6]^{4-}$
Hexacyanoferrate(4-) ion (Ewens-Bassett system)

1.22 - A neutral or electropositive coordination entity is named by adding, as prefixes, the names of all ligands, in alphabetical order, to the unmodified name of the central atom.

Examples :

Amido(chloro)methoxooxophosphorus * $[P(NH_2)Cl(OCH_3)O]$
Hexaamminecobalt(III) ion $[Co(NH_3)_6]^{3+}$

D-1.3 - Substitutive nomenclature

When substitutive nomenclature is used in Section D, the principal characteristic group is not always expressed by a suffix to the name of a parent compound. Nevertheless, in extending organic

---

* Although the use of parentheses around simple ligands is not required, it can avoid the possibility of misinterpretation.

chemical nomenclature to compounds containing characteristic groups other than those of Section C, it is also necessary to extend Table II of Rule C-10.1 which gives general classes of compounds in the order in which the characteristic groups have decreasing priority for citation as principal group. The extended version of this Table is given in Rule D-1.32 .

Choice of the functional class of a characteristic group

1.31 - When there is a choice for the functional class of a characteristic group, this is made according to the order of precedence given in Rule D-1.32 .

Examples

Ethylphosphonamidic acid
(Acids take precedence over amides) $C_2H_5P(O)\!\!<^{NH_2}_{OH}$

O-Methyl hydrogen ethylphosphonimidate *
(Acids and esters take precedence over $C_2H_5P(NH)\!\!<^{OCH_3}_{OH}$
imides)

Ethylphosphonamidic chloride (Rule D-5.63) $C_2H_5P(O)\!\!<^{NH_2}_{Cl}$
(Acyl halides take precedence over amides)

Choice between characteristic groups bound to the same structure for citation as the principal group

1.32 - (see Rule C-10.3). When there is a choice between possible principal groups, the group corresponding to the functional class cited first in the following list is selected.

General classes of compounds in decreasing order of priority for naming and choosing a principal group :

1a. Anionic centers in the order of corresponding acids.

1b. "Onium" and similar cations (see Subsection C-0.8).

2a. Acids : in the order COOH, CO.OOH, then successively their S and Se derivatives, followed by sulfonic, sulfinic acids.., selenonic, seleninic acids..,phosphonic,phosphinic, phosphonous, phosphinous acids, etc. † .

---

* Although the 1970 Inorganic Nomenclature Rules recommend the joining of hydrogen to the following name of the anion (see Rule 6.2), the previously used format in Section A-C is followed here, i. e., hydrogen is written as a separate word.

† see footnote † on page 329.

2b. Acidic derivatives of inorganic acids in which the central atom of the acid residue is linked through one or more heteroatoms to the carbon skeleton of the organic part of the molecule * † .

3. Derivatives of acids belonging to class 2a , in the order : anhydrides. esters, acyl halides, amides, hydrazides, imides, amidines, etc. * † .

4. Nitriles (cyanides), then isocyanides .

5. Aldehydes; then successively their S and Se analogues; then their derivatives .

6. Ketones; then their analogues and derivatives, in the same order as for aldehydes .

7. Alcohols and phenols; then their S, Se and Te analogues; then neutral esters of alcohols and phenols with inorganic acids except hydrogen halides, in the same order *† .

8. Hydroperoxides .

9a. Amines, imines, hydrazines, etc. ; then phosphines, arsines, etc. .

9b. Other derivatives of inorganic acids which cannot be considered either as acids (class 2b) or as esters (class 7) * .

10. Ethers; then successively their S and Se analogues .

11. Peroxides .

---

* For supplementary information concerning seniority inside these classes, see Rule D-1.33 .

† The nomenclature of esters of inorganic acids is subject to discussion because two points of view are in conflict .

In Section C, the organic part of the molecule is considered as the more important and seniority rules relate mainly to the substituents of this organic part . As a consequence, neutral esters of inorganic acids, nitrates for instance, are considered after the corresponding alcohols or phenols (class 7) .

In Section D, to the contrary, the inorganic part of the molecule is often considered as the more important and the name is built accordingly . For instance, esters of phosphoric acid are generally named as phosphates and not as derivatives of organic parent compounds . Then the seniority rules of class 3 are applied to the inorganic parent compound .

Examples :

4-Phosphonocyclohexanecarbox-
   ylic acid
(Carboxylic acids take precedence
 over phosphonic acids)

$(HO)_2P(O)$ —⬡— CO_2H

[4-(Ethoxycarbonyl)cyclohexyl]-
  phosphonic acid
(Acids take precedence over esters)

$(HO)_2P(O)$ —⬡— $CO-OC_2H_5$

2-(Dimethoxyphosphoryl)benzene-
  sulfonic acid
(Acids take precedence over esters)

⬡ with SO_3H and $PO(OCH_3)_2$

2-Phosphinoethylamine
(Amines take precedence over phosphines)

$H_2P-CH_2-CH_2-NH_2$

(Arsinomethyl)diphenylphosphine
(Phosphines take precedence over arsines)

$H_2As-CH_2-P(C_6H_5)_2$

1.33 – If a choice has to be made inside classes 2a, 2b, 7 and 9b of Rule D-1.32 for derivatives of inorganic acids, the following criteria are applied successively, in the order listed, until a decision is reached :

(a) Central atom of the characteristic group appearing first in the sequence of Table I given in Appendix , **p. 459, except that carbon as the central atom has always priority.**

(b) Highest oxidation number of the central atom of the characteristic group.

(c) Greatest number of atoms linked to the central atom and appearing as early as possible in Table I , **p. 459,** *with the exception that,* according to existing practice, halogens rank between the oxygen family and nitrogen .

Examples :

$$(HO)_2As(O)-CH_2-CH_2-CH_2-CH_2-P(O)(OH)_2$$

(4-Arsonobutyl)phosphonic acid (P takes precedence over As)

$$HO-As-⬡-As-OH$$ (with OH groups and O on the right As)

4-[(Dihydroxyarsino)phenyl]arsonic acid
    (arsenic(V) takes precedence over arsenic(III))

$$Br_2P(O)-CH_2-CH_2-P(O)Cl_2$$

[2-(Dibromophosphoryl)ethyl]phosphonic dichloride
   (Cl takes precedence over Br)

Choice of a parent between two directly bound characteristic groups

1.34 - When two atoms of different elements are directly bound and when the choice for parent compound is between those two elements, the parent compound is the one containing the element cited later in Table I., p. 459. All other groups are designated by a prefix.

Examples :

(Methylphosphino)silane	$CH_3 PH-SiH_3$
(Diphenylphosphino)borane	$(C_6H_5)_2 P-BH_2$
Acetylborane	$CH_3 CO-BH_2$

Rule 1.35 of the provisional version has been deleted.

1.36- As an exception to Rule D-1.34, compounds in which carbon atoms are directly linked to the phosphorus, arsenic, antimony or bismuth atoms may be named as derivatives of parent phosphorus, arsenic, antimony or bismuth hydrides.

Examples :

Acetylphosphine	$CH_3-CO-PH_2$
Carbamoylphosphine	$NH_2-CO-PH_2$

D-1.4 - Radicofunctional nomenclature

In Section D, radicofunctional nomenclature is mainly applied to two important classes of compounds : esters and acyl halides . In both cases, the "functional group" is a notional anion : phosphate, arsenate, silicate, halide, azide, etc.

The general seniority rules for these compounds are given in Subsection C-0.2 . Inside each class, seniority is defined by the following rules .

1.41 – The choice of the names of complex anions is directed by the principles of Rule D-1.31. Esters and acid esters derived from complex anions are named according to radicofunctional nomenclature, other ligands of the complex anions being cited inside the name of the anion.

Example :

Dimethyl phosphorochloridate
(Esters take precedence over
acyl halides)

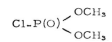

1.42 – For acyl halides and similar compounds, the order of seniority is determined first by the acyl radical, according to the order of seniority of the corresponding acid. If an order of seniority has then to be based upon the anionic name, this is done according to Rule C-22 in the order : fluoride, chloride, bromide, iodide, ... cyanide ..., azide, etc..

Examples :

2-(Dichlorophosphoryl)ethane-
    sulfonyl chloride $Cl_2 P(O)-CH_2-CH_2-SO_2-Cl$

2-(Chlorosulfonyl)ethane-
    sulfonyl fluoride $Cl-SO_2-CH_2-CH_2-SO_2-F$

D.1.5 – Additive nomenclature

1.51 – Additive nomenclature involves naming atoms added to the structure denoted by the rest of the name. This may be done in two ways :

(a) by adding a prefix "hydro-" to the parent name (Rule C-32),

(b) by adding a functional class name after the parent name (Rule C-33).

For the nomenclature of addition compounds, see Rule D-1.55.

1.52 – The prefix "hydro-" is used as described in Rules A-23.1, B-1.1, B-1.2, C-16 and C-32.1 as denoting the addition of one hydrogen atom. Multiplying prefixes are used as needed.

Example :

1,2,3,4-Tetrahydroarsinoline

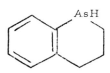

1.53 . Bonding numbers higher than those listed in Rule D-1.61 may be indicated by hydro prefixes ; however, **the designation given by Rule D-1.62 is preferred.**
Examples :

2,2,4,4,6,6-Hexaethyl-2,2,4,4,6,6-
hexahydro-1,3,5,2,4,6-triaza-
triphosphorine

2-Ethoxy-2,2-dihydro-2,2-dime-
thyl-1,3,2-dioxaphospholane

1.54 - Important functional class names involved in additive nomenclature are :

Class name	Added atom or group
Oxide	O
Sulfide	S
Selenide	Se
Telluride	Te
Imide	NH

Examples :

Trimethylarsine sulfide	$(CH_3)_3 AsS$
Triphenylarsine oxide	$(C_6H_5)_3 AsO$
Trimethylphosphine ethylimide	$(CH_3)_3 P(NC_2H_5)$

1.55 - The names of addition compounds are formed by citing those of the individual compounds, separated by a rule (long hyphen) . Boron compounds and water are always cited last and in that order . Other molecules are cited in order of increasing number; any which occur in equal numbers are cited in alphabetical order . The numbers of each individual molecule are indicated by Arabic numerals separated by a solidus, the total expression being enclosed in parentheses *.

Examples :

Aluminium chloride—ethanol (1/4)	$AlCl_3 \cdot 4C_2H_5OH$
Methanol—boron trifluoride (2/1)	$2CH_3OH \cdot BF_3$

* A colon may be used instead of the solidus.

D-1.6 - Replacement nomenclature

1.60 - The system of naming organic compounds in which carbon atoms have been replaced by heteroatoms has been described in Section B for cyclic compounds : Rule B-4, "a" Nomenclature, and in Section C for acyclic compounds : Subsection C-0.6, Replacement nomenclature .

In Section D, this system is broadened by extending Table I of Rule B-1 to include most of the elements . The corresponding "a" terms are given in Table I (Appendix, p. 459). This seniority order of elements is based upon the **element sequence given in Table IV** of "Nomenclature of Inorganic Chemistry, 1970, p. 104", and supersedes Rule B-1.4 .

1.61 - The affixes listed in Table I in decreasing order of priority are used according to the principles stated in Subsection C-0.6 for acyclic structures and in Rule B-4 for heterocyclic systems. These prefixes are used alone when the heteroatoms considered in the structure have the normal **bonding** number. The accepted normal **bonding** numbers are the following :

Elements	Normal **bonding** number
F, Cl, Br, I, At	1
O, S, Se, Te, Po	2
N, P, As, Sb, Bi	3
C, Si, Ge, Sn, Pb	4
B, Al, Ga, In, Tl	3
Be, Mg, Zn, Cd, Hg	2

For elements **and bonding numbers not listed here see Rule D-1.62.**

Examples :

1-Methyl-1-aluminacyclohexane

$$\begin{array}{c} \quad\quad CH_2 - CH_2 \\ H_2C \diagup \quad\quad\quad\quad \diagdown Al-CH_3 \\ \quad\quad\diagdown CH_2 - CH_2 \diagup \end{array}$$

Methyl 3-phospha-5-silahexanoate

$CH_3-SiH_2-CH_2-PH-CH_2-CO-OCH_3$

1.62 - When the bonding number of a neutral heteroatom is other than that given in Rule D-1.61 or if the heteroatom is not listed under that rule, the bonding number is expressed by an Arabic number cited as a superscript to the Greek letter λ (see also D-0.3) which follows the locant for the heteroatom. If no locant for the heteroatom normally appears in the name for the parent compound, the lambda symbol and an appropriate locant are prefixed to that part of the name which designates the heteroatom.

Note - In some cases, bonding numbers higher than those listed in Rule D-1.61 may be indicated by hydro prefixes according to Rule D-1.53 .

Examples :

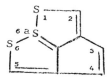

2,2,4,4,6,6-Hexaethyl-1,3,5,2λ^5,4λ^5,6λ^5-triazatriphosphorine

1,6,6aλ^4-Trithiacyclopenta[*c d*]pentalene

1$\lambda^4\delta^2$,2,4,6,3,5-Thiatriazadiphosphorine
(for the use of δ^2, see D-0.3)

2-Ethoxy-2,2-dimethyl-1,3,2λ^5-dioxaphospholane

1.63 - Ionic heteroatoms are expressed, whenever possible, by the usual terms such as "-onium", "-ium", "-ide", "-onia", "-ata", with their usual meaning (see Subsections C-0.8 and D-7.63) associated with normal bonding numbers.

Examples :

Tetramethylphosphonium bromide $[(CH_3)_4P]^+ Br^-$

Pyridinium perchlorate $[C_5H_5NH]^+[ClO_4]^-$

Sodium 4a-borataanthracene Na^+

D-1.7 - Fundamental heterocyclic systems

Section B deals with the general principles of the nomenclature of fundamental heterocyclic systems and accordingly Section D is restricted to systems not already covered by Section B and to appropriate extensions of the rules contained in Section B .

Rule B-1 (Extended Hantzsch-Widman system) already covers some heteroatoms other than oxygen, sulfur, selenium and nitrogen . Further extension may be accomplished by using Rule D-1.61 and Table I of the Appendix , p. 459.

Rule B-2 gives trivial and semitrivial names for heterocyclic systems but many other names, not given in this Rule, are commonly used for naming various heterocycles containing selenium, tellurium, phosphorus, arsenic, antimony and mercury . Table IV of the Appendix, p. 466, lists those names which are approved.

Rule B-3, concerning fused heterocyclic systems, may be also applied to component heterocycles named in Table IV of the Appendix , p. 466.

Replacement nomenclature, described in Rule B-4 for heterocyclic systems, may be generalized by use of the principles given in Subsection D-1.6 .

Special nomenclature rules concerning rings with regular patterns of heteroatoms, are given in Subsection D-4 .

D-2. COORDINATION COMPOUNDS

D-2.0 - Introduction

The following Rules are a summary of Section 7 of "Nomenclature of Inorganic Chemistry, 1970", issued by the IUPAC Commission on Nomenclature of Inorganic Chemistry.

The purpose of this summary is to enable an organic chemist to understand a coordination-type name ; however, should he wish to name a coordination structure he must consult "Nomenclature of Inorganic Chemistry, 1970". The appropriate inorganic rule number is therefore given in parentheses after the D-rule number.

D-2.1 (7.1) - Definitions

In its oldest sense the term *coordination entity* generally refers to molecules or ions in which there is an atom (A) to which are attached other atoms (B) or groups (C) to a number in excess of that corresponding to the classical or stoicheiometric valence of the atom (A). However, the system of nomenclature originally evolved for these compounds within this narrow definition has proved useful for a much wider class of compounds and for the purposes of nomenclature the restriction "in excess of stoicheiometric valence" is to be omitted. Any compound formed by addition of one or several ions and/or molecules to one or more ions or/and molecules may be named according to this system.

The effect of this definition is to bring many simple and well-known compounds under the same nomenclature rules as those previously accepted for coordination compounds. Thus, the diversity of names is reduced and also many controversial issues **are** avoided. It is not intended to imply the existence of any constitutional analogy between different compounds merely because they are named under a common system of nomenclature. The system may be extended to many addition compounds.

In the rules which follow certain terms are used in the senses here indicated : the atom referred to above as (A) is known as the *nuclear* or *central* atom, and all other atoms which are directly attached to (A) are known as *coordinating* or *ligating* atoms. Each central atom (A) has a characteristic *coordination number* or *ligancy* which is the number of atoms directly attached to it. Atoms (B) and groups (C) are called *ligands* . A group containing more than one potential coordinating atom is termed a *multidentate* ligand, the number of potential coordinating atoms being indicated by the terms *unidentate, bidentate,* etc. A *chelate* ligand is a ligand attached to one central atom through two or more coordinating atoms, while a *bridging group* is attached to more than one centre of coordination. The whole assembly of one or more central atoms with their attached ligands is referred to as a *coordinating entity,* which may be a cation, an anion or an uncharged molecule. A *polynuclear* entity is one which contains more than one nuclear or central atom, their number being designated by the terms *mononuclear, dinuclear,* etc.

D-2.2 (7.2) - Principles of coordination nomenclature

Formulae

2.21 (7.21) - In *formulae* the usual practice is to place the symbol(s) for the central atom(s) first (except in formulae which are primarily structural), with the ionic and then the neutral ligands following, and the formula for the whole **coordination entity** is enclosed in **square brackets [ ].**

The normal nesting order for enclosing marks is $\{[( )]\}$. However, in the formulae of coordination compounds, square brackets are used to enclose a complex ion or a neutral coordination entity. Enclosing marks are then nested with the square brackets as follows : $[( )]$, $[\{( )\}]$, $[\{[( )]\}]$, $[\{[( )]\}]$, etc., and a space is left between the outer square brackets with their associated subscript or superscript symbol if any, and the remainder of the formula.

Order within each class of ligand should be the alphabetical order of the symbols of the ligating atoms.

Examples :

$$K_3 \left[Co(C_2O_4)_3 \right]$$
$$\left[Co\{SC[N(CH_3)_2]_2\}_4 \right] \left[NO_3 \right]_2$$

Names

2.22 (7.21 and 7.25) - In *names* the central atom(s) should be placed after the ligands . The ligands are listed in alphabetical order, regardless of the number of each . A compound ligand is treated as a single unit .

Two kinds of multiplying prefixes are used within the complete name of the coordination entity. The prefixes di-, tri-, etc. are used to indicate a set of identical unsubstituted radicals or parent compounds. The prefixes bis-, tris-, etc., are used to indicate a set of identical radicals or parent compounds each substituted in the same way or to avoid ambiguity. Enclosing marks are used with multiplying prefixes as in organic nomenclature (Sections A, B and C. pp. 80-1).

Indication of oxidation number [*] and proportion of constituents

2.23 (7.22) - The names of coordination entities have always been intended to indicate the charge of the central atom (ion) from which the entity is derived . Since the charge on the coordination entity is the algebraic sum of the charges of the constituents, the necessary information may be supplied by giving either the oxidation number, according to Stock's system (formal charge on the central atom, i.e. the oxidation number, given in Roman numerals, the cipher 0 being used to indicate an oxidation state of zero) or the Ewens-Bassett number (charge on the entire ion, given in Arabic numerals) . In using Ewens-Bassett numbers, zero is omitted . Alternatively, the proportion of constituents may be given by means of multiplying prefixes .

Examples :

Potassium hexacyanoferrate(III) $K_3 [Fe(CN)_6]$
Potassium hexacyanoferrate(3-)
Tripotassium hexacyanoferrate

Potassium hexacyanoferrate(II) $K_4 [Fe(CN)_6]$
Potassium hexacyanoferrate(4-)
Tetrapotassium hexacyanoferrate

Note : In examples to the following rules, often only one name is given, either with Stock or Ewens-Bassett number . The choice only serves illustrative purposes and does not indicate a preference .

Terminations

2.24 (7.24) - Anions are given the terminations -ide, -ite, or -ate . Cations and neutral molecules are not given any distinguishing alphabetic termination.

Structural prefixes

2.25 (7.23) - Structural information may be given in formulae and names by prefixes such as *cis-*, *trans-* , *fac-*, *mer-* , etc. in conformity with Rule D-2.61 and Table III of the Appendix , p. 464.

---

[*] The oxidation number is the net charge of the entire coordination entity minus the sum of the charges of the ligands (see Rules D-2.33 (7.313) and D-2.37 (7.323). For full definitions and discussions, refer to "IUPAC - Nomenclature of Inorganic Chemistry, 2nd Edition, Butterworths, London (1971), p. 5.

D-2.3 (7.3) - Ligands

Anionic ligands

2.31 (7.311) - The names of anionic ligands, whether inorganic or organic, end in -o [see, however, Rule D-2.33 (7.313)] . In general, if the anion name ends in -ide, -ite or -ate, the final -e is replaced by -o, giving -ido, -ito or -ato respectively [see also Rule D-2.34 (7.314)] . Enclosing marks are required for inorganic anionic ligands the names of which contain numerical prefixes, as (triphosphato), and for thio, seleno and telluro analogues of oxo anions containing more than one oxygen atom, as (thiosulfato) .

Examples :

Acetato	CH_3-COO^-
Methyl sulfito	CH_3O-SOO^-
Dimethylamido	$(CH_3)_2N^-$
Acetamido	CH_3-$CONH^-$
Hydrido *	H^-

2.32 (7.312) - The anions listed below do not follow exactly Rule D-2.31 . Modified forms have become established in some cases along with the regular forms .

Formula	Ion name	Ligand name
F^-	Fluoride	Fluoro
Cl^-	Chloride	Chloro
Br^-	Bromide	Bromo
I^-	Iodide	Iodo
O^{2-}	Oxide	Oxo
H^-	Hydride	Hydrido or Hydro *
HO^-	Hydroxide	Hydroxo
O_2^{2-}	Peroxide	Peroxo
HO_2^-	Hydrogenperoxide †	Hydrogenperoxo
S^{2-}	Sulfide	Thio ‡

.../...

* Both hydrido and hydro are used for coordinated hydrogen, but the latter term is usually restricted to boron and carbon compounds, the regular name hydrido being generally preferred in other cases.

† For organic radicofunctional nomenclature the name hydroperoxide is used to describe the -OOH group and the substituent prefix is hydroperoxy (Rule C-218.1).

‡ However, for the disulfide anion S_2^{2-}, the regular ligand name, disulfido, is used .

(continued)

Formula	Ion name	Ligand name
HS⁻	Hydrogensulfide *	Mercapto
NC⁻	Cyanide	Cyano
CH₃O⁻	Methoxide or **Methanolate**	Methoxo or **Methanolato**
CH₃S⁻	Methanethiolate	Methylthio or Methanethiolato

2.33 (7.313) - The presence of hydrocarbon radicals in coordination entities is indicated by the customary radical names even though they are considered as anions in computing the oxidation number .

Examples

Potassium tetraphenylborate(1-)
Potassium tetraphenylborate(III)   K $[B(C_6H_5)_4]$

Potassium pentachloro(phenyl)antimonate(1-)† K $[SbCl_5(C_6H_5)]$
Potassium pentachloro(phenyl)antimonate(V)†

Potassium triethynylcuprate(2-)
Potassium triethynylcuprate(I)   K₂ $[Cu(C_2H)_3]$

Potassium tetrakis(phenylethynyl)niccolate(4-) K₄ $[Ni(C_2C_6H_5)_4]$
Potassium tetrakis(phenylethynyl)niccolate(0)

2.34 (7.314) - Ligands derived from organic compounds by the loss of a proton and not covered by Rules D-2.31-2.33(7.311-7.313) are given the ending -ato . Enclosing marks are used to set off all such ligands regardless of whether or not they are substituted, e.g. (benzoato), (p-chlorophenolato), [2-(chloromethyl)-1-naphtholato] .

Examples :

Bis(2,3-butanedione dioximato)nickel
Bis(2,3-butanedione dioximato)nickel(II) $[Ni(C_4H_7N_2O_2)_2]$

Bis(2,4-pentanedionato)copper
Bis(2,4-pentanedionato)copper(II) $[Cu(C_5H_7O_2)_2]$

Bis(8-quinolinolato)silver
Bis(8-quinolinolato)silver(II)

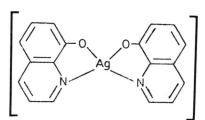

---

* For organic radicofunctional nomenclature the name hydrosulfide is used for the HS- group and the substituent prefix is mercapto (Rules C-511.1 and C-511.5).

† Although the use of parentheses around simple ligands such as phenyl is not required, it can avoid the possibility of misinterpretation.

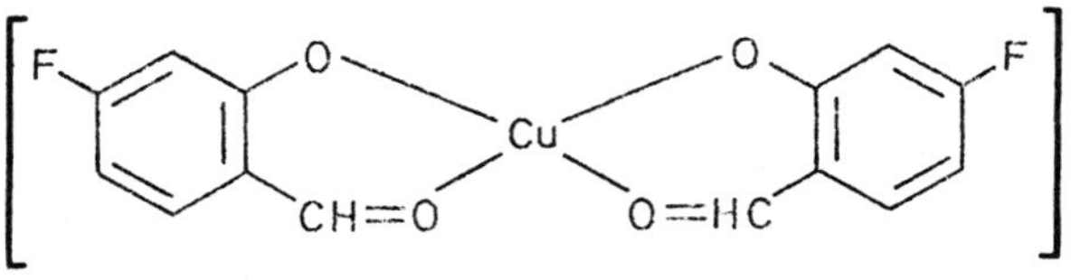

Bis(4-fluorosalicylaldehydato) copper(II)

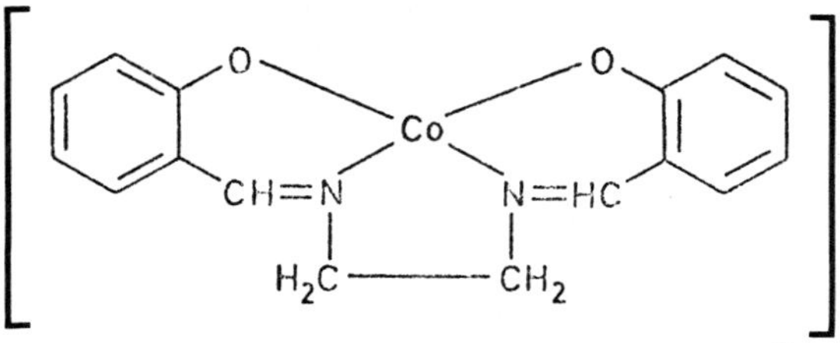

[2,2'-Ethylenebis(nitrilomethylidyne)diphenolato]cobalt(II)

Note : Where a neutral organic compound forms ligands with loss of different numbers of protons, the charge is indicated in parentheses after the name of the ligand .

Examples :

Tartrato(3-)

$$CH(OH)\text{-}COO^-$$
$$|$$
$$CH(O^-)\text{-}COO^-$$

Tartrato(2-)

$$CH(OH)\text{-}COO^-$$
$$|$$
$$CH(OH)\text{-}COO^-$$

Neutral and cationic ligands

2.35 (7.321) - The name of the coordinated molecule or cation is to be used without change, except for the special cases provided for in Rules D-2.36 (7.322) and D-2.37 (7.323) . The names of all neutral ligands, other than those of Rules D-2.36 and D-2.37, are set off with enclosing marks .

Examples :

Bis(2,3-butanedione dioxime)dichloro- $[CoCl_2(C_4H_8N_2O_2)_2]$
   cobalt(II)
   [cf. nickel derivative in Rule D-2.34 (7.314).]

Dichlorobis(triethylphosphine)platinum(II) $[PtCl_2(PEt_3)_2]$

Dichlorobis(methylamine)copper(II) $[CuCl_2(CH_3NH_2)_2]$

Tetrakis(pyridine)platinum(2+) $[Pt(py)_4]$ $[PtCl_4]$
 tetrachloroplatinate(2-)

Tris(2,2'-bipyridine)iron(2+) chloride $[Fe(bpy)_3]$ Cl_2

Tris(ethylenediamine)cobalt(3+) sulfate $[Co(en)_3]_2$ $[SO_4]_3$

$$[Zn\{NH_2-CH_2-CH(NH_2)-CH_2-NH_2\}_2]\ I_2$$
Bis(1,2,3-propanetriamine)zinc(2+) iodide

Potassium trichloro(ethylene)platinate(II)* $K[PtCl_3(C_2H_4)]$

Hexakis(phenyl isocyanide)chromium $[Cr(C_6H_5NC)_6]$

$$[PtCl_2\{NH_2-CH_2-CH(NH_2)-CH_2-NH_3\}]\ Cl$$
Dichloro[(2,3-diaminopropyl)ammonium]platinum(1+) chloride

2.36 (7.322) - Water and ammonia as neutral ligands in coordination entities are called "aqua" and "ammine" respectively .

Examples :

Hexaaquachromium(3+) chloride $[Cr(H_2O)_6]$ Cl_3
Hexaaquachromium trichloride

Pentaaquahydroxoaluminium(2+) ion $[Al(OH)(H_2O)_5]^{2+}$

$$[CoCl_3(NH_3)_2\{(CH_3)_2NH\}]$$
Diamminetrichloro(dimethylamine)cobalt
Diamminetrichloro(dimethylamine)cobalt(III)

2.37 (7.323) - The groups NO and CO, when linked directly to a metal atom, are called nitrosyl and carbonyl respectively . In computing the oxidation number these ligands are treated as neutral.

Examples :

Dicarbonylbis(triphenylphosphine)nickel(0) $[Ni(CO)_2(PPh_3)_2]$

Sodium pentacyanonitrosylferrate(III) $Na_2[Fe(CN)_5(NO)]$

$$[Fe(en)_3]\ [Fe(CO)_4]$$
Tris(ethylenediamine)iron(2+) tetracarbonylferrate(2-)

---

* Although the use of parentheses around simple ligands is not required, it can avoid the possibility of misinterpretation.

Different modes of linkage of some ligands

2.38 (7.33) - The different points of attachment of a ligand may be denoted by adding, at the end of the name of the ligand, the symbol(s), in italic letters, for the atom(s) through which attachment occurs .

Thus the 1,2-dithiooxalato(2-) anion may be attached through S or O by any one of four ways distinguished as follows :

$$1,2\text{-dithiooxalato } -S,S'; \ -O,O' \ ; \ -O,S \ ; \ -O,S'$$

The points of unsymmetrical attachment are cited in the order of Table I of the Appendix , p. 459, that atom occuring earliest being assigned the lowest locant.

In some cases different names are already in use for different modes of attachment, as for example, thiocyanato (-SCN) and isothiocyanato (-NCS), nitro (-NO$_2$) and nitrito (-ONO). In the absence of structural knowledge about the linkage actually present, thiocyanato and nitrito should be used.

Examples :

Potassium bis(1,2-dithio-oxalato-*S,S'*)niccolate(2-)

Dichloro[*N,N*- dimethyl-2,2'-thiobis(ethylamine)-*S,N*']platinum

There are some ligands which occasionally coordinate in a manner quite different from the normal mode. The mode of coordination can be indicated by the name of the ligand or by designating, by italicized symbol(s), the atom(s) of the ligands through which coordination occurs.

Examples :

Glycinato-*O,N* $-NH_2-CH_2-CO-O^-$

Glycine-*N* $-NH_2-CH_2-CO-OH$

Glycinato-O $NH_2-CH_2-CO-O^-$_

Glycine-O $^+NH_3-CH_2-CO-O^-$_

$$[Pt(CH_3)_3\{CH(COCH_3)_2\}(bpy)]$$

(1-Acetylacetonyl)(2,2'-bipyridine)trimethylplatinum

(2,2'-Bipyridine)(diacetylmethyl)trimethylplatinum

(2,2'-Bipyridine)trimethyl(2,4-pentanedionato-C^3)platinum

Designation of active coordination site(s) among several possibilities

2.39 (7.34) - In some cases, several possible coordination sites may be involved. The different possible locations may be indicated by the particular atoms through which coordination occurs: cysteinato-S,N ; cysteinato-O,N ; etc. If the same element is involved in the different possibilities, the position in the chain or ring to which the **coordinating atom(s) is(are) attached is indicated by** numerical superscripts :

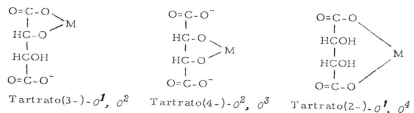

Tartrato(3-)-O^1, O^2 Tartrato(4-)-O^2, O^3 Tartrato(2-)-O^1, O^4

D-2.4 (7.35) - Abbreviations

General principles

2.41 - Abbreviations are used extensively for ligand names, and especially in formulae . A list of common abbreviations is given in Rules D-2.42 and D-2.43 . The following principles should govern the use of abbreviations :

(a) Each paper should explain the abbreviations used in it.

(b) Abbreviations should be short, generally not more than four letters .

(c) Abbreviations should be such as not to cause confusion with the commonly accepted abbreviations used for organic radicals : Me, methyl ; Et, ethyl ; Ph, phenyl, etc. .

(d) All abbreviations for ligands, except L, the general abbreviation for ligand, those referred to in (c) and those shown with capital letters under Rule D-2.42 should be in lower-case letters, e.g. : en, pn, tren, bpy, etc.   M should be the general abbreviation for metal .

(e) Abbreviations should not involve hyphens, e.g. : phen and not *o*-phen for *o*-phenanthroline (or 1,10-phenanthroline).

(f) The abbreviations for the neutral compound and the ligand ion derived from it must be clearly differentiated, for instance :

Hacac Acetylacetone **(2,4-Pentanedione)**

acac **Acetylacetonato**

H_2dmg Dimethylglyoxime (2,3-Butanedione dioxime)

Hdmg Dimethylglyoximato(1-)

dmg **Dimethylglyoximato(2-)**

H_4edta Ethylenediaminetetraacetic acid
     ((Ethylenedinitrilo)tetraacetic acid)

Hedta or edta Coordinated ions derived from H_4edta

(g) Care should be taken to ensure that there is no confusion between abbreviations and symbols. Abbreviations should be separated from symbols by a space or enclosed in parentheses :

$$[Co(en)_3]^{3+} \quad or \quad [Co\ en_3]^{3+} \ ; \quad not \quad [Coen_3]^{3+}$$

Anionic groups

2.42 - The following is a list of commonly used abbreviations, the parent **compound being given.**

Hacac Acetylacetone (2,4-Pentanedione)
    $CH_3-CO-CH_2-CO-CH_3$

Hbg Biguanide
   $NH_2-C(=NH)-NH-C(=NH)-NH_2$

H_2dmg Dimethylglyoxime (2,3-Butanedione dioxime)
    $CH_3-C(=NOH)-C(=NOH)-CH_3$

H_4edta Ethylenediaminetetraacetic acid
    ((Ethylenedinitrilo)tetraacetic acid)

    $(HO-CO-CH_2)_2N-CH_2-CH_2-N(CH_2-CO-OH)_2$

H_2ox Oxalic acid
   $HO-CO-CO-OH$

Neutral groups

2.43 - The following is a list of commonly used abbreviations.

bpy 2,2'-Bipyridine
   (2,2'-Bipyridyl)

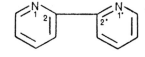

diars   **o**-Phenylenebis(dimethylarsine)
$(CH_3)_2As-C_6H_4-As(CH_3)_2$

dien   **N-(2-Aminoethyl)ethylenediamine**
(Bis(2-aminoethyl)amine**)**(Diethylenetriamine)
$NH_2-CH_2-CH_2-NH-CH_2-CH_2-NH_2$

diphos   Ethylenebis(diphenylphosphine)
$Ph_2P-CH_2-CH_2-PPh_2$

en   Ethylenediamine
$NH_2-CH_2-CH_2-NH_2$

phen   1,10-Phenanthroline

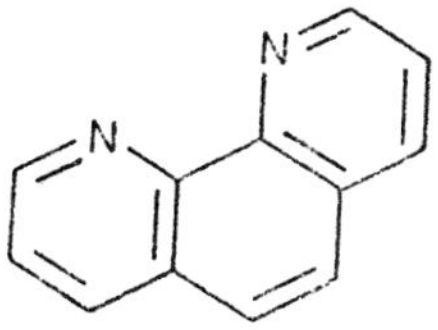

pn   **1,2-Propanediamine**   (Propylenediamine)
$NH_2-CH(CH_3)-CH_2-NH_2$

py   Pyridine

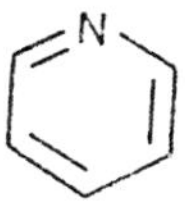

tren   *N,N*-Bis(2-Aminoethyl)ethylenediamine
(Tris(2-aminoethyl)amine)
$(NH_2-CH_2-CH_2)_3N$

trien   *N,N'*-Bis(2-aminoethyl)ethylenediamine
(Triethylenetetramine)
$NH_2-CH_2-CH_2-NH-CH_2-CH_2-NH-CH_2-CH_2-NH_2$

ur   Urea
$NH_2-CO-NH_2$

D-2.5 (7.4) - Coordination entities with unsaturated molecules or groups

A wide variety of unsaturated hydrocarbon-metal compounds is known . In many of these the metal atom is bonded to two or more contiguous atoms of the ligand rather than to a specific atom . The name π-complexes has been introduced for this class of compounds but the use of π in the naming of individual compounds is not recommended .

Designation of stoicheiometric composition only

2.51 (7.41) - The name of the ligand group is given as in Subsection D-2.3.

Examples :

Tricarbonyl(cyclooctatetraene)iron $[Fe(CO)_3(C_8H_8)]$

Tetracarbonyl(2-methylallyl)-
   manganese $[Mn(CO)_4\{CH_2=C(CH_3)-CH_2\}]$

Designation of structure

2.52 (7.421) - When all the atoms in a chain or ring are bound
to the central atom, the name of the ligand group is given as **in Rule D-2.3**
but with a prefix η which may be read as eta or hapto. Hapto
comes from the Greek ἄπτειν, to fasten.

Examples :

Bis(η-cyclopentadienyl)-
   hydridorhenium $[ReH(C_5H_5)_2]$

Bis(η-benzene)chromium $[Cr(C_6H_6)_2]$

Tricarbonyl(η-cycloheptatrien-
   ylium)molybdenum(1+) ion

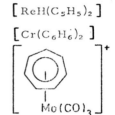

2.53 (7.422) - When all the multiple-bonded ligand atoms are
bound to the central atom, the names are derived as in Rule
D-2.52 (7.421).

Example :

(η-Bicyclo[2.2.1]hepta-
   2,5-diene)tricarbonyliron

2.54 (7.423) - When some, but not all, ligand atoms in a chain
or ring or some, but not all, ligand atoms involved in double bonds
are bound to the central atom, locants, to indicate the atoms involved, are inserted before the prefix η . When a sequence of atoms
is involved, this is indicated by the locants of the first and the
last such atom.

When it is desired to stress that a ligand is bound by a bond to a single atom, the prefix σ may be used.

Examples :

$[1 — 3 - \eta - 2 - \text{Butenyl}]$
tricarbonylcobalt

Tetracarbonyl$[1 — 3 - \eta -$
$(2 - \text{methylallyl})]$manganese

$(\eta - \text{Cyclopentadienyl})(1 — 3 - \eta - \text{cyclopenta-}$
dienyl$)[\sigma - (2,4 - \text{cyclopentadien} - 1 - \text{yl})] -$
nitrosylmolybdenum

Dicyclopentadienyl coordination compounds of metals : metallocenes

2.55 (7.43) - The general term for bis(η-cyclopentadienyl) coordination compounds of metals and their derivatives is metallocenes. The trivial name for bis (η-cyclopentadienyl)iron, $[Fe(C_5H_5)_2]$, is ferrocene, and for the cation $[Fe(C_5H_5)_2]^+$, ferrocene (1+) or ferrocenium. Similarly coined names such as nickelocene and cobaltocene are not recommended and purely trivial names, such as cymanthrene and cytizel are disapproved.

Examples :

Bis(η-cyclopentadienyl)iron
Bis(η-cyclopentadienyl)iron(II) $Fe(C_5H_5)_2$
Ferrocene

Bis(η-cyclopentadienyl)iron(1+)
tetrafluoroborate(1-) $[Fe(C_5H_5)_2]\ [BF_4]$
Bis(η-cyclopentadienyl)iron(III) tetrafluoroborate
Ferrocene(1+) tetrafluoroborate(1-)
Ferrocenium tetrafluoroborate

2.56 (7.431) - Derivatives of ferrocene are named by use
of the ordinary prefixes and suffixes of organic substitutive or
conjunctive nomenclature. Since all of the positions of the two
cyclopentadienyl rings are equivalent, substituents are given the
lowest possible locants without regard to the attachment of the
iron atom (see also D-2.57).

Examples :

1,1'-Dichloroferrocene

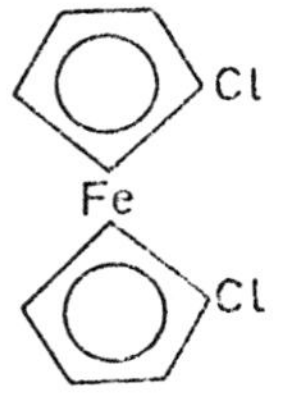

Ferrocenecarbaldehyde
Formylferrocene

$Fe(C_5H_5)(C_5H_4CHO)$

Ferrocenecarboxylic acid
Carboxyferrocene

$Fe(C_5H_5)(C_5H_4COOH)$

2-Ferrocenylethanol
(2-Hydroxyethyl)ferrocene
Ferroceneethanol

$Fe(C_5H_5)(C_5H_4CH_2CH_2OH)$

Ethylferrocenium chloride
Ethylferrocene(1+) chloride

$[Fe(C_2H_5C_5H_4)(C_5H_5)]$ Cl

2.57 (7.432) - Radicals formed from ferrocene by loss of one,
two, three or more hydrogen atoms are named ferrocenyl, ferro-
cenediyl, ferrocenetriyl, etc. : locants are used as necessary to
name ferrocene derivatives.

Examples :

3-Ferrocenylalanine

$[Fe(C_5H_5)\{C_5H_4CH_2CH(NH_2)COOH\}]$

Ferrocenyldiphenylarsine
(Diphenylarsino)ferrocene

$[Fe(C_5H_5)\{C_5H_4As(C_6H_5)_2\}]$

3,5-(1,1'-Ferrocenediyl)-
valeric acid

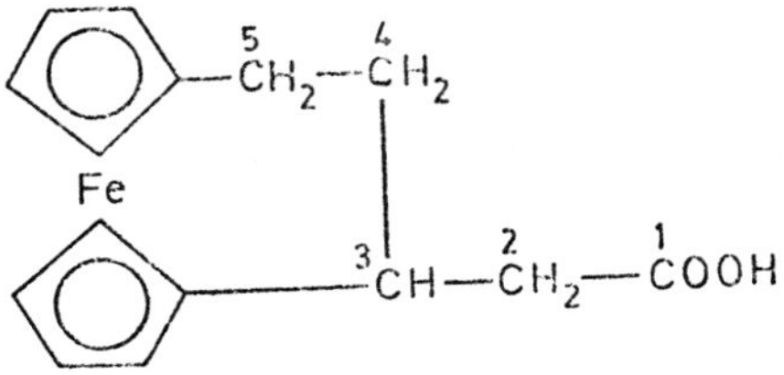

2.58 (7.432) - For coordination compounds of the metallocene type derived from fused-ring systems or systems other than cyclopentadiene, the word "ferrocene" should not be used as a basis for fusion-type names.

Example * :

Bis(1—3a, 7a-η-indenyl)iron

not Benzoferrocene

not Dibenzoferrocene

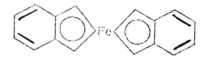

2.59 (7.433 and 7.434) - The stereoisomers and conformations of ferrocene derivatives are specified according to the methods of Section E, pp. 486-489.

D-2.6 (7.5) - Designation of isomers

2.60 - Isomerism may arise in coordination compounds in a number of ways :

(a) Coordination to a central atom may occur through different atoms of a ligand (see Rules D-2.38 and D-2.39) .

(b) Ligands may themselves be isomeric . This is indicated by their names, e.g. :

1,2-Propanediamine $NH_2-CH(CH_3)-CH_2-NH_2$

N-Methylethylenediamine $CH_3-NH-CH_2-CH_2-NH_2$

(c) Interchange of coordinated and non-coordinated ions may occur . This is indicated by the names, e.g. :

Pentaamminesulfatocobalt(III)
bromide $[Co(SO_4)(NH_3)_5]$ Br

Pentaamminebromocobalt(III)
sulfate $[CoBr(NH_3)_5]$ $[SO_4]$

---

* The locants 1—3a,7a mean 1,2,3,3a,7a .

(d) The geometrical arrangement of two or more kinds of ligand
     in the coordination sphere may differ.

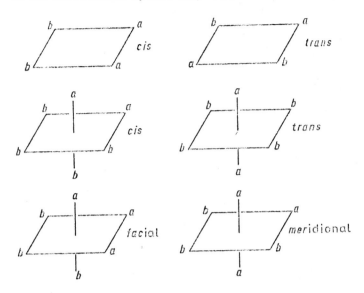

(e) The arrangement of ligands in the coordination sphere may be
     chiral.

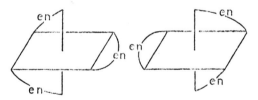

$$en = H_2N-CH_2-CH_2-NH_2$$

(f) Asymmetry of an atom in a ligand may result from coordi-
     nation.
     Examples :

$H_2N---CH_2$ Cl_4Pt $H_2N---\overset{*}{CH}$ CH_2NH_2	The carbon atom marked with the asterisk is asymmetric in the coordination entity.

CH_3 CH_3 B N B Cl Cl CH_3 H	The nitrogen atoms are asymmetric in the coordination entity.

352

Geometrical isomerism

2.61 (7.511) - The prefixes *cis-*, *trans-*, *fac-* and *mer-* are used where they are sufficient to differentiate specific isomers.

Examples :

Planar configuration

cis-Diiodobis(triethylstibine)-
   platinum

$$Et_3Sb \diagdown \diagup I \\ Pt \\ Et_3Sb \diagup \diagdown I$$

trans-Bis(dipropyl sulfide)-
   dinitroplatinum

$$Pr_2S \diagdown \diagup NO_2 \\ Pt \\ O_2N \diagup \diagdown SPr_2$$

trans-Bis[(ethylthio)-
   acetato −*O,S*]platinum

$$O=C-O \quad \overset{Et}{\underset{}{S}}-CH_2 \\ | \qquad Pt \qquad | \\ H_2C-S \quad \underset{Et}{\overset{}{O}}-C=O$$

Octahedral configuration

cis-Bis(ethylenediamine)-
   difluorocobalt(1+) ion

trans-Tetraamminedi-
   chlorochromium(1+) ion

$$H_3N \qquad NH_3 \\ H_3N \qquad NH_3$$

fac-Trichlorotris(pyridine)-
   ruthenium

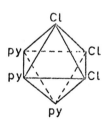

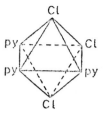

mer-Trichlorotris(pyridine)-
ruthenium

(In the above and in some of the following examples the central
metal atom has been omitted from the formulae for the sake of
clarity).

2.62 (7.512) - Italicized letters are used as locants for specifying spatial positions in various configurations. The assignments
for the square planar and octahedral configurations are :

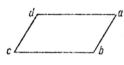

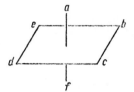

Examples :

a-Ammine-*b*-(hydroxylamine)-*d*-
nitro-*c*-(pyridine)platinum(1+)
chloride

There are two more isomers of this composition.

a b-Diammine-*cf*-diaqua-*de*-
bis(pyridine)cobalt(3+) ion

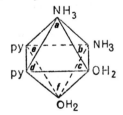

There are four more possible isomers of this composition
which are achiral. This compound has an enantiomer.

2.63 (7.513) - The detailed rules for the application of these locants to chelate ligands are given in Inorganic Rule 7.513 .

Chiral isomerism

2.64 (7.52) - Locants may distinguish enantiomeric forms of a coordination compound . Contrary to the situation common in organic chemistry, the attribution of locants may specify a precise enantiomer . This is for instance the case with the octahedral configuration given in Rule D-2.62 *.

In such cases, when the absolute configuration is not known, locants may still be used, but the complete name should be prefixed by X . When the absolute configuration is known, the principles laid down in Inorganic Rule 7.8 should be followed .

The symbol X , applied to locant designators, has a role similar to the one of the symbol *rel* applied in the *RS* system (Section E, Rule **E-4.10**).

D-2.7 (7.6) - Dinuclear and polynuclear compounds with bridging groups

Compounds with bridging atoms or groups

2.71 (7.611) - Bridging atoms or groups are indicated as follows :

(a) A bridging group is indicated by adding the Greek letter μ immediately before the name of the group and separating that name from that of the rest of the **coordination entity by hyphens.**

(b) Two or more bridging groups of the same kind are indicated by di-μ- (or bis-μ-), etc. .

(c) The bridging groups are listed with the other groups in alphabetical order *unless the symmetry of the molecule permits simpler names by the use of multiplicative prefixes.*

(d) Where the same ligand is present as a bridging ligand and as a non-bridging ligand, it is cited first as a bridging ligand (See third example in Subsection D-3. 0).

Bridging groups between two centres of coordination are of two types : (1) the two centres are attached to the same atom of the bridging group and (2) the two centres are attached to different atoms of the bridging group .

For bridging groups of the first type it is often desirable to indicate the bridging atom . This is done by adding the italicized symbol for the atom at the end of the name of the ligand as in Rule D-2.38 (7.33) . For bridging groups of the second type, the symbols of all coordinated atoms of the ligand are added.

---

* There is objection to this practice by some who prefer that locants for enantiomers be identical. This matter is under careful consideration by the Commission.

Examples :

μ-Hydroxo-bis[penta-
  amminechromium(III)] $[(NH_3)_5Cr-OH-Cr(NH_3)_5]$ Cl_5
  chloride

Di-μ-chloro-bis{[bis(picolinaldehyde oximato -*N,N'*)platinum -*O,O'*]-
  copper}(2+) ion

Di-μ-chloro-bis{[bis(picolinaldehyde oximato -*N,N'*)platinum(II)-
  -*O,O'*]copper(II)} ion

Bis(μ-nonafluorovalerato -*O,O'*)-disilver
Bis(μ-nonafluorovalerato -*O,O'*)-disilver(I)

Di-μ-chloro-(η-1,5-cyclooctadiene)(η-ethylene)dirhodium
Di-μ-chloro-(η-1,5-cyclooctadiene)(η-ethylene)dirhodium(I)

2.72 (7.612) - If the number of central atoms bound by one bridging group exceeds two, the number is indicated by adding a subscript numeral to the μ.

Examples :

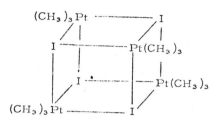

Tetra-μ$_3$-iodo-tetrakis(trimethylplatinum)

Tetra-μ$_3$-iodo-tetrakis[trimethylplatinum(IV)]

Hexa(μ-acetato −O,O′)-μ$_4$-oxo-tetraberyllium

Hexa(μ-acetato −O,O′)-μ$_4$-oxo-tetraberyllium(II)

$$[Be_4O(CH_3COO)_6]$$

Hexa(μ-acetato −O,O′)-μ$_3$-oxo-trichromium(1+) chloride

Hexa(μ-acetato −O,O′)-μ$_3$-oxo-trichromium(III) chloride

$$[Cr_3O(CH_3COO)_6] Cl$$

μ$_4$-Thio-tetrakis(methyl-mercury)(2+) ion

μ$_4$-Thio-tetrakis[methyl-mercury(II)] ion

$$[(CH_3Hg)_4S]^{2+}$$

2.73 (7.614) - When the bridged polynuclear coordination entity contains more than one kind of nuclear atom the locants will be assigned according Inorganic Rules 7.613 and 7.614.

Extended structures

2.74 (7.62) - Where bridging causes indefinite extension of the structure compounds are best named on the basis of the repeating unit and use of the prefix *catena* .

Chains may also be formed by the attachment of two centres of coordination to different atoms of the bridging group. Italicized symbols of the coordinating atoms are then appended to the name of the bridging ligand.

Example :

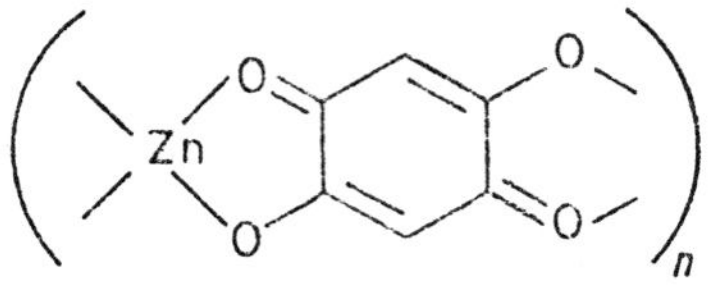

catena-$\left[\mu-\left[2,5\text{-Dihydroxy-1,4-benzoquinonato}(2-)-O^1,O^2:O^4,O^5\right]\text{zinc}\right]$

D-2.8 (7.7) - Dinuclear and polynuclear compounds without bridging groups

Direct linking between centres of coordination

2.81 (7.711) - Compounds containing metal-metal bonds, when symmetrical, are named by the use of multiplicative prefixes ; when unsymmetrical, one central atom and its attached ligands are together treated as a ligand on the other central atom. The primary central atom is the last encountered in Table I of the Appendix, p. 459.

The names of radicals with the free valence on the metal, e.g., chloromercurio and dimethylarsenio, are constructed by prefixing the names of the organic and inorganic radicals to the modified name of the metal given in Table II of the Appendix, p. 461.

Examples :

Bis(tricarbonyl-η-cyclopenta-
  dienylmolybdenum) $\left[\eta\text{-}C_5H_5(CO)_3Mo\text{-}Mo(CO_3\text{-}\eta\text{-}C_5H_5)\right]$

Pentacarbonyl(tetracarbonyl-
  cobaltio)rhenium $\left[(CO)_4Co\text{-}Re(CO)_5\right]$

Pentacarbonyl[(triphenyl-
  arsine)aurio]manganese $\left[\{(C_6H_5)_3As\{Au\{Mn(CO)_5\}\}\right]$

Tricarbonyl-η-cyclopenta-
  dienyl(tricarbonyl-η-cyclo-
pentadienylmolybdenio)tungsten $\left[\eta\text{-}C_5H_5(CO)_3W\text{-}Mo(CO)_3\text{-}\eta\text{-}C_5H_5\right]$

2.82 (7.712) - Where there are bridging groups as well as the metal-metal bond between the same pair of atoms, the compound is named as a bridged compound. The existence of the metal-metal bond may be indicated by the italicized symbols in parentheses at the end of the name.

Examples :

μ$_3$-(Iodomethylidyne)-*cyclo*-

tris(tricarbonylcobalt)(3 *Co-Co*)

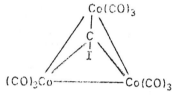

Hexacarbonyl-μ-(η-3-hexyne)-

dicobalt *(Co-Co)*

Homoatomic aggregates

2.83 (7.72) - There are several instances of a finite group of metal atoms with bonds directly between the metal atoms but also with some nonmetal atoms of groups (ligands) intimately associated with the cluster. The geometrical shape of the cluster is designated by *triangulo, quadro , tetrahedro, octahedro* , etc., and the nature of the bonds to the ligands by the conventions for bridging bonds and simple bonds. Numbers are used as locants as in Subsections D-4 and D-7 .

Example :

$$[Mo_6Cl_8Cl_3\{(C_6H_5)_2PCH_2CH_2P(C_6H_5)_2\}(py)]Cl$$

Octa-μ$_3$-chloro-trichloro[ethylenebis(diphenylphosphine)]-
(pyridine)-*octahedro*-hexamolybdenum(II) chloride

D-3. ORGANOMETALLIC COMPOUNDS

D-3.0 - Introduction

In these Rules the term "organometallic compound" includes all compounds in which a carbon atom is joined directly to **an atom of** any element other than hydrogen, carbon, nitrogen, oxygen, fluorine, chlorine, bromine, iodine, and astatine, although the above elements may also be present in the molecule.

For expediency, in this **Subsection of these rules the term "metal atom" includes the atoms of all the elements except those named above.**

There are special rules for naming organic compounds containing sulfur, selenium and tellurium (Section C), phosphorus, arsenic and in some cases antimony and bismuth (Subsection D-5), silicon (Subsection D-6) and boron (Subsection D-7) but the following rules may also be applied .

Although many organometallic compounds exist in associated molecular forms and contain structural solvent, their names are generally based on the stoicheiometric compositions of the compounds, the solvent, if any, being ignored.

Examples :

Butyllithium $(C_4H_9Li)_n$

Triethylaluminium $[(C_2H_5)_3Al]_2$

Where the structure is known and it is desired to emphasise this, the compounds may be named as coordination compounds (see Subsection D-2) .

Examples :

Di-μ-ethyl-tetraethyldialuminium

or

Di-μ-ethyl-bis(diethylaluminium)

catena-(Di-μ-methyl-beryllium)

Bis(diethyl ether)iodo-
(methyl)magnesium *

$$[MgI(CH_3)\{(C_2H_5)_2O\}_2]$$

D-3.1 - Organometallic compounds in which the metal is bound to organic radicals and hydrogen only

3.11 - Compounds consisting solely of individual metal atoms joined to the carbon atom of one or more organic radicals and/or to one or more hydrogen atoms are named by citing such radicals or hydrogen atoms in *alphabetical* order, followed by the name of the metal . The presence of hydrogen attached to the metal atom must aways be indicated by the prefix "hydrido" † . No break is made between the radical names and that of the metal .

Examples :

Phenylarsenic	$(C_6H_5As)_n$
Methyllithium	$(CH_3Li)_n$
Vinylsodium	$CH_2=CHNa$
Diéthyltellurium	$(C_2H_5)_2Te$
Tribenzylarsenic	$(C_6H_5CH_2)_3As$
Tetraphenyltellurium	$(C_6H_5)_4Te$
Tetrabutylgermanium	$(C_4H_9)_4Ge$
Penta-2-naphthylantimony	$(C_{10}H_7)_5Sb$

---

* See footnote on page 327.

† Hydrogen attached to carbon or boron is named only when necessary and then by the prefix "hydro".

Examples:

(Dihydronaphthyl)dihydridorhenium	$(C_{10}H_9)ReH_2$
Difluorohydro(pyridine)boron	$(C_5H_5N)BHF_2$

Butylethylmethylpropyltin $CH_3(C_2H_5)(C_3H_7)(C_4H_9)Sn$

Ethynyltriphenyltin $(C_6H_5)_3SnC\equiv CH$

Ethylhydridoberyllium C_2H_5BeH

Dihydridophenylarsenic $C_6H_5AsH_2$

Hydridomethylpropylantimony $(CH_3)(C_3H_7)SbH$

Trihydridophenylgermanium $C_6H_5GeH_3$

Benzylhydridophenylgold $(C_6H_5)(C_6H_5CH_2)AuH$

3.12 (Partly alternative to D-3.11) - Compounds of tervalent phosphorus, arsenic, antimony and bismuth, and of quadrivalent silicon, germanium, tin and lead may also be named as substitution derivatives of phosphine, PH_3, arsine, AsH_3, stibine, SbH_3, bismuthine, BiH_3, silane, SiH_4, germane, GeH_4, stannane, SnH_4, and plumbane, PbH_4, respectively * . In using this alternative nomenclature the compounds are named by citing the organic radicals in alphabetical order followed by the name of the hydride.

Examples :

Triethylarsine $(C_2H_5)_3As$

Ethylmethyldiphenylgermane $(C_6H_5)_2(C_2H_5)(CH_3)Ge$

Tetraphenylplumbane $(C_6H_5)_4Pb$

Ethylarsine $C_2H_5AsH_2$

Diphenylstannane $(C_6H_5)_2SnH_2$

---

* In naming hydrides the ending "-ane" should only be used for hydrides of B, C, Si, Ge, Sn, Pb and, according to Rule D-4.11, for hydrides of unbranched chains of identical atoms of P, As, Sb, Bi, S, Se and Te.

D-3.2 - Organometallic compounds having uncharged ligands

3.21 - Organometallic compounds containing uncharged ligands are named by coordination nomenclature, all ligands (anions, radicals and molecules) being cited in alphabetical order before the name of the metal.

Examples :

(Pyridine)tris(trifluoromethyl)antimony $[Sb(CF_3)_3(C_5H_5N)]$

trans-Chloro(methyl)bis(triethylphosphine)platinum *

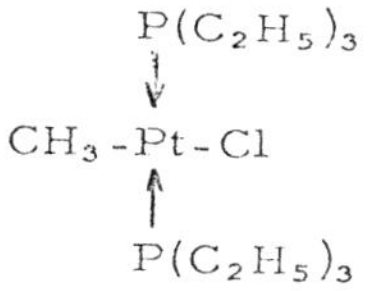

$$P(C_2H_5)_3$$
$$\downarrow$$
$$CH_3-Pt-Cl$$
$$\uparrow$$
$$P(C_2H_5)_3$$

trans-Bis[diethyl(phenyl)phosphine]-
   bis(pentachlorophenyl)iron *

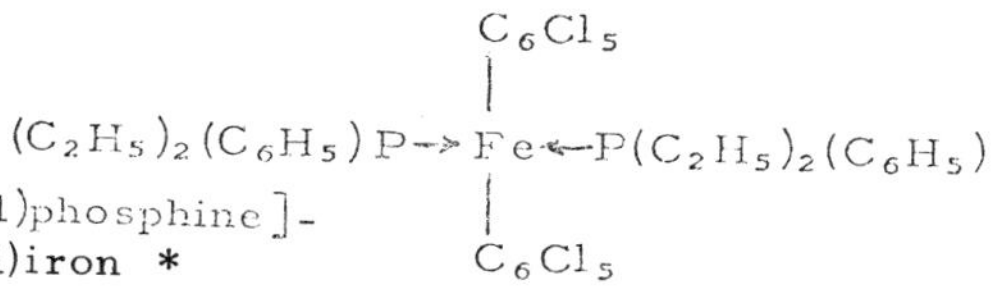

$$C_6Cl_5$$
$$|$$
$$(C_2H_5)_2(C_6H_5)P \rightarrow Fe \leftarrow P(C_2H_5)_2(C_6H_5)$$
$$|$$
$$C_6Cl_5$$

Bis(2,5-dimethyl-2,5-diphospha-
   hexane)hydridophenyl-
   ruthenium
or
Bis[ethylenebis[dimethylphosphine]]-
   hydridophenylruthenium

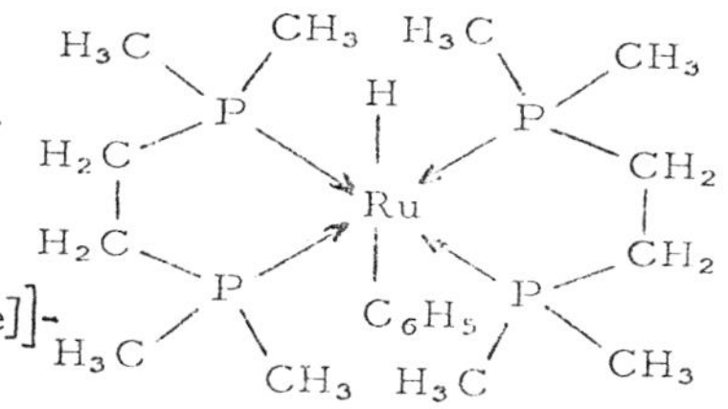

(2,5-Dimethyl-2,5-diphospha-
   hexane)(2,5-dimethyl-2,5-
   diphosphahexyl)hydridoruthenium
or
[[[2-(Dimethylphosphino)ethyl]-
   methylphosphino]methyl]-P, P', C]-
   [ethylenebis[dimethylphosphine]]-
   hydridoruthenium

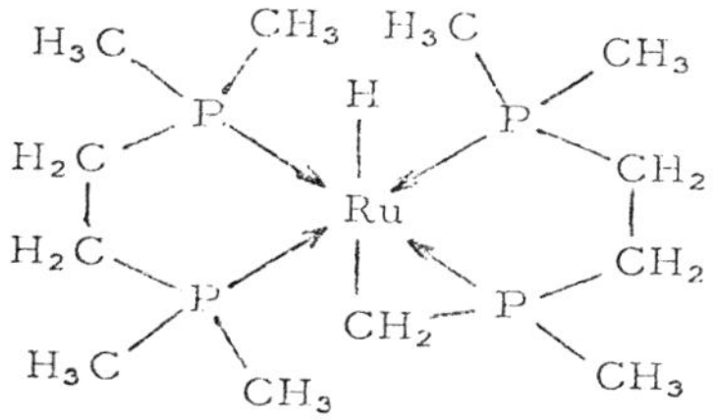

* See footnote on page 327.

D-3.3 - Organometallic compounds having anionic ligands

3.31 (Partly alternative to D-3.12) - Organometallic compounds containing anionic ligands are named by methods a, b or c :

(a) by stating the name(s) of the organic radical(s) followed by the name of the metallic element and the anion name(s), note being taken of the stoicheiometric composition only. The names of the radicals and anions are each cited in alphabetical order,

(b) by adding, as prefixes, all the radical and anion names to the name of the metal, note being taken of the molecular structure. The radical and anion names are cited together in alphabetical order . This method may be used for monomolecular compounds, when the structure is known, and for polymeric compounds by addition of the prefix poly, with further addition of information about bridging (see Rule D-2.71) .

(c) where applicable, according to Rule D-3.12 .

Examples :

(a) Methyltin chloride dihydride CH_3SnH_2Cl

(b) Chlorodihydridomethyltin

(c) Chloro(methyl)stannane   *

(a) Methylmagnesium iodide $(CH_3MgI)_n$

(b) Poly[iodo(methyl)magnesium]   *

(a) Phenylmercury hydroxide C_6H_5HgOH

(b) Hydroxo(phenyl)mercury *

(a) Diethylaluminium chloride $\{(C_2H_5)_2AlCl\}_2$

(b) Di-μ-chloro-bis(diethylaluminium)

(a) Di-p-tolylgermanium oxide $(p\text{-}CH_3C_6H_4)_2GeO$

(b) Oxodi-p-tolylgermanium   †

(c) Oxodi-p-tolylgermane

(a) Bis(phenylgermanium) trioxide $(C_6H_5Ge)_2O_3$

(b) Trioxodiphenyldigermanium ‡

---

* Although the use of parentheses around simple groups like methyl and phenyl is not required, it can avoid the possibility of misinterpretation.

† When the polymeric nature of such compounds is indicated as $[(p\text{-}CH_3C_6H_4)_2GeO]_n$, the (b) name may be modified as follows :

Poly [μ-oxo-di-p-tolylgermanium].

‡ Since the structure indicated does not show the relationship between the oxygen atoms, a (c) name is not possible. Similarly, a polymer name including the citation of bridging groups cannot be written ; the polymeric nature of this example must be shown just by prefixing "Poly" to name (b).

(a) Phenylaluminium phenylimide $\{C_6H_5Al(NC_6H_5)\}_4$

(b) Tetrakis-μ_3-(phenylimido)
tetrakis(phenylaluminium)

(a) Diphenylantimony chloride $(C_6H_5)_2SbCl$

(b) Chlorodiphenylantimony

(c) Chlorodiphenylstibine

(a) Triphenylbismuth chloride $(C_6H_5)_3BiCl(OH)$
hydroxide

(b) **Chlorohydroxotriphenylbismuth**

3.32 – Compounds of the general formula $R_xM(O)_y(OH)_z$, where M is a metal atom, are designated by functional names of the type hydroxide oxide, unless other functional names, such as phosphonic acid and seleninic acid are permitted in special nomenclature systems (See Subsections C-6.4, C-7, D-5, D-6 and D-7). However, the use of these systems for organoantimony, organotellurium and their extension to organogermanium, tin and lead compounds must be carefully considered since most of these compounds are usually amphoteric polymers.

Examples:

Phenylantimony dihydroxide oxide $C_6H_5SbO(OH)_2$

Methyltin hydroxide oxide $CH_3SnO(OH)$

Diphenylantimony hydroxide oxide $(C_6H_5)_2SbO(OH)$

Ethyltellurium hydroxide oxide $C_2H_5TeO(OH)$

Triethyltellurium hydroxide oxide $(C_2H_5)_3TeO(OH)$

D-3.4 – Organometallic radicals

3.41 – Radicals derived from organometallic compounds and having the free valence (or valences) on the metal atom, are named by prefixing the names of the organic and inorganic ligands to the modified name of the metal given in Table II (Appendix, p. 462).

Examples :

Methyltellurio	CH_3Te-
Hydridodimethylstannio	$(CH_3)_2HSn-$
Dimethylstannio	$(CH_3)_2Sn\lneq$
Methylstannio *	$CH_3Sn\leqq$
Methylstannio *	CH_3Sn-
Triphenylstannio	$(C_6H_5)_3Sn-$
Dichlorodimethylantimonio	$(CH_3)_2Cl_2Sb-$
Chloro(methyl)antimonio †	$(CH_3)ClSb-$

3.42 (Partly alternative to D-3.11) – Compounds containing univalent organometallic radicals attached to carbon may be named by citing their radical names before the name of the parent compound. The radical name takes its place in alphabetical order with those of the other substituents.

Examples :

1,1,2,2-Tetraphenyl-1,2-disodioethane	$(C_6H_5)_2C-C(C_6H_5)_2$
	$\qquad\quad\overset{|}{Na}\ \ \overset{|}{Na}$
1-Cupriopropyne	$CH_3C\equiv CCu$
Lithiotriphenylmethane	$(C_6H_5)_3CLi$
(Methylmagnesio)benzene	$CH_3MgC_6H_5$

3.43 – Names of univalent radicals may be derived from the hydrides phosphine, arsine, stibine, bismuthine, silane, germane, stannane and plumbane as follows :

---

* The use of "multivalent" organometallic radicals as substituents in organic nomenclature is not completely free from ambiguity and is under study by both the Inorganic and Organic Nomenclature Commissions.

† See footnote on page 327.

Phosphino	$-PH_2$	Silyl	$-SiH_3$
Arsino	$-AsH_2$	Germyl	$-GeH_3$
Stibino	$-SbH_2$	Stannyl	$-SnH_3$
Bismuthino	$-BiH_2$	Plumbyl	$-PbH_3$

Examples :

Dimethylstannyl	$-SnH(CH_3)_2$
Methylphenylarsino	$-As(CH_3)(C_6H_5)$
Triphenylplumbyl	$-Pb(C_6H_5)_3$

3.44. - Multivalent radicals derived from the hydrides in Rule D-3.43 are name systematically by adding diyl, triyl, etc., to the name of the parent hydride.

Examples :

| Methylstibinediyl | $>SbCH_3$ |
| Methylstannanetriyl | $\geq SnCH_3$ |

D-3.5 - Organic radicals containing characteristic groups

3.51 - An organometallic compound may be named by adding the name of the organic part as a prefix to that of the metal atom regardless of the nature of the organic characteristic group present.

Examples :

Bis(4-carboxyphenyl)mercury	$(4-HOOCC_6H_4)_2Hg$
Triethyl(2-formylethyl)tin	$(C_2H_5)_3SnCH_2CH_2CHO$
(Carboxymethyl)triphenylgermanium	$(C_6H_5)_3GeCH_2COOH$
(4-Carboxyphenyl)phenylmercury	$4-HOOCC_6H_4HgC_6H_5$
(4-Carboxyphenyl)methyltellurium	$4-CH_3TeC_6H_4COOH$
Diphenyl(4-sulfophenyl)bismuth	$4-HO_3SC_6H_4Bi(C_6H_5)_2$
(1,2-Dicarboxyethyl)hydridodimethyltin	$(CH_3)_2HSnCHCOOH$
	$\quad\quad\quad CH_2COOH$

3.52 (Partly alternative to D-3.51) - An organometallic compound containing an organic characteristic group that can be expressed as a suffix may be named according to the Rules of Sections A, B and C, hydrogen atoms of the organic parent compound being replaced by metal-containing radicals formed according to Rules D-3.41, D-3.43 and D-3.44.

Examples :

3-(Triethylstannio)propanal $\qquad$ $(C_2H_5)_3SnCH_2CH_2CHO$

3-(Triethylstannyl)propanal

(Triphenylgermanio)acetic acid $\qquad$ $(C_6H_5)_3GeCH_2COOH$

(Triphenylgermyl)acetic acid

4-(Phenylmercurio)benzoic acid $\qquad$ $4-(C_6H_5-Hg)C_6H_4-COOH$

4-(Diphenylbismuthio)benzene-
sulfonic acid $\qquad$ $4-[(C_6H_5)_2Bi]C_6H_4-SO_3H$

4-(Diphenylbismuthino)benzene-
sulfonic acid

4-(Methyltellurio)benzoic acid $\qquad$ $4-(CH_3Te)C_6H_4-COOH$

4-(Methyltelluro)benzoic acid

(Hydridodimethylstannio)-
succinic acid $\qquad$ $(CH_3)_2SnH-CH-COOH$

(Dimethylstannyl)succinic acid $\qquad\qquad\qquad\qquad | \atop CH_2-COOH$

Tris(hydroxomercurio)acetic acid $\qquad$ $(HOHg)_3C-COOH$

$$[4-\{(CH_3)_2SbCl_2\}C_6H_4\{N(CH_3)_3\}]\ Cl$$

[4-(Dichlorodimethylantimonio)phenyl]trimethylammonium
chloride

Bis(triphenylstannio)carbodiimide $\qquad$ $(C_6H_5)_3Sn-N=C=N-Sn(C_6H_5)_3$

Bis(triphenylstannyl)carbodiimide

3.53 (Partly alternative to D-3.51) - If two or more identical organic structures, each containing a characteristic group which can be designated by a suffix, are each linked to one metal atom by a single bond, the resulting organometallic compound may be named according to the principles of Rule C-72.

The metal is designated, in a prefix, by a radical name formed by citing other groups and radicals as prefixes to the modified name of the metal as given by Table II (Appendix, p. 461). This radical name is followed by the appropriate numerical affix, then by the name of the organic parent compound.
Multivalent radicals derived according to Rule D-3.44 may also be used.

Examples :

4,4'-Mercuriodibenzoic acid $(4\text{-}HOOCC_6H_4)_2Hg$

2,2'-(Ethylmethylstannio)-
dipropionic acid

2,2'-(Ethylmethylstannanediyl)-
dipropionic acid

$$
\begin{array}{c}
CH_3 \\
| \\
CH_3 \quad CH\!-\!\!COOH \\
\diagdown Sn \diagup \\
C_2H_5 \quad CH\!-\!\!COOH \\
| \\
CH_3
\end{array}
$$

D-3.6 – Organometallic compounds with two or more metal atoms

3.61 – Compounds in which rings are linked by metal atoms may be named :

(a) by using "io" radical names (Rule D-3.41 and Table II , p. 461) as prefixes to the name of a cyclic compound chosen according to the seniority order defined in Rule C-14.1 ,

(b) by coordination nomenclature, the metal encountered last in Table I (p. 459) being cited as nuclear atom.

If there is a further choice for the parent cyclic compound (method a) or for the nuclear atom (method b), this is made so that the substituents (method a) or the ligands (method b) are named with the smallest possible number of successive substitutions .

Examples :

$$
C_6H_5\text{-}Hg-\!\!\!\underset{}{\bigcirc}\!\!\!-Sb\begin{array}{c} \nearrow C_6H_5 \\ \searrow C_6H_5 \end{array}
$$

(a) 1-(Diphenylantimonio)-4-(phenylmercurio)benzene
     1-(Diphenylstibino)-4-(phenylmercurio)benzene

(b) [4-(Diphenylantimonio)phenyl]phenylmercury(II)

    [4-(Diphenylstibino)phenyl]phenylmercury(II)

Note : these names are preferred to alternative names, such as phenyl[[4-(phenylmercurio)phenyl]antimonio]benzene , which express more substitutions in the ligand.

(a) 4-(Dimethylarsenio)-3-[{5-(hydroxomercurio)-2-thienyl}-mercurio]pyridine

4-(Dimethylarsino)-3-[(5-hydroxomercurio)-2-thienyl]-mercurio|pyridine

(b) [4-(Dimethylarsenio)-3-pyridyl][5-(hydroxomercurio)-2-thienyl]mercury(II)

[4-(Dimethylarsino)-3-pyridyl][5-(hydroxomercurio)-2-thienyl]mercury (II)

D-3.7 - Organometallic ions

3.71 - Organometallic anions may be named by modifying the ending of the appropriate organometallic radical to end in "-ate". The oxidation number of the metal may be indicated by the Stock system (a), or the charge on the anion by the Ewens-Bassett system (b) (see Subsection D-2).

Examples :

(a) Hexaethynylchromate(III) $[Cr(C{\equiv}CH)_6]^{3-}$
(b) Hexaethynylchromate(3-)

(a) Pentahydroxophenyltellurate(IV) $[C_6H_5Te(OH)_5]^{2-}$
(b) Pentahydroxophenyltellurate(2-)

(a) Triphenylstannate(II) $[Sn(C_6H_5)_3]^-$
(b) Triphenylstannate(1-)

(a) Triethylzincate(II) $[Zn(C_2H_5)_3]^-$
(b) Triethylzincate(1-)

(a) Triethylhydridoaluminate(III) $[AlH(C_2H_5)_3]^-$
(b) Triethylhydridoaluminate(1-)

(a) Trihydroxotris(trifluoromethyl)- $[(CF_3)_3Sb(OH)_3]^-$
    antimonate(V)
(b) Trihydroxotris(trifluoromethyl)antimonate(1-)

(a) Dioxophenylgermanate(IV) $[C_6H_5GeO_2]^-$
(b) Dioxophenylgermanate(1-)

(a) Methyltrioxoantimonate(V) $[CH_3SbO_3]^{2-}$
(b) Methyltrioxoantimonate(2-)

(a) Ethyltrioxoarsenate(V) $[C_2H_5AsO_3]^{2-}$

(b) Ethyltrioxoarsenate(2-)

3.72 - Organometallic cations may be named according to Rule D-3.31 and the oxidation number of the metal or the charge of the cation is indicated .

Examples :

(a) Tricyclohexyltin(IV) $[(C_6H_{11})_3Sn]^+$

(b) Tricyclohexyltin(1+)

(a) Bis(cyclopentadienyl)iron(III) $[(C_5H_5)_2Fe]^+$

(b) Bis(cyclopentadienyl)iron(1+)

(a) Hydridotrimethylarsenic(V) $[AsH(CH_3)_3]^+$

(b) Hydridotrimethylarsenic(1+)

3.73 (Partially alternative to D-3.72) - Cations of the types $[MR_4]^+$ (with M = P, As Sb or Bi) or $[TeR_3]^+$, where R is a hydrogen atom or an organic radical, may be named as substitution derivatives of the following parent cations (see also Rule C-82.1) :

Phosphonium $[PH_4]^+$

Arsonium $[AsH_4]^+$

Stibonium $[SbH_4]^+$

Bismuthonium $[BiH_4]^+$

Telluronium $[TeH_3]^+$

Examples :

Tetramethylarsonium $[As(CH_3)_4]^+$

Dimethylarsonium $[AsH_2(CH_3)_2]^+$

Tetra-_p_-tolylbismuthonium $[Bi(4-CH_3C_6H_4)_4]^+$

Triphenyltelluronium $[Te(C_6H_5)_3]^+$

3.74 - Organometallic salts are designated by citing the name(s) of the cation(s) followed by the name(s) of the anion(s) .

Examples :

Sodium trihydroxotris(trifluoro-   Na $[(CF_3)_3Sb(OH)_3]$
  methyl)antimonate(V)

Potassium dioxo(phenyl)germanate(IV) *   K $[C_6H_5GeO_2]$

Tricyclohexyltin(IV) tetrafluoroborate $[(C_6H_{11})_3Sn]$ $[BF_4]$

---

* Although the use of parentheses around such simple ligands is not required, it can avoid the possibility of misinterpretation.

3.75 - When an ion is not named as a constituent of a salt, it is generally advisable to add "ion", "cation" or "anion" to the names formed according to Rules D-3.71, D-3.72 and D-3.73 . This is recommended when Stock numbers are used .

Examples :

Hexamethylchromate(III) ion (or anion)

Tricyclohexyltin(IV) ion (or cation)

Tetramethylarsonium ion (or cation)

D-4. CHAINS AND RINGS WITH REGULAR PATTERNS

OF HETEROATOMS

D-4.0 - Introduction

These rules codify and extend the best features of established practices to provide consistent names for compounds containing regular patterns of heteroatoms . The basis of this nomenclature is the use of the "a" terms given in Table I (Appendix, p. 459). Boranes are an exception to this set of rules (see Subsection D-7).

D-4.1 - Homogeneous chains

4.11 - A compound having an unbranched chain of identical heteroatoms saturated with hydrogen atoms to the extent required by the standard bonding numbers given in Rule D-1.61 is named by citing successively :

(a) the appropriate multiplying prefix : "di", "tri", "tetra", etc. The terminal vowel is *not* elided even if the "a" term begins with the same vowel,

(b) the appropriate "a" term given in Table I (Appendix, p. 459) denoting the heteroatom; except for chains of sulfur, selenium and tellurium atoms for which the "a" terms are "sulfa", "sela" and "tella",

(c) the termination "ne".
If the bonding number (see Preamble, under Bonding number, p. 325) is that indicated in Rule D-1.61 as normal bonding number, then the appropriate "a" term is used unmodified. If, however, the bonding number of the heteroatom differs from the normal bonding number or is not given in Rule D-1.61, then the "a" term is always modified by affixing a λ^n symbol, according to Rule D-1.62.

Any valence not involved in a chain-linkage will be considered as satisfied by a hydrogen atom.

Examples :

Pentasilane	$H_3Si-SiH_2-SiH_2-SiH_2-SiH_3$
Tetrasulfane	$HS-S-S-SH$
Nonaazane	$H_2N-[NH]_7-NH_2$
Tetra-λ^5-phosphane	$H_4P-PH_3-PH_3-PH_4$

4.12 - The name of a compound having an unbranched chain of identical heteroatoms with fewer hydrogen atoms than required to satisfy the standard bonding numbers given in Rule D-1.61 is derived from the name of the corresponding saturated chain by changing the ending "-ane" to "-ene" or "-yne" and adding any necessary locants and multipliers .

Example :

1, 1, 6, 6 Tetraphenyl-2, 4-hexa-
   azadiene

$$\overset{6}{(C_6H_5)_2N}-\overset{5}{N}=\overset{4}{N}-\overset{3}{N}=\overset{2}{N}-\overset{1}{N(C_6H_5)_2}$$

4.13 - A compound having an unbranched chain of identical heteroatoms with associated hydrogen atoms is numbered by Arabic numerals from one terminal heteroatom to the other . That heteroatom is numbered 1 as will give the lowest locants * to :

(a) unsaturation, according to Rule A-3.3 ,

(b) substituents .

These priorities will be considered in sequence until a unique numbering is obtained .

Examples :

2-Pentaazene
(Priority : 2 before 3)

$$\overset{5}{H_2N}-\overset{4}{NH}-\overset{3}{N}=\overset{2}{N}-\overset{1}{NH_2}$$

2-Chloropentasilane
(Priority : 2 before 4)

$$\overset{5}{H_3Si}-\overset{4}{SiH_2}-\overset{3}{SiH_2}-\overset{2}{SiHCl}-\overset{1}{SiH_3}$$

4.14 - A radical consisting of an unbranched chain of identical heteroatoms is named by adding the endings "-yl", "-diyl", "-triyl", etc. to the name of the corresponding heteroatom chain, coined according to Rules D-4.11 and D-4.12, eliding the final "e" of the name of the heteroatom chain before "y". The name is preceded by the locant(s) of the heteroatom(s) carrying the free valence † .

The chain of the radical is numbered from one end to the other so that the lowest possible locants * are given to :

(a) free valence(s) ,

(b) unsaturation, according to Rule A-3.3 ,

(c) substituents .

The abbreviated forms silyl, germyl, stannyl and plumbyl are used instead of silanyl, etc. , to distinguish between disilyl, i. e. two silyl groups (2 SiH_3), and disilanyl, i.e. one disilanyl group (Si_2H_5), etc.

---

* When series of locants containing the same number of terms are compared term by term, that series is "lowest" which contains the lowest number on the occasion of the first difference [see Rule C-13.11(e), footnote] .

† Note that, in contrast to the Rules given in Section A, this rule does not require that a free valence terminates the chain (cf. Rules A-2.25, A-3.5). Note also that the suffixes "-ylidene", "-ylidyne" and "-ylene" codified in Sections A and B are not included here. The Commissions are continuing to study these areas.

Examples :

1-Trisilanyl $\qquad$ $H_3Si-SiH_2-SiH_2-$

1,2,3-Trisilanetriyl $\qquad$ $-SiH_2-SiH-SiH_2-$
$\qquad\qquad\qquad\qquad\qquad\quad |$

1,1-Distibanediyl $\qquad$ $H_2Sb-Sb\big\langle$

1,1,2-Tristibanetriyl $\qquad$ $H_2Sb-Sb-Sb\big\langle$
$\qquad\qquad\qquad\qquad\qquad\qquad |$

4.15 - A compound having a branched chain of identical hetero-atoms is named by substituting heteroatomic radicals (see Rule D-4.14) into the longest unbranched chain, named and numbered according to Rules D-4.11, D-4.12 and D-4.13 .

Example :

3-Digermanylpentagermane $\qquad$ $H_3Ge-GeH_2-GeH-GeH_2-GeH_3$
$\qquad\qquad\qquad\qquad\qquad\qquad\qquad\qquad |$
$\qquad\qquad\qquad\qquad\qquad\qquad\qquad\quad GeH_2$
$\qquad\qquad\qquad\qquad\qquad\qquad\qquad\qquad |$
$\qquad\qquad\qquad\qquad\qquad\qquad\qquad\quad GeH_3$

D-4.2 - Homogeneous rings

4.21 - A compound having a single saturated ring of identical heteroatoms is named by adding the prefix "cyclo-" to the name of the saturated unbranched chain containing the same number of identical heteroatoms. The ring is numbered to give the lowest possible locants to substituents, using the conventions of Sections A, B and C.

Examples :

Cyclooctasilane

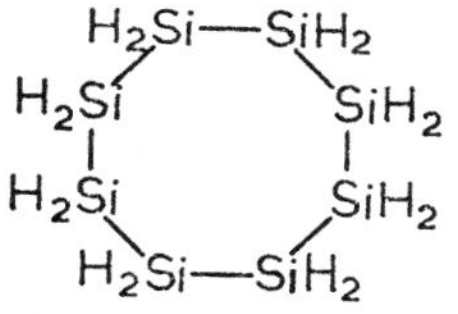

$$
\begin{array}{c}
H_2Si\!-\!\!SiH_2 \\
H_2Si \qquad\quad SiH_2 \\
| \qquad\qquad | \\
H_2Si \qquad\quad SiH_2 \\
H_2Si\!-\!\!SiH_2
\end{array}
$$

Cyclopentaazane

$$
\begin{array}{c}
HN\!-\!\!NH \\
HN \qquad NH \\
\diagdown\ N \diagup \\
H
\end{array}
$$

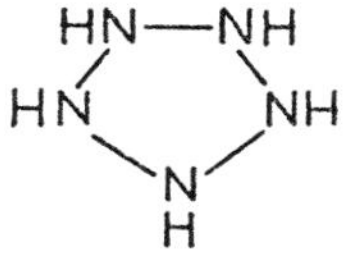

4.22 - The name of a compound having a single ring of identical heteroatoms with fewer hydrogen atoms than required to satisfy the standard bonding numbers given in Rule D-1.61 is derived from the name of the corresponding saturated ring by changing "ane" to

"ene" or "yne" and adding any necessary locants and multipliers .
The origin and the direction of enumeration are chosen so that the
lowest locants are given to unsaturation according to Rule A-3.3 .

Example :

Cyclopentaazene

4.23 - A radical consisting of a ring of identical heteroatoms is
named by adding the endings "yl", "diyl", "triyl", etc. to the name
of the corresponding ring coined according to Rules D-4.21 and
D-4.22 with elision of the final "e" of the name of the ring before "y".
The name is preceded by the locant(s) of the heteroatom(s) carrying
the free valence(s).

The ring of the radical is numbered so that the lowest locants
are given to :

(a) the free valence(s),

(b) unsaturation (according to Rule A-3.3),

(c) substituents.

Example :

2-Cyclopentaazen-1-yl
(not 1-Cyclopentaazen-3-yl)

D-4.3 – Chains of alternating atoms limited by identical atoms

4.31 - When a compound contains an unbranched chain of
alternating atoms, its parent chain is considered to be that chain
(abababa) which is limited by the two identical atoms of the element
coming last in Table I (Appendix, p. 459). It is then named by citing
successively :

(a) the prefix "di", "tri", "tetra"... denoting the number of atoms
of the limiting element, followed by the "a" term of that element,

(b) the "a" term denoting the other element of the chain,

(c) the ending "ne" if the chain is fully saturated

If the chain is not fully saturated, the syllable "ane" is changed
to "ene", "yne", "adiene", etc. preceded by the appropriate
locant(s) .

The chain is numbered from one terminal atom to the other, so
as to give the lowest possible locant(s) to unsaturation, if any .

Examples :

Disilazane	$H_3Si-NH-SiH_3$
1,3-Tricarbazadiene	$H_3C-N=CH-N=CH_2$
Tetrasilathiane	$H_3Si-S-SiH_2-S-SiH_2-S-SiH_3$
Trigermasilane	$H_3Ge-SiH_2-GeH_2-SiH_2-GeH_3$
1,3-Disilyldigermasilane	$H_3Si-GeH_2-SiH_2-GeH_2-SiH_3$
Triphosphazane	$H_2P-NH-PH-NH-PH_2$
Tetrastannoxane	$H_3Sn-O-SnH_2-O-SnH_2-O-SnH_3$

D-4.4 - Chains of repeating units *

4.41 - A compound having an unbranched chain of repeating units, each consisting of two (ababab) or more atoms (abcabc etc.) is named by citing successively :

(a) the prefix "catena" followed by a numerical affix : "di", "tri", "tetra", etc. denoting the number of repeating units ;

(b) the "a" terms of the atoms of the repeating unit, placed in parentheses and cited in the order in which they occur in the repeating unit, with elision of "a" before "a" or "o" . If there is a choice, the citation begins by the terminal atom of the unit cited *last* in Table I of the Appendix , p. 459;

(c) within the parentheses, a suffix "ane", "ene", "yne",... specifying the state of hydrogenation of the repeating unit †. These units should be linked head to tail by single bonds .

The numbering of the chain starts with the terminal atom cited *first* in Table I of the Appendix , p. 459.

Examples :

Catenadi(phosphazane)	$H_2N-PH-NH-PH_2$
Catenatri(phosphazene)	$HN=P-N=P-N=PH$
Catenatri(carbathiazane)	$\overset{9\ \ 8\ \ 7\ \ \ 6\ \ \ \ 5\ 4\ \ \ 3\ \ \ \ \ 2\ 1}{H_3C-S-NH-CH_2-S-NH-CH_2-S-NH_2}$

4.42 - If double or triple bonds are between repeating units, or are irregularly scattered on diverse units, the appropriate endings "ene", "diene", "yne" ... are added to the name, formed according to Rule D-4.41, of the compound in which these double or triple bonds are replaced by single bonds, the "a" terms being enclosed in parentheses. The lowest possible locants are allocated to double, and then to triple bonds if the numbering has not been decided by a terminal atom of the chain .

---

* This rule is an alternative to other procedures. For example, some of the compounds of this type can be named as substitution derivatives of parent compounds named by Rule D-4.3 ; others may be easily named by regular replacement nomenclature (see Subsection C-0.6).

† The final "a" of the "a" term is elided before the "ane", "ene" or "yne".

Examples :

Catenatri(carbathiacarbane)-
3, 6-diene

$$\overset{9\ 8\ 7}{H_3C}-S-\overset{6}{CH}=\overset{5\ 4}{CH}-S-\overset{3}{CH}=\overset{2\ 1}{CH}-S-CH_3$$

9 8 7 6 5 4 3 2 1
H₃C-S-CH=CH-S-CH=CH-S-CH₃

Catenatri(azacarbazane)-
3, 5-diene

9 8   7   6 5 4 3 2   1
H₂N-CH₂-NH-N=CH-N=N-CH₂-NH₂

D-4.5 - Rings of repeating units

4.51 - Monocyclic systems consisting of repeating units, each of two or more atoms, are named by citing successively :

(a) the prefix "cyclo" followed by a numerical affix : "di", "tri", "tetra"... denoting the number of repeating units ,

(b) the "a" terms of the atoms of the repeating unit, cited in the order in which they occur in the repeating unit, with elision of "a" before "a" or "o" . The citation begins by the terminal atom of the unit cited *last* in Table I of the Appendix (p. 459) and is made, as far as possible, in the *reverse* order of this table, in contradistinction to the numbering method which starts from the atom first cited in Table I, p. 459,

(c) a suffix "ane", "ene",*... specifying the state of hydrogenation of the repeating unit .

The above defined names apply to structures in which the repeating units are linked head to tail by single bonds . The numbering starts at a terminal atom of a repeating unit as *high* as possible in Table I of the Appendix , p. 459, and the lowest possible locants are allocated to the atoms *highest* in this Table.

When the repeating unit is unsaturated, the terms b and c are enclosed together within parentheses . It is also generally advisable to use parentheses when the repeating unit contains more than two atoms .

Examples:

Cyclotriboraphosphane

Cyclotri(silazene)

---

* See second footnote on page 377.

Cyclotetraazathiane

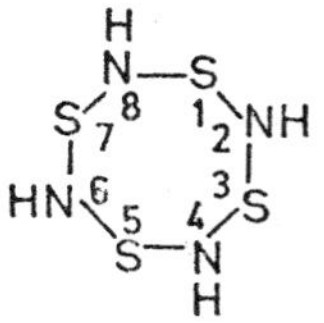

$$\begin{array}{c} H \\ N{-}S \\ S^{8}\;\;^{1}NH \\ S_{7}\;\;\;\;\;\;\;\;^{2} \\ | \\ HN_{6}\;\;\;_{3}S \\ S_{5}\;\;_{4} \\ S{-}N \\ H \end{array}$$

Cyclotri(silazoxane)

$$\begin{array}{c} {-}NH \\ O_{4}\;\;^{5}\;\;_{6}SiH_{2} \\ H_{2}Si_{3}\;\;\;\;\;\;\;\;_{7}O \\ | \\ HN^{2}\;\;\;\;\;_{8}NH \\ O{-}SiH_{2} \end{array}$$

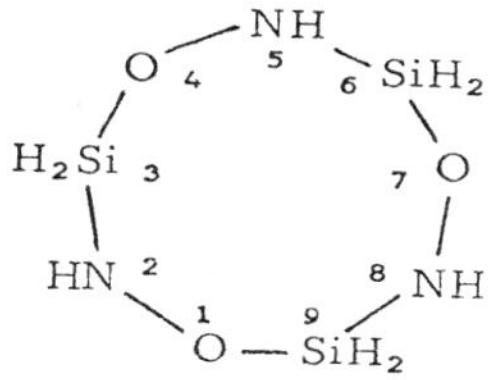

4.52 - If double bonds exist between repeating units, or are irregularly scattered on the different units, the appropriate endings "ene", "diene", ..., are added to the name given, according to Rule D-4.51, to the compound in which these double bonds are replaced by single bonds, the "a" terms being enclosed in parentheses . The lowest possible locants are allocated to double bonds, so far as Rule D-4.51 leaves a choice .

Example :

Cyclodi(boradiazan)-1-ene

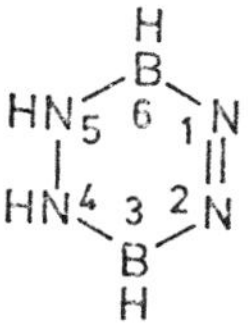

$$\begin{array}{c} H \\ B \\ HN_{5}\;\;^{6}\;\;_{1}N \\ | \;\;\;\;\;\;\;\;|| \\ HN_{4}\;\;_{3}\;\;^{2}N \\ B \\ H \end{array}$$

4.53 - The presence of cationic and anionic centers in a formally neutral chain or ring structure in some cases may be indicated by the appropriate "onia" and "ata" terms.

Example :

Catenadi(boratazoniane) $H_3\overset{-}{B}{-}\overset{+}{N}H_2{-}\overset{-}{B}H_2{-}\overset{+}{N}H_3$

D-4.6 - Radicals from chains and rings of repeating units

4.61 - The names of radicals conceptually derived by loss of hydrogen atom(s) from any chain or ring atom are formed by adding "yl", "diyl", "triyl", ..., to the name of the chain or ring, with elision of "e" before "y" *. The numbering of the chain or ring is retained. If there is a choice, the lowest possible locant(s) are allocated to free valence(s).

Examples:

Catenatri(phosphazan)-2-yl

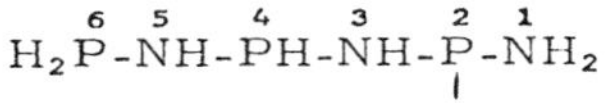

$$\overset{6}{H_2}\overset{5}{P}-\overset{}{NH}-\overset{4}{PH}-\overset{3}{NH}-\overset{2}{P}-\overset{1}{NH_2}$$

2, 2, 4-Cyclotristannoxanetriyl

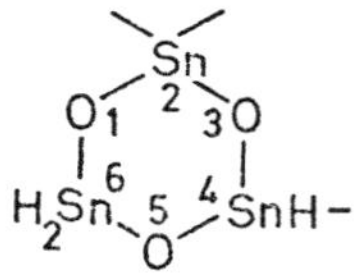

D-4.7 - Substituents

4.71 - All substituents are cited as prefixes, in alphabetical order. If the fixed numbering of the ring or chain, defined by the preceding rules, leaves any choice, that numbering is chosen as gives lowest locants (see Rule D-4.13, footnote) to the substituents.

Examples :

4-Ethyl-2,2-dimethylcyclodisilazane

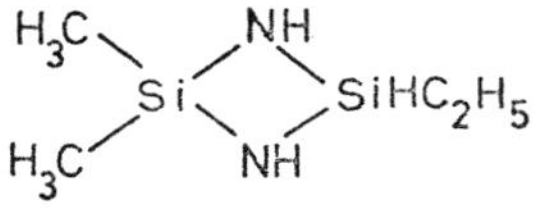

2,4-Dihydroxycyclotrisilathiane

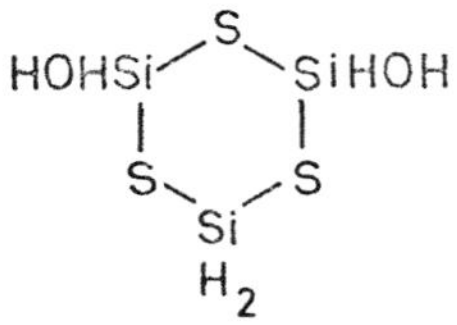

4.72 - If it is desirable to include substituents as part of the repeating unit as may be the case for long chains, the numerical affixes "bis", "tris", "tetrakis", ..., are used for the repeating unit, and the name of the complex unit is cited in parentheses.

---

* See also footnote † on page 374

Examples :

1-Methyl-6-(methylthio)catenatris-
(methoxyoxo-λ^5-phosphathiane)

2, 4, 6-Trimethoxy-1-methyl-6-(methylthio)- 2, 4, 6-trioxotri(λ^5-phos-
phathiane)

S- Methyl $-P-$ (methylthio)catenapoly-
(methoxyoxo-λ^5-phosphathiane)

D-5. ORGANIC COMPOUNDS CONTAINING PHOSPHORUS, ARSENIC, ANTIMONY OR BISMUTH

D-5.0 - Introduction

Organic compounds containing tervalent phosphorus, arsenic, antimony or bismuth show analogies to the corresponding nitrogen compounds and their nomenclature can therefore, to some extent, be patterned on that of the corresponding nitrogen compounds .

The chemistry of compounds containing quinquevalent phosphorus, arsenic, antimony or bismuth, on the other hand, is often quite different from the chemistry of the nitrogen derivatives with similar stoicheiometric composition so that a nomenclature similar to that of the nitrogen compounds cannot always be used . Nevertheless, the organic compounds of quinquevalent phosphorus and arsenic generally have similar structures and accordingly can be named by rules covering both classes of compounds .

The following nomenclature systems are of particular use in the naming of phosphorus and arsenic derivatives :

(a) Substitutive nomenclature (Subsections C-0.1 and D-1.3) covering both substitution by an organic group of hydrogen in an inorganic hydride or other parent compound, such as PH_3, PH_4^+, PH_5, $HP(O)(OH)_2$; and substitution of hydrogen in a parent organic compound by a phosphorus, arsenic, etc., group, such as $-PH_2$, $-AsH_2$, $-P(O)(OH)_2$, etc.

(b) Additive nomenclature (Subsections C-0.3 and D-1.5), e.g. phosphine oxide ,

(c) Replacement nomenclature (Subsections C-0.6 and D-1.6) also known as "a" nomenclature ,

(d) Use of "a" terms in the nomenclature of polyheteroatomic chains and rings (Subsection D-4) ,

(e) A special infix nomenclature based on phosphorus or arsenic oxo acids, in which replacement of oxygen atoms or hydroxyl groups is indicated by the infixes given in the following list :

Infix :	Replaced atom(s) :	Replacing atom(s) :
-peroxo	-OH	-OOH
-thio-	=O -OH	=S -SH
-seleno-	=O -OH	=Se -SeH

Infix :	Replaced atom(s) :	Replacing atom(s) :
-telluro-	$\begin{cases} =O \\ -OH \end{cases}$	$=Te$ $-TeH$
-amid(o)-	$-OH$	$-NH_2$
-hydrazid(o)-	$-OH$	$-NHNH_2$
-imid(o)-	$=O$	$=NH$
-nitrid(o)-	$=O$ and $-OH$	$-N\diagdown$
-fluorid(o)-	$-OH$	$-F$
-chlorid(o)-	$-OH$	$-Cl$
-bromid(o)-	$-OH$	$-Br$
-iodid(o)-	$-OH$	$-I$
-cyanid(o)-	$-OH$	$-C \equiv N$
-cyanatid(o)-	$-OH$	$-O-C \equiv N$
-thiocyanatid(o)- *	$-OH$	$-S-C \equiv N$
-isocyanatid(o)-	$-OH$	$-N=C=O$
-isothiocyanatid(o)- *	$-OH$	$-N=C=S$

The appropriate infix (in alphabetical order if more than one) is inserted before the suffix (-ic, -oic, -ous, -ate, -oate, or -ite) in the parent name. Multiplying an infix does not change its place in the alphabetical order.

(f) Replacement of oxygen in the derivatives of **oxo acids containing** quinquevalent phosphorus may also be indicated by prefixes (Section 5, Nomenclature of Inorganic Chemistry). This leads to names such as thiophosphoric acid for H_3PO_3S,

(g) Compounds of phosphorus, arsenic, antimony or bismuth may also be named by coordination nomenclature (Subsection D-2), e.g. $(C_6H_5)_5P$, pentaphenylphosphorus.

The organic compounds of quinquevalent antimony and bismuth are often polymeric or pseudo-ionic and are therefore different in character from apparently analogous phosphorus and arsenic compounds, so that it is better to fashion their nomenclature on that of coordination compounds (Subsection D-2) and of organometallic compounds (Subsection D-3). These systems of nomenclature are sometimes suitable also for organic compounds of tervalent antimony and bismuth.

Throughout this Subsection D-5, chemical formulae are generally written according to common practice of organic chemistry and the normal nesting order for enclosing marks is used :

$$\{[(\quad)]\}$$

However, when a name is coined according to the Nomenclature of Coordination Compounds (Subsection D-2), in the corresponding formula enclosing marks should be nested according to Rule D-2.21. This convention has been applied in Subsection D-5.8 where coordination nomenclature is the only one used.

---

*   Selenium and tellurium analogues are similarly included.

D-5.1 - Phosphines, their analogues and their metal

derivatives (Coordination number 3)

5.11 - The hydrides MH_3 of tervalent phosphorus, arsenic, antimony and bismuth are named phosphine, arsine, stibine, and bismuthine * .

Organic derivatives of these hydrides are named generically phosphines, arsines, stibines and bismuthines ; alkyl, aryl, etc. derivatives PH_2R^1, PHR^1R^2 and $PR^1R^2R^3$ are named primary, secondary, and tertiary phosphines respectively and are named as substituted phosphines by use of appropriate prefixes. Arsines, stibines and bismuthines are classified and named similarly.

Examples :

Ethylphosphine $CH_3-CH_2-PH_2$

Cyclohexylarsine

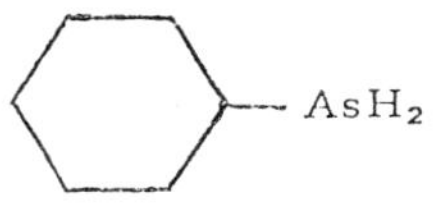

2-Naphthylphosphine

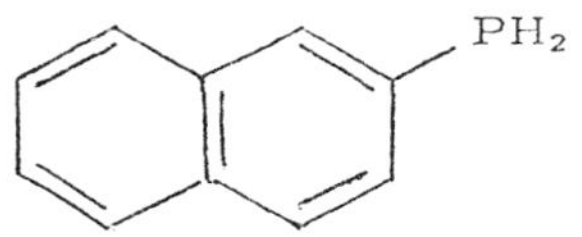

2-Benzofuranylphosphine

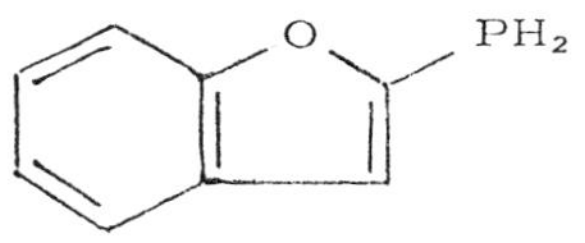

Triphenylphosphine $P(C_6H_5)_3$
Trivinylstibine $Sb(CH=CH_2)_3$
Trimethylbismuthine $Bi(CH_3)_3$

(1-Chloroethyl)(2-chloroethyl)arsine

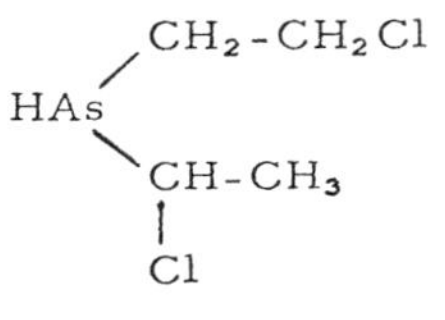

---

* The systematic names for these hydrides are phosphane, arsane, stibane, and bismuthane, as codified by Rule 2.3 in the 1970 Inorganic Nomenclature Rules, but because of long and firmly established usage, the trivial names phosphine, arsine, etc., are retained in these Rules.

Ethyl(phenyl)propylphosphine *

$$P \overset{\displaystyle C_6H_5}{\underset{\displaystyle C_3H_7}{-C_2H_5}}$$

1,2,4-Butanetriyltris(phosphine)

$$H_2P-CH_2-CH-CH_2-CH_2-PH_2$$
$$\underset{\displaystyle PH_2}{|}$$

o-Phenylenebis(arsine)

(3,7-Dibenzofurandiyl-
   bismethylene-
   bis(phosphine))

H_2P-CH_2 ... CH_2-PH_2

Note : In these **three last examples, the multiplying prefixes** "bis", "tris", etc. are used instead of "di", "tri", etc. to avoid confusion with names such as "diphosphine" and "diarsine" which have been previously used to denote respectively the compounds H_2P-PH_2 and $H_2As-AsH_2$. The latter compounds are now named "diphosphane" and "diarsane" according to Rule D-4.11.

Similarly "bismethylene" is used because "dimethylene" could be $-CH_2-CH_2-$.

5.12 - When the characteristic group $-MH_2$ does not constitute the principal group, it is designated by one of the prefixes shown in the following list and used in the same way as the prefix "amino", in conformity with Rule C-811.3. The univalent substituents $-MHR$, $-MR_2$ and $-MRR'$ are denoted as substituted "phosphino", "arsino", etc., groups.

Substituent :	Prefix :
$-PH_2$	Phosphino
$-AsH_2$	Arsino
$-SbH_2$	Stibino
$-BiH_2$	Bismuthino

Examples :

Phosphinoacetic acid H_2P-CH_2-COOH

* See footnote on page 327.

4-Phosphinopentanoic acid $CH_3-CH-CH_2-CH_2-COOH$

$\qquad\qquad\qquad\qquad\qquad\qquad\qquad |$

$\qquad\qquad\qquad\qquad\qquad\qquad\qquad PH_2$

3-Arsino-1-propanol $H_2As-CH_2-CH_2-CH_2OH$

5.13 - A compound containing, bound to the same structure, more than one type of the groups cited in Rule D-5.12 is named as a derivative of the hydride first named in Rule D-5.11 . If the compound also contains an amino group or a nitrogen heterocycle, these have priority for citation as principal group, according to Rule D-1.32 .

Examples :

2-Phosphinoethylamine $H_2P-CH_2-CH_2-NH_2$

(4-Stibinophenyl)arsine

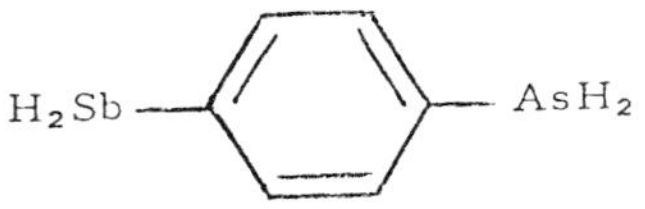

H_2Sb —⟨ ⟩— AsH_2

(Arsinomethyl)diphenylphosphine $H_2As-CH_2-P(C_6H_5)_2$

4-Arsinoquinoline

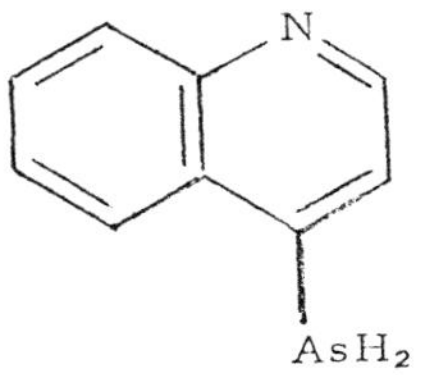

4-(Phosphinomethyl)imidazole

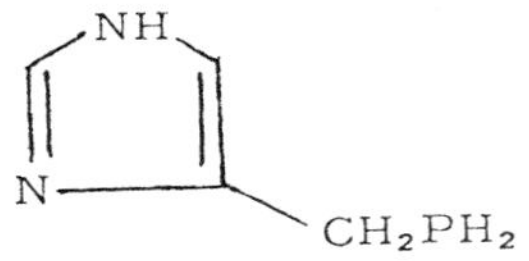

1-(Ethylmethylarsino)acridine

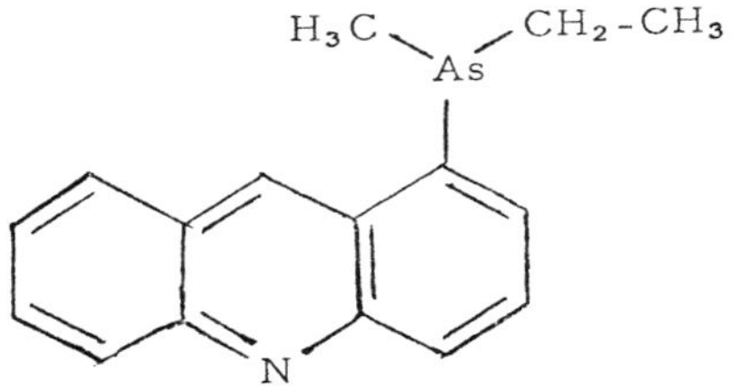

5.14 - When they do not constitute the principal group, bivalent substituents -MH- receive names shown in the following list :

Bivalent substituent :	Prefix :
-PH-	Phosphinediyl
-AsH-	Arsinediyl
-SbH-	Stibinediyl
-BiH-	Bismuthinediyl

Examples :

4,4'-Phosphinediyl-
   dibenzoic acid

$$HOOC - \langle\!\!\!\bigcirc\!\!\!\rangle - PH - \langle\!\!\!\bigcirc\!\!\!\rangle - COOH$$

3,3'-Phosphinediyldipropionic acid $HP(CH_2-CH_2-COOH)_2$

5.15 - When they do not constitute the principal group, atoms of phosphorus, arsenic, antimony or bismuth carrying three substituents receive the names shown in the following list.

Tervalent substituent :	Prefix :
-P<	Phosphinetriyl
-As<	Arsinetriyl
-Sb<	Stibinetriyl
-Bi<	Bismuthinetriyl

Examples :

Phosphinetriyltriacetic acid $P(CH_2COOH)_3$

[o-Phenylenebis(arsinetriyl)]-
   tetraacetic acid

$$\langle\!\!\!\bigcirc\!\!\!\rangle \begin{array}{l} As(CH_2-COOH)_2 \\ \\ As(CH_2-COOH)_2 \end{array}$$

5.16 - Groups -MH- forming a bridge between two carbon atoms of a cyclic system may be designated by the prefixes shown in Rule D-5.14 . In this case, these prefixes are non-detachable .

Example :

1,2,3,4-Tetrahydro-1,4-
arsinediylnaphthalene

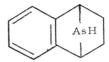

5.17 - Compounds in which there is a chain containing several atoms of phosphorus, arsenic, antimony and/or bismuth are named by replacement nomenclature, using the prefixes : "phospha", "arsa", "stiba" and "bisma" listed in Table I (Appendix, p. 459) in this order and with locants to indicate the positions of the hetero-atoms in the chain in agreement with Rule C-62.

Example :

$$\overset{12}{CH_3}-\overset{11}{CH_2}-\underset{\underset{CH_3}{|}}{\overset{10}{P}}-\overset{9}{CH_2}-\overset{8}{CH_2}-\underset{\underset{C_2H_5}{|}}{\overset{7}{As}}-\overset{6}{CH_2}-\overset{5}{AsH}-\overset{4}{CH_2}-\overset{3}{CH_2}-\underset{\underset{CH_3}{|}}{\overset{2}{P}}-\overset{1}{CH_3}$$

7-Ethyl-2,10-dimethyl-2,10-diphospha-5,7-diarsadodecane

5.18 - Phosphorus, arsenic, antimony and bismuth compounds analogous to amides, ureas, etc., are named as acyl derivatives of phosphines, arsines, stibines and bismuthines.

Examples :

Acetyldiethylphosphine	$CH_3-CO-P(C_2H_5)_2$
Carbamoylphosphine	$H_2N-CO-PH_2$
(Phenylsulfonyl)phosphine	$C_6H_5-SO_2-PH_2$
Carbonylbis(phosphine)	$OC(PH_2)_2$

Note : concerning the use of "bis" in the last example, see note to examples of Rule D-5.11.

5.19 - Derivatives of PH_3, AsH_3, SbH_3 or BiH_3 in which one or more hydrogen atoms have been replaced by a metal are named phosphides, arsenides, antimonides and bismuthides.

Examples :

Sodium diethylphosphide	$[(C_2H_5)_2P]Na$
Calcium phenylarsenide	$[C_6H_5As]Ca$

D-5.2 - Phosphines and their analogues containing
electronegative substituents (halogen, OH, SH, SeH, NH_2, etc.)
and their derivatives (Coordination number 3)

5.21 - Compounds derived from parent structures H_2MX, HMX_2
or MX_3, in which X is an electronegative substituent (halogen, OH,
SH, SeH, NH_2, etc.), may be named by three methods :

(a) as substitution products of phosphine PH_3, arsine AsH_3,
stibine SbH_3, or bismuthine, BiH_3,

(b) as derivatives of the parent compounds :

phosphinous acid H_2POH or arsinous acid H_2AsOH,

phosphonous acid $HP(OH)_2$ or arsonous acid $HAs(OH)_2$,

phosphorous acid $P(OH)_3$ or arsenous acid $As(OH)_3$,

(c) by coordination nomenclature (Subsection D-2) as compounds of
phosphorus, arsenic, antimony or bismuth. Generally it is not
necessary to include the oxidation number but it may be affixed,
as in the first and third example, if desired.

Method (b) is not applied to antimony or bismuth.

Examples :

Derivatives of parent structures H_2MX

(a) Methoxydiphenylphosphine $(C_6H_5)_2P\text{-}OCH_3$
(b) Methyl diphenylphosphinite
(c) Methoxodiphenylphosphorus(III)

(a) Chlorodiphenylphosphine $(C_6H_5)_2P\text{-}Cl$
(b) Diphenylphosphinous chloride
(c) Chlorodiphenylphosphorus

(b)   Sodium dibenzylphosphinite $(C_6H_5-CH_2)_2P-O^-Na^+$

(c)   Sodium dibenzyloxophosphate(1-)
Sodium dibenzyloxophosphate(III)

(a)   [(Diethylthiocarbamoyl)thio]- $[(CH_3)_2CH]_2P-S-CS-N(C_2H_5)_2$
diisopropylphosphine

(b)   Diethylthiocarbamic diisopropyl-
phosphinous thioanhydride (cf.
Rule C- 543.5)

(c)   (Diethyldithiocarbamato)diisopropyl-
phosphorus

(a)   Hydroxydiphenylarsine $(C_6H_5)_2AsOH$

(b)   Diphenylarsinous acid

(c)   Hydroxodiphenylarsenic

(a)   Isocyanatodiphenylstibine $(C_6H_5)_2SbNCO$

(c)   Isocyanatodiphenylantimony

Derivatives of parent structures HMX_2

(a)   Dihydroxy(phenyl)phosphine   * $C_6H_5P(OH)_2$

(b)   Phenylphosphonous acid

(c)   Dihydroxophenylphosphorus

(a)   Dichloro(phenyl)phosphine   * $C_6H_5PCl_2$

(b)   Phenylphosphonous dichloride

(c)   Dichloro(phenyl)phosphorus *

(a)   Chloro(ethylthio)methylphosphine

(b)   Ethyl methylphosphonochloridothioite

(c)   Chloro(ethylthio)methylphosphorus

$$CH_3P\begin{cases}Cl\\SC_2H_5\end{cases}$$

(a)   Methoxy(methyl)(methylamino)phosphine   *

(b)   Methyl N,P-dimethylphosphonamidite

(c)   Methoxomethyl(methylamido)phosphorus

$$CH_3P\begin{cases}OCH_3\\NHCH_3\end{cases}$$

Note on (a) - This compound is named as a phosphine according to
Rule D-1.34 .

---

*   Although the use of parentheses around such simple ligands as phenyl
or methyl is not required, it can avoid the possibility of misinterpretation.

(a) Methylenebis(diacetoxyarsine)

(b) Methylenebis(arsonous) tetraacetic
    tetraanhydride

$$H_2C \begin{array}{c} As(OCOCH_3)_2 \\ As(OCOCH_3)_2 \end{array}$$

(c) μ-Methylene-bis(diacetatoarsenic)

(a) Dichloro(phenyl)bismuthine *

(c) Dichloro(phenyl)bismuth   *

$$C_6H_5BiCl_2$$

Derivatives of parent structure MX$_3$

(a) Trimethoxyphosphine

(b) Trimethyl phosphite

$$P(OCH_3)_3$$

(c) Trimethoxophosphorus

(a) Chloro(dimethylamino)methoxyphosphine

(b) Methyl dimethylphosphoramidochloridite

(c) Chloro(dimethylamido)methoxophosphorus

$$P \begin{array}{c} Cl \\ OCH_3 \\ N(CH_3)_2 \end{array}$$

(b) Sodium O,O'-diethyl phosphorothioite

(c) Sodium diethoxothiophosphate(1-)

$$P \begin{array}{c} OC_2H_5 \\ OC_2H_5 \\ SNa \end{array}$$

(a) Tripiperidinophosphine

(c) Tripiperididophosphorus

$$P \left(-N \bigcirc \right)_3$$

5.22 - When more than one acid group each containing one atom
of tervalent phosphorus or arsenic is present as principal acid
group in an organic compound, the parent acid may be named,
according to method (b), as phosphinous, phosphonous, arsinous or
arsonous acids substituted by multivalent hydrocarbon or heterocyclic radicals . This method applies also to the derivatives of
these acids .

Examples :

(a) P,P,P',P'-Tetrahydroxy-
    1,5-naphthylenebis(phosphine)

(b) 1,5-Naphthylenebis(phosphonous acid)

(c) μ-1,5-Naphthylene-bis-
    (dihydroxophosphorus)

$$P(OH)_2 \quad \text{(naphthalene with two } P(OH)_2 \text{ groups)} \quad P(OH)_2$$

---

*   See footnote on page 390.

(a) P,P,P',P'--Tetrachloro-
   1,5-naphthylenebis(phosphine)

(b) 1,5-Naphthylenebis-
   (phosphonous dichloride)

(c) μ-1,5-Naphthylene-bis-
   (dichlorophosphorus)

D-5.3 - Onium compounds (Coordination number 4)

5.31 - Salts and hydroxides containing an atom of quadricoordinate phosphorus, arsenic, antimony or bismuth of the type $R_4M^+ X^-$, where the groups R may be identical or different, are called "phosphonium", "arsonium", "stibonium" and "bismuthonium" compounds, respectively.

5.32 - When the group containing the quadricoordinate phosphorus, arsenic, antimony or bismuth atom is the principal group, the compound is named as a salt or hydroxide of substituted phosphonium, arsonium, stibonium or bismuthonium. The names of the substituent radicals are prefixed to the word "phosphonium", "arsonium", etc., and the name of the anion is placed, as a separate word, after the name of the cation thus obtained (see Rule C-816.1).

Examples :

Benzyltriphenylphosphonium
   hydroxide

$$\left[C_6H_5-CH_2-\overset{+}{P}(C_6H_5)_3\right] OH^-$$

Tetramethylarsonium iodide

$$\left[\overset{+}{As}(CH_3)_4\right] I^-$$

5.33 - When several types of cationic centres are present in a single structure and when the group containing the quadricoordinate phosphorus, arsenic, antimony or bismuth atom does not have priority for designation as suffix, the prefix "phosphonio", "arsonio", "stibonio" or "bismuthonio" is used in conformity with the principles laid down in Rules C-85.1 and C-85.2.

The parent "onium" ions PH_4^+, AsH_4^+, SbH_4^+, and BiH_4^+ are placed in that order, immediately after NH_4^+ in Table V of Rule C-82.1.

Examples :

$$\left[(CH_3)_3 \overset{+}{As} - (CH_2)_6 - \overset{+}{N}(CH_3)_3\right] 2\ Cl$$

Trimethyl[6-(trimethylarsonio)hexyl]ammonium dichloride

1-Methyl-4-(trimethylphosphonio)pyridinium dichloride

5.34 – When replacement nomenclature is used, groups containing quadricoordinate atoms of phosphorus, arsenic, antimony or bismuth are designated by the prefixes "phosphonia", "arsonia", "stibonia" or "bismuthonia" respectively .

Example :

1-Methyl-1-phosphonianaphthalene chloride

D-5.4 – Phosphine oxides and their analogues

(Coordination number 4)

5.41 – Compounds R_3PX and R_3AsX in which X is a bivalent atom (O, S or Se) or a bivalent group (NH, CH_2, etc.) may be named :

(a) as oxides, sulfides, selenides or imides of phosphines or arsines when X = O, S, Se or NR . The alkylidene compounds are not named by this system ,

(b) as substitution products of the hydrides PH_5 or AsH_5 ,

(c) as coordination compounds of phosphorus or arsenic . This method is also suitable for antimony and bismuth compounds .

Examples:

(a) Triphenylphosphine oxide $(C_6H_5)_3PO$

(b) Oxotriphenylphosphorane

(c) Oxotriphenylphosphorus

(a) Triphenylphosphine phenylimide   * $(C_6H_5)_3P(NC_6H_5)$
    Tetraphenylphosphine imide

(b) Triphenyl(phenylimino)phosphorane

(c) Triphenyl(phenylimido)phosphorus

(b) Methylenetriphenylphosphorane $(C_6H_5)_3PCH_2$

(c) Methylenetriphenylphosphorus

(b) (Dicyanomethylene)triphenylarsorane $(C_6H_5)_3As[C(CN)_2]$

(c) (Dicyanomethylene)triphenylarsenic

(a) Triphenylstibine selenide $(C_6H_5)_3SbSe$

(c) Triphenylselenoantimony

    5.42 - (Alternative to part of Rule D-5.41) - Individual "ylides" of the type $[R_3\overset{+}{P}\text{-}\overset{-}{N}Y \longleftrightarrow R_3P{=}NY]$ and $[R_3\overset{+}{P}\text{-}\overset{-}{C}Y_2 \longleftrightarrow R_3P{=}CY_2]$ in which NY or CY_2 is replaced by sulfur or selenium, or those in which phosphorus is replaced by arsenic, are named as ionic derivatives when it is desired to draw attention to the dipolar limiting structures (see Rule C-87).

Examples :

(Triethylphosphonio)cyclopentadienide $(C_2H_5)_3\overset{+}{P}$

Phenyl(triphenylphosphonio)amide $(C_6H_5)_3\overset{+}{P}\text{-}\overset{-}{N}C_6H_5$

(Triphenylphosphonio)anilide (see Rule
    C-87)

(Triethylarsonio)sulfide $(C_2H_5)_3\overset{+}{As}\text{-}\overset{-}{S}$

(Triphenylphosphonio)methanide $(C_6H_5)_3\overset{+}{P}\text{-}\overset{-}{C}H_2$

---

*   According to Rules C-661.1 and C-661.3, both methods of indicating substituted amides and imides are allowed.

D-5.5 - Oxo acids containing quinquevalent phosphorus and arsenic
(Coordination number 4)

5.51 - When containing one or more phosphorus or arsenic atom(s) being part of a principal group attached through P or As to carbon, organic derivatives of oxo acids of quinquevalent phosphorus or arsenic are named as substitution derivatives of the following parent acids :

Phosphinic acid	$H_2PO(OH)$	Arsinic acid	$H_2AsO(OH)$
Phosphonic acid	$HPO(OH)_2$	Arsonic acid	$HAsO(OH)_2$

When multiplying affixes are needed for the oxoacid residues, "bis", "tris", "tetrakis", etc. are always used to avoid ambiguity with inorganic polyacids such as diphosphonic acid, $H_2P_2H_2O_5$ (Nomenclature of Inorganic Chemistry, 1970, Rules 4.12 and 5.214).

Examples:

Diphenylphosphinic acid	$(C_6H_5)_2PO(OH)$
Ethylphosphonic acid	$C_2H_5PO(OH)_2$
Dimethylarsinic acid	$(CH_3)_2AsO(OH)$
(4-Aminophenyl)arsonic acid	$(4)-H_2NC_6H_4AsO(OH)_2$

1,5-Naphthylenebis(phosphonic acid)

1,3,5-Naphthalenetriyl-tris(phosphonic acid)

Note : the names phosphinic, phosphonic, arsinic or arsonic can only be used when phosphorus or arsenic is attached to atoms of hydrogen, carbon or other atom of a parent compound, such as N, As or Si. Thus, $C_6H_5PO(Cl)(OH)$ is phenylchlorophosphonic acid or phenylphosphonochloridic acid and not chloro(phenyl)phosphinic acid ; $C_5H_{10}N-P(O)(Cl)(OH)$ is piperidinochlorophosphonic acid or piperidinophosphonochloridic acid and not chloropiperidinophosphinic acid ; and $ClP(O)(OH)_2$ is chlorophosphoric acid or phosphorochloridic acid and not chlorophosphonic acid.

5.52 - When the unsubstituted acid groups mentioned in Rule D-5.51 do not constitute the principal group in an organic compound, they are designated by the following prefixes : *

phosphono † $-PO(OH)_2$

phosphonato	$-PO(O^-)_2$

phosphinico $>PO(OH)$

phosphinato	$>POO^-$

arsono $-AsO(OH)_2$

arsonato	$-AsO(O^-)_2$

arsinico $>AsO(OH)$

arsinato	$>AsOO^-$

Examples :

Phosphonoacetic acid $(HO)_2OP-CH_2COOH$

4,4'-Arsinicodibenzoic acid $(HO)AsO$

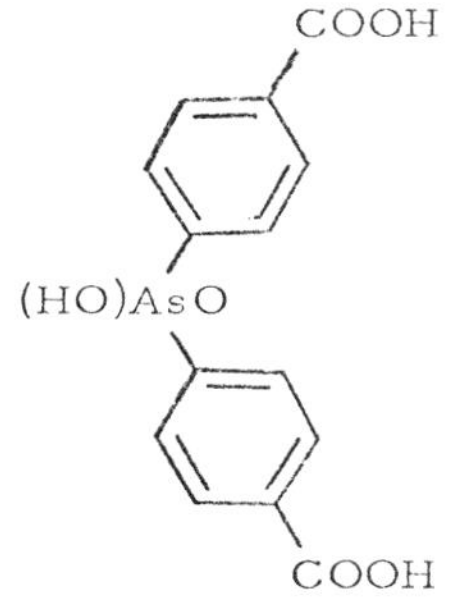

2-Phosphonatobenzoate

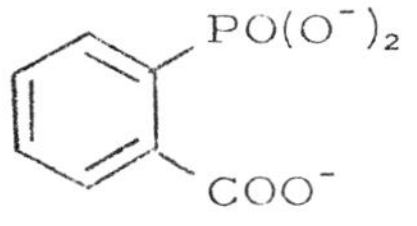

---

*   Alternatively they may be named as organometallic radicals, as shown in Rule D-5.69 .

†   Common prefix for biochemical usage : phospho- when linkage is to a heteroatom, but not when it is to carbon (Nomenclature of phosphorus-containing compounds of Biochemical Importance; IUPAC Information Bulletin, No 66, December 1977).

5.53 - Analogues of the parent **oxo** acids named in Rule D-5.51, of phosphoric acid, $PO(OH)_3$, and of arsenic acid, $AsO(OH)_3$, in which oxygen atoms and/or hydroxyl groups are replaced by other groups, as given in Subsection D-5.0, can be named in the following ways :

(a) as substitution derivatives of the acids named according to Section 5 of Nomenclature of Inorganic Chemistry (Definitive Rules 1970) (thiophosphoric acid, thiophosphonic acid, amido-phosphoric acid, etc.) and substituting hydrogen atoms bound to P or As only with radicals having the free valence on carbon,

(b) by use of infix nomenclature (see Subsection D-5.0),

(c) as "hydrogen salts", the anion being named by coordination nomenclature ; all ligands including organic radicals and hydrogen (designated by hydrido) are cited in alphabetical order, followed by "phosphate" or "arsenate".

When the acids are named by methods (a) or (b), the anions are named by changing "-ic acid" to "-ate" (except that -or- is deleted in naming derivatives of phosphoric acid).

Examples :

(a) Diethyldithiophosphinic acid $(C_2H_5)_2PS(SH)$

(b) Diethylphosphinodithioic acid

(c) Hydrogen diethyldithiophosphate(V)

(a) Dimethylamidophosphoric acid $(CH_3)_2NPO(OH)_2$

(b) *N,N*- Dimethylphosphoramidic acid *

(c) Dihydrogen (dimethylamido)-
    trioxophosphate (V)

(a) Diphenyl(phenylimido)phosphinic
    acid † $(C_6H_5)_2P(NC_6H_5)(OH)$

(b) *N,P,P*- Triphenylphosphinimidic
    acid   *

(c) Hydrogen oxodiphenyl(phenylimido)-
    phosphate (V)

(a) Phenyl(chloro)(phenylimido)-
    thiophosphonic acid †‡ $C_6H_5P(NC_6H_5)(Cl)(SH)$

(b) *N,P*- Diphenylphosphonochloridimido-
    thioic acid   *

(c) Hydrogen chloro(phenyl)(phenylimido)-
    thiophosphate (V) ‡

---

* Although locants are not really needed here, they are included for clarity.

† See footnote on page 394.

‡ See footnote on page 327.

(a) Phenylnitridophosphonic acid $C_6H_5P(N)OH$

(b) Phenylphosphononitridic acid

(c) Hydrogen nitridooxophenylphosphate(V)

(a) Ethylselenophosphonic acid $C_2H_5PSe(OH)_2$

(b) Ethylphosphonoselenoic acid

(c) Dihydrogen ethyldioxoselenophosphate(V)

(a) Phenyl(methylamido)thiophosphonic
    acid   * $C_6H_5PS(NHCH_3)(OH)$

(b) *N*-Methyl-*P*-phenylphosphonamidothioic
    acid

(c) Hydrogen (methylamido)oxo(phenyl)-
    thiophosphate (V) †

(a) (Dimethylamido)(isothiocyanato)-
    thiophosphoric acid   * $PO[N(CH_3)_2](NCS)(SH)$

(b) *N*,*N*-Dimethylphosphoramidoisothio-
    cyanatidothioic acid ‡

(c) Hydrogen (dimethylamido)-
    isothiocyanatooxothiophosphate (V)

(a) (Dimethylamido)(phenylimido)-
    (thiocyanato)phosphoric acid * $P(NC_6H_5)[N(CH_3)_2](SCN)(OH)$

(b) *N*,*N*-Dimethyl-*N*′-phenylphosphoramidimido-
    thiocyanatidic acid

(c) Hydrogen (dimethylamido)oxo(phenylimido)-
    thiocyanatophosphate (V)

(a,b) Ethylphosphinic acid $(C_2H_5)HPO(OH)$

(c) Hydrogen ethylhydridodioxophosphate(V)

(a) Phenylchlorophosphonic acid $C_6H_5PO(Cl)(OH)$

(b) Phenylphosphonochloridic acid

(c) Hydrogen chlorodioxophenylphosphate(V)

    5.54 - (Alternative to part of Rule D-5.53) - Generally the point
of attachment of ionisable hydrogen is not specified in naming acids
because they generally exist in solution as a mixture of tautomeric
and ionised forms and in the solid the hydrogen is usually bonded to
more than one anion.

    However, if it is desired to specify a particular tautomer this can
be done according to the procedure given in Rule C-541, for the
names coined according to methods (a) and (b). With coordination
nomenclature (c), the compound is considered as a neutral (non-
ionised) molecule.

---

* See footnote on page 394.

† See footnote on page 327.

‡ See first footnote on page 397.

Examples :

(a) Ethylselenophosphonic O,O'-acid $C_2H_5PSe(OH)_2$
(b) Ethylphosphonoselenoic O,O'-acid
(c) Ethyldihydroxoselenophosphorus(V)

(a) Ethylselenophosphonic O,Se-acid $C_2H_5PO(OH)(SeH)$
(b) Ethylphosphonoselenoic O,Se-acid
(c) Ethyl(hydrogenselenido)hydroxooxophosphorus(V) *

D-5.6 - Derivatives of oxo acids containing quinquevalent
phosphorus and arsenic (coordination number 4)

5.60 - The concept of a "derivative" is peculiar to organic chemistry and the names of esters, amides, etc., can be formed according to Section C on the basis of the names of the parent acids. By coordination nomenclature all ligands are named in alphabetical order followed by the name of the central atom, phosphorus or arsenic, and oxidation number or charge of the ion if desired.

5.61 - Esters of the acids considered in Subsection D-5.5 are named by changing the ending "-ic acid" to "-ate", except that "-or" is deleted in naming derivatives of phosphoric acid, [methods (a) and (b)] or by coordination nomenclature [method (c)].

Examples :

(a)(b) Trimethyl phosphate $PO(OCH_3)_3$
(c) Trimethoxooxophosphorus

(a)(b) Ethyl methyl phenyl phosphate $PO(OC_2H_5)(OCH_3)(OC_6H_5)$
(c) Ethoxomethoxooxophenoxophosphorus

(a)(b) Methyl dihydrogen phosphate † $PO(OCH_3)(OH)_2$
(c) Dihydroxomethoxooxophosphorus

(a)(b) Sodium diethyl phosphate $PO(OC_2H_5)_2ONa$
(c) Sodium diethoxodioxophosphate(1-)

(a) O-Methyl phenylthiophosphinate $C_6H_5PHS(OCH_3)$
(b) O-Methyl phenylphosphinothioate
(c) Hydridomethoxophenylthiophosphorus

---

* For organic radicofunctional nomenclature the name hydroselenide is used for the -SeH group and the substituent prefix is hydroseleno (see Rule C-701).

† See the first footnote on page 328.

(a,b) Dimethyl phosphonate $\qquad$ $HPO(OCH_3)_2$

(c) Hydridodimethoxooxophosphorus

(a) Ethyl diethyldithiophosphinate $\qquad$ $(C_2H_5)_2PS(SC_2H_5)$

(b) Ethyl diethylphosphinodithioate

(c) Diethyl(ethylthio)thiophosphorus

(a) S-Methyl dimethylthioarsinate $\qquad$ $(CH_3)_2AsO(SCH_3)$

(b) S-Methyl dimethylarsinothioate

(c) Dimethyl(methylthio)oxoarsenic

(a,b) Diethyl methylphosphonate $\qquad$ $CH_3PO(OC_2H_5)_2$

(c) Diethoxo(methyl)oxophosphorus *

(a) S-Ethyl O-methyl phenyl-
     thiophosphonate $\qquad$ $C_6H_5PO(OCH_3)(SC_2H_5)$

(b) S-Ethyl O-methyl phenylphosphonothioate

(c) (Ethylthio)methoxooxophenylphosphorus

(a) Methyl phenylchlorophosphonate $\qquad$ $C_6H_5POCl(OCH_3)$

(b) Methyl phenylphosphonochloridate

(c) Chloro(methoxo)oxophenylphosphorus *

(a) Dimethyl (dimethylamido)phosphate $\qquad$ $PO[(CH_3)_2N](OCH_3)_2$

(b) Dimethyl dimethylphosphoramidate

(c) (Dimethylamido)dimethoxooxophosphorus

(a) Ethyl (diethylamido)-
     (thiocyanato)phosphate $\qquad$ $PO(NCS)[(C_2H_5)_2N](OC_2H_5)$

(b) Ethyl diethylphosphoramido-
     (thiocyanatidate)

(c) (Diethylamido)ethoxooxo-
     (thiocyanato)phosphorus

(a) Diethyl fluoroarsenate $\qquad$ $AsOF(OC_2H_5)_2$

(b) Diethyl arsenofluoridate

(c) Diethoxofluorooxoarsenic

(a) S-Ethyl difluorothioarsenate $\qquad$ $AsOF_2(SC_2H_5)$

(b) S-Ethyl arsenodifluoridothioate

(c) (Ethylthio)difluorooxoarsenic

---

* See footnote on page 327.

5.62 - Amides derived from the acids considered in Subsection D.5.5 [methods (a) and (b)] are named by changing "acid" to "amide". Imides and hydrazides are named analogously. N-substituted amides are named as substitution products. Alternatively they may be named by coordination nomenclature [method (c)].

Examples :

(a)   Dimethylphosphinic dimethylamide * $(CH_3)_2PO[N(CH_3)_2]$

(a) N,N,P,P- Tetramethylphosphinic amide †

(c)   (Dimethylamido)dimethyloxophosphorus

(a)   Phenylphosphonic bis(ethylamide) * $C_6H_5PO(NHC_2H_5)_2$

(a) N,N'-Diethyl-P-phenylphosphonic diamide

(c)   Bis(ethylamido)oxophenylphosphorus

(a) Phenylthiophosphonic bis(dimethylamide)* $C_6H_5PS[N(CH_3)_2]_2$

(b) N,N,N',N'- Tetramethyl-P-phenyl-
        phosphonothioic diamide

(c) Bis(dimethylamido)phenylthiophosphorus

(a) N,N-Diethyl-N',N'-dimethyl- $C_6H_5AsS[N(CH_3)_2][N(C_2H_5)_2]$
        (phenyl)thioarsonic diamide * ‡

(b) N,N- Diethyl-N',N'-dimethyl-As-
        phenylarsonothioic diamide

(c) (Diethylamido)(dimethylamido)-
        (phenyl)thioarsenic ‡

(a) (Phenylsulfonylimido)phosphoric $(C_6H_5SO_2N)P(NHC_6H_5)_3$
        trianilide

(b) N, N',N''-Triphenyl-N'''-(phenylsulfonyl)-
        phosphorimidic triamide

(c) Tris(phenylamido)(phenylsulfonylimido)-
        phosphorus

    Note : As shown in name (a), "anilide" may replace "phenyl-amide" .

---

*   See footnote on page 394.

†   See first footnote on page 397.

‡   See footnote on page 327.

(a) Thiophosphoric tris(2,2-dimethylhydrazide) $PS[NH-N(CH_3)_2]_3$

(b) 2, 2, 2', 2', 2'', 2''-Hexamethylphosphorothioic trihydrazide

(c) Tris(2,2-dimethylhydrazido)thiophosphorus

Note : If a compound contains more than one amide group, these are treated in the same way . For example, $RPO(NH_2)_2$ is a phosphonic diamide, not an amidophosphonic or a phosphonamidic amide .

5.63 - Acyl halides derived from the acids considered in Subsection D-5.5 [methods (a) and (b)] are named by replacing the name "acid" by the name of the appropriate halide . Alternatively they may be named by coordination nomenclature .

Examples :

(a) Diethyl(phenylimido)phosphinic chloride * $(C_2H_5)_2PCl(NC_6H_5)$

(b) P ,P-Diethyl-N-phenylphosphinimidic chloride

(c) Chloro(diethyl)(phenylimido)phosphorus

(a) Diethylthiophosphinic chloride $(C_2H_5)_2PSCl$

(b) Diethylphosphinothioic chloride

(c) Chlorodiethylthiophosphorus

(a,b) Phenylphosphonic dichloride $C_6H_5POCl_2$

(c) Dichlorooxophenylphosphorus

(a) Ethyl(diethylamido)phosphonic chloride * $C_2H_5POCl[N(C_2H_5)_2]$

(b) N ,N ,P-Triethylphosphonamidic chloride †

(c) Chloro(diethylamido)(ethyl)oxophosphorus ‡

(a) (Dimethylamido)isocyanatophosphoric chloride $POCl(NCO)[N(CH_3)_2]$

(b) N,N-Dimethylphosphoramidoisocyanatidic chloride †

(c) Chloro(dimethylamido)isocyanatooxo-phosphorus

Note : The use of the radical names listed in Rule D-5.66 to designate acyl halides is not recommended .

5.64 - Mixed anhydrides of phosphorus acids and carboxylic acids may be named in the following ways :

(a) according to Rule C-491.3 ,

---

* See footnote on page 394.

† See first footnote on page 397.

‡ See footnote on page 327.

(b) as acyl derivatives of the phosphorus acid ,

(c) by coordination nomenclature .

Examples :

(a) Acetic phosphoric monoanhydride $\quad$ $CH_3CO\text{-}O\text{-}PO(OH)_2$

(b) Monoacetyl phosphate

(c) (Acetato)dihydroxooxophosphorus

Note: For name (b) , common biochemical usage is to omit the prefix "mono" and to name this "acetyl phosphate"

(a) Acetic dimethylphosphinic anhydride $\quad$ $CH_3CO\text{-}O\text{-}PO(CH_3)_2$

(b) Acetyl dimethylphosphinate

(c) (Acetato)dimethyloxophosphorus

5.65 - Derivatives of stoicheiometric composition RPO_2, $ROPO_2$, $ROPO(NR')$, $(RO)_2PN$, etc. and their arsenic analogues, being polymeric, may be named as metaphosphites, metaphosphates, imidometaphosphates, phosphonitriles, etc. when their structures and molecular weights are unknown .

Examples :

Methyl trithiometaphosphate $\quad$ $(CH_3SPS_2)_n$

Poly(phosphonitrile dichloride) $\quad$ $(Cl_2PN)_n$

Note: Compounds with known chain or ring structures are named according to Subsection D-4 .

5.66 - When another group is also present that has priority for citation as principal group, radicals derived from oxo acids of quinquevalent phosphorus or arsenic are named as follows :

Phosphinoyl *	$H_2P(O)-$
Phosphonoyl *	$HP(O)\!\!<$
Phosphoryl *	$P(O)\!\!\underset{\sim}{<}$
Arsinoyl	$H_2As(O)-$
Arsonoyl	$HAs(O)\!\!<$
Arsoryl †	$As(O)\!\!\underset{\sim}{<}$

---

* The names "phosphinyl", "phosphinylidene", "phosphinylidyne" have also been used but are not recommended .

† "Arsoryl" is preferred to "arsenyl" for clarity and consistency . See also Rule D-5.71, Note p. 406.

Examples :

Phosphinoylacetic acid $H_2P(O)-CH_2COOH$

Phosphoryltriacetic acid $P(O)(CH_2COOH)_3$

5.67 - Acid radicals derived from the acids described in Rule
D-5.53 are preferably named by replacing the termination "-ic
acid" by "-oyl" in the corresponding acid name coined according to
method (b).

As an alternative, radical names may be derived from acid
names coined according to method (a) but this procedure is usually
restricted to replacement by sulfur, selenium or tellurium.

In the following examples, letters (a) and (b) relate to the
corresponding methods described in Rule D-5.53.

Examples :

(a) Thiophosphinoyl $H_2P(S)-$
(b) Phosphinothioyl

(a) Selenophosphonoyl $HP(Se)\big\langle$
(b) Phosphonoselenoyl

(a) Thiophosphoryl $P(S)\big\langle$
(b) Phosphorothioyl

(b) Phosphonimidoyl $HP(NH)\big\langle$

(b) Phosphonamidoyl $HP(O)(NH_2)-$

(b) Phosphonamidimidoyl $HP(NH)(NH_2)-$

(b) Phosphononitridoyl $HP(N)-$

(b) Phosphorimidoyl $P(NH)\big\langle$

(b) Phosphoronitridoyl $P(N)\big\langle$

5.68 - Compound radicals may also be named from the radicals
phosphoryl, arsoryl and their derivatives.

Examples :

4-(Diethoxyphosphoryl)-
   benzoic acid $(C_2H_5O)_2P(O)-$$-COOH$

4,4'-(Methoxyphosphoryl)-
   dibenzoic acid

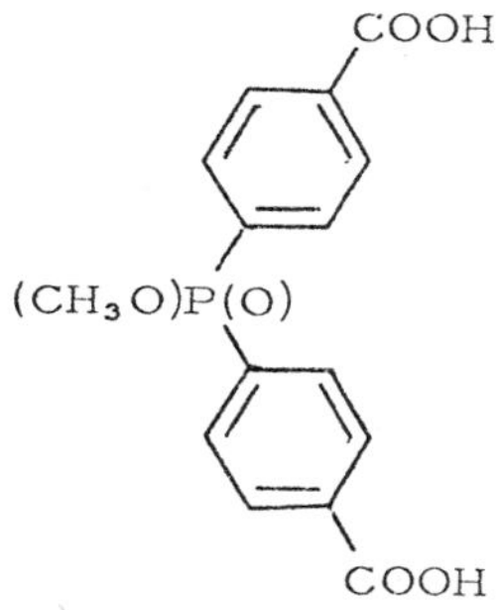

$(CH_3O)P(O)$

COOH

COOH

(Dichlorophosphoryl)acetyl chloride $Cl_2P(O)-CH_2COCl$

4-[Dimethyl(thiophosphoryl)]-
   benzoic acid

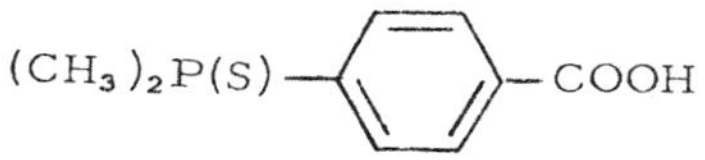

$(CH_3)_2P(S)$ —⟨ ⟩— COOH

Methyl (dimethoxyphosphor-
   imidoyl)acetate

$(CH_3O)_2P(NH)-CH_2COOCH_3$

4-(Methoxyphosphoronitridoyl)-
   benzenesulfonic acid

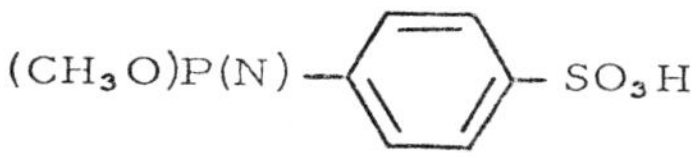

$(CH_3O)P(N)$ —⟨ ⟩— SO_3H

    5.69 - (Partly alternative to Rules D-5.52, D-5.66, D-5.67 and D-5.68) - Radicals having the free valence(s) on antimony or bismuth are named as organometallic radicals according to Rule D-3.41 . This method is also applicable to phosphorus and arsenic radicals .

Examples :

Dihydroxooxophosphorio $(HO)_2(O)P-$

Hydroxooxoarsenio $(HO)(O)As<$

Oxophosphorio $OP\lesssim$

Hydridomethoxothiophosphorio $H(CH_3O)(S)P-$

Dimethoxo(phenylimido)phosphorio $(CH_3O)_2(C_6H_5N)P-$

D-5.7 - Phosphoranes and their analogues

(Coordination number 5)

5.71 - Alkyl or aryl, etc., derivatives of PH_5 and AsH_5 are called generically "phosphoranes" and "arsoranes" *. These compounds may be designated :

(a) as substitution derivatives of these hydrides,

(b) as coordination compounds.

Examples :

(a) Pentaphenylphosphorane $P(C_6H_5)_5$

(b) Pentaphenylphosphorus

(a) Pentaphenylarsorane $As(C_6H_5)_5$

(b) Pentaphenylarsenic

5.72 - Derivatives of five-coordinate phosphorus, arsenic, antimony or bismuth containing one or more electronegative univalent substituents are named according to one of the following two methods :

(a) as substitution products of MH_5 (only for phosphorus and arsenic compounds),

(b) as coordination compounds

Examples :

(a) Dichlorotriphenylphosphorane $(C_6H_5)_3PCl_2$

(b) Dichlorotriphenylphosphorus

(a) Bis(dimethylamino)triphenylarsorane $(C_6H_5)_3As[N(CH_3)_2]_2$

(b) Bis(dimethylamido)triphenylarsenic

---

* Although the syllable "-or-" belongs only to phosphorus, it is introduced in the name arsorane because arsenane is the name of a six-membered ring with one arsenic atom (see Rule B-1.1).

(b) Chlorohydroxotriphenylbismuth $(C_6H_5)_3BiCl(OH)$

5.73 - If the structure of derivatives of five-coordinate phosphorus, arsenic, antimony or bismuth is unknown, a name formed according to organometallic rules (Subsection D-3) or by coordination nomenclature should be used (Subsection D-2).

Example :

Benzyltriphenoxophosphorus iodide $(C_6H_5O)_3PI(CH_2-C_6H_5)$
Benzyliodotriphenoxophosphorus

5.74 - When they do not constitute the principal group, radicals derived from phosphorane PH_5 or arsorane AsH_5 may be named in two ways :

(a) by the use of the following prefixes :

H_4P-	Phosphoranyl		H_4As-	Arsoranyl
$H_3P\langle$	Phosphoranediyl		$H_3As\langle$	Arsoranediyl
$H_2P\langle$	Phosphoranetriyl		$H_2As\langle$	Arsoranetriyl

(b) as organometallic radicals according to Rule D-3.41 and Table II (Appendix, p. 461). This method is also applicable to antimony and bismuth radicals.

Example :

(a) 4-(Tetraphenylphosphoranyl)- $(C_6H_5)_4P—$⟨benzene ring⟩$-$ COOH
benzoic acid

(b) 4-(Tetraphenylphosphorio)-
benzoic acid

Note : According to coordination nomenclature (Subsection D-2), the above compound can also be named :

(4-Carboxyphenyl)tetraphenylphosphorus

D-5.8 -   Anions (Coordination number 6)

5.81 - Anions in which the phosphorus, arsenic, antimony or bismuth atom has coordination number 6 are named by placing the names of all ligands, radicals, or anions as prefixes to the word "phosphate", "arsenate", "antimonate" or "bismuthate". The name is completed according to Rule D-2.23 by indication of either :

(a) the oxidation state of the central atom (Stock system),

(b) the charge of the entire ion (Ewens-Bassett system) .

Examples :

(a) Sodium hexaphenylphosphate(V) $Na\left[P(C_6H_5)_6\right]$

(b) Sodium hexaphenylphosphate(1-)

(a) Ammonium tris(2,2'- biphenylylene)arsenate(V)

(b) Ammonium tris(2,2'- biphenylylene)arsenate(1-)

(a) Lithium hexamethylantimonate(V) $Li\left[Sb(CH_3)_6\right]$

(b) Lithium hexamethylantimonate(1-)

(a) Sodium trichlorotriphenylbismuthate(V) $Na\left[BiCl_3(C_6H_5)_3\right]$

(b) Sodium trichlorotriphenylbismuthate(1-)

D-6 - ORGANOSILICON COMPOUNDS

D-6.0 - Introduction

The Rules for naming organosilicon compounds in Subsection D-6.1 to D-6.5 are essentially the same as those of 1949 (IUPAC Comptes rendus of the 15th Conference, pp. 127-132) except as follows:

(a)   "silthiane", etc. has been changed to "silathiane", etc.,

(b)   it is now recommended that the names of all organosilicon radicals containing two or more silicon atoms be formed in a regular fashion (without contractions; e.g. "disilanyloxy" rather than "disilanoxy"),

(c)   naming of silasesquioxanes and silasesquithianes has been added, and

(d)   naming of multivalent radicals has been modified (see Rules D-6.12 and D-6.51).

D-6.1 - Chains and rings of silicon atoms

6.11 - The name of the compound SiH_4 is "silane". According to the number of silicon atoms present, compounds having the general formula $H_3Si[SiH_2]_nSiH_3$ are named "disilane", "trisilane", etc. The generic name of the branched or unbranched silicon hydrides is "silane".

Example:

Tetrasilane $H_3Si-SiH_2-SiH_2-SiH_3$

6.12 - A radical consisting of an unbranched acyclic chain of silicon atoms is named by adding the endings "-yl", "-diyl", "-triyl", etc. to the name of the corresponding silicon chain (with elision of "e" before "y") coined according to Rule D-6.11. The suffix is preceded by the locant(s) of the silicon atom(s) carrying the free valence(s).
Example:

Disilanyl $H_3Si-SiH_2-$

The chain of the radical is numbered from one end to the other so that the lowest possible permissible locants are given to:

(a) free valence(s),
(b) substituents .

The abbreviated form silyl is used instead of silanyl, to distinguish between disilyl, i.e. two silyl groups (2 SiH_3), and disilanyl, i.e. one disilanyl group (Si_2H_5), etc. (see Rule D-4.14).

Examples:

Disilanyl	$H_3Si-SiH_2-$
Silanediyl	$-SiH_2-$
1,3-Trisilanediyl	$-SiH_2-SiH_2-SiH_2-$
1,1,1-Disilanetriyl	$SiH_3-Si\langle$
1,1,4,4-Tetrasilanetetrayl	$\rangle SiH-SiH_2-SiH_2-SiH\langle$
1,1,3-Trisilanetriyl	$-SiH_2-SiH_2-SiH\langle$
1,2,2,3-Trisilanetetrayl	$-SiH_2-\overset{\mid}{\underset{\mid}{Si}}-SiH_2-$

Note : Multivalent radicals derived from the silanes are named differently by this Rule than by either the 1949 IUPAC Silicon Rules or by the principles in Rule A-4. A free valence need not terminate the chain and the suffixes "-ylidene", "-ylidyne", and "-ylene" are not included here (see also footnote † on page 374). The following table compares the two methods of naming.

Present Rules	Former Rules
Silanediyl	Silylene
2-Trisilanyl	1-Silyldisilanyl
1,3-Trisilanediyl	Trisilanylene
1,1,1-Trisilanetriyl	Trisilanylidyne
1,2,2,3-Trisilanetetrayl	1,3-Trisilanediyl-2-ylidene
1,1,1,3,3,3-Trisilanehexayl	1,3-Trisilanediylidyne

6.13 - A branched silane is named by adding the names of the radicals corresponding to the side chains to the name of the longest chain of silicon atoms present in the molecule. Numbering and citation of side chains proceed according to the principles set forth in Subsection A-2.

Example:

$$\overset{5}{}\overset{4}{}\overset{3}{}\overset{2}{}\overset{1}{}$$

3-Disilanyl-2-silylpentasilane $H_3Si-SiH_2-SiH-SiH-SiH_3$

with branches SiH_2 and SiH_3 / SiH_3

6.14 - Cyclic compounds with the general formula $\overset{\frown}{[SiH_2]_n}$ are named "cyclotrisilane", "cyclotetrasilane", etc., according to the number of silicon atoms present. The generic name of these compounds is "cyclosilane". Radicals derived from cyclosilanes are named by adding "-yl", "-diyl", "-triyl", etc., to the name of the cyclosilane with elision of final "e" of the silane name before a suffix "-yl". *

Examples:

Cyclohexasilane

$$H_2Si \underset{SiH_2}{\overset{SiH_2}{}} SiH_2 \quad H_2Si \quad SiH_2$$

Cyclotrisilanyl

$$H_2Si \!-\! SiH\!- \quad \diagdown\!\diagup \quad SiH_2$$

D-6.2 - Chains and rings of silicon and oxygen atoms

6.21 - Compounds in which the skeleton is composed of silicon and oxygen atoms in such a way that each silicon atom is separated from its nearest silicon atoms by single oxygen atoms, while each oxygen atom is separated from its nearest oxygen atoms by single silicon atoms are called "siloxanes".

6.22 - Compounds with the general formula $H_3Si-[O-SiH_2]_n-O-SiH_3$ are named "disiloxane", "trisiloxane", etc., according to the number of silicon atoms present. The chain is numbered from one end to the other starting with a terminal silicon atom and numbering each oxygen and silicon atom in turn. Branched open-chain siloxanes are named according to the principles set forth in Rule D-6.13.

---

* See also footnote † on page 374 and the note to Rule D-6.12 on page 410.

Examples:

Tetrasiloxane

$$\overset{1}{}\;\overset{2}{}\;\overset{3}{}\quad\overset{4}{}\;\overset{5}{}\quad\overset{6}{}\;\overset{7}{}$$
$$H_3Si-O-SiH_2-O-SiH_2-O-SiH_3$$

5-(Disiloxanyloxy)pentasiloxane $H_3Si-O-SiH_2-O-SiH-O-SiH_2-O-SiH_3$
$$\qquad\qquad\qquad\qquad\qquad\qquad\qquad\quad |$$
$$\qquad\qquad\qquad\qquad\qquad\qquad\qquad O-SiH_2-O-SiH_3$$

1,1,1-Trichloro-5,5,5-
    trimethoxytrisiloxane $(CH_3O)_3Si-O-SiH_2-O-SiCl_3$

6.23 - Monocyclic siloxanes with general formula $\left(OSiH_2\right)_n$ are named cyclotrisiloxane, cyclotetrasiloxane, etc., according to the number of silicon atoms present. Numbering is assigned according to the principles of Rule B-1.53.

Example:

2,2-Dimethylcyclotrisiloxane

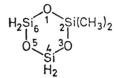

6.24 - Bi- and polycyclic siloxanes are named by a prefix defining the ring structure, such as "bicyclo[3.3.1]" (cf. Rule A-31 and A-32) or "spiro[4.5]" (cf. Rule A-41), followed by an infix giving the number of silicon atoms present in the molecule, the whole being attached to "siloxane". Numbering of these cyclosiloxanes is assigned according to the principles set forth in Rules A-31, A-32 or A-41 for the ring structure and Rules B-1 and B-3.4 for heteroatoms.

Examples:

3,3-Dimethylbicyclo[3.3.1]tetrasiloxane

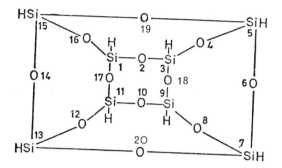

2, 2-Dimethylspiro[5. 7]hexasiloxane

Pentacyclo[9.5.1.1$^{3, 9}$.1$^{5, 15}$.1$^{7, 13}$]octasiloxane
(an octasilasesquioxane by Rule D-6.61)

Tetracyclo[5.5.1.1$^{3, 11}$.1$^{5, 9}$]hexasiloxane
(a hexasilasesquioxane by Rule D-6.61)

D-6.3 - Chains and rings of silicon and sulfur atoms

6.31 - Compounds having the formula $H_3Si-[S-SiH_2]_n-S-SiH_3$ are named "disilathiane", "trisilathiane", etc., according to the number of silicon atoms present; they have the generic name "silathiane".

Example:

Trisilathiane $H_3Si-S-SiH_2-S-SiH_3$

6.32 - Cyclic compounds having the general formula $(S-SiH_2)_n$ are named "cyclotrisilathiane", "cyclotetrasilathiane" etc., according to the number of silicon atoms present; they have the generic name "cyclosilathiane". Locants are assigned according to the principles of Rule B-1.53.

Example:

2-Methylcyclotrisilathiane

6.33 - Bi- and poly-cyclic silathianes are named according to the principles set forth in Rule D-6.24 for the cyclosiloxane analogues (cf. Rule D-6.61).

Examples:

Tricyclo[5.1.1.1^{3,5}]tetrasilathiane

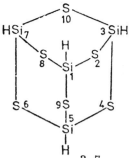

Tricyclo[3.3.1.1^{3,7}]tetrasilathiane

6.34 - Selenium and tellurium analogs of siloxanes are named in the same way as silathianes by using "selen" or "tellur" in place of "thi".

Example:

Disilaselenane H_3Si-Se-SiH_3

D-6.4 - Chains and rings of silicon and nitrogen atoms

6.41 - Compounds with the general formula H_3Si-[NH-SiH_2]$_n$-NH-SiH_3 are named "disilazane", "trisilazane", etc., according to the number of silicon atoms present; they have the generic name "silazane".

Example:

Trisilazane H_3Si-NH-SiH_2-NH-SiH_3

6.42 - Cyclic compounds having the general formula $\overbrace{[SiH_2-NH]}_n$ are named "cyclotrisilazane", "cyclotetrasilazane", etc., according to the number of silicon atoms present; they have the generic name "cyclosilazane". Numbering is assigned according to the principles of Rule B-1.53.

Example:

1-Methylcyclotrisilazane

6.43 - Bi- and polycyclic silazanes are named by the principles set forth in Rule D-6.24. The prefixes *1Si*- and *1N*- are used when necessary to indicate the atom at the bridgehead to be numbered 1.

Examples:

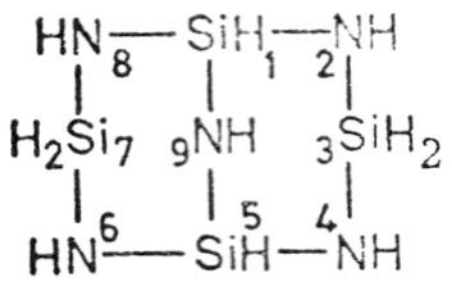

$$HN_8 \text{—} SiH_1 \text{—} NH_2$$

$$H_2Si_7 \quad {}_9NH \quad {}_3SiH_2$$

$$HN_6 \text{—} SiH_5 \text{—} NH_4$$

Bicyclo[3.3.1]tetrasilazane

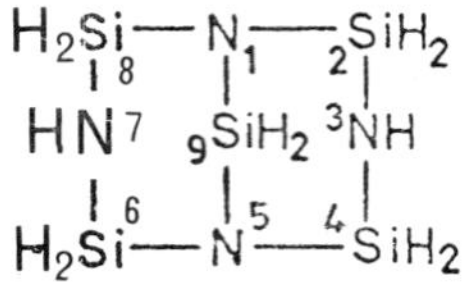

$$H_2Si_8 \text{—} N_1 \text{—} SiH_2{}_2$$

$$HN_7 \quad {}_9SiH_2 \quad {}^3NH$$

$$H_2Si_6 \text{—} N_5 \text{—} SiH_2{}_4$$

Bicyclo[3.3.1]pentasilazane

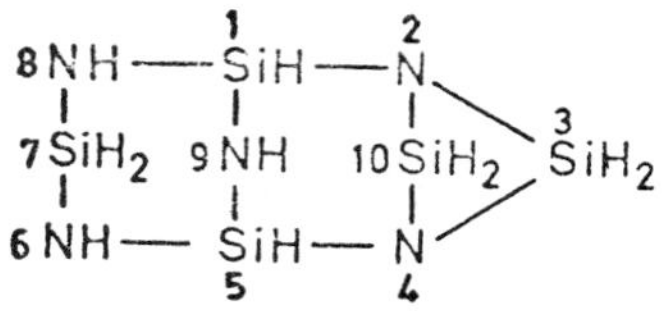

$${}_8NH \text{—} SiH_1 \text{—} N_2$$

$${}_7SiH_2 \quad {}_9NH \quad {}_{10}SiH_2 \quad {}_3SiH_2$$

$${}_6NH \text{—} SiH_5 \text{—} N_4$$

1Si-Tricyclo[3.3.1.1^{2,4}]pentasilazane

$$H_2Si_8 \text{—} N_1 \text{—} SiH_2$$

$$HN_7 \quad {}_9SiH_2 \quad {}_{10}NH \quad NH_3$$

$$H_2Si_6 \text{—} N_5 \text{—} SiH_4$$

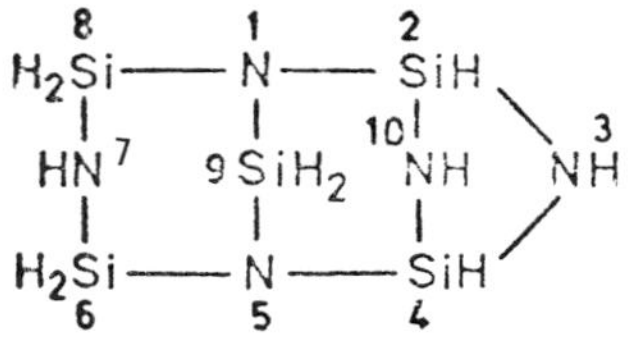

1N- Tricyclo[3.3.1.1^{2,4}]pentasilazane

D-6.5 - Radicals and compounds having more than one chain

6.51 - Radicals derived from acyclic and cyclic siloxanes, silathianes and silazanes are named by adding "-yl", "-diyl", "-triyl", etc., to the name of the siloxane, silathiane or silazane with elision of final "e" before the suffix "-yl". Numbering of acyclic and cyclic silicon radicals proceeds so that free valences have locants as low as possible. *

Examples:

1-Trisilazanyl	$\overset{5}{H_3}\text{Si}-\overset{4}{\text{NH}}-\overset{3}{\text{SiH}_2}-\overset{2}{\text{NH}}-\overset{1}{\text{SiH}_2}-$
2-Trisilazanyl	$H_3\text{Si}-\text{NH}-\text{SiH}_2-\underset{\mid}{\text{N}}-\text{SiH}_3$
1,3-Disiloxanediyl	$-\text{SiH}_2-\text{O}-\text{SiH}_2-$

2,2-Cyclotrisilathianediyl

$$H_2\overset{}{\text{Si}_6}\overset{S}{\diagup}\overset{1}{}\overset{}{\diagdown}{}_2\text{Si}{<}$$
$$\overset{}{S}^5\,{}_4\,{}^3\overset{}{S}$$
$$\diagdown\text{SiH}_2\diagup$$

1,2,4,6-Cyclotrisilazanetetrayl

$$-\text{HSi}_6\overset{\mid}{\diagup}\overset{N}{}\overset{1}{}\diagdown{}_2\text{SiH}-$$
$$\text{HN}^5\,{}_4\,{}^3\text{NH}$$
$$\diagdown\overset{\mid}{\text{SiH}}\diagup$$

Bicyclo[3.3.1]tetrasilazan-3-yl
(see Rule B-14.1)

$$\overset{8}{\text{HN}}\;-\;\overset{1}{\text{SiH}}\;-\;\overset{2}{\text{NH}}$$
$$H_2\text{Si}_7 \quad {}_9\text{NH} \quad {}_3\text{SiH}-$$
$$\text{HN}\;-\;\text{SiH}\;-\;\text{NH}$$
$$\overset{}{6}\qquad\overset{}{5}\qquad\overset{}{4}$$

---

* See also footnote † on page 374   and the note in Rule D-6.12.

6.52 - Acyclic silicon compounds, containing more than one chain as defined in Rules D-6.1 — D-6.4, are named from the parent having the largest number of silicon atoms.

Examples:

3-Siloxytrisilathiane

$$\overset{5\quad\ 4\ 3\quad\ 2\ 1}{H_3Si-S-\underset{\underset{O-SiH_3}{|}}{SiH}-S-SiH_3}$$

1,1,1,3,3,5,5-Heptamethyl-5-(trimethylsilyl)trisilazane

$$\overset{\quad\ CH_3\quad\ CH_3\quad\ CH_3\ CH_3}{\underset{\quad\ CH_3\quad\ CH_3\quad\ CH_3\ CH_3}{H_3C-Si-NH-Si-NH-Si-Si-CH_3}}$$

6.53 - When there is a choice between two parent polysilicon chains having the same number of silicon atoms, the order of precedence is as follows, listed in order of decreasing priority: siloxanes, silathianes, silazanes, and silanes.

Examples:

1-(Silylthio)disiloxane

$$H_3Si-O-SiH_2-S-SiH_3$$

2-Disilanyl-1,3-disilyldisilazane

$$H_3Si-SiH_2-N-SiH_2-SiH_3$$
$$|$$
$$SiH_2$$
$$|$$
$$SiH_3$$

D-6.6 - Silasesquioxanes, silasesquithianes, etc.

6.61 - (Partly alternative to Rules D-6.24, D-6.33, and D-6.43). Compounds in which each silicon atom is linked to three oxygen atoms and in which every oxygen atom is linked to two silicon atoms are named generically silasesquioxanes. In the name of an individual compound this generic name is preceded by a prefix indicating the number of silicon atoms present. The silasesquioxanes have the general formula $Si_{2n}H_{2n}O_{3n}$. The names tetrasilasesquioxane (n = 2), hexasilasesquioxane (n=3), octasilasesquioxane (n = 4), tetrasilasesquithiane, tetrasilasesquiazane, etc., are class names indicating $Si_{2n}H_{2n}X_{3n}$ where X = O, S or NH.

Individual compounds are named and numbered according to Rules D-6.24, D-6.33, and D-6.43.

D-6.7 - General nomenclature systems

Organosilicon compounds that cannot be named as silanes, siloxanes, etc. (Subsection D-6.1) may be named as indicated in the following Rules: (a) by replacement nomenclature [Subsection C-0.6 (cf. Rule B-4,] (b) by extended Hantzsch-Widman nomenclature (cf. Rule B-1), (c) by coordination nomenclature (cf. Subsection D-2), or (d) as organometallic compounds (cf. Subsection D-3).

Replacement nomenclature

6.71 - Acyclic organosilicon compounds may be named by the Rules of Subsection C-0.6. Cyclic organosilicon compounds may be named by Rules B-4, B-10, B-11, B-14 and B-15.

Examples:

3-Thia-6,7-diaza-
  2,4,8-trisilanonane

$$\overset{9}{CH_3}-\overset{8}{SiH_2}-\overset{7}{NH}-\overset{6}{NH}-\overset{5}{CH_2}-\overset{4}{SiH_2}-\overset{3}{S}-\overset{2}{SiH_2}-\overset{1}{CH_3}$$

2,2-Dimethyl-1-oxa-4-thia-2-
  silacyclohexan-6-one

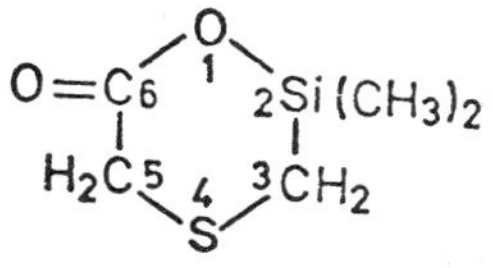

1-Methyl-2,8,9-trioxa-5-aza-
  1-silabicyclo[3.3.3]undecane
(cf. Rules A-31, A-32, A-41
and B-14).

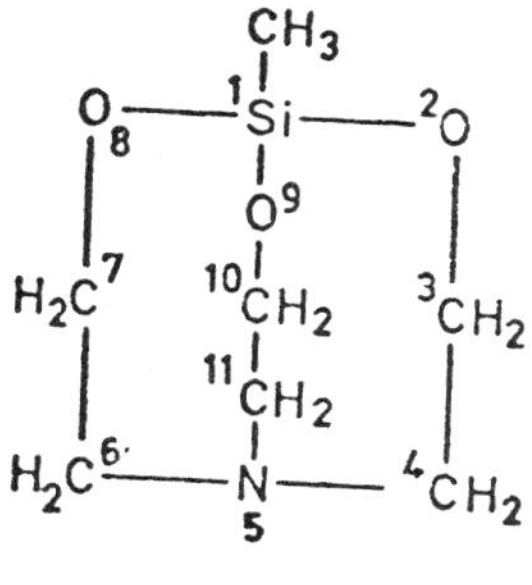

9-Sila-9,9'-spirobifluorene
(cf. Rules B-10 and B-11)

Extended Hantzsch-Widman nomenclature

6.72 - Monocyclic heterocyclic systems containing silicon atoms may also be named according to the extended Hantzsch-Widman system described in Rule B-1.

Example:

2,2-Dimethyl-1,3-bis(trimethylsilyl)-1,3,2-diazasilolidine

Coordination nomenclature

6.73 - Anions, in which silicon is the central atom, which are named by the rules for coordination compounds, are given the termination "silicate" (see Rule D-2.24).

Examples:

Lithium bis[1,2-ethanediolato(2-)]-
methoxosilicate(1-)

Tetrafluorodimethylsilicate(2-) $[(CH_3)_2SiF_4]^{2-}$

6.74 - When organosilicon radicals function as ligands by attachment through silicon they are indicated by their customary radical names.

Examples:

Triphenyl(triphenylsilyl)-
     borate(1-)

$$\left[B(C_6H_5)_3\{(C_6H_5)_3Si\}\right]^-$$

Dicarbonyl(η-cyclopentadienyl)-
    (trimethylsilyl)iron

$$[Fe(C_5H_5)\{(CH_3)_3Si\}(CO)_2]$$

6.75 - The name of a cation having silicon as central atom is formed by adding the names of its ligands as prefixes to "silicon" (cf. Rule D-2.35).

Example:

(2,2'-Bipyridine -N,N'')triphenyl-
    silicon(1+) iodide

Compounds having a silicon-metal bond

6.76 - (Alternative to Rule D-6.78). Compounds consisting of an atom of an element that occurs later than silicon in Table I (p. 459) joined to silicon radicals or to silicon and organic radicals may be named by citing those radicals in alphabetical order followed by the name of the element (cf. Rule D-3.11).

Examples:

(Pentaphenyldisilanyl)potassium	$(C_6H_5)_3Si-Si(C_6H_5)_2K$
Triphenyl(trimethylsilyl)lead	$(CH_3)_3Si-Pb(C_6H_5)_3$
(Trimethylsilyl)sodium	$(CH_3)_3SiNa$

6.77 - Compounds consisting of an atom of an element that occurs earlier than silicon in Table I (p. 459) joined to silicon radicals or silicon and organic radicals may be named as substituted silanes, the radicals being named according to Subsection D-3.4).

Examples:

(Diphenylarsino)trimethylsilane	$(CH_3)_3Si-As(C_6H_5)_2$
Trimethyl(methyltellurio)silane	$(CH_3)_3Si-TeCH_3$

6.78 - (Partly alternative to Rule D-6.76). The presence of a metal atom bonded to silicon may also be indicated by its radical name ending in "-io" (Subsection D-3.4 and Table II, p. 461).

Examples:

9-Lithio-9-methyl-9-silafluorene

or

5-Lithio-5-methyl-5H-dibenzosilole

Trimethylsodiosilane $(CH_3)_3SiNa$

Compound radicals

6.79 - The following are examples of compound radicals.

Examples:

1-Disilazanylthio	$H_3Si-NH-SiH_2-S-$
1-Pentasilanyloxy	$H_3Si-SiH_2-SiH_2-SiH_2-SiH_2-O-$
Disilanylamino	$H_3Si-SiH_2-NH-$
3-Aminodisiloxanyl	$H_2N-SiH_2-O-SiH_2-$
Silylthio	H_3Si-S-

Exception:

Siloxy	H_3Si-O-

Note: Contractions such as disilanoxy, disilazanoxy, etc., were permitted in the 1949 Rules but are no longer recommended.

D-6.8 - Substituting groups

6.81 - Silicon compounds containing characteristic groups which can only be designated by prefixes according to Table I of Rule C-10.1 are named by prefixing the names of the characteristic groups, together with those of any other group being cited as prefixes, to the name of the parent silicon compound.

Examples:

Trichloropropoxysilane	$Cl_3Si-O-CH_2-CH_2-CH_3$
3-Methoxy-1-methyltrisilazane	$H_3Si-NH-SiH(OCH_3)-NH-SiH_2-CH_3$

6.82 - Silicon compounds containing characteristic groups which may be expressed either by a prefix or by a suffix according to Table III of Rule C-10.4 or Rule C-833.1, Table XIV, are named
(a) by suffixing the names(s) of the principal group(s) to that of the parent silicon compound, or

(b) by adding, as prefix(es), the name(s) of the corresponding substituent(s) to that of the parent silicon compound.

Examples:

(a)   1, 3, 5-Trisiloxanetriol

$$\overset{5}{}\quad\overset{4}{}\overset{3}{}\quad\overset{2}{}\overset{1}{}$$
$$HO\text{-}SiH_2\text{-}O\text{-}SiH\text{-}O\text{-}SiH_2\text{-}OH$$
$$\underset{OH}{\big|}$$

(b)   1, 3, 5-Trihydroxytrisiloxane

(a)   Dimethylsilanedicarbonitrile

$$(CH_3)_2Si(CN)_2$$

(b)   Dicyanodimethylsilane

(a)   Disilathianeselenol

$$\overset{3}{H_3}\overset{2}{Si}\text{-}S\text{-}\overset{1}{SiH_2}\text{-}SeH$$

(b)   (Hydroseleno)disilathiane

(a)   Triphenylsilanecarboxylic acid

$$(C_6H_5)_3Si\text{-}COOH$$

(b)   Carboxytriphenylsilane

6.83 - Compounds containing silicon radicals linked to organic molecules containing a principal group (see Rule C-10.3) may be named
(a) according to the Rules of Subsection C-0.1, or
(b) by the use of replacement nomenclature (see Subsection C-0.6).

Examples:

(a)   (3-Ethyldisiloxanyl)methanethiol

$$CH_3\text{-}CH_2\text{-}SiH_2\text{-}O\text{-}SiH_2\text{-}CH_2\text{-}SH$$

(b)   3-Oxa-2, 4-disilahexane-1-thiol

$$[(CH_3)_3Si\text{-}O\text{-}Si(CH_3)_2\text{-}O\text{-}Si(CH_3)_2\text{-}CH_2]_3N$$

(a)   Tris[(heptamethyl-1-trisiloxanyl)methyl]amine   or
       1, 1', 1''-Tris(heptamethyl-1-trisiloxanyl)trimethylamine

(b)   2, 2, 4, 4, 6, 6, 10, 10, 12, 12, 14, 14-Dodecamethyl-8-[(heptamethyl-trisiloxanyl)methyl]-3, 5, 11, 13-tetraoxa-8-aza-2, 4, 6, 10, 12, 14 hexasilapentadecane

$$\text{[benzene with OH, OH]}-SiH_2-CH_2-[CH_2]_8-CH_2-CHO$$

(a)   11-[(2,3-Dihydroxyphenyl)silyl]undecanal

6.84 - The presence of one or more amine or substituted amine groups attached to one or more silicon atoms may be indicated
(a) by prefixing the name of the parent radical to the functional class name "-amine",
(b) by the prefix "amino-", or
(c) by adding "amine", "diamine", "triamine", etc., to the name of the parent compound (cf. Rules C-811.3, C-812.1 and C-813.1).

Examples:

(a) N,N- Diethyl-(3-methyl-1-tri-
           silanyl)amine

$$CH_3-SiH_2-SiH_2-SiH_2-N(C_2H_5)_2$$

(b)   1-(Diethylamino)-3-methyltri-
           silane

(a)   Silylamine

(b)   Aminosilane

$$H_3Si-NH_2$$

$$[(CH_3)_2CH-O]_2\underset{\underset{NH_2}{|}}{Si}-NH-\underset{\underset{NH_2}{|}}{Si}[O-CH(CH_3)_2]_2$$

(c)   1,1,3,3-Tetraisopropoxy-1,3-disilazanediamine

$$CH_3-CH_2-\underset{\underset{CH_3-CH_2}{|}}{\overset{\overset{NH_2}{|}}{Si}}-CH_2-\underset{\underset{CH_2-CH_3}{|}}{\overset{\overset{CH_2-CH_3}{|}}{Si}}-CH_2-CH_3$$

(c)   3,5,5-Triethyl-3,5-disilaheptan-3-amine

6.85 - Names of anions of silanols, siloxanols, silathianols, and silazanols and their sulfur, selenium and tellurium analogues (silanethiols, silaneselenols, silanetellurols)may be formed
(a) by changing the ending "-ol" to "-olate" (see Rules C-206.1 and C-511.3),
(b) by changing the prefixes "hydroxy", "mercapto" and "hydroseleno" to "oxido", "sulfido" and "selenido", respectively (see Rules C-86.2, C-511.4 and C-701.1), or
(c) by the nomenclature for coordination compounds.

Examples:

(a)   Sodium trimethylsilanethiolate(1-)

(b)   Sodium trimethylsulfidosilane $(CH_3)_3SiSNa$

(c)   Sodium trimethylthiosilicate(1-)

(a)   Sodium pentaphenyldisilanolate(1-)

(b)   Sodium oxidopentaphenyldisilane $(C_6H_5)_3Si\text{-}Si(C_6H_5)_2ONa$

(c)   Sodium oxopentaphenyldisilicate(1-)

6.86 - Silazane cations are named by applying Rule C-82.2.

Example:

2,4-Trisilazanedium dichloride $H_3Si\text{-}NH_2^+\text{-}SiH_2NH_2^+\text{-}SiH_3 \ 2Cl^-$

6.87 - Silazane anions may be named by applying Inorganic Rule 7.24.

Example :

Disodium 2,4-trisilazanediide $2\,Na^+ \ H_3Si\text{-}N^-\text{-}SiH_2\text{-}N^-\text{-}SiH_3$

D-6.9 - Esters

6.91 - Esters derived from silanols, siloxanols, silathianols and silazanols and their sulfur, selenium and tellurium analogues may be named
(a) as esters (see Rules C-463.1 and C-543.1),
(b) as acyloxy, acylthio, acylseleno or acyltelluro derivatives of silanes, siloxanes, silathianes or silazanes,
(c) by the nomenclature for coordination compounds, or
(d) as siloxy, silylthio, silylseleno or silyltelluro derivatives of hydrides

Examples:

(a)   Trimethylsilyl benzoate

(b)   (Benzoyloxy)trimethylsilane $(CH_3)_3Si\text{-}OOC\text{-}C_6H_5$

(c)   (Benzoato)trimethylsilicon

(a)   Tris(trimethylsilyl)
             phosphite

(d)   Tris(trimethylsiloxy)- $[(CH_3)_3SiO]_3P$
             phosphine

6.92 - Esters of orthosilicic acid, R_4SiO_4, may be named by citing the names of the radical(s) before "orthosilicate" (cf. Rule C-463).

Examples:

Tetraethyl orthosilicate

$(C_2H_5O)_4Si$

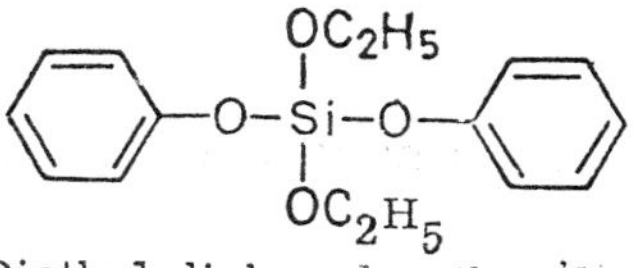

Diethyl diphenyl orthosilicate

6.93 - Cyclic esters of orthosilicic acid may be named
(a) by citing the names of the multivalent organic radical(s) in
    alphabetical order before "orthosilicate" or
(b) as heterocyclic compounds.

Example:

(a)   Diethylene orthosilicate

(b)   1,4,6,9-Tetraoxa-5-
             silaspiro[4.4]nonane

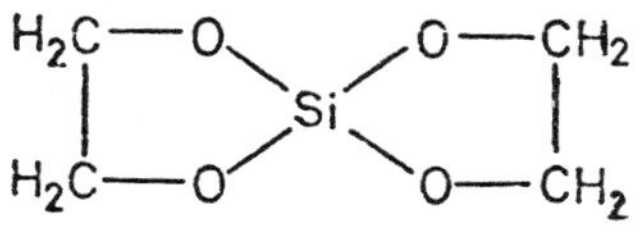

6.94 - Orthosilicates in which one or more oxygen atoms are replaced by one or more sulfur atoms are called thioorthosilicates, dithioorthosilicates, etc., according to the number of sulfur atoms present. Thioorthosilicates are named according to the principles of Rules D- 6.92 and D-6.93 using o and/or S as necessary, to denote attachment to oxygen and sulfur, respectively (see Rule C-651).

Examples:

Tetramethyl tetrathioorthosilicate
    (cf. Rule D- 6.92)
          or
Tetrakis(methylthio)silane

$(CH_3S)_4Si$

Di-o-phenylene tetrathio-
    orthosilicate
       or
2, 2'-Spirobi[1, 3, 2-benzodithia-
    silole]
    (cf. Rule D- 6.93)

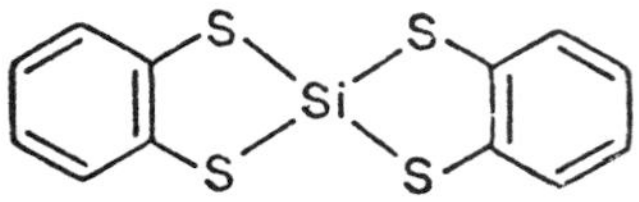

o,o',o"- Tripropyl hydrogen thio-
    orthosilicate

$(CH_3-CH_2-CH_2O)_3Si-SH$

6.95 - Acyl derivatives of orthosilicic acids may be named (a) by applying Rule D-6.92, or (b) by applying the rules for the nomenclature of coordination compounds, or (c) as acyloxy derivatives of silane.

Examples:

(a)   Tetraacetyl orthosilicate

(b)   Tetra(acetato) silicon $(CH_3-COO)_4Si$

(c)   Tetraacetoxysilane

D-7. ORGANOBORON COMPOUNDS

D-7.0 - Introduction

The nomenclature of the organoboron compounds is based on the names of the boron hydrides, their skeletal replacement derivatives, and radicals derived from these compounds. Inorganic boron compounds are named in IUPAC Nomenclature of Inorganic Chemistry, 2nd edition (Definitive Rules 1970), Section 11, summarised here. For more details, "Nomenclature of Inorganic Boron Compounds" (*Pure and Applied Chemistry* , 1972, Vol. 30, pages 683 - 710) should be consulted.

D-7.1 - Boron hydrides

7.11 - The molecular hydrides of boron are called boranes. They are named by citing the number of boron atoms in the molecule as a numerical prefix to the term borane (except for the prefix mono- which is omitted, BH_3 being named borane). These numerical prefixes are the same as those used in hydrocarbon nomenclature (Rule A-1.1,*) The number of hydrogen atoms in the molecule is designated by an arabic numeral enclosed in parentheses immediately following the name derived as above. The numerical designation of the number of hydrogen atoms may be omitted for BH_3 and other cases where no ambiguity arises.

Examples:

Diborane(6)	B_2H_6
Pentaborane(9)	B_5H_9
Pentaborane(11)	B_5H_{11}
Decaborane(14)	$B_{10}H_{14}$

The numbering of these polyboranes is given in the following structural diagrams.

---

* Note that the prefix for twenty is spelled "icosa' not "eicosa" (see foot- note on page 5)

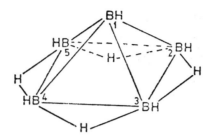

B_2H_6 Diborane(6)

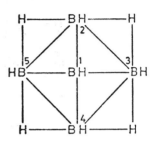

B_5H_9 Pentaborane(9)   **planar projection**

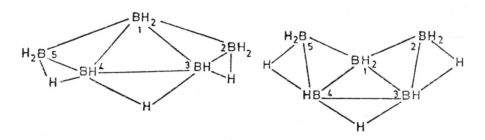

B_5H_{11} Pentaborane(11)   **planar projection**

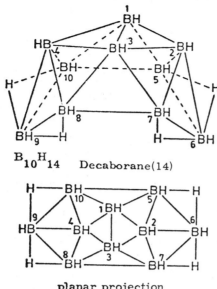

$B_{10}H_{14}$ Decaborane(14)

planar projection

7.12 - The polyboranes are of two general classes: (1) closed structures (i.e., structures with boron skeletons that are polyhedra having only triangular faces) and (2) non-closed structures. The members of the first class are designated by the italic prefix *"closo "*. Some members of the second class have structures very similar to a closed structure. These may be differentiated from other open structures by the italic prefix *"nido "*.

Examples:

nido- Pentaborane(9) B_5H_9

closo- Icosaborane(16) $B_{20}H_{16}$

7.13 - Polyboranes that may be considered to be formed by fusion or joining of simpler boranes may also be named as derivatives of the latter. *

---

* The prefixes *"iso"* and *"neo"* have been used to distinguish isomers of unknown structures. However, once structures are known, a structural name is preferred.

Examples:

1,1'-Bipentaborane(9)	$B_{10}H_{16}$
Decaborano(14)[6',7':5,6]decaborane(14) [*]	$B_{18}H_{22}$
Decaborano(14)[6',7':6,7]decaborane(14) [*]	

D-7.2 - Polyboranes with skeletal replacement

7.21 - Polyboranes with skeletal replacement. Compounds in which one or more boron atoms in a polyborane skeleton have each been replaced by another atom are termed heteroboranes. Classes of such compounds are named according to the heteroatoms, e.g. carbaboranes, thiaboranes. The compounds are named by a modification of organic replacement nomenclature. The replacement of a boron atom by another atom is denoted by use of the corresponding prefix listed in Table I, p. 459, regardless of the valence of the replacing heteroatom. The number of hydrogen atoms in the heteroborane is indicated by an Arabic numeral enclosed in parentheses and immediately following the name.

Example:

Dicarbadodecaborane(12) [†]	$B_{10}C_2H_{12}$

7.22 - The heteroboranes are numbered so as to give the lowest numbers to the heteroatoms, consistent with the skeletal numbering of the parent polyborane. The prefixes *closo* and *nido* may also be used, as appropriate, when structures are known.

[*] These names are fusion names formed in a similar manner as for hydrocarbons and heterocyclic compounds (see Rule A-21).

[†] Dicarbadodecaborane(12) isomers are very stable and form a large number of derivatives. They are "dicarba" derivatives of the unknown $B_{12}H_{12}$ and may be considered as derived from the stable $[B_{12}H_{12}]^{2-}$ ion by the replacement of two skeletal boron atoms with two skeletal carbon atoms.

Examples:

1,2-Dicarba-*closo*-dodecaborane(12)

1,7-Dicarba-*closo*-dodecaborane(12) $B_{10}C_2H_{12}$

1,12-Dicarba-*closo*-dodecaborane(12)

7-Thia-*nido*-undecaborane(12) $B_{10}SH_{12}$

D-7.3 - Boron radicals

7.31 - Radicals derived from the boron hydrides are named by adding the endings yl, diyl, triyl, etc., to the name of the boron hydride with elision of "e" before "y".

Examples:

Boryl (not boranyl) * H_2B-

Boranediyl $HB{<}$

Boranetriyl $B{\leqq}$

Diboran(6)yl

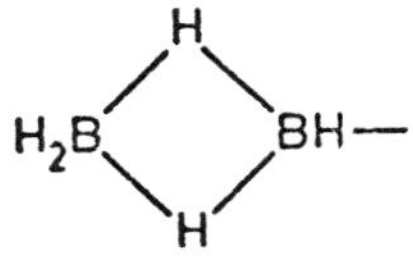

1,2-Diborane(4)diyl $-HB-BH-$

2,2-Tetraborane(10)diyl

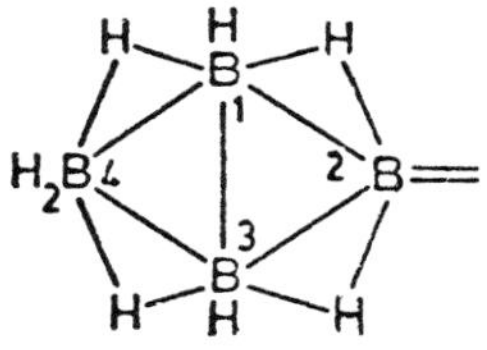

6,9-Decaborane(14)diyl $6,9-(B_{10}H_{12}){<}$

---

* The abbreviated form boryl is used instead of boranyl to distinguish between two boryl groups (2 BH_2) and one diboranyl group (see also Rule D-4.14).

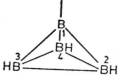

closo-Tetraboran(4)yl

1,2-Dicarba-*closo*-dodecarboran(12)-1-yl $B_{10}C_2H_{11}{}^-$

(1-2)Diboran(6)yl

7.4 - Derivatives of the boron hydrides

7.41 - Organic derivatives of boron hydrides may be named from a boron-containing parent or from an organic parent. In a compound containing more than one non-metal, boron takes precedence for choice of the part to be named as parent.

7.42 - Derivatives of borane BH_3 are named as substitution products, the substituents being denoted by prefixes, cited in alphabetical order.

Examples:

Trimethylborane $(CH_3)_3B$

Dichloro(4-chlorophenyl)borane

Bromodimethylborane $(CH_3)_2BBr$

Trimethoxyborane * $(CH_3O)_3B$

Tris(methylthio)borane * $(CH_3S)_3B$

---

* These compounds have been named as esters, trimethyl borate and trimethyl trithioborate respectively.

Bis(dimethylamino)methylborane $[(CH_3)_2N]_2B(CH_3)$

Difluoro(trimethylsilyl)borane $(CH_3)_3SiBF_2$

(Dimethylstibino)dimethoxyborane $(CH_3O)_2BSb(CH_3)_2$

Methoxydimethylborane * $(CH_3O)B(CH_3)_2$

Dihydroxymethylborane * $CH_3B(OH)_2$

1,3,5-Hexanetriyltris-
(dihydroxyborane) *

$$\underset{\underset{CH_3-CH-CH_2-CH-CH_2-CH_2-B(OH)_2}{\mid \qquad\qquad\quad \mid}}{B(OH)_2 \qquad B(OH)_2}$$

Oxybis(dimethylborane) † $(CH_3)_2BOB(CH_3)_2$

Ethylenebis(dichloroborane) $Cl_2BCH_2CH_2BCl_2$

Triacetoxyborane ‡ $(CH_3CO_2)_3B$

Tris(2,2-dimethylhydrazino)borane $[(CH_3)_2NNH]_3B$

$[(CH_3-CH_2)_2N]_2B-O-CH_2CH_2-O-B[N(CH_2CH_3)_2]_2$

(Ethylenedioxy)bis[bis(diethylamino)borane] #

---

* These compounds have been named as borinic acids, $R_2B(OH)$, and boronic acids, $RB(OH)_2$; for example, methyl dimethylborinate and methylboronic acid.

† This compound may also be named Tetramethyldiboroxane according to Rule D-4.3.

‡ This compound may also be named Tri(acetato)boron by coordination nomenclature.

This compound may also be named by replacement nomenclature as 4,9-bis(diethylamino)-3,10-diethyl-5,8-dioxa-3,10-diaza-4,9-dibora-dodecane (see Subsection C-0.6).

7.43 – Organoboron compounds that can be regarded as derived from a polyborane by substitution of a hydrogen atom may be named as derivatives of a polyborane or heteroborane as described in Rules D-7.42 and D-7.21.

When a non-bridging hydrogen is replaced, locants are used in the customary way. When a bridging hydrogen is replaced a letter μ is used as necessary, preceded by the locants of attachment of the bridge (see Rule D-2.71).

Examples:

Tetramethoxydiborane(4) $(CH_3O)_2B-B(OCH_3)_2$

1-Methyldiborane(6)

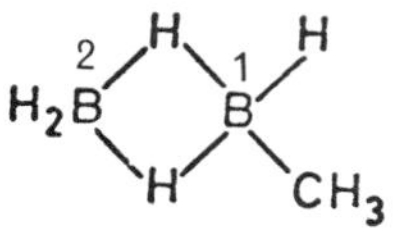

2-Ethyl-3-methylpentaborane(9)

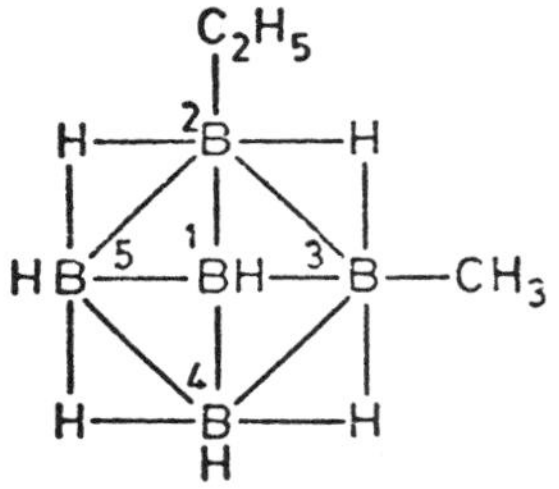

1,3,4,5,6,7,8,9,10,11,12-
Undecachloro-1,2-dicarba-
closo-dodecaborane(12)

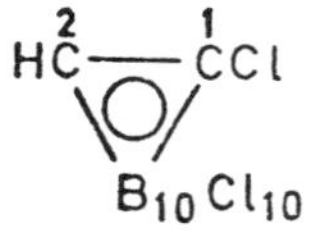

2-Methyl-2,4-dicarba-*closo*-
heptaborane(7)

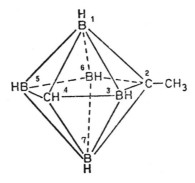

cis-1,2-Dimethyldiborane(6)

1,1-Tetramethylenediborane(6)

μ-(Dimethylamino)diborane(6)

1,2-μ-Anilinotetraborane(10)

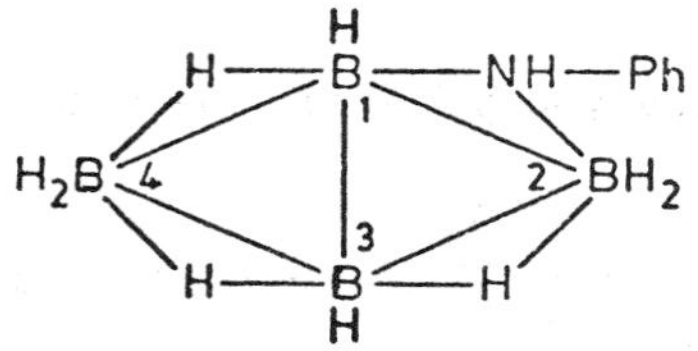

Note 1　　In the absence of structural information, non-bridging substituents may be denoted by the Greek letter τ(tau).

Note 2　　The hydrogen atoms of a BH_2 group, when stereochemically different, may be differentiated by the prefixes *exo-* and *endo-*.

Example:

exo- 2-Methyltetraborane(10)

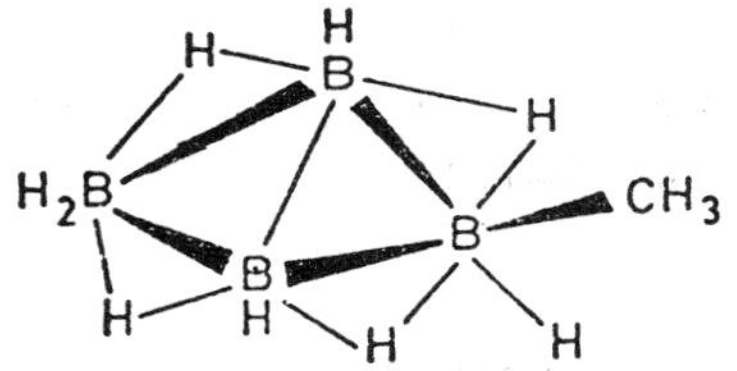

7.44 - (Partly alternative to Rules D-7.42 and D-7.43). Derivatives of the boron hydrides containing groups that are conveniently named by means of suffixes (Rule C-10.3) or by an accepted trivial name may be named by the use of boron radical names (see Rule D-7.31).

Examples:

4-(Dimethylboryl)phenol

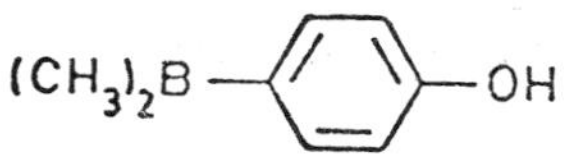

　　　or

(4-Hydroxyphenyl)-
　　dimethylborane (D-7.42)

4-(Dihydroxyboryl)-2-nitrobenzoic
    acid
         or
(4-Carboxy-3-nitrophenyl)dihydroxy-
    borane (D-7.42)

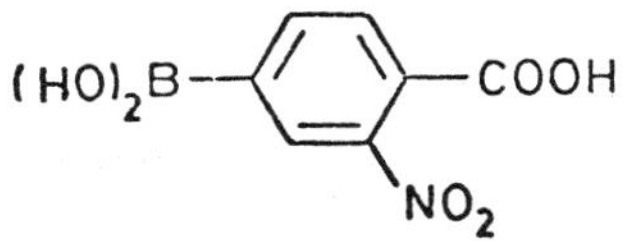

$$(HO)_2B-\underset{NO_2}{\bigodot}-COOH$$

4,4',4''-Boranetriyltrianiline
         or

Tris(4-aminophenyl)borane (D-7.42)

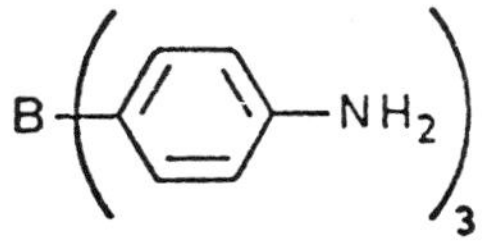

$$B\left(\!\!\left(\bigodot\right)-NH_2\right)_3$$

2,2'-(Methoxyboranediyl)di-1-
    propanol
         or

Bis(2-hydroxy-1-methylethyl)methoxy-
    borane (D-7.42)

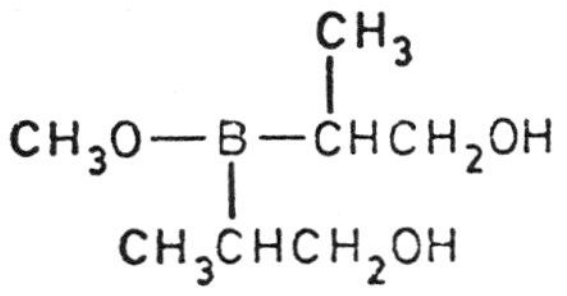

$$CH_3O-\underset{CH_3CHCH_2OH}{\overset{CH_3}{\underset{|}{\overset{|}{B}}}}-CHCH_2OH$$

N,N'-Bis(diphenylboryl)-
    1,4-butanediamine
         or

(Tetramethylenediimino)bis-
    (diphenylborane) (D-7.42)

$$(C_6H_5)_2B-NH[CH_2]_4NH-B(C_6H_5)_2$$

7.45 - Derivatives of the carbaboranes in which an organic group **is** attached to carbon are treated according to the rules of Section C.

Examples:

2-Methyl-1,2-dicarba-*closo*-
    dodecaborane(12)-1-carboxylic
    acid

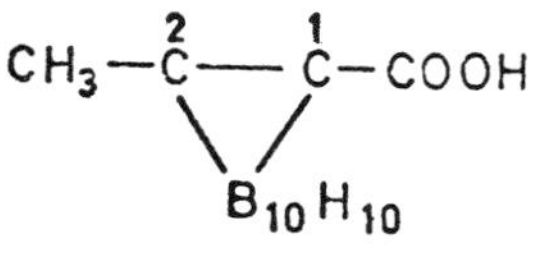

$$CH_3-\overset{2}{C}-\overset{1}{C}-COOH$$
$$B_{10}H_{10}$$

1-(Chloromethyl)-1,2-dicarba-
    closo-dodecaborane(12)

$$H\overset{2}{C}-\overset{1}{C}-CH_2Cl$$
$$B_{10}H_{10}$$

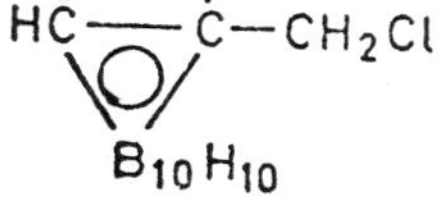

7.5 - Heterocyclic boron compounds

Organoboron heterocyclic compounds with uncharged boron atoms
named according to the general patterns established for other organic heterocycles. Choice among the various systems is made according to the type of compound and the requirements of each system.
Numbering of the atom in the rings and the use of locants follows the
usual practices described in Section B. *

Note :   The prefix "borazaro" is not recommended. The
             mesomeric structure of the isoelectronic (isoconjugate)

             group $-\bar{B}H=\overset{+}{N}H-$ is equivalent to -BH-NH- and the name is
             based on the uncharged structure . Rule B-4.1 provides for
             systematic replacement nomenclature.

Example:

1,2-Dihydro-1-aza-2-boranaphthalene
(not 2,1-Borazaronaphthalene)

7.51 - Heterocyclic compounds containing uncharged boron atoms
may be named by replacement nomenclature (see also Subsection B-4).

Examples

1,2,3,4-Tetrahydro-2,3-
   dimethyl-1,4-diaza-2,3
   diboranaphthalene

1,4a,8-Triaza-8a-
   boranaphthalene

---

*   The practice of using letters to designate positions in heterocyclic
       boron compounds is not included here.
       Example :
       2,4,6-Trimethylborazine

       (not *B, B, B,*-trimethylborazine)

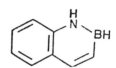

1,4-Diboracyclohexa-
2,5-diene

Perhydro-9b-boraphenalene
Dodecahydro-9b-boraphenalene

2,5,7,10,11,14-Hexaoxa-1,6-
diborabicyclo[4.4.4]tetradecane

3,9-Dihydroxy-2,4,8,10-tetraoxa-
3,9-diboraspiro[5.5]undecane

10,11-Dihydro-5 *H*-**5**-boradibenzo-
[*a,d*]cycloheptene (see Rule A-21.5)

7.52 - Three- to ten- membered monocyclic rings containing un-charged boron atoms may be named by the extended Hantzsch-Widman system (see Subsection B-1).

Examples:

1,3,2-Dioxaboretane

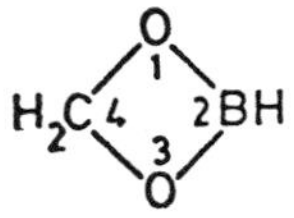

H_2C_4 $_2BH$ with ring: O_1, O_3

1H-1,3,2-Diazaborole

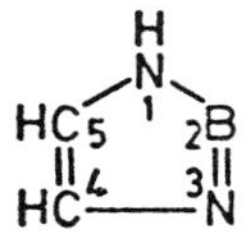

Ring: N (1, H), B (2), N (3), HC_4, HC_5

Δ⁴-1,3,2-Diazaboroline *

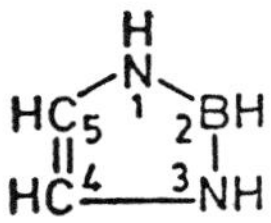

Ring: N (1, H), $_2BH$, $_3NH$, HC_4, HC_5

1,3,2-Diazaborolidine

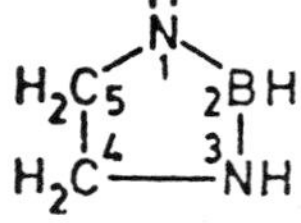

Ring: N (1, H), $_2BH$, $_3NH$, H_2C_4, H_2C_5

Borolane

Ring: B (1, H), $_2CH_2$, $_3CH_2$, H_2C_4, H_2C_5

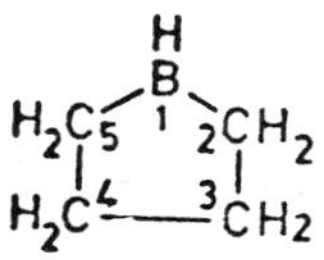

* For use of the Greek letter **Δ** see the footnote to Rule B-1.2.

1-Propylborinane *

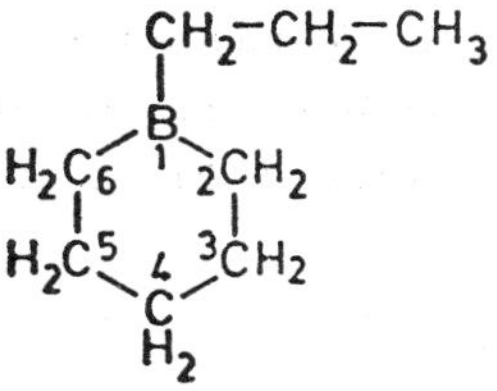

$$CH_2-CH_2-CH_3$$

2-Phenyl-4H-1,3,2-dioxaborin

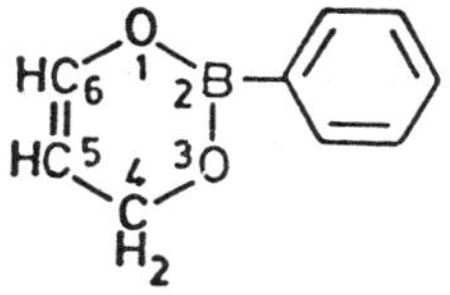

1 H-Borepin

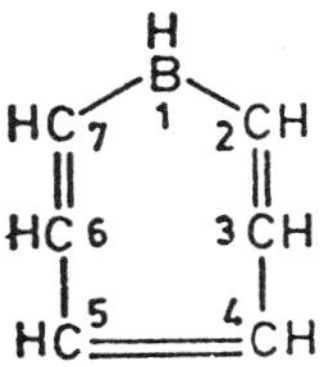

1,5-Diborocane

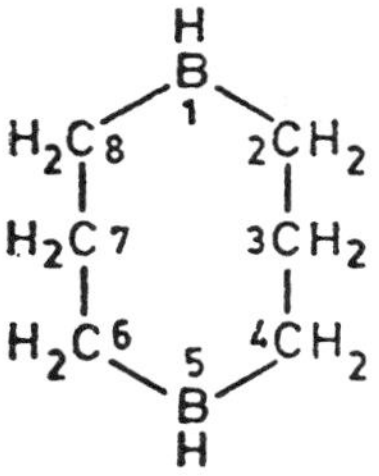

2,4-Bis(methylthio)-1,3,2,4-dithiadiboretane

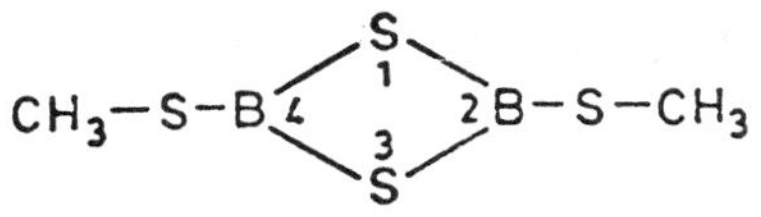

$$CH_3-S-B \quad \quad B-S-CH_3$$

2-(Methylthio)-1,3,2-oxathiaborepin

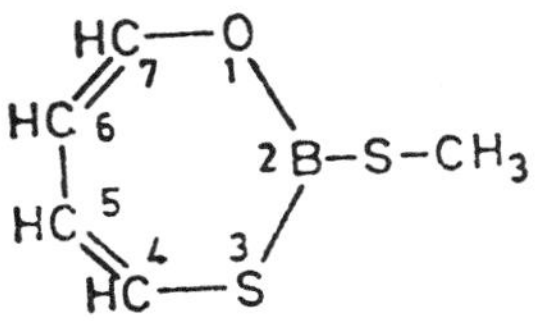

$$B-S-CH_3$$

---

*   The suffix "-inane" is used by analogy with phosphorinane and for the
same reason (see Subsection B-1).

7.53 - Fused heterocyclic systems containing uncharged boron atoms may be named by fusion nomenclature (see Subsection B-3).

Examples:

1,2,3,4-Tetrahydro-2,3-dimethyl-
1,4,2,3-benzodiazadiborine

[1,3,2]Diazaborino[1,2-a][1,3,2]-
diazaborine

5H-Dibenzobcrole

2-(Benzyloxy)-1,2-dihydro-4H-3,1,2-
benzoxazaborin-4-one

2-Phenyl-4H-1,3,2-dioxaborolo-
[4,5-d]imidazole

7.54 - Heterocyclic compounds containing uncharged boron atoms alternating with another heteroatom may be named according to the rules for rings of repeating units (see Rule D-4.51). *

Examples:

Cyclodiborathiane

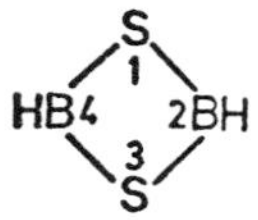

Cyclotetraborazane

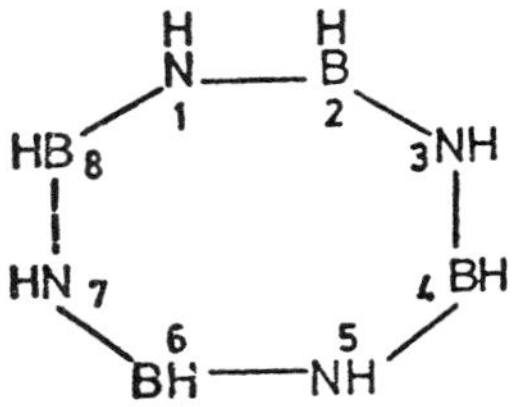

Bicyclo[4.4.0]pentaborazane

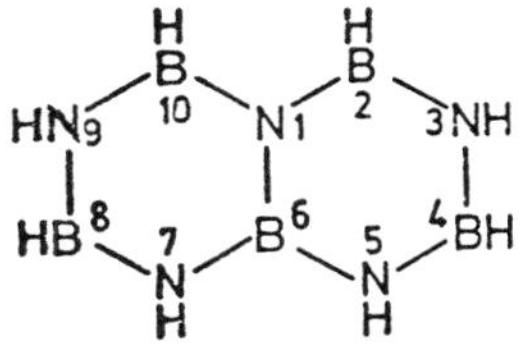

The following trivial names may be retained for organoboron heterocycles of this type.

Borazine (cyclotriborazane)

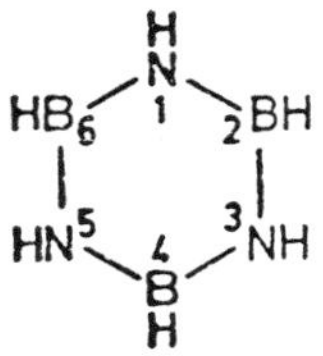

Boroxin (cyclotriboroxane)

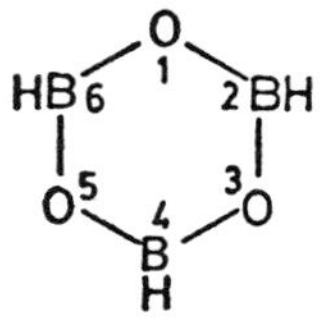

$$
\begin{array}{c}
\text{O} \\
\text{HB}\,6\quad 1\quad 2\,\text{BH} \\
\text{O}\,5\quad 4\quad 3\,\text{O} \\
\text{B} \\
\text{H}
\end{array}
$$

Borthiin (cyclotriborathiane)

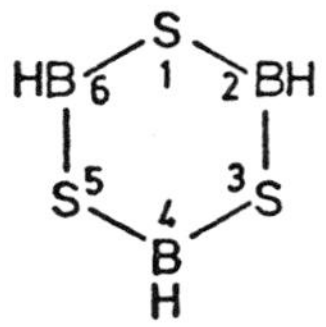

$$
\begin{array}{c}
\text{S} \\
\text{HB}\,6\quad 1\quad 2\,\text{BH} \\
\text{S}\,5\quad 4\quad 3\,\text{S} \\
\text{B} \\
\text{H}
\end{array}
$$

Note : The names borazole, boroxole, and borthiole for these compounds should not be used since they imply five-membered rings according to the principles of the Hantzsch-Widman system.

7.6 - Organoboron coordination compounds

7.61 - Organoboron coordination compounds may be named **according** to the general rules for coordination compounds (see Subsection D-2) except that the ligand name for H is hydro instead of **hydrido**. The position of the attachment of the ligands to boron may **be** indicated according to Subsection 7.3 of IUPAC Nomenclature of Inorganic Chemistry.

7.62 - Organoboron coordination compounds in which the boron **atom** has coordination number 4 and in which one or more boron **atoms** is charged are named as coordination compounds (see Subsection D-2), using the name borate, if the coordination entity is negatively **charged**, or the name boron, if the coordination entity is positively charged. The charge is given by the appropriate Ewens-Bassett number enclosed in parentheses following the name.

Examples:

Tetramethylborate(1-) $[(CH_3)_4B]^-$

Hexaphenyldiborate(2-) $[(C_6H_5)_3B-B(C_6H_5)_3]^{2-}$

Hydropentamethyldiborate(2-) $[(CH_3)_5B_2H]^{2-}$

Tridecahydro(pyridine)decaborate(1-) $[(C_5H_5N)B_{10}H_{13}]^-$

Trifluoro[(triphenylphosphonio)-
  methyl]borate $(C_6H_5)_3\overset{+}{P}-CH_2-\overset{-}{B}F_3$

Dihydrobis(pyridine)boron(1+) $[(C_5H_5N)_2BH_2]^+$

Bis(2,4-pentanedionato)boron(1+) chloride

Bis(pyridine)[3-(2-pyridyl)propyl]boron(2+)

[1,3,5-Cyclohexanetriolato(3-)]hydroxoborate(1-)

Methoxo[methyltris(2-oxidoethyl)-ammonium-O,O',O'']borate

7.63 - (Partly alternative to Rule D-7.62). Organoboron compounds in which a charged boron atom is in a ring system may also be named by means of replacement nomenclature. Anionic boron atoms are designated by the replacement prefix "borata" and cationic boron atoms by the replacement prefix "borylia".

Examples:

1,1-Dimethyl-1-boratacyclohexane

1-Methoxy-1,3-dimethyl-1-borata-indene

(2, 2'-Bipyridine)(oxydi-*o*-phenylene)boron(1+) perchlorate

[2, 2'-[1, 8-naphthylenebis(nitrilomethylidyne)] diphenolato-
O,N,N', O']boron (1+) perchlorate

7.64 - Neutral organoboron coordination entities may be named coordination nomenclature. Alternatively, such compounds may be named as addition compounds of boranes or organoboron heterocycles, as given in Rule D-7.7 .

Examples:

Trimethyl(pyridine)boron

or

Pyridine—trimethylborane

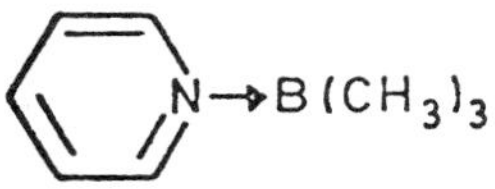

(Dimethyl sulfide)decahydro-
  (trimethylamine)dodecaboron

  or

Dimethyl sulfide — trimethylamine —
  dodecaborane(10)

$$B_{12}H_{10}[(CH_3)_3N][(CH_3)_2S]$$

Difluoro[2, 4-pentanedionato-O, O']boron

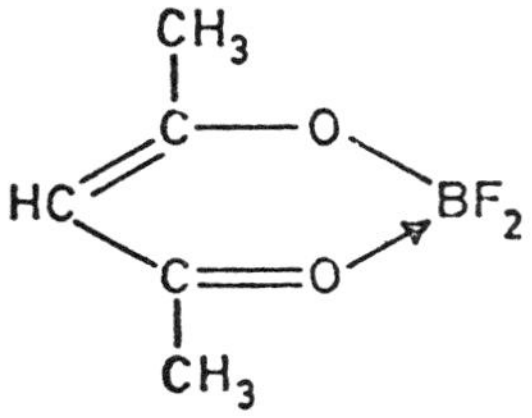

Dimethyl [8-quinolinato-O, N]boron

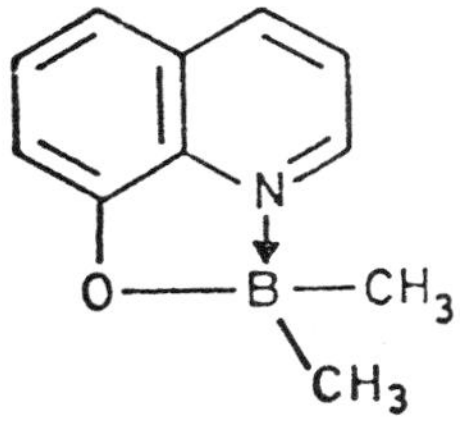

451

(Glycinato-*O, N*)methoxophenylboron

Chloro[1, 3-propanediolato(2-)-*O, O′*]
   (pyridine)boron

or

Pyridine—2-chloro-1, 3, 2-dioxa-
   borinane

(2-Aminobenzyl-*N*)-
   [pyrocatecholato(2-)-*O, O′*]boron

[2, 2', 2"-nitrilotriethanolato(3-)-O, O', C'', N] -
   boron

Bis[[2-(2-hydroxybenzylidene)amino]-
   benzoato(2-)-O, O', N]-μ-oxo-diboron

Rule D-7.65 of the Provisional Version has been incorporated into Rule
D-7.64 of this version.

7.66 - Anionic ligands formed from boranes or carboboranes have the
ending "borato".

Examples:

Diethyl(tetrahydroborato)-
   aluminium

$(CH_3CH_2)_2AlBH_4$

Tetrakis(trihydromethylborato)-
   uranium

$U(CH_3BH_3)_4$

Dimethyl(tetramethylborato)-
   thallium

$(CH_3)_2Tl[B(CH_3)_4]$

Bis[η-undecahydro-7,8-
   dicarbaundecaborato(2-)]-
   ferrate(1-)

$[(B_9C_2H_{11})_2Fe]^-$

D-7.7 - Organoboron addition compounds

7.71 - Many neutral organoboron compounds may be considered formally as the union of a neutral Lewis base (electron-pair donor or nucleophile) with a boron compound (electron-pair acceptor or electrophile). This union may be depicted by means of an arrow bond. This is a formal device designed to avoid specifying a particular electronic structure of the bond (see also Rule C-612). These compounds may also be considered as derivatives of a boron anion formed by replacement of a hydride ion of the boron anion by a neutral compound. Such compounds may be named as neutral coordination derivatives of boron (see Rule D-7.64).

7.72 - Addition compounds between neutral Lewis bases and boron compounds are named by connecting the names of each component by a rule. The number of molecules of each component, when greater than unity, is denoted by appropriate numerical prefixes. It is customary to cite the boron component last in the name, except for addition compounds containing water (see IUPAC Nomenclature of Inorganic Chemistry, Section 8).

Examples:

Carbon monoxide — borane	$OC \cdot BH_3$
Trimethylamine — trichloroborane [*]	$(CH_3)_3N \cdot BCl_3$
Diethyl sulfide — methylborane	$(C_2H_5)_2S \cdot BH_2(CH_3)$
Ethylenediamine — bis(borane)	$NH_2-CH_2-CH_2-NH_2 \cdot 2BH_3$
Bis(ethylamine) — pentaborane(9)	$2\ C_2H_5NH_2 \cdot B_5H_9$
Dimethyl sulfide — trimethylamine — dodecaborane(10)	$(CH_3)_2S \cdot (CH_3)_3N \cdot B_{12}H_{10}$
Trimethylamine — triborane(7)	$(CH_3)_3N \cdot B_3H_7$

7.73 - The points of attachment of the donor molecule to the boron component may be indicated by italicized atomic symbols joined by a rule, enclosed in parentheses, and cited between the name of each component of the addition compound. Each atomic symbol refers to the component nearest to it. The components and the atomic symbols are to be cited in the sequence donor-acceptor.

Examples:

O-Methylhydroxylamine *(N— B)*-- borane	$CH_3ONH_2 \cdot BH_3$
Aminoborane *(N —B)* trimethyl- borane	$BH_2NH_2 \cdot B(CH_3)_3$
(Dimethylamino)difluorophos- phine *(P— B)* tetraborane(8)	$(CH_3)_2NPF_2 \cdot B_4H_8$

---

[*] May also be named Trimethylamine — boron trichloride (see IUPAC Nomenclature of Inorganic Chemistry, Subsection 2.2).

Note: The locant of a component may be added as a superscript
            to the appropriate atomic symbol.

Example:

1-Phenylurea $(N^3 —B)$--
borane

7.74 - An intramolecular addition compound between a group
acting as a Lewis base and a boron-containing group in the same
molecule may be denoted by atomic symbol pairs in the order donor-
acceptor as described in Rule D-7.73, but cited in front of the
complete name of the compound . Such intramolecular addition compounds may also be named by coordination nomenclature (See D-7.64).

Examples:

$(N — B)$-(2-Aminoethoxy)dimethyl-
       borane

            or

$(N — B)$-2-[(Dimethylboryl)oxy]-
       ethylamine

$(O — B)$-Dichloro(2-nitrophenoxy)-
       borane

(*N* — *B*)- 1 - Aza-5-borabicyclo[3.3.3]-
   undecane

(*O* — *B*)-Difluoro[(methylnitrosoamino)oxy]-
   borane

                    or

(*O* — *B*)-*O*-(Difluoroboryl)-*N*-methyl-
   N- nitrosohydroxylamine

(N^3—*B*)- 2-[2-(dihydroxyboryl)phenyl]-
   benzimidazole

APPENDIX

Table I – Seniority list of elements and "a" terms used in replacement nomenclature, in decreasing order of priority

Element	"a" Term	Element	"a" Term
F	fluora	Au	aura
Cl	chlora	Ni	nickela
Br	broma	Pd	pallada
I	ioda	Pt	platina
At	astata	Co	cobalta
O	oxa	Rh	rhoda
S	thia	Ir	irida
Se	selena	Fe	ferra
Te	tellura	Ru	ruthena
Po	polona	Os	osma
N	aza	Mn	mangana
P	phospha	Tc	techneta
As	arsa	Re	rhena
Sb	stiba	Cr	chroma
Bi	bisma	Mo	molybda
C	carba	W	tungsta (wolframa)
Si	sila	V	vanada
Ge	germa	Nb	nioba
Sn	stanna	Ta	tantala
Pb	plumba	Ti	titana
B	bora	Zr	zircona
Al	alumina	Hf	hafna
Ga	galla	Sc	scanda
In	inda	Y	yttra
Tl	thalla	La	lanthana
Zn	zinca	Ce	cera
Cd	cadma	Pr	praseodyma
Hg	mercura	Nd	neodyma
Cu	cupra	Pm	prometha
Ag	argenta	Sm	samara

TABLE I (Continued)

Element	"a" Term	Element	"a" Term
Eu	europa	Am	america
Gd	gadolina	Cm	cura
Tb	terba	Bk	berkela
Dy	dysprosa	Cf	californa
Ho	holma	Es	einsteina
Er	erba	Fm	ferma
Tm	thula	Md	mendeleva
Yb	ytterba	No	nobela
Lu	luteta	Lr	lawrenca
Ac	actina	Be	berylla
Th	thora	Mg	magnesa
Pa	protactina	Ca	calca
U	urana	Sr	stronta
Np	neptuna	Ba	bara
Pu	plutona	Ra	rada

TABLE II - Element radical names *

Element name	Radical name	Element name	Radical name
Actinium	Actinio	Fermium	Fermio
Aluminium	Aluminio	Fluorine	Fluorio
Americium	Americio	Francium	Francio
Antimony	Antimonio	Gadolinium	Gadolinio
Argon	Argonio	Gallium	Gallio
Arsenic	Arsenio	Germanium	Germanio
Astatine	Astatio	Gold(Aurum)	Aurio
Barium	Bario	Hafnium	Hafnio
Berkelium	Berkelio	Helium	Helio
Beryllium	Beryllio	Holmium	Holmio
Bismuth	Bismuthio	Hydrogen	-
Boron	Borio	Indium	Indio
Bromine	Bromio	Iodine	Iodio
Cadmium	Cadmio	Iridium	Iridio
Caesium	Caesio	Iron(Ferrum)	Ferrio
Calcium	Calcio	Krypton	Kryptonio
Californium	Californio	Lanthanum	Lanthanio
Carbon	-	Lawrencium	Lawrencio
Cerium	Cerio	Lead(Plumbum)	Plumbio
Chlorine	Chlorio	Lithium	Lithio
Chromium	Chromio	Lutetium	Lutetio
Cobalt	Cobaltio	Magnesium	Magnesio
Copper(Cuprum)	Cuprio	Manganese	Manganio
Curium	Curio	Mendelevium	Mendelevio
Deuterium	Deuterio	Mercury	Mercurio
Dysprosium	Dysprosio	Molybdenum	Molybdenio
Einsteinium	Einsteinio	Neodymium	Neodymio
Erbium	Erbio	Neon	Neonio
Europium	Europio	Neptunium	Neptunio

---

* Element radical names should not be confused with cationic substituent prefixes defined in Rules C-82 and D-5.33 . These names are compared in the following list , on page 463 .

TABLE II - (Continued)

Element name	Radical name	Element name	Radical name
Nickel	Nickelio	Silicon	Silicio
Niobium	Niobio	Silver(Argentum)	Argentio
Nitrogen	-	Sodium	Sodio(Natrio)
Nobelium	Nobelio	Strontium	Strontio
Osmium	Osmio	Sulfur	Sulfurio
Oxygen	-	Tantalum	Tantalio
Palladium	Palladio	Technetium	Technetio
Phosphorus	Phosphorio	Tellurium	Tellurio
Platinum	Platinio	Terbium	Terbio
Plutonium	Plutonio	Thallium	Thallio
Polonium	Polonio	Thorium	Thorio
Potassium	Potassio(Kalio)	Thulium	Thulio
Praseodymium	Praseodymio	Tin(Stannum)	Stannio
Promethium	Promethio	Titanium	Titanio
Protactinium	Protactinio	Tritium	Tritio
Radium	Radio	Tungsten	Tungstenio
Radon	Radonio	(Wolfram)	(Wolframio)
Rhenium	Rhenio	Uranium	Uranio
Rhodium	Rhodio	Vanadium	Vanadio
Rubidium	Rubidio	Xenon	Xenonio
Ruthenium	Ruthenio	Ytterbium	Ytterbio
Samarium	Samario	Yttrium	Yttrio
Scandium	Scandio	Zinc	Zincio
Selenium	Selenio	Zirconium	Zirconio

Element radical names :

Element	Name
N	–
P	Phosphorio
As	Arsenio
Sb	Antimonio
Bi	Bismuthio
O	–
S	Sulfurio
Se	Selenio
Cl	Chlorio
Br	Bromio
I	Iodio

Cationic substituent names :

Substituent	Name	Rule
H_3N^+–	Ammonio	C-82
H_3P^+–	Phosphonio	D-5.33
H_3As^+–	Arsonio	D-5.33
H_3Sb^+–	Stibonio	D-5.33
H_3Bi^+–	Bismuthonio	D-5.33
H_2O^+–	Oxonio	C-82
H_2S^+–	Sulfonio	C-82
H_2Se^+–	Selenonio	C-82
HCl^+–	Chloronio	C-82
HBr^+–	Bromonio	C-82
HI^+–	Iodonio	C-82

TABLE III - Affixes used in inorganic nomenclature

Multiplying affixes

(a) mono, di, tri, tetra, penta, hexa, hepta, octa, nona (ennea), deca, undeca (hendeca), dodeca, etc. used by direct joining without hyphens

(b) bis, tris, tetrakis, pentakis, etc. used by direct joining without hyphens but usually with enclosing marks around each whole expression to which the prefix applies

Structural affixes

These affixes are italicized and separated from the rest of the name by hyphens.

antiprismo	eight atoms bound into a rectangular antiprism
asym	asymmetrical
catena	a chain structure ; often used to designate linear polymeric substances
cis	two groups occupying adjacent positions ; sometimes used in the sense of *fac*
closo	a cage or closed structure, especially a boron skeleton that is a polyhedron having all triangular faces
cyclo	a ring structure. Cyclo here is used as a modifier indicating structure and hence is italicized. In organic nomenclature, cyclo is considered to be part of the parent name since it changes the molecular formula and therefore is not italicized.
dodecahedro	eight atoms bound into a dodecahedron with triangular faces
fac	three groups occupying the corners of the same face of an octahedron
hexahedro	eight atoms bound into a hexahedron (e.g., cube)
hexaprismo	twelve atoms bound into a hexagonal prism

icosahedro	twelve atoms bound into a triangular icosahedron
mer	meridional ; three groups on an octahedron in such a relationship that one is *cis* to the two others which are themselves *trans*
nido	a nest-like structure, especially a boron skeleton that is almost closed
octahedro	six atoms bound into an octahedron
pentaprismo	ten atoms bound into a pentagonal prism
quadro	four atoms bound into a quadrangle (e.g., square)
sym	symmetrical
tetrahedro	four atoms bound into a tetrahedron
trans	two groups directly across a central atom from each other ; i.e., in the polar positions on a sphere
triangulo	three atoms bound into a triangle
triprismo	six atoms bound into a triangular prism
η *(eta or hapto)*	signifies that two or more contiguous atoms of a group are attached to a metal
μ *(mu)*	signifies that a group so designated bridges two centres of coordination
σ *(sigma)*	signifies that one atom of a group is attached to a metal by a sigma bond
λ *(lambda)*	signifies, with its superscript, the bonding number, i.e., the sum of the number of skeletal bonds and the number of hydrogen atoms associated with an atom in a parent compound
δ *(delta)*	signifies, with its superscript, the number of skeletal cumulative double bonds terminating at a heteroatom in a cyclic compound.

TABLE IV - Trivial names for heterocyclic systems

The following list of structures and trivial names of heterocyclic systems is an extended version of the list contained in Rule B-2.11. For compounds which already appear in Rule B-2.11, only the names and numbers are given under which they are listed there. Rings are cited in increasing order of preference for selection as base compounds in fusion names, i.e. the ring which comes later in the list takes precedence. In two cases, compounds (43) and (45), marked by an asterisk, this list updates the one in Rule B-2.11 with respect to the order of precedence.

Compounds (2), (3), (36) and (46) from B-2.11 have been omitted since they are not used as base compounds in fusion names. Except for names of the type phenothiazine the names given in the list should not be used as models for the coinage of new names. Replacement (Rule B-4), **Hantzsch-Widman (Rule B-1) and fusion (Rule B-3) names are preferred.**

Phenomercurin

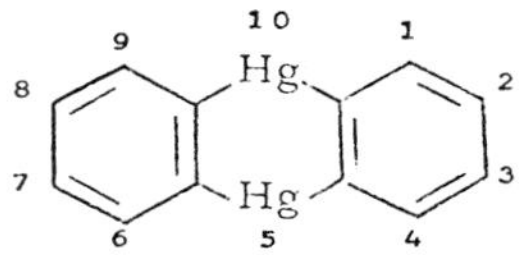

Isoarsindole

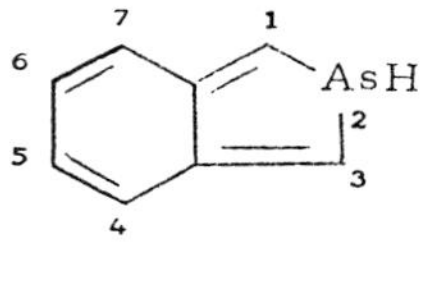

Arsindole

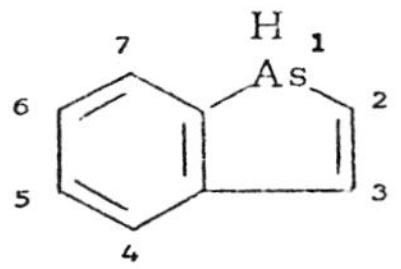

Isoarsinoline

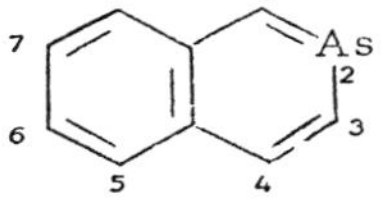

Arsinoline

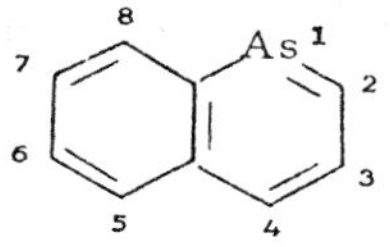

Arsanthridine

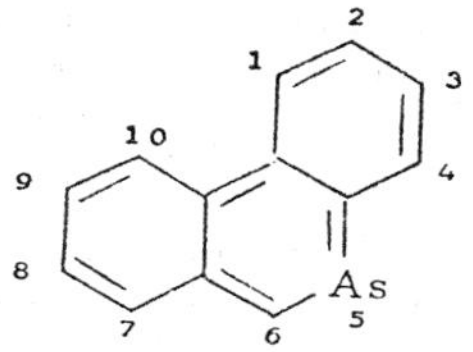

Acridarsine

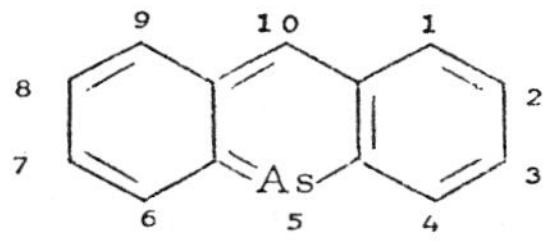

Arsanthrene

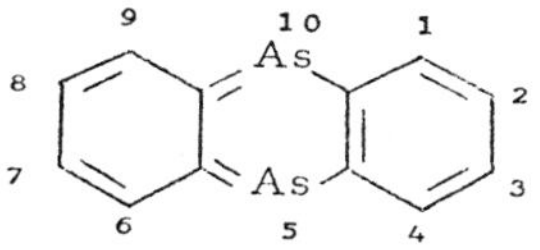

Isophosphindole

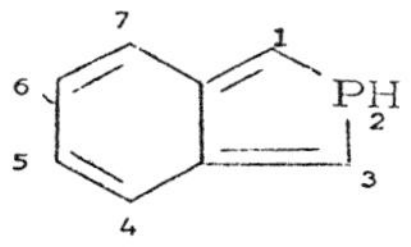

Phosphindole

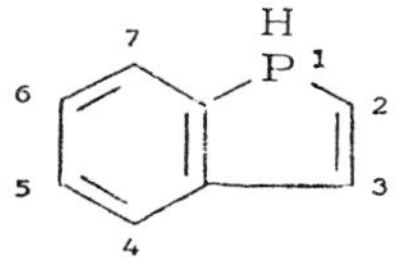

Isophosphinoline

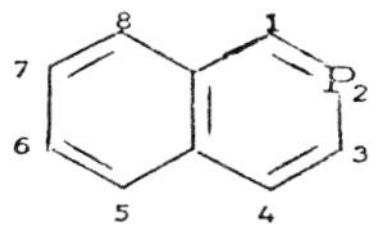

Phosphinoline

Phosphanthrene

Tellurophene

Selenophene

Selenanthrene

Thiophene (1)
Thianthrene (4)

Phenothiarsine

Furan (5)
Pyran (6)
Isobenzofuran (7)
Chromene (8)
Xanthene (9)

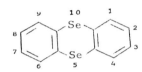

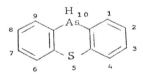

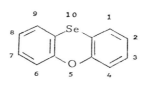

Phenoxantimonin

Phenoxarsine

Phenoxaphosphine

Phenoxatellurin

Phenoxaselenin

Phenoxathiin (10)
Pyrrole (11, 12)
Imidazole (13)
Pyrazole (14)
Isothiazole (43)*
Isoxazole (45)*
Pyridine (15)
Pyrazine (16)
Pyrimidine (17)
Pyridazine (18)

Pyrrolizine (1*H*-form shown)

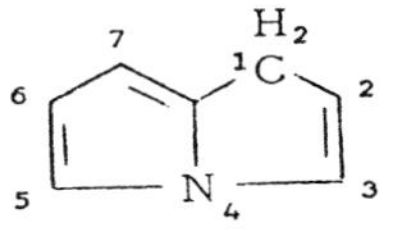

Indolizine (19)
Isoindole (20)
Indole (21, 22)
Indazole (23)
Purine (24)
Quinolizine (25)
Isoquinoline (26)
Quinoline (27)
Phthalazine (28)
Naphthyridine (29)
Quinoxaline (30)
Quinazoline (31)
Cinnoline (32)
Pteridine (33)
Carbazole (34, 35)
Phenanthridine (37)
Acridine (38)
Perimidine (39)
Phenanthroline (40)
Phenazine (41)

Phenomercazine

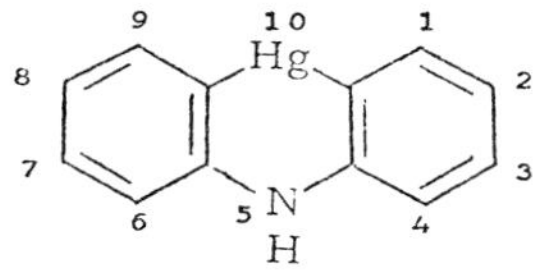

Phenarsazine (42)

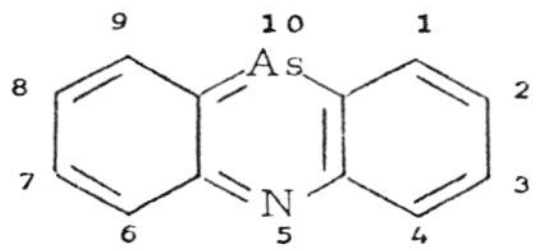

Phenophosphazine

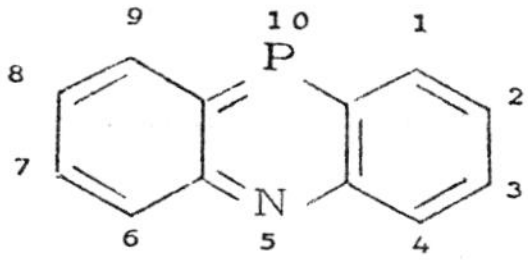

Phenotellurazine

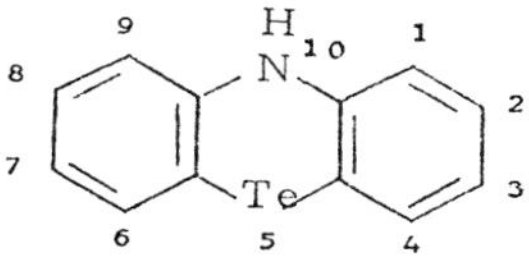

Phenoselenazine

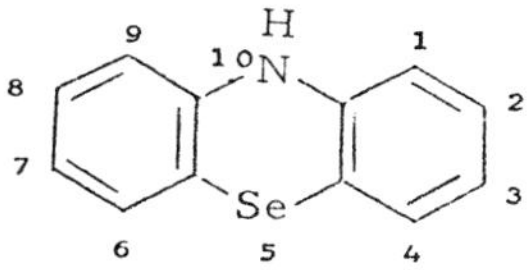

Phenothiazine (44)
Phenoxazine (47)

Section E: Stereochemistry (Recommendations 1974)

INTRODUCTION

This Section of the IUPAC Rules for Nomenclature of Organic Chemistry differs from previous Sections in that it is here necessary to legislate for words that describe concepts as well as for names of compounds.

At the present time, concepts in stereochemistry (that is, chemistry in three-dimensional space) are in the process of rapid expansion, not merely in organic chemistry, but also in biochemistry, inorganic chemistry, and macromolecular chemistry. The aspects of interest for one area of chemistry often differ from those for another, even in respect of the same phenomenon. This rapid evolution and the variety of interests have led to development of specialized vocabularies and definitions that sometimes differ from one group of specialists to another, sometimes even within one area of chemistry.

The Rules in this Section deal only with the main principles of stereochemistry; work on further aspects is under way. The present rules have two objects: to prescribe, for basic concepts, terms that may provide a common language in all areas of stereochemistry; and to define the ways in which these terms may, so far as necessary, be incorporated into the names of individual compounds. Many of these Rules do little more than codify existing practice, often of long standing; however, others extend old principles to wider fields, and yet others deal with nomenclatures that are still subject to controversy. The Commission recognizes that specialized nomenclatures are required for local fields; in some cases, such as carbohydrates, amino acids, peptides and proteins, and steroids, international rules already exist; for other fields, study is in progress.

RULES

Rule E-0

The steric structure of a compound is denoted by an affix or affixes to the name that does not prescribe the stereochemistry; such affixes, being additional, do not change the name or the numbering of the compound. Thus, enantiomers, diastereoisomers, and *cis-trans*-isomers receive names that are distinguished only by means of different stereochemical affixes. The only exceptions are those trivial names that have stereochemical implications (for example, fumaric acid, cholesterol).

Note: In some cases (see Rules E-2.2.3 and E-2.3.1) stereochemical relations may be used to decide between alternative numberings that are otherwise permissible.

E-1 Types of isomerism

E-1.1. The following non-stereochemical terms are relevant to the stereochemical nomenclature given in the Rules that follow:

(a) The term structure may be used in connexion with any aspect of the organization of matter.

Hence: structural (adjectival)

(b) Compounds that have identical molecular formulae but differ in the nature or sequence of bonding of their atoms or in arrangement of their atoms in space are termed isomers.

Hence: isomeric (adjectival)

isomerism (phenomenological)

Examples:

$H_3C-O-CH_3$ is an isomer of H_3C-CH_2-OH

473

(In this and other Rules a broken line denotes a bond projecting behind the plane of the paper. and a thickened line denotes a bond projecting in front of the plane of the paper. In such cases a line of normal thickness denotes a bond lying in the plane of the paper.)

(c) The constitution of a compound of given molecular formula defines the nature and sequence of bonding of the atoms. Isomers differing in constitution are termed constitutional isomers.

Hence: constitutionally isomeric (adjectival)
        constitutional isomerism (phenomenological)

Example:

H₃C—O—CH₃   is   a   constitutional   isomer   of
H₃C—CH₂—OH

Note: Use of the term "structural" with the above connotation is abandoned as insufficiently specific.

E-1.2. Isomers are termed stereoisomers when they differ only in the arrangement of their atoms in space.
Hence: stereoisomeric (adjectival)
        stereoisomerism (phenomenological)

Examples:

E-1.3. Stereoisomers are termed *cis-trans*-isomers when they differ only in the positions of atoms relative to a specified plane in cases where these atoms are, or are considered as if they were, parts of a rigid structure.
Hence: *cis-trans*-isomeric (adjectival)
        cis-trans-isomerism (phenomenological)

Examples:

E-1.4. Various views are current regarding the precise definition of the term "configuration". (a) Classical interpretation: The configuration of a molecule of defined constitution is the arrangement of its atoms in space without regard to arrangements that differ only as after

*Determination of absolute configuration became possible through work by J. M. Bijvoet, A. F. Peerdeman and A. J. van Bommel, Nature **168**, 271 (1951); cf. J. M. Bijvoet, Proc. Kon. Ned. Akad. Wet., Amsterdam **52**, 313 (1949).

rotation about one or more single bonds. (b) This definition is now usually limited so that no regard is paid also to rotation about π-bonds or bonds of partial order between one and two. (c) A third view limits the definition further so that no regard is paid to rotation about bonds of any order, including double bonds.

Molecules differing only in configuration are termed configurational isomers.

Hence: configurational isomerism

Notes: (1) Contrast conformation (Rule E-1.7). (2) The phrase "differ only as after rotation" is intended to make the definition independent of any difficulty of rotation. in particular independent of steric hindrance to rotation. (3) For a brief discussion of views (a-c) see Appendix 1, p. 486.

Examples:

The following pairs of compounds differ in configuration:

(i)

(ii)

(iii)

(iv)   These isomers (iv) are configurational in view (a) but are conformational (see Rule E-1.7) in view (b) or (c)

E-1.5. The terms relative stereochemistry and relative configuration are used to describe the positions of substituents on different atoms in a molecule relative to one another.

E-1.6. The terms absolute stereochemistry and absolute configuration* are used to describe the three-dimensional arrangement of substituents around a chiral element. (For a description of chiral element, see Appendix 2.)

E-1.7. Various views are current regarding the precise definition of the term "conformation". (a) Classical interpretation: The conformations of a molecule of defined configuration are the various arrangements of its atoms in space that differ only as after rotation about single bonds. (b) This is usually now extended to include rotation about π-bonds or bonds of partial order between one and two. (c) A third view extends the definition further to include also rotation about bonds of any order, including double bonds.

Molecules differing in conformation are termed conformational isomers.

Hence: conformational isomerism

Notes: All the Notes to Rule E-1.4 apply also to E-1.7.

Examples:

Each of the following pairs of formulae represents a

compound in the same configuration but in different conformations.

(a,b,c)

(a,b,c)

(a,b,c)

(b,c)   Fe   Fe

(c)   See Example (iv) to Rule E–1.4.

E–2 cis-trans-*Isomerism* *
Preamble:

The prefixes *cis* and *trans* have long been used for describing the relative positions of atoms or groups attached to non-terminal doubly bonded atoms of a chain or attached to a ring that is considered as planar. This practice has been codified for hydrocarbons by IUPAC. There has, however, not been agreement on how to assign *cis* or *trans* at terminal double bonds of chains or at double bonds joining a chain to a ring. An obvious solution was to use *cis* and *trans* where doubly bonded atoms formed the backbone and were non-terminal and to enlist the sequence-rule preferences to decide other cases; however, since the two methods, when generally applied, do not always produce analogous results, it would then be necessary to use different symbols for the two procedures. A study of this combination showed that both types of symbol would often be required in one name and, moreover, it seemed wrong in principle to use two symbolisms for essentially the same phenomenon. Thus it seemed to the Commission wise to use only the sequence-rule system, since this alone was applicable to all cases. The same decision was taken independently by *Chemical Abstracts Service*† who introduced *Z* and *E* to correspond more conveniently to *seqcis* and *seqtrans* of the sequence rule.

It is recommended in the Rules below that these designations *Z* and *E* based on the sequence rule shall be used in names of compounds; but *Z* and *E* do not always

*These Rules supersede the Tentative Rules for olefinic hydrocarbons published in the *Comptes rendus of the 16th IUPAC Conference*, New York, 1951, pp. 102–103.

†J. E. Blackwood, C. L. Gladys, K. L. Loening, A. E. Petrarca and J. E. Rush, *J. Amer. Chem. Soc.* **90**, 509 (1968); J. E. Blackwood, C. L. Gladys, A. E. Petrarca, W. H. Powell and J. E. Rush, *J. Chem. Docum.* **8**, 30 (1968).

correspond to the classical *cis* and *trans* which show the steric relations of like or similar groups that are often the main point of interest. So the use of *Z* and *E* in names is not intended to hamper the use of *cis* and *trans* in discussions of steric relations of a generic type or of groups of particular interest in a specified case (see Rule E–2.1 and its Examples and Notes, also Rule E–4.11).

It is also not necessary to replace *cis* and *trans* for describing the stereochemistry of substituted monocycles (see Rule E–2.3). For cyclic compounds the main problems are usually different from those around double bonds; for instance, steric relations of substituents on rings can often be described either in terms of chirality (see Subsection E–4) or in terms of *cis–trans*-relationships, and, further, there is usually no single relevant plane of reference in a hydrogenated polycycle. These matters are discussed in the Preambles to Rule E–2.3 and Rule E–3.

E–2.1. Definition of cis-trans. Atoms or groups are termed *cis* or *trans* to one another when they lie respectively on the same or on opposite sides of a reference plane identifiable as common among stereoisomers. The compounds in which such relations occur are termed *cis–trans*-isomers. For compounds containing only doubly bonded atoms the reference plane contains the doubly bonded atoms and is perpendicular to the plane containing these atoms and those directly attached to them. For cyclic compounds the reference plane is that in which the ring skeleton lies or to which it approximates. When qualifying another word or a locant, *cis* or *trans* is followed by a hyphen. When added to a structural formula, *cis* may be abbreviated to *c*, and *trans* to *t* (see also Rule E–2.3.3).

Examples:
[Rectangles here denote the reference planes and are considered to lie in the plane of the paper.]

cis   *trans*

trans   *cis*

The groups or atoms a,a are the pair selected for designation but are not necessarily identical; b,b are also not necessarily identical but must be different from a,a.

cis or *trans* according as a or b is taken as basis of comparison

cis   *trans*

Notes: The formulae above are drawn with the reference plane in the plane of the paper, but for doubly bonded compounds it is customary to draw the formulae so that this plane is perpendicular to that of the paper; atoms attached directly to the doubly bonded atoms then lie in the plane of the paper and the formulae appear as, for instance:

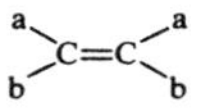

cis

Cyclic structures, however, are customarily drawn with the ring atoms in the plane of the paper, as above. However, care is needed for complex cases, such as:

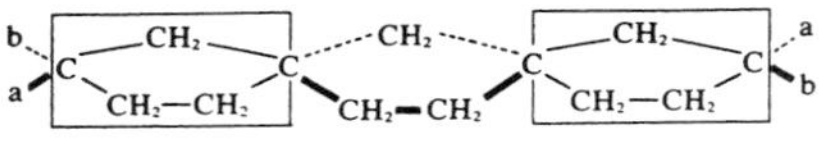

The central five-membered ring lies (approximately) in a plane perpendicular to the plane of the paper. The two a groups are *trans* to one another; so are the b groups; the outer cyclopentane rings are *cis* to one another with respect to the plane of the central ring.

cis or *trans* (or *Z* or *E*: see Rule E–2.2–1) may also be used in cases involving a partial bond order when a limiting structure is of sufficient importance to impose rigidity around the bond of partial order. An example is:

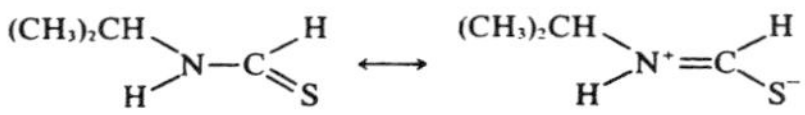

trans (or *E*)

E-2.2. cis–trans-*Isomerism around double bonds.*

E–2.2.1. In names of compounds steric relations around one double bond are designated by affixes *Z* and/or *E*, assigned as follows. The sequence-rule-preferred* atom or group attached to one of a doubly bonded pair of atoms is compared with the sequence-rule-preferred atom or group attached to the other of that doubly bonded pair of atoms: if the selected pair are on the same side of the reference plane (see Rule E–2.1) an italic capital letter *Z* prefix is used; if the selected pair are on opposite sides an italic capital letter *E* prefix is used.† These prefixes, placed in parentheses and followed by a hyphen, normally precede the whole name: if the molecule contains several double bonds, then each prefix is immediately preceded by the lower or less primed locant of the relevant double bond.

Examples:

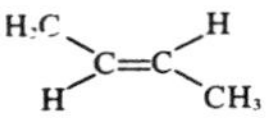

(*E*)-2-Butene

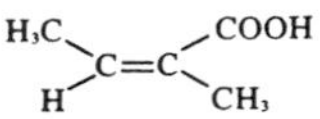

(*Z*)-2-Methyl-2-butenoic acid‡
or (*Z*)-2-Methylisocrotonic acid
(see Exceptions below)

---

*For sequence-rule preferences see Appendix 2, p. 486.
†These prefixes may be rationalized as from the German *zusammen* (together) and *entgegen* (opposite).
‡The name angelic acid is abandoned because it has been associated with the designation *trans* with reference to the methyl groups.
§The name tiglic acid is abandoned because it has been associated with the designation *cis* with reference to the methyl groups.
‖Systematic names are recommended for derivatives of these compounds formed by substitution on carbon.

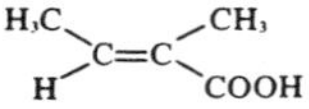

(*E*)-2-Methyl-2-butenoic acid§
or (*E*)-2-Methylcrotonic acid
(see Exceptions below)

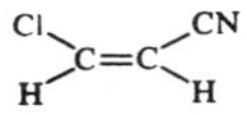

(*Z*)-3-Chloroacrylonitrile

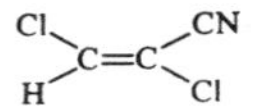

(*E*)-2,3-Dichloroacrylonitrile

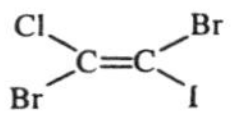

(*Z*)-1,2-Dibromo-1-chloro-2-iodoethylene (by the sequence rule, Br is preferred to Cl, but I to Br)

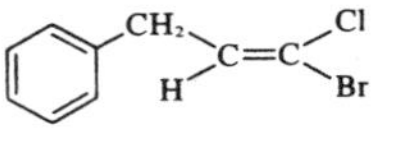

(*E*)-(3-Bromo-3-chloroallyl)-benzene

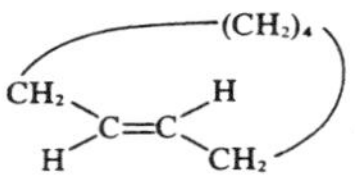

(*E*)-Cyclooctene

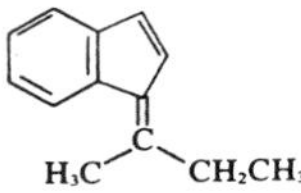

(*E*)-1-*sec*-Butylideneindene

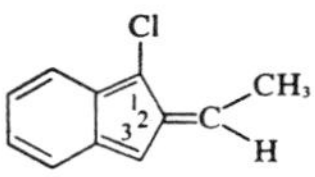

(*Z*)-1-Chloro-2-ethylidene-2*H*-indene

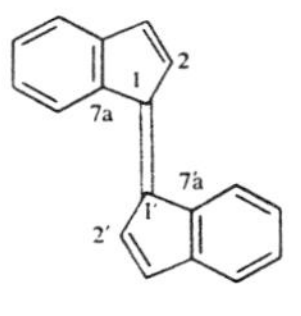

(*E*)-1,1'-Biindenylidene

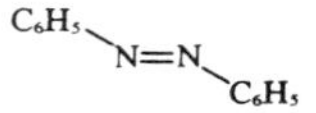

(*E*)-Azobenzene

Exceptions to Rule E–2.2.1:
The following are examples of accepted trivial names in which the stereochemistry is prescribed by the name and is not cited by a prefix:

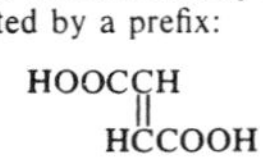

Fumaric acid

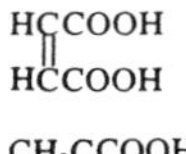

Maleic acid

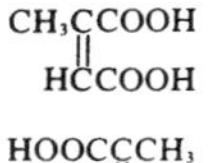

Citraconic acid‖

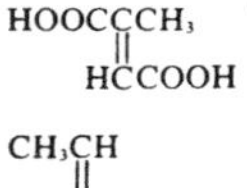

Mesaconic acid‖

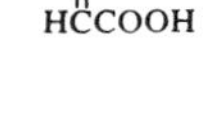

Crotonic acid

HCCH₃
‖
HCCOOH   Isocrotonic acid

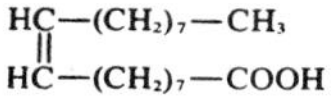

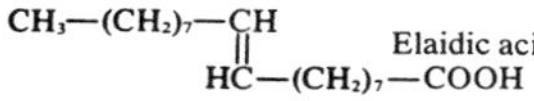

HC—(CH₂)₇—CH₃
‖
HC—(CH₂)₇—COOH   Oleic acid

CH₃—(CH₂)₇—CH
‖
HC—(CH₂)₇—COOH   Elaidic acid

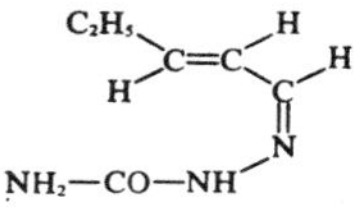

(Z)-[(E)-2-Pentenal semicarbazone]

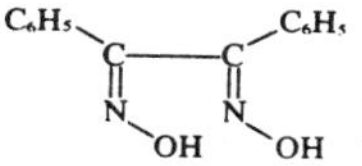

(Z,E)-(Benzil dioxime)

E–2.2.2. (a) When a molecule contains more than one double bond, each E or Z prefix has associated with it the lower locant of the double bond concerned.

(b) The E, Z prefixes are given at the beginning of the complete name, unless the prefix is related to a double bond within a substituent: it then forms part of the name of the substituent.

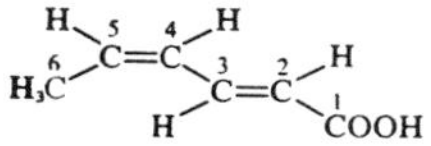

(2E,4Z)-2,4-Hexadienoic acid

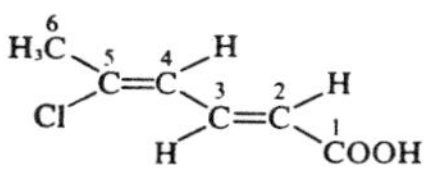

(2E,4Z)-5-Chloro-2,4-hexa-dienoic acid

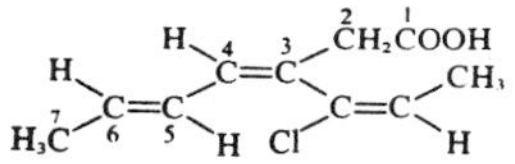

3-[(E)-1-Chloropropenyl]-(3Z,5E)-3,5-heptadienoic acid

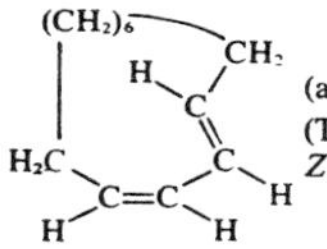

(a) (1Z,3E)-1,3-Cyclododecadiene
(The lower locant is assigned to the Z double bond.)

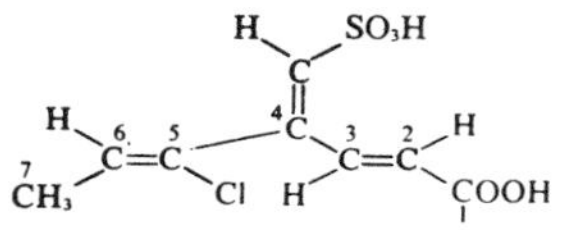

(2E,5Z)-5-Chloro-4-(E-sulfomethylene)-2,5-heptadienoic acid

[In application of the sequence rule, the relation of the SO₃H to CCl (rather than to C-3), and of the CH₃ to Cl, are decisive.]

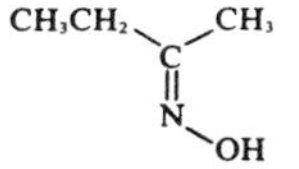

(E)-(Butanone oxime)*

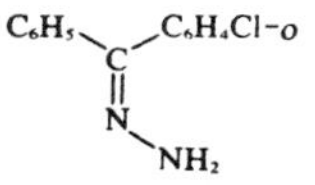

(Z)-(2-Chlorobenzophenone hydrazone)

E–2.2.3. When Rule C–13.1 permits alternatives, preference for lower locants and for inclusion in the principal chain is allotted as follows, in the order stated, so far as necessary: Z over E groups; cis over trans cyclic groups; if the nature of these groups is not decisive, then the lower locant for such a preferred group at the first point of difference.

Examples:

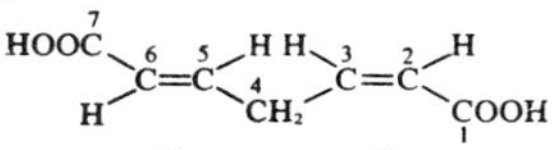

(2Z,5E)-2,5-Heptadienedioic acid
[The lower numbers are assigned to the Z double bond.]

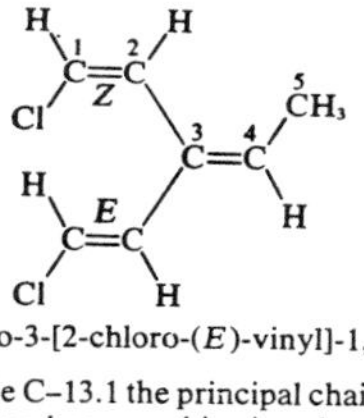

(1Z,3Z)-1-Chloro-3-[2-chloro-(E)-vinyl]-1,3-pentadiene

[According to Rule C–13.1 the principal chain must include the C=C–CH₃ group because this gives lower numbers to the double bonds (1,3 rather than 1,4); then the Cl-containing Z group is chosen for the remainder of the principal chain in accord with Rule E–2.2.3.]

E–2.3. *Relative stereochemistry of substituents in monocyclic compounds*†.

Preamble:

cis- and trans-Prefixes are commonly used to designate the positions of substituents on rings relative to one another; when a ring is, or is considered to be, rigidly planar or approximately so and is placed horizontally, these prefixes define which groups are above and which below the (approximate) plane of the ring. This differentiation is often important, so this classical terminology is retained in Subsection E–2.3; since the difficulties inherent in end-groups do not arise for cyclic compounds it is unnecessary to resort to the less immediately informative E/Z symbolism.

When the cis–trans-designation of substituents is applied, rings are considered in their most extended form; reentrant angles are not permitted; for example:

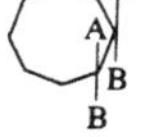

and not

cis   apparently *trans*

---

*The terms *syn*, *anti*, and *amphi* are abandoned for such compounds.

†Formulae in Examples to this Rule denote relative (not absolute) configurations.

The absolute stereochemistry of optically active or racemic derivatives of monocyclic compounds is described by the sequence-rule procedure (see Rule E–4.9 and Appendix 2).* The relative stereochemistry may be described by a modification of sequence-rule symbolism as set out in Rule E–4.10. If either of these procedures is adopted, it is then superfluous to use also *cis* or *trans* in the names of individual compounds.

E–2.3.1. When alternative numberings of the ring are permissible according to the Rules of Section C, that numbering is chosen which gives a *cis* attachment at the first point of difference; if that is not decisive, the criteria of Rule E–2.2.3 are applied. *cis-* and *trans-* may be abbreviated to *c-* and *t-*, respectively, in names of compounds when more than one such designation is required.

Examples:

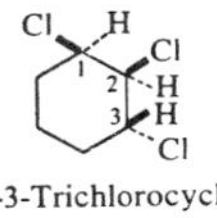

1,*c*-2,*t*-3-Trichlorocyclohexane

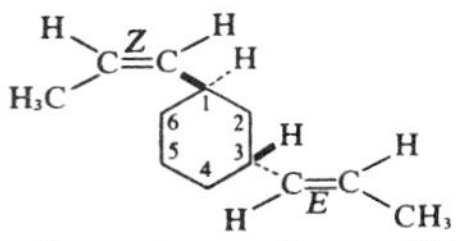

1-[(Z)-1-Propenyl]-*trans*-3[(E)-1-propenyl]cyclohexane

E–2.3.2. When one substituent and one hydrogen atom are attached at each of two positions of a monocycle, the steric relations of the two substituents are expressed as *cis* or *trans*, followed by a hyphen and placed before the name of the compound.

Examples:

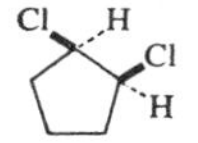

cis-1,2-Dichlorocyclopentane

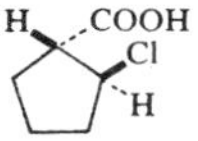

trans-2-Chloro-1-cyclopentane-carboxylic acid

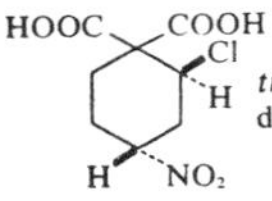

trans-2-Chloro-4-nitro-1,1-cyclohexane-dicarboxylic acid

E–2.3.3. When one substituent and one hydrogen atom are attached at each of more than two positions of a monocycle, the steric relations of the substituents are expressed by adding *r* (for *reference* substituent), followed by a hyphen, before the locant of the lowest-numbered of these substituents and *c* or *t* (as appropriate), followed by a hyphen, before the locants of the other substituents to express their relation to the reference substituent.

Examples:

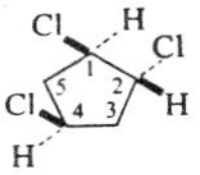

r-1,*t*-2,*c*-4-Trichlorocyclopentane (not *r*-1, *t*-2, *t*-4 which would follow from the alternative direction of numbering; see Rule E–2.3.1)

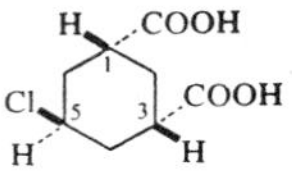

t-5-Chloro-*r*-1,*c*-3-cyclohexanedi-carboxylic acid

E–2.3.4. When two different substituents are attached at the same position of a monocycle, then the lowest-numbered substituent named as suffix is selected for designation as reference group in accordance with Rule E–2.3.2 or E–2.3.3; or, if none of the substituents is named as suffix, then of the lowest-numbered pair that one preferred by the sequence rule is selected as reference group; and the relation of the sequence-rule preferred group at each other position, relative to the reference group, is cited as *c* or *t* (as appropriate).

Examples:

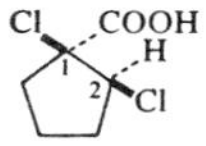

1,*t*-2-Dichloro-*r*-1-cyclopentanecarboxylic acid

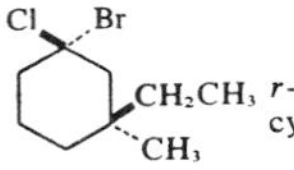

r-1-Bromo-1-chloro-*t*-3-ethyl-3-methyl-cyclohexane

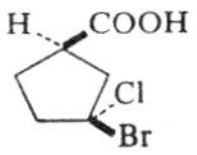

c-3-Bromo-3-chloro-*r*-1-cyclopentane-carboxylic acid

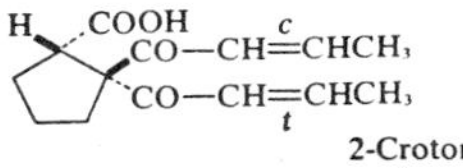

2-Crotonoyl-*t*-2-isocrotonoyl-*r*-1-cyclopentanecarboxylic acid

E–3 Fused rings

Preamble:

In simple cases the relative stereochemistry of substituted fused-ring systems can be designated by the methods used for monocycles. For the absolute stereochemistry of optically active and racemic compounds the sequence-rule procedure can be used in all cases (see Rule E–4.9 and Appendix 2);* for relative configurations of such compounds the procedure of Rule E–4.10 can be applied. Sequence-rule methods are, however, not descriptive of geometrical shape. There is as yet no generally acceptable system for designating in an immediately interpretable manner the configuration of bridged polycyclic compounds (the *endo–exo* nomenclature, which should solve part of the problem, has been used in different ways). These matters will be considered in a later document.

E–3.1. Steric relations at saturated bridgeheads common to two rings are denoted by *cis* or *trans*, followed by a hyphen and placed before the name of the ring system,

---

* p.486.

* p.486.

according to the relative positions of the exocyclic atoms or groups attached to the bridgeheads. Such rings are said to be *cis*-fused or *trans*-fused.

Examples:

cis-Decalin   1-Methyl-*trans*-bicyclo[8.3.1]tetradecane

E-3.2. Steric relations at more than one pair of saturated bridgeheads in a polycyclic compound are denoted by *cis* or *trans*, each followed by a hyphen and, when necessary, the corresponding locant of the lower-numbered bridgehead and a second hyphen, all placed before the name of the ring system. Steric relations between the nearest atoms* of *cis*- or *trans*-bridgehead pairs may be described by affixes *cisoid* or *transoid*, followed by a hyphen and, when necessary, the corresponding locants and a second hyphen, the whole placed between the designations of the *cis*- or *trans*-ring junctions concerned. When a choice remains amongst nearest atoms, the pair containing the lower-numbered atom is selected. *cis* and *trans* are not abbreviated in such cases. In complex cases, however, designation may be more simply effected by the sequence rule procedure (see Appendix 2)‡ Note: the terms *syn* and *anti* were formerly used for *cisoid* and *transoid*.

Examples:

cis-cisoid-trans-Perhydro-
phenanthrene

cis-4a- cisoid-4a,10a-trans- 10a.–
Perhydroacridine *or*
rel-(4aS,8aS,9aS,10aR)- Perhydro-
acridine†

trans-3a-cisoid-3a,4a-cis-4a-Per-
hydrobenz[f]indene or rel-(3aR,
4aS,8aR,9aR)-Perhydrobenz[f]-
indene

E-4. Chirality

E-4.1. The property of non-identity of an object with its mirror image is termed chirality. An object, such as a molecule in a given configuration or conformation, is termed chiral when it is not identical with its mirror image; it is termed achiral when it is identical with its mirror image.

Notes: (1) Chirality is equivalent to handedness, the term

---

*The term "nearest atoms" denotes those linked together through the smallest number of atoms, irrespective of actual separation in space. For instance, in the second Example to this Rule, the atom 4a is "nearer" to 10a than to 8a.

†For the designation *rel*- see Rule E–4.10.

‡ p. 486.

being derived from the Greek $\chi\epsilon\iota\rho$ = hand. Note: The term dissymetry was formerly used.

(2) All chiral molecules are molecules of optically active compounds, and molecules of all optically active compounds are chiral. There is a 1:1 correspondence between chirality and optical activity.

(3) In organic chemistry the discussion of chirality usually concerns the individual molecule or, more strictly, a model of the individual molecule. The chirality of an assembly of molecules may differ from that of the component molecules, as in a chiral quartz crystal or in an achiral crystal containing equal numbers of dextrorotatory and laevorotatory tartaric acid molecules.

(4) The chirality of a molecule can be discussed only if the configuration or conformation of the molecule is specifically defined or is considered as defined by common usage. In such discussions structures are treated as if they were (at least temporarily) rigid. For instance, ethane is configurationally achiral although many of its conformations, such as (A), are chiral; in fact, a configuration of a mobile molecule is chiral only if all its possible conformations are chiral; and conformations of ethane such as (B) and (C) are achiral.

(A)   **(B)**   **(C)**

Examples;

(D)   **(E)**

(D) and (E) are mirror images and are not identical, not superposable. They represent chiral molecules. They represent (D) dextrorotatory and (E) laevorotatory glyceraldehyde.

(F)

(F) is identical with its mirror image. It represents an achiral molecule, namely, a molecule of 1,2,3-propanetriol (glycerol).

E-4.2. The term asymmetry denotes absence of any symmetry. An object, such as a molecule in a given configuration or conformation, is termed asymmetric if it has no element of symmetry.

Notes: (1) All asymmetric molecules are chiral, and all compounds composed of them are therefore optically active; however, not all chiral molecules are asymmetric since some molecules having axes of rotation are chiral.

(2) Notes (3) and (4) to Rule E-4.1 apply also in discussions of asymmetry.

Examples:

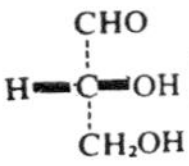

CHO
|
H—C—OH　　has no element of symmetry and represents
|　　　　　a molecule of an optically active compound.
CH₂OH

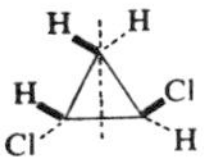

has a C_2 axis of rotation; it is chiral although not asymmetric, and therefore represents a molecule of an optically active compound.

C_2 axis

E-4.3. (a) An asymmetric atom is one that is tetrahedrally bonded to four different atoms or groups, none of the groups being the mirror image of any of the others. *Note:* One 'group' may be a lone-pair of electrons, as in sulfoxides.

(b) An asymmetric atom may be said to be at a chiral centre since it lies at the centre of a chiral tetrahedral structure. In a general sense, the term "chiral centre" is not restricted to tetrahedral structures; the structure may, for instance, be based on an octahedron or tetragonal pyramid.

(c) When the atom by which a group is attached to the remainder of a molecule lies at a chiral centre the group may be termed a chiral group.

Notes: (1) The term "asymmetric", as applied to a carbon atom in rule E-4.3(a), was chosen by van't Hoff because there is no plane of symmetry through a tetrahedron whose corners are occupied by four atoms or groups that differ in scalar properties. For differences of vector sense between the attached groups see Rule E-4.8.

(2) In Subsection E-4 the word "group" is used to denote the series of atoms attached to one bond.

For instance, in （structure） the groups attached to C*

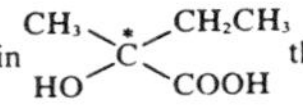

are —CH₃, —OH, —CH₂CH₃, and —COOH;

in （structure） they are —CH₃, —OH,

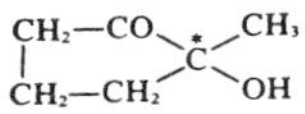

—COCH₂CH₂CH₂, and —CH₂CH₂CH₂CO.

(3) For the chiral axis and chiral plane (which are less common than the chiral centre) see Appendix 2, p.436.

(4) There may be more than one chiral centre in a molecule and these centres may be identical, or structurally different, or structurally identical but of opposite chirality; however, presence of an equal number of structurally identical chiral groups of opposite chirality, and no other chiral group, leads to an achiral molecule. These statements apply also to chiral axes and chiral planes. Identification of the sites and natures of the various factors involved is essential if the overall chirality of a molecule is to be understood.

(5) Although the term "chiral group" is convenient for use in discussions it should be remembered that chirality attaches to molecules and not to groups or atoms. For instance, although the *sec*-butyl group may be termed chiral in dextrorotatory 2-*sec*-butylnaphthalene, it is not chiral in the achiral compound (CH₃CH₂)(CH₃)CH-CH₃.

Examples:

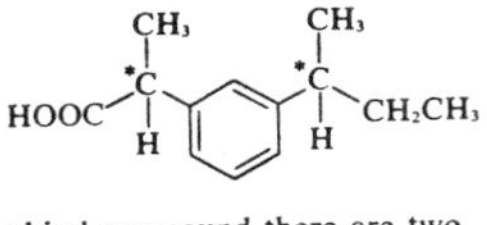

In this chiral compound there are two asymmetric carbon atoms, marked C*, each lying at a chiral centre. These atoms form parts of different chiral groups, namely, —CH(CH₃)COOH and —CH(CH₃)CH₂CH₃.

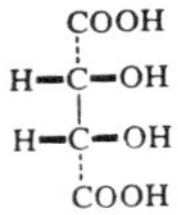

COOH
|
H—C—OH　　In this molecule (*meso*-tartaric acid) the two
|　　　　　central carbon atoms are asymmetric atoms
H—C—OH　　and each is part of a chiral group
|　　　　　—CH(OH)COOH. These groups, however,
COOH　　　although structurally identical, are of opposite chirality, so that the molecule is achiral.

E-4.4. Molecules that are mirror images of one another are termed enantiomers and may be said to be enantiomeric. Chiral groups that are mirror images of one another are termed enantiomeric groups.

Hence: enantiomerism (phenomenological)

Note: Although the adjective enantiomeric may be applied to groups, enantiomerism strictly applies only to molecules [see Note (5) to Rule E-4.3].

Examples:

The following pairs of molecules are enantiomeric.

(i)
CHO　　　　　　　CHO
|　　　　　　　　|
H—C—OH　　　HO—C—H
|　　　　　　　　|
CH₂OH　　　　　CH₂OH

(ii) 
COOH　　　　　　COOH
|　　　　　　　　|
H—C—OH　　　HO—C—H
|　　　　　　　　|
HO—C—H　　　H—C—OH
|　　　　　　　　|
COOH　　　　　　COOH

(iii)

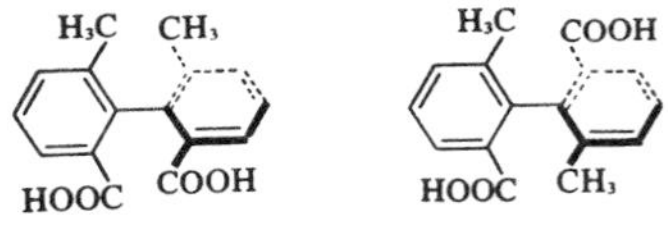

(iv) 
(*E*)-Cyclooctene

(v)

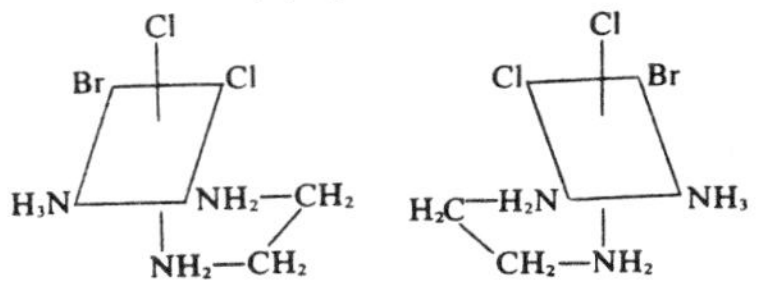

(vi)

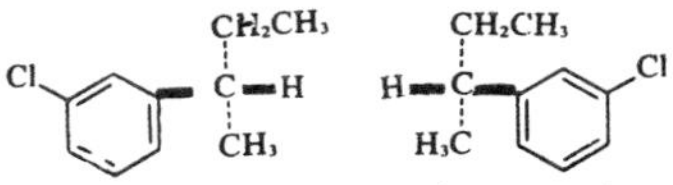

The *sec*-butyl groups in (vi) are enantiomeric.

Enantiomers whose absolute configurations are not known may be differentiated as dextrorotatory (prefix +) or laevorotatory (prefix −) depending on the direction in which, under specified experimental conditions, they rotate the plane of polarized light. The use of d instead of + and l instead of − is deprecated.

E-4.5. When equal amounts of enantiomeric molecules are present together, the product is termed racemic, independently of whether it is crystalline, liquid, or gaseous. A homogeneous solid phase composed of equimolar amounts of enantiomeric molecules is termed a racemic compound. A mixture of equimolar amounts of enantiomeric molecules present as separate solid phases is termed a racemic mixture. Any homogeneous phase containing equimolar amounts of enantiomeric molecules is termed a racemate.

Examples:

The mixture of two kinds of crystal (mirror-image forms) that separate below 28°C from an aqueous solution containing equal amounts of dextrorotatory and laevorotatory sodium ammonium tartrate is a racemic mixture.

The symmetrical crystals that separate from such a solution above 28°C, each containing equal amounts of the two salts, provide a racemic compound.

E-4.6. Stereoisomers that are not enantiomeric are termed diastereoisomers.

Hence: diastereoisomeric (adjective)
diastereoisomerism (phenomenological)

Note: Diastereoisomers may be chiral or achiral.

Examples:

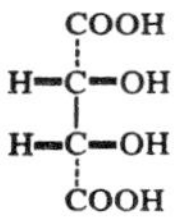

COOH
H—C—OH
H—C—OH   and
COOH

COOH
H—C—OH
HO—C—H
COOH

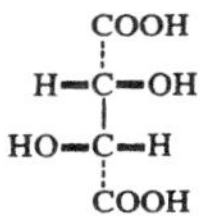

are diastereoisomers, the former is achiral, and the latter is chiral

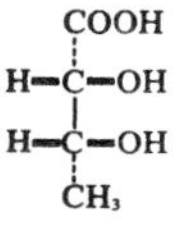

COOH
H—C—OH
H—C—OH   and
CH₃

COOH
H—C—OH
HO—C—H
CH₃

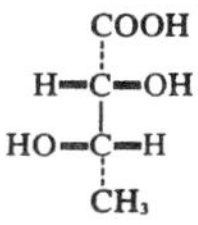

are diastereoisomers; both are chiral.

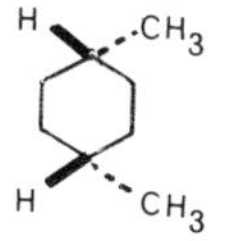

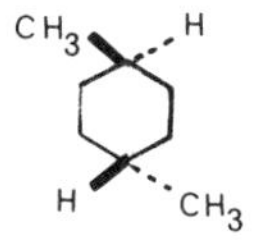

are diastereoisomers and *cis-trans* isomers; both are achiral

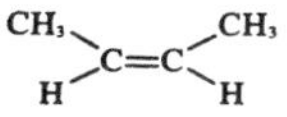

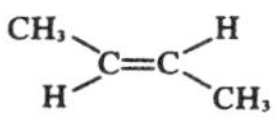

are diastereoisomers and *cis-trans* isomers; both are achiral

E-4.7. A compound whose individual molecules contain equal numbers of enantiomeric groups, identically linked, but no other chiral group, is termed a *meso*-compound.

Example:

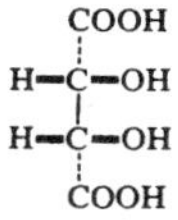

COOH
H—C—OH
H—C—OH
COOH
meso-Tartaric acid

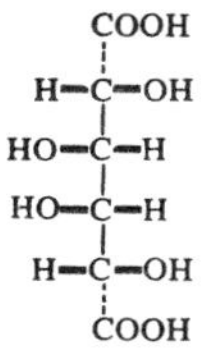

COOH
H—C—OH
HO—C—H
HO—C—H
H—C—OH
COOH
Galactaric acid

E-4.8. An atom is termed pseudoasymmetric when bonded tetrahedrally to one pair of enantiomeric groups (+)-a and (−)-a and also to two atoms or achiral groups b and c that are different from each other.

Examples:

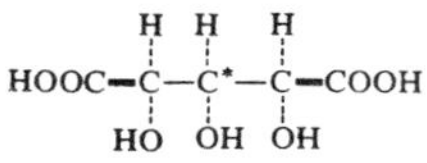

HOOC—C—C*—C—COOH with H H H and HO OH OH
(A)

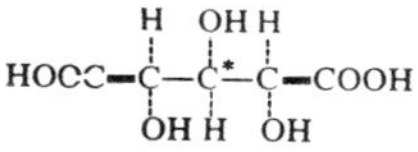

HOOC—C—C*—C—COOH with H OH H and OH H OH
(B)

C* are pseudoasymmetric

Notes: (1) The molecular structure around a pseudoasymmetric atom gives on reflexion an identical (superimposable) structure.

(2) Compounds differing at a pseudo-asymmetric atom belong to the larger class of diastereoisomers. Structures (A) and (B) are interconverted by interchange of the H and OH on C*. (A) and (B) are achiral diastereoisomers (see Rule E-4.6).

E-4.9. Names of chiral compounds whose absolute configuration is known are differentiated by prefixes R, S, etc., assigned by the sequence-rule procedure (see Appendix 2),* preceded when necessary by the appropriate locants.

Examples:

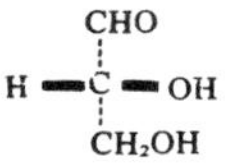

CHO
H—C—OH
CH₂OH
(R)-Glyceraldehyde

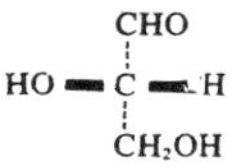

CHO
HO—C—H
CH₂OH
(S)-Glyceraldehyde

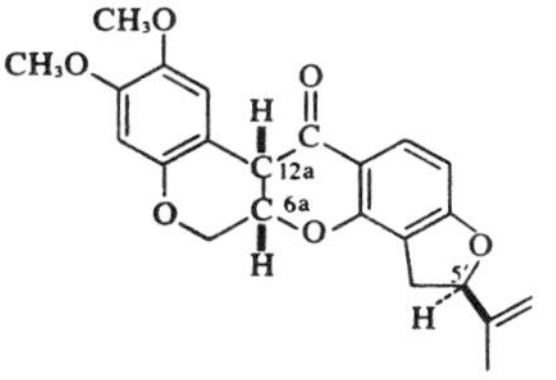

(6aS,12aS,5'R)-Rotenone

Methyl phenyl (R)-sulfoxide

* p.486.

E-4.10. (a) Chiral centres, of which the relative but not the absolute configuration is known, may be labelled arbitrarily by prefixes R^*, S^* (spoken R star, S star), preceded when necessary by the appropriate locants. These prefixes are assigned by the sequence-rule procedure (see Appendix 2)* on the arbitrary assumption that the centre of chirality with the lowest locant has chirality R.

(b) In complex cases the stars may be omitted and, instead, the whole name is prefixed by *rel-* (for *relative*).

(c) When a compound contains chiral centres with known absolute configurations and a sterically unrelated set of chiral centres with known relative configurations, then R^* and S^* must be used to designate the latter. The prefix *rel-* cannot be used.

This rule (E–4.10) does not form part of the Sequence Rule procedure formulated in the original papers (see Appendix 2)*.

Examples:

(a)

(1R^*,3S^*)-1-Bromo-3-chlorocyclohexane

(b)

rel-(1R,3R,5R)-1-Bromo-3-chloro-5-nitrocyclohexane

(c)

(1R^*,3R^*,5S^*)-[(1S)-*sec*-Butoxy]-3-chloro-5-nitro-cyclohexane

E-4.11. When it is desired to express relative or absolute configuration with respect to a class of compound, specialized local systems may be used. The sequence rule may, however, be used additionally for positions not amenable to treatment by the local system.

Examples:
erythro, threo, arabino, gluco, etc., combined when necessary with D or L for carbohydrates and their derivatives (see IUPAC/IUB Tentative Rules for Carbohydrate Nomenclature, *IUPAC Information Bull.* Appendix No. 7, 1970).

D, L for amino acids and peptides [see IUPAC/IUB Nomenclature of α-Amino-acids, Appendix No. 46 (September 1975) to *IUPAC Information Bull.*].

D, L and a series of other prefixes and trivial names for cyclitols and their derivatives [see IUPAC/IUB 1973 Recommendations for the Nomenclature of Cyclitols, *Pure and Applied Chem.* **37**, 285–297 (1974)].

α, β, and a series of trivial names for steroids and related compounds [see IUPAC/IUB, 1971 Recommendations for the Nomenclature of Steroids, *Pure and Applied Chem.* **31**, 283–322 (1972)].

* p.486.

The α, β system for steroids can be extended to other classes of compound such as terpenes and alkaloids when their absolute configurations are known; it can also be combined with stars or the use of a prefix *rel-* when only the relative configurations are known.

In spite of the Rules of Subsection E–2, *cis* and *trans* are used when the arrangement of the atoms constituting an unsaturated backbone is the most important factor, as, for instance, in polymer chemistry and for carotenoids. When a series of double bonds of the same stereochemistry occurs in a backbone, the prefix all-*cis* or all-*trans* may be used.

E-4.12. (a) An achiral object having at least one pair of features that can be distinguished only by reference to a chiral object or to a chiral reference frame is said to be prochiral, and the property of having such a pair of features is termed prochirality. A consequence is that if one of the features of the pair in a prochiral object is considered to differ from the other feature the resultant object is chiral.

(b) In a molecule an achiral centre or atom is said to be prochiral if it would be held to be chiral when two attached atoms or groups, that taken in isolation are indistinguishable, are considered to differ.

Notes: (1) For a tetrahedrally bonded atom prochirality requires a structure Xaabc (where none of the groups a, b, or c is the enantiomer of another).

(2) For a fuller exploration of the concept of prochirality, which is of particular importance to biochemists and spectroscopists, and for its extension to axes, planes, and unsaturated compounds, see K. R. Hanson, *J. Amer. Chem. Soc.* **88**, 2731 (1966); H. Hirschmann and K. R. Hanson, *J. Org. Chem.* **36**, 3293 (1971); *Europ. J. Biochem.* **22**, 301 (1971).

Examples:

In both examples (A) and (B) the methylene carbon atom is prochiral; in both cases it would be held to be a chiral centre if one of the methylene hydrogen atoms were considered to differ from the other. An actual replacement of one of these protium atoms by, say, deuterium would produce an actual chiral centre at the methylene carbon atom; as a result compound (A) would become chiral, and compound (B) would be converted into one of two diastereoisomers.

E-4.13. Of the identical pair of atoms or groups in a prochiral compound, that one which leads to an (R)-compound when considered to be preferred to the other by the sequence rule (without change in priority with respect to other ligands) is termed *pro-R*, and the other is termed *pro-S*.

Example:

H^1 is *pro-R*.
H^2 is *pro-S*.

E-5. Conformations

E-5.1. A molecule in a conformation into which its atoms return spontaneously after small displacements is termed a conformer.

Examples:

are different conformers

E-5.2. (a) When, in a six-membered saturated ring compound, atoms in relative positions 1, 2, 4, and 5 lie in one plane, the molecule is described as in the chair or boat conformation according as the other two atoms lie, respectively, on opposite sides or on the same side of that plane.

Examples:

Chair   Boat

Note: These and similar representations are idealized, minor divergences being neglected.

(b) A molecule of a monounsaturated six-membered ring compound is described as being in the half-chair or boat conformation according as the atoms not directly bound to the doubly bonded atoms lie, respectively, on opposite sides or on the same side of the plane containing the other four (adjacent) atoms.

Examples:

Half-chair   Boat*

(c) A median conformation through which one boat form passes during conversion into the other boat form is termed a twist conformation. Similar twist conformations are involved in conversion of a chair into a boat form or vice versa.

Example:

Boat   Twist   Boat

E-5.3. (a) Bonds to a tetrahedral atom in a six-membered ring are termed equatorial or axial according as they or their projections make a small or a large angle, respectively, with the plane containing a majority of the

---

*The term 'half-boat' has been used here.
†The terms axial, equatorial, pseudo-axial, and pseudo-equatorial [see Rule E-5.3(b)] may be used also in connexion with other than six-membered rings if, but only if, their interpretation is then still beyond dispute.

---

ring atoms†. Atoms or groups attached to such bonds are also said to be equatorial or axial, respectively.
Notes: (1) See, however, pseudo-equatorial and pseudo-axial [Rule E-5.3(b)].

(2) The terms equatorial and axial may be abbreviated to e and a when attached to formulae: these abbreviations may also be used in names of compounds and are there placed in parentheses after the appropriate locants, for example, 1(e)-bromo-4(a)-chlorocyclohexane.

Examples:

E-5.4

(b) Bonds from atoms directly attached to the doubly bonded atoms in a monounsaturated six-membered ring are termed pseudo-equatorial or pseudo-axial according as the angles that they make with the plane containing the majority of the ring atoms approximate to those made by, respectively, equatorial or axial bonds from a saturated six-membered ring. Pseudo-equatorial and pseudo-axial may be abbreviated to e' and a', respectively, when attached to formulae: these abbreviations may also be used in names, then being placed in parentheses after the appropriate locants.

Example:

E-5.4. Torsion angle: In an assembly of attached atoms X–A–B–Y, where neither X nor Y is collinear with A and B, the smaller angle subtended by the bonds X–A and Y–B in a plane projection obtained by viewing the assembly along the axis A–B is termed the torsion angle (denoted by the Greek lower-case letter theta θ or omega ω). The torsion angle is considered positive or negative according as the bond to the front atom X or Y requires to be rotated to the right or left, respectively, in order that its direction may coincide with that of the bond to the rear selected atom Y or X. The multiplicity of the bonding of the various atoms is irrelevant. A torsion angle also exists if the axis for rotation is formed by a collinear set of more than two atoms directly attached to each other.
Notes: (1) It is immaterial whether the projection be viewed from the front or the rear.

(2) For the use of torsion angles in describing molecules see Rule E-5.6.

Examples:
(For construction of Newman projections, as here, see Rule E-6.2):

$\theta = \sim +60°$ $\theta = \sim 0°$

$\theta = \sim 180°$

Newman projections of
propionaldehyde
$\theta = \sim -60°$ $\theta = \sim -120°$

Newman projection of
hydrogen peroxide
$\theta = \sim 180°$

E-5.5. If two atoms or groups attached at opposite ends of a bond appear one directly behind the other when the molecule is viewed along this bond, these atoms or groups are described as eclipsed, and that portion of the molecule is described as being in the eclipsed conformation. If not eclipsed, the atoms or groups and the conformation may be described as staggered.

Examples:

Eclipsed conformation.
The pairs a/a', b/b', and c/c' are eclipsed.

Staggered conformation.
All the attached groups are staggered.
In an ideal case the torsion angles are all 60°.

Projection of CH_3CH_2CHO.
The CH_3 and the H of the CHO are eclipsed.
The O and the H's of CH_2 in CH_2CH_3 are staggered.

E-5.6. Conformations are described as synperiplanar (*sp*), synclinal (*sc*), anticlinal (*ac*), or antiperiplanar (*ap*) according as the torsion angle is within ±30° of 0°, ±60°, ±120°, or ±180°, respectively; the letters in parentheses are the corresponding abbreviations. Atoms or groups are selected from each set to define the torsion angle according to the following criteria: (1) if all the atoms or groups of a set are different, that one of each set that is preferred by the sequence rule; (2) if one of a set is unique, that one; or (3) if all of a set are identical, that one which provides the smallest torsion angle.

---

*The terms *cis*, *gauche*, and *trans* (or their initial letters) have been used, especially in polymer chemistry, to indicate the approximate torsion angles shown below.

cis	c	0°
gauche	g	60°
trans	t	180°

Gauche may have + and − signs as superscripts (g^+, g^-). Since *cis* and *trans* are used in so many other ways, the Commission does not recommend their use in describing conformations. However, '*gauche*' may sometimes be convenient.
†The lone pair of electrons (represented by two dots) on the nitrogen atoms are the unique substituents that decide the description of the conformation (these are the "phantom atoms" of the sequence-rule symbolism).

Examples:

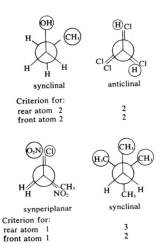

antiperiplanar   anticlinal   synclinal   synperiplanar
(*trans*)   (*gauche* or *skew*)   (*cis*)*

In the above conformations, all CH_2Cl-CH_2Cl, the two Cl atoms decide the torsion angle.

synclinal   anticlinal

Criterion for:

rear atom 2		2
front atom 2		2

synperiplanar   synclinal

Criterion for:

rear atom 1		3
front atom 1		2

The *ringed* atoms or groups are those used for determining the torsion angles in each case.

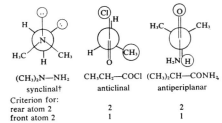

$(CH_3)_2N-NH_2$ CH_3CH_2-COCl $(CH_3)_2CH-CONH_2$
synclinal†   anticlinal   antiperiplanar

Criterion for:

rear atom 2	2	2
front atom 2	1	1

E-6. Stereoformulae

E-6.1. In a Fischer projection the atoms or groups attached to a tetrahedral centre are projected on to the plane of the paper from such an orientation that atoms or groups appearing above or below the central atom lie behind the plane of the paper and those appearing to left and right of the central atom lie in front of the plane of the paper, and that the principal chain appears vertical with the lowest-numbered chain member at the top.

Example:

CHO   1   CHO   CHO

H—C—OH 2   H——C——OH or H————OH

CH₂OH 3   CH₂OH   CH₂OH

Orientation   Fischer projection

Notes: (1) The first of the two types of Fischer projection should be used whenever convenient.

(2) If a formula in the Fischer projection is rotated through 180° in the plane of the paper, the upward and downward bonds from the central atom still project behind the plane of the paper, and the sideways bonds project in front of that plane. If, however, the formula is rotated through 90° in the plane of the paper, the upward and downward bonds now project in front of the plane of the paper and the sideways bonds project behind that plane. In the latter orientation it is essential to use thickened and dashed lines to avoid confusion.

E-6.2. To prepare a Newman projection a molecule is viewed along the bond between two atoms; a circle is used to represent these atoms, with lines from outside the circle towards its centre to represent bonds to other atoms; the lines that represent bonds to the nearer and the further atom end at, respectively, the centre and the circumference of the circle. When two such bonds would be coincident in the projection, they are drawn at a small angle to each other.*

Examples:

Perspective   Newman   Perspective   Newman
               projection   projection

E-6.3. General note. Formulae that display stereochemistry should be prepared with extra care so as to be unambiguous and, whenever possible, self-explanatory. It is inadvisable to try to lay down rules that will cover every case, but the following points should be borne in mind.

A thickened line (▬) denotes a bond projecting from the plane of the paper towards an observer, a broken line (----) denotes a bond projecting away from an observer, and, when this convention is used, a full line of normal thickness (——) denotes a bond lying in the plane of the paper. A wavy line (∿) may be used to denote a bond whose direction cannot be specified or, if it is explained in the text, a bond whose direction it is not desired to specify in the formula. Dotted lines (······) should preferably not be used to denote stereochemistry, and never when they are used in the same paper to denote mesomerism, intermediate states, etc. Wedges should not be used as complement to broken lines (but see below). Single large dots have sometimes been used to denote atoms or groups

attached at bridgehead positions and lying above the plane of the paper, with open circles to denote them lying below the plane of the paper, but this practice is strongly deprecated.

Hydrogen or other atoms or groups attached at sterically designated positions should never be omitted.

In chemical formulae, rings are usually drawn with lines of normal thickness, that is, as if they lay wholly in the plane of the paper even though this may be known not to be the case. In a formula such as (I) it is then clear that the H atoms attached at the A/B ring junction lie further from the observer than these bridgehead atoms, that the H atoms attached at the B/C ring junction lie nearer to the observer than those bridgehead atoms, and that X lies nearer to the observer than the neighbouring atom of ring C.

(I)   (II)

(III)

However, ambiguity can then sometimes arise, particularly when it is necessary to show stereochemistry within a group such as X attached to the rings that are drawn planar. For instance, in formula (II), the atoms O and C*, lying above the plane of the paper, are attached to ring B by thick bonds; but then, when showing the stereochemistry at C*, one finds that the bond *from C* to* ring B projects away from the observer and so should be a broken line. Such difficulties can be overcome by using wedges in places of lines, the broader end of the wedge being considered nearer to the observer, as in (III).

In some fields, notably for carbohydrates, rings are conveniently drawn as if they lay almost perpendicular to the plane of the paper, as shown in (IV); however, conventional formulae such as (V), with the lower bonds considered as the nearer to the observer, are so well established that it is rarely necessary to elaborate this to form (IV).

CH₂OH   CH₂OH

HO   O OH   HO   O OH

(IV)   (V)

By a similar convention, in drawings such as (VI) and (VII), the lower sets of bonds are considered to be nearer than the upper to the observer. In (VII), note the gaps in the rear lines to indicate that the bonds crossing them pass in front (and thus obscure sections of the rear bonds). In some cases, when atoms have to be shown as lying in

---

*Cf. M. S. Newman. *Chem. Progr. Kreskge-Hooker Sci. Lab. Rep.* **13**, 111 (1952); *J. Chem. Educ.* **32**, 344 (1955); *Steric Effects in Organic Chemistry.* John Wiley, New York 1956, p. 6. A similar projection was used earlier by J. Böeseken and R. Cohen, *Rec. Trav. chim.* **47**, 839 (1928).

several planes, the various conventions may be combined, as in (VIII).

(VI)　　　　　　　(VII)　　　　　　　(VIII)

In all cases the overriding aim should be clarity.

APPENDIX 1

Configuration and Conformation

See Rules E-1.4 and E-1.7.

Various definitions have been propounded to differentiate configurations from conformations.

The original usage was to consider as conformations those arrangements of the atoms of a molecule in space that can be interconverted by rotation(s) around a single bond, and as configurations those other arrangements whose interconversion by rotation requires bonds to be broken and then re-formed differently. Interconversion of different configurations will then be associated with substantial energies of activation, and the various species will be separable; but interconversion of different conformations will normally be associated with less activation energy, and the various species, if separable, will normally be more readily interconvertible. These differences in activation energy and stability are often large.

Nevertheless, rigid differentiation on such grounds meets formidable difficulties. Differentiation by energy criteria would require an arbitrary cut in a continuous series of values. Differentiation by stability of isolated species requires arbitrary assumptions about conditions and half-lives. Differentiation on the basis of rotation around single bonds meets difficulties connected both with the concept of rotation and with the selection of single bonds as requisites; and these need more detailed discussion here.

Enantiomeric biaryls are nowadays usually considered to differ in conformation, any difficulty in rotation about the 1,1'-bond due to steric hindrance between the neighbouring groups being considered to be overcome by bond bending and/or bond stretching, even though the movements required must closely approach bond breaking if these substituents are very large. Similar doubts about the possibility of rotation occur with a molecule such as (A), where rotation of the benzene ring around the oxygen-to-ring single bonds affords easy interconversion if x is large but appears to be physically impossible if x is small; and no critical size of x can be reasonably established. For reasons such as these, Rules E-1.4 and E-1.7 are so worded as to be independent of whether rotation appears physically feasible or not (see Note 2 to those Rules).

(A)　　　　　　　　　(B)

The second difficulty arises in the many cases where rotation is around a bond of fractional order between one and two, as in the helicenes, crowded aromatic molecules, metallocenes, amides, thioamides, and carbene-metal coordination compounds (such as B). The term conformation is customarily used in these cases and

---

*The ligancy of an atom refers to the number of neighbouring atoms bonded to it, irrespective of the nature of the bonds.

that appears a reasonable extension of the original conception, though it will be wise to specify the usage if the reader might be in doubt.

When interpreted in these ways, Rules E-1.4 and E-1.7 reflect the most frequent usage of the present day and provide clear distinctions in most situations.

Nevertheless, difficulties remain and a number of other usages have been introduced.

It appears to some workers that, once it is admitted that change of conformation may involve rotation about bonds of fractional order between one and two, it is then illogical to exclude rotation about classical double bonds because interconversion of open-chain *cis-trans*-isomers depends on no fundamentally new principle and is often relatively easy, as for certain alkene derivatives such as stilbenes and for azo compounds, by irradiation. This extension is indeed not excluded by Rules E-1.4 and E-1.7 but if it is applied that fact should be explicitly stated.

A further interpretation is to regard a stereoisomer possessing some degree of stability (that is, one associated with an energy hollow, however shallow) as a configurational isomer, the other arrangements in space being termed conformational isomers; the term conformer (Rule E-5.1) is then superfluous. This definition, however, requires a knowledge of stability (energy relations) that is not always available.

In another view a configurational isomer is any stereoisomer that can be isolated or (for some workers) whose existence can be established (for example, by physical methods); all other arrangements then represent conformational isomers. But it is then impossible to differentiate configuration from conformation without involving experimental efficiency or conditions of observation.

Yet another definition is to regard a conformation as a precise description of a configuration in terms of bond distances, bond angles, and torsion angles.

In none of the above views except the last is attention paid to extension or contraction of the bond to an atom that is attached to only one other atom, such as –H or =O. Yet such changes in interatomic distance due to non-bonded interactions may be important, for instance in hydrogen bonding, in differences due to crystal form, in association in solution, and in transition states. This area may repay further consideration.

Owing to the circumstances outlined above, the Rules E-1.4 and E-1.7 have been deliberately made imprecise, so as to permit some alternative interpretations; but they are not compatible with all the definitions mentioned above. The time does not seem ripe to legislate for other than the commoner usages or to choose finally between these.

It is, however, encouraging that no definition in this field has (yet) involved atomic vibrations for which, in all cases, only time-average positions are considered.

Finally it should be noted that an important school of thought uses conformation with the connotation of "a particular geometry of the molecule, i.e. a description of atoms in space in terms of bond distances, bond angles, and dihedral angles", a definition much wider than any discussed above.

APPENDIX 2

Outline of the Sequence Rule Procedure

The sequence rule procedure is a method of specifying the absolute molecular chirality (handedness) of a compound, that is, a method of specifying in which of two enantiomeric forms each chiral element of a molecule exists. For each chiral element in the molecule it provides a symbol, usually R or S, which is independent of nomenclature and numbering. These symbols define the chirality of the specific compound considered; they may not be the same for a compound and some of its derivatives, and they are not necessarily constant for chemically similar situations within a chemical or a biogenetic class. The procedure is applied directly to a three-dimensional model of the structure, and not to any two-dimensional projection thereof.

The method has been developed to cover all compounds with ligancy up to 4 and with ligancy 6,* and for all configurations and conformations of such compounds. The following is an outline confined to the most common situations; it is essential to study the original papers, especially the 1966 paper,† before using the sequence rule for other than fairly simple cases.

General basis. The sequence rule itself is a method of arranging atoms or groups (including chains and rings) in an order of precedence, often referred to as an order of preference; for discussion this order can conveniently be generalized as a > b > c > d, where > denotes "is preferred to".

The first step, however, in considering a model is to identify the nature and position of each chiral element that it contains. There are three types of chiral element, namely, the chiral centre, the chiral axis, and the chiral plane. The chiral centre, which is very much the most commonly met, is exemplified by an asymmetric carbon atom with the tetrahedral arrangement of ligands, as in (1). A chiral axis is present in, for instance, the chiral allenes such as

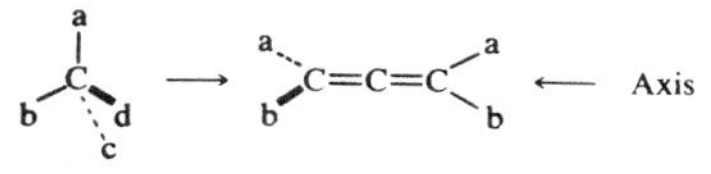

(1)   (2)

(2) or the chiral biaryl derivatives. A chiral plane is exemplified by the plane containing the benzene ring and the bromine and oxygen atoms in the chiral compound (3), or by the underlined atoms in the cycloalkene (4). Clearly, more than one type of chiral element

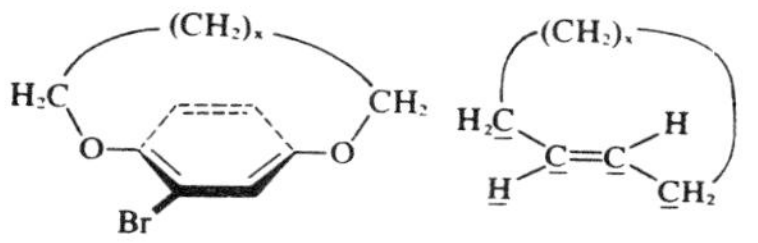

(3)   (4)

may be present in one compound; for instance, group "a" in (2) might be a *sec*-butyl group which contains a chiral centre.

The chiral centre. Let us consider first the simplest case, namely, a chiral centre (such as carbon) with four ligands, a, b, c, d which are all different atoms, tetrahedrally arranged, as in CHFClBr. The four ligands are arranged in order of preference by means of the sequence rule; this contains five sub-rules, which are applied in succession so far as necessary to obtain a decision. The first sub-rule is all that is required in a great majority of actual cases; it states that ligands are arranged in order of decreasing atomic number, in the above case (a) Br> (b) Cl> (c) F> (d) H. There would be two (enantiomeric) forms of the compound and we can write these as (5) and (6). In the sequence rule procedure

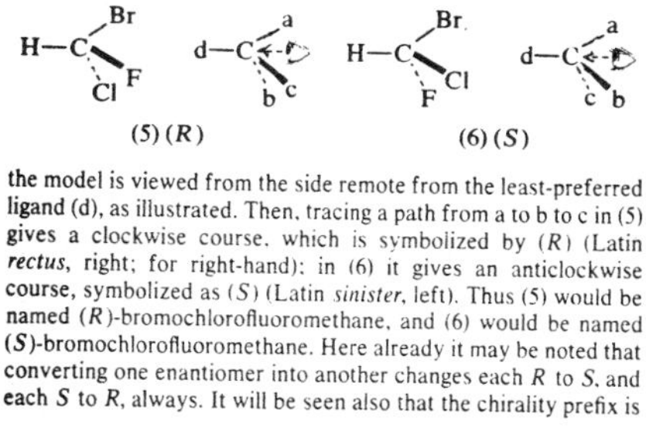

(5) (R)   (6) (S)

the model is viewed from the side remote from the least-preferred ligand (d), as illustrated. Then, tracing a path from a to b to c in (5) gives a clockwise course, which is symbolized by (R) (Latin *rectus*, right; for right-hand); in (6) it gives an anticlockwise course, symbolized as (S) (Latin *sinister*, left). Thus (5) would be named (R)-bromochlorofluoromethane, and (6) would be named (S)-bromochlorofluoromethane. Here already it may be noted that converting one enantiomer into another changes each R to S, and each S to R, always. It will be seen also that the chirality prefix is

---

†R. S. Cahn, (Sir) Christopher Ingold, V. Prelog, *Angew. Chem. intern. Edit.* **5**, 385 (1966) (in English); errata, *ibid.*, p. 511; *Angew. Chem.* **78**, 413 (1966) (in German). Earlier papers: R. S. Cahn, C. K. Ingold, *J. Chem. Soc. (London)* 612 (1951); R. S. Cahn, (Sir) Christopher Ingold, V. Prelog, *Experientia* **12**, 81 (1956). For a partial, simplified account see R. S. Cahn, *J. Chem. Educ.* **41**, 116 (1964).

the same whether the alphabetical order is used, as now recommended, for naming the substituents or whether this is done by an order of complexity (giving fluorochlorobromomethane).

Next, suppose we have H_3C-CHClF. We deal first with the atoms directly attached to the chiral centre; so the four ligands to be considered are Cl > F > C (of CH_3) > H. Here the H's of the CH_3 are not concerned, because we do not need them in order to assign our symbol.

However, atoms directly attached to a centre are often identical, as for example the underlined C's in H_3C–CHCl–CH_2OH. For such a compound we at once establish a preference (a) Cl > (b,c) C,C > (d) H. Then to decide between the two C's we work outwards, to the atoms to which they in turn are directly attached and we then find:

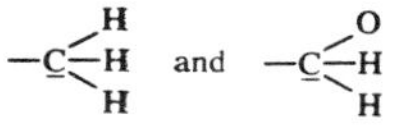

which we can conveniently write as C(H,H,H) and C(O,H,H). We have to compare H,H,H with O,H,H, and since oxygen has a higher atomic number than hydrogen we have O > H and thence the complete order Cl > C (of CH_2OH) > C (of CH_3) > H, so that the chirality symbol can then be determined from the three-dimensional model.

We must next meet the first complication. Suppose that we have a molecule (7):

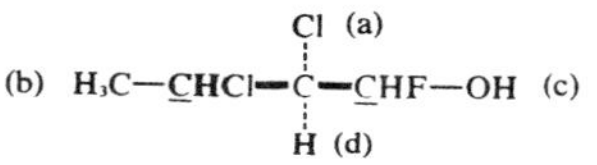

(7)   (S)

To decide between the two C's we first arrange the atoms attached to them in *their* order of preference, which gives C(Cl,C,H) on the left and C(F,O,H) on the right. Then we compare the preferred atom of one set (namely, Cl) with the preferred atom (F) of the other set; and as Cl > F we arrive at the preferences a > b > c > d shown in (7) and chirality (S). If, however, we had a compound (8):

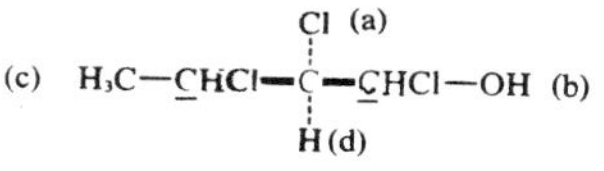

(8) (R)

we should have met C(Cl,C,H) and C(Cl,O,H) and, since the atoms of first preference are identical (Cl) we should have had to make the comparisons with the atoms of second preference, namely, O > C, which leads to the different chirality (R) as shown in (8).

Branched ligands are treated similarly. Setting them out in full gives a picture that at first sight looks complex but the treatment is in fact simple. For instance, in compound (9) a first quick glance again shows (a) Cl > (b,c) C,C > (d) H.

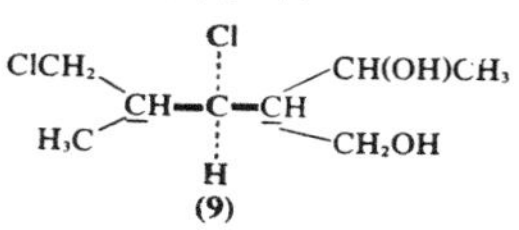

(9)

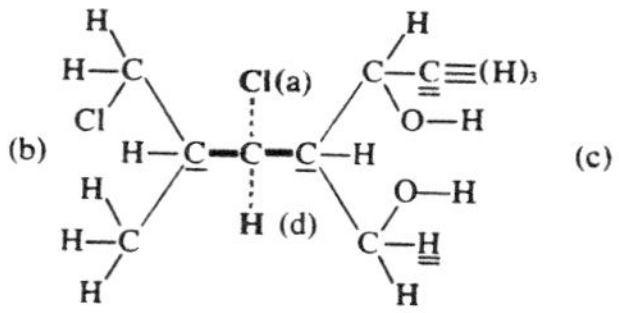

(9)   (S)

When we expand the two C's we find they are both C(C,C,H), so we continue exploration. Considering first the left-hand ligand we arrange the branches and their sets of atoms in order thus: C(Cl,H,H) > C(H,H,H); and on the right-hand side we have C(O,C,H) > C(O,H,H) (because C > H). We compare first the preferred of these branches from each side and we find C(Cl,H,H) > C(O,C,H) because Cl > O, and that gives the left-hand branch preference over the right-hand branch. That is all we need to do to establish chirality (S) for this highly branched compound (9). Note that it is immaterial here that, for the lower branches, the right-hand C(O,H,H) would have been preferred to the left-hand C(H,H,H); we did not need to reach that point in our comparisons and so we are not concerned with it; but we should have reached it if the two top (preferred) branches had both been the same CH₂Cl.

Rings, when met during outward exploration, are treated in the same way as branched chains.

With these simple procedures alone, quite complex structures can be handled; for instance, the analysis alongside formula (10) for natural morphine explains why the specification is as shown. The reason for considering C-12 as C(CCC) is set out in the next paragraphs.

Thus in D-glyceraldehyde (11) the CHO group is treated as C(O,(O),H) and is thus preferred to the C(O,H,H) of the CH₂OH group, so that the chirality symbol is (R).

$$\begin{array}{ccc} & CHO & (b) \\ (d) & H-C-OH & (a) \\ & CH_2OH & (c) \end{array}$$

D-Glyceraldehyde (11) (R)

Only the doubly bonded atoms themselves are duplicated, and not the atoms or groups attached to them; the duplicated atoms may thus be considered as carrying three phantom atoms (see below) of atomic number zero. This may be important in deciding preferences in certain complicated cases.

Aromatic rings are treated as Kekulé structures. For aromatic hydrocarbon rings it is immaterial which Kekulé structure is used because "splitting" the double bonds gives the same result in all cases; for instance, for phenyl the result can be represented as (12a) where "(6)" denotes the atomic number of the duplicate representations of carbon.

(12)   (13)   (12a)

For aromatic hetero rings, each duplicate is given an atomic number that is the mean of what it would have if the double bond were located at each of the possible positions. A complex case is illustrated in (13). Here C-1 is doubly bonded to one or other of the nitrogen atoms (atomic number 7) and never to carbon, so its added duplicate has atomic number 7; C-3 is doubly bonded either to C-4 (atomic number 6) or to N-2 (atomic number 7), so its added duplicate has atomic number 6½; so has that of C-8; but C-4a may be doubly bonded to C-4, C-5, or N-9, so its added duplicate has atomic number 6 ⅓.

One last point about the chiral centre may be added here. Except for hydrogen, ligancy, if not already four, is made up to four by adding "phantom atoms" which have atomic number zero (0) and are thus always last in order of preference. This has various uses but perhaps the most interesting is where nitrogen occurs in a rigid skeleton, as for example in α-isosparteine (14); here the phantom atom can be placed where the nitrogen lone pair of electrons is; then N-1 appears as shown alongside the formula, and the chirality (R) is the consequence; the same applies to N-16. Phantom atoms are similarly used when assigning chirality symbols to chiral sulfoxides (see example to Rule E-4.9).

(10)   (5R,6S,9R,13S,14R)-Morphine

Now, using the sequence rule depends on exploring along bonds. To avoid theoretical arguments about the nature of bonds, simple classical forms are used. Double and triple bonds are split into two and three bonds respectively. A >C=O group is treated as >C-O, where the (O) and the (C) are duplicate representations of the atoms at the other end of the double bond. -C≡CH is treated as

$$\begin{array}{cccc} -C & -CH \\ (C) & (C) & (C) & (C) \end{array}$$

and —C≡N is treated as

$$\begin{array}{cc} -C & N \\ (N)(N) & (C)(C) \end{array}$$

(14)   (1R,6R,7S,9S,11R,16R)-Sparteine

Symbolism. In names of compounds. the *R* and *S* symbols, together with their locants. are placed in parentheses. normally in front of the name. as shown for morphine (10) and sparteine (14): but this may be varied in indexes or in languages other than English. Positions within names are required. however. when more than a single series of numerals is used. as for esters and amines. When relative stereochemistry is more important than absolute stereochemistry. as for steroids or carbohydrates. a local system of stereochemical designation may be more useful and sequence rule symbols need then be used only for any situations where the local system is insufficient.

Racemates containing a single centre are labelled (*RS*). If there is more than one centre the first is labelled (*RS*) and the others are (*RS*) or (*SR*) according to whether they are *R* or *S* when the first is *R*. For instance. the 2,4-pentanediols CH_3–CH(OH)–CH_2–CH(OH)–CH_3 are differentiated as:

one chiral form	(2*R*,4*R*)-
other chiral form	(2*S*,4*S*)-
meso-compound	(2*R*,4*S*)-
racemic compound	(2*RS*,4*RS*)-

Finally the principles by which some of the least rare of other situations are treated will be very briefly summarized.

Pseudoasymmetric atoms. A sub-rule decrees that *R* groups have preference over *S* groups and this permits pseudo-asymmetric atoms. as in Cab(*c-R*)(*c-S*) to be treated in the same way as chiral centres: but as such a molecule is achiral (not optically active) it is given the lower-case symbol *r* or *s*.

Chiral axis. The structure is regarded as an elongated tetrahedron and viewed along the axis—it is immaterial from which end it is viewed: the nearer pair of ligands receives the first two positions in the order of preference, as shown in (15) and (16).

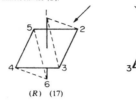

(15)

(16)

Chiral plane. The sequence-rule-preferred atom directly attached to the plane is chosen as "pilot atom". In compound (3)

(p. 487) this is the C on the left-hand CH_2 group. Now this is attached to the left-hand oxygen atom in the plane. The sequence rule-preferred path from this oxygen atom is then explored in the plane until a rotation is traced which is clockwise (*R*) or anticlockwise (*S*) when viewed from the pilot atom. In (3) this path is O→C→C(Br) and it is clockwise (*R*).

Other sub-rules. Other sub-rules cater for new chirality created by isotopic labelling (higher mass number preferred to lower) and for steric differences in the ligands. Isotopic labelling rarely changes symbols allotted to other centres.

Octahedral structures. Extensions of the sequence rule enable ligands arranged octahedrally to be placed in an order of preference. including polydentate ligands, so that a chiral structure can then always be represented as one of the enantiomeric forms (17) and (18). The face 1–2–3 is observed from the side remote from the face 4–5–6 (as marked by arrows). and the path 1→2→3 is observed; in (17) this path is clockwise (*R*), and in (18) it is anticlockwise (*S*).

(*R*) (17)   (18) (*S*)

Conformations. The torsion angle between selected bonds from two singly bonded atoms is considered. The selected bond from each of these two atoms is that to a unique ligand. or otherwise to the ligand preferred by the sequence rule. The smaller rotation needed to make the front ligand eclipsed with the rear one is noted (this is the rotatory characteristic of a helix); if this rotation is right-handed it leads to a symbol *P* (plus); if left-handed to *M* (minus). Examples are:

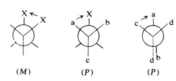

(*M*)   (*P*)   (*P*)

Details and complications. For details and complicating factors the original papers should be consulted. They include treatment of compounds with high symmetry or containing repeating units (e.g cyclitols), also π-bonding (metallocenes. etc.). mesomeric compounds and mesomeric radicals, and helical and other secondary structures.

Some common groups in order of sequence·rule preference
Note: ANY alteration to structure. or substitution. etc.. may alter the order of preference.

A. *Alphabetical order:* higher number denotes greater preference

64 Acetoxy	75 Bromo	46 Ethylamino
36 Acetyl	42 *tert*-Butoxycarbonyl	21 Ethynyl
48 Acetylamino	5 Butyl	68 Fluoro
	16 *sec*-Butyl	35 Formyl
10 Allyl	19 *tert*-Butyl	63 Formyloxy
43 Amino	38 Carboxy	62 Glycosyloxy
44 Ammonio ⁺H_3N-	74 Chloro	7 Hexyl
37 Benzoyl	17 Cyclohexyl	1 Hydrogen
49 Benzoylamino	52 Diethylamino	57 Hydroxy
65 Benzoyloxy	51 Dimethylamino	76 Iodo
50 Benzyloxycarbonylamino	34 2,4-Dinitrophenyl	9 Isobutyl
13 Benzyl	28 3,5-Dinitrophenyl	8 Isopentyl
60 Benzyloxy	59 Ethoxy	20 Isopropenyl.
41 Benzyloxycarbonyl	40 Ethoxycarbonyl	14 Isopropyl
	3 Ethyl	

69 Mercapto
58 Methoxy
39 Methoxycarbonyl
2 Methyl
45 Methylamino
71 Methylsulfinyl
66 Methylsulfinyloxy
72 Methylsulfonyl
67 Methylsulfonyloxy
70 Methylthio
11 Neopentyl
56 Nitro

27 m-Nitrophenyl
33 o-Nitrophenyl
24 p-Nitrophenyl
55 Nitroso
6 Pentyl
61 Phenoxy
22 Phenyl
47 Phenylamino
54 Phenylazo
18 1-Propenyl
4 Propyl

29 1-Propynyl
12 2-Propynyl
73 Sulfo
25 m-Tolyl
30 o-Tolyl
23 p-Tolyl
53 Trimethylammonio
32 Trityl
15 Vinyl
31 2,6-Xylyl
26 3,5-Xylyl

B. *Increasing order of sequence-rule preference*

1 Hydrogen
2 Methyl
3 Ethyl
4 Propyl
5 Butyl
6 Pentyl
7 Hexyl
8 Isopentyl
9 Isobutyl
10 Allyl
11 Neopentyl
12 2-Propynyl
13 Benzyl
14 Isopropyl
15 Vinyl
16 *sec*-Butyl
17 Cyclohexyl
18 1-Propenyl
19 *tert*-Butyl
20 Isopropenyl
21 Ethynyl
22 Phenyl
23 p-Tolyl
24 p-Nitrophenyl
25 m-Tolyl
26 3,5-Xylyl

27 m-Nitrophenyl
28 3,5-Dinitrophenyl
29 1-Propynyl
30 o-Tolyl
31 2,6-Xylyl
32 Trityl
33 o-Nitrophenyl
34 2,4-Dinitrophenyl
35 Formyl
36 Acetyl
37 Benzoyl
38 Carboxy
39 Methoxycarbonyl*
40 Ethoxycarbonyl*
41 Benzyloxycarbonyl*
42 *tert*-Butoxycarbonyl*
43 Amino
44 Ammonio ⁺H₃N–
45 Methylamino
46 Ethylamino
47 Phenylamino
48 Acetylamino
49 Benzoylamino
50 Benzyloxycarbonylamino
51 Dimethylamino
52 Diethylamino

53 Trimethylammonio
54 Phenylazo
55 Nitroso
56 Nitro
57 Hydroxy
58 Methoxy
59 Ethoxy
60 Benzyloxy
61 Phenoxy
62 Glycosyloxy
63 Formyloxy
64 Acetoxy
65 Benzoyloxy
66 Methylsulfinyloxy
67 Methylsulfonyloxy
68 Fluoro
69 Mercapto HS–
70 Methylthio CH₃S–
71 Methylsulfinyl
72 Methylsulfonyl
73 Sulfo HO₃S–
74 Chloro
75 Bromo
76 Iodo

*These groups are RO—C—
‖
O

490

SECTION F : GENERAL PRINCIPLES FOR THE NAMING OF NATURAL PRODUCTS AND RELATED COMPOUNDS

(Provisional Recommendations 1976)

PREAMBLE

In the field of natural products, three levels of nomenclature usage are recognized. A new compound, isolated from a natural source, is generally given a *trivial* name. By common usage, these trivial names are normally related to the biological origin of the material, but frequently not in any rational way, since the available structural information is insufficient.

Since such trivial names often carry either no structural information or misleading indications and their multiplicity constitutes a strain on information processes, both human and mechanical, they should preferably be considered as ephemeral and replaced for chemical purposes by *semisystematic* names, created according to systematic procedures, as soon as the structural class and details of functional groups, and so on, become available. Alternatively, if the structures are sufficiently simple, *systematic* names based on Sections A — C and Section D should be used.

Examples:

Thus flemichapparin-C, flemichapparin-B and medigol are all closely related coumestans, whose structures are completely known and which have been interconverted. The names used should give the information that the compounds have the following structures:

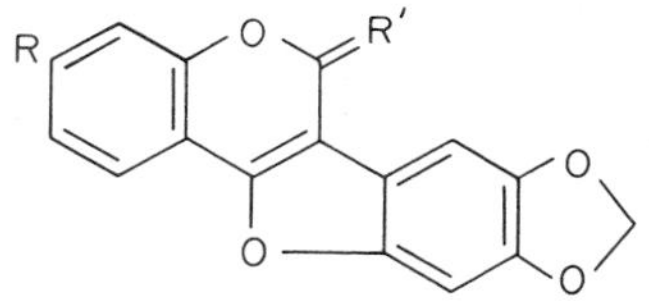

I

Flemichapparin-C I; R = OMe, R′ = O
Flemichapparin-B I; R = OMe, R′ = H_2
Medigol I; R = OH, R′ = O

Sclarene (2) is labda-8(17),13(16),14-triene

Opuntiol (3) is 6-(hydroxymethyl)-4-methoxy-2*H*-pyran-2-one
(Rule C-314)

The structure of any naturally occurring compound can be
defined in terms of
    (a)   a fundamental or slightly modified acyclic, homo-
        cyclic, or heterocyclic parent structure,
    (b)   the degree of hydrogenation of the natural product
        compared with the parent structure,
    (c)   the groups attached to the parent structure,
    (d)   the stereochemical configuration at all chiral
        centres or other chiral elements
Most known naturally occurring compounds belong to one of a
limited number of structural classes. Such classes are generally
well defined and each group can be characterized by a set of
parent structures that are closely related structurally.
Treatment of features (b) - (d) is standard and independent
of the natural product class; indeed much is common to all
organic nomenclature. These rules are therefore in three parts:
    (a)   general principles for the coining and use of biolog-
        ically based trivial names,

(b)   rules for the selection, naming and numbering of parent
      structures that will form the basis of the semisystem-
      atic nomenclature (these parent structures will be de-
      fined in separate publications. For special rules
      for steroids, carotenoids, carbohydrates, etc., see
      '"Biochemical Nomenclature and Related Documents,
      1978", The Biochemical Society, London*.

(c)   rules of general applicability for the derivation of
      semisystematic names for natural products.

RULES

Rule F-1   Biologically Based Trivial Names

F-1.1   When a compound is isolated from a natural source and
its structure or novelty is unknown, it may be accorded a trivial name, based on the family or genus or species name of the
biological material from which it has been isolated. The biological name will be the binominal Latin one recognized or authorized by the appropriate international authority (for example,
Index Kewensis, Index Animalium, Nomenclator Zoologicus)

Examples:
      Opuntiol from *Opuntia eliator*
      Laurene from *Laurencia glandulifera*
      Sativene from *Helminthosporium sativum*

F-1.2   The trivial name should not give a false implication as
to structure or identity of principal group(s) present.

F-1.3   The following groups of letters have significance as
terminations in general organic chemistry and should not be used
as terminations to the trivial names coined for natural products
unless
      (a)   it is known that the natural product contains the
            group usually designated by that termination,
      (b)   that group is the principal group present (see Rules
            C-10.2 ─ C-10.5).

-	-	al	-	am	an	ane	-	ate	-
-	-	-	-	-	en	ene	et	ete	-
ic	ide	-	ile	-	in	ine	-	-	ium
-	-	ol	ole	-	-	one	-	-	-
-	yde	yl	-	-	-	yne	-	-	-

---

*Available from Biochemical Society Depot, PO Box 32, Commerce
Way, Colchester, Essex CO2 8HP, UK.

Example:

> A newly isolated natural product in which an oxo group
> [$>$C(:O)] contains the only heteroatom may be given the
> termination -one.

F-1.4   Preferably, the trivial name of a natural product of unknown structure should end in: un, une or ur, since these are
terminations having no structural implications.

F-1.5   If it is subsequently found that the structure of the
compound is identical with or can be derived from a known structure, the trivial name accorded later should be abandoned in
favour of the previously recorded one or a name should be derived
systematically from the existing name of the known structure.

F-1.6   If subsequent study confirms the novelty of the compound,
its trivial name should be replaced by the semisystematic or
systematic name as soon as sufficient detail of the structure is
known.

Rule F-2   Parent Structures for Semisystematic Names

F-2.1   The semisystematic name of a naturally occurring compound, or of a synthetic derivative or artefact, is based on the
name of an appropriate parent structure. [Parent structures and
their names will be listed in separate publications (to be
published later, after full consultation with workers in the
fields concerned].

To the name of the parent structure chosen as basis are
added indications of any modification of that parent structure,
of the state of hydrogenation, of any functional groups present,
and of any stereochemical configuration not already implied.
Unless specified to the contrary, the principles of Sections A —
E are followed.

F-2.2   The fundamental parent structures are usually chosen to
relate to the structure of the key, or biogenetically simple,
compounds in a given class of natural product, by

    (a)   removal of all characteristic, functional, or non-
hydrocarbon groups or radicals, but including heterocyclic rings as appropriate;

    (b)   retaining hydrocarbon groups attached to the residual
skeleton if their presence results in asymmetry at
their points of attachment to the skeleton;

(c) expressing the residual skeleton usually in either the
highest or the lowest possible state of hydrogenation.
The former is preferable because it permits the maximum amount of stereochemistry implied in the parent
structure.

F-2.3   The semisystematic name corresponding to a parent structure is derived from the trivial name of the key compound to
which the parent structure is related, by changing the root of
the trivial name to a termination of the following type:
    If the parent structure is hydrocarbon or carbocyclic, to
'ane' for a saturated system,
'ene' for an unsaturated system.
    If the parent structure contains a heteroatom(s), to
'an' for a saturated system,
'en' for an unsaturated system.
    For parent structures containing nitrogen, the following
have been used but are *not* recommended:
'anine' for a saturated system,
'enin' or 'enine' for an unsaturated system.

Examples:

Lupeol (4), a trivial name, provides lupane (5)
        as the semisystematic parent name

F-2.4   For many parent structures corresponding to the trivial
name there are well-established numberings, based either on biogenetic considerations or on application of the principles of
Sections A — C.

Where no numbering system, or more than one exists, it is
recommended that the principles of Sections A - C be followed.
It is essential to give an individual locant to every ring position, including valley positions and heteroatoms. Of the two
groups at a *gem*-disubstituted position, the lower locant is
given to the α-group (see Rule F-6.2).

F-2.5   Certain modifications of parent structures require the
use of ring-designators rather than atom locants. The letter *A*
is assigned to the ring that contains the lowest-numbered carbon
atom.

Rule F-3   Degree of Saturation or Unsaturation

Normally the parent structure used as basis of the name of
the natural product will be either fully saturated or will contain the maximum number of non-cumulative double bonds.

F-3.1   If the parent structure contains the maximum number of
non-cumulative double bonds (*cf*. Rule A-21), any other state of
hydrogenation can generally be indicated by use of the prefix
dihydro, tetrahydro, etc., together with the necessary locants,
according to Rule A-23.

F-3.2   If the parent structure is fully saturated and its name
ends in 'ane' or 'an', unsaturation in the natural product is
indicated by replacing the termination 'ane' or 'an' by 'ene',
'en', or 'adiene', 'adien', etc.

F-3.3   If the parent structure is fully saturated and its name
does *not* end in 'ane' or 'an', unsaturation in the natural
product requires use of the prefix didehydro, tetradehydro, etc.
together with necessary locants, according to Rule C-41.2.

F-3.4   When 'anine' is used as a termination for a parent
structure containing nitrogen (Rule F-2.3), unsaturation is indicated by the terminations 'enine', 'adienine', etc.

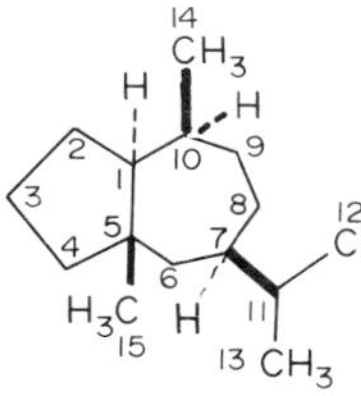

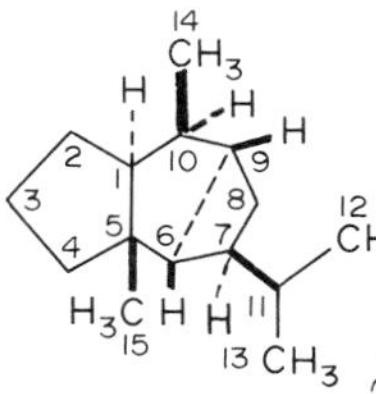

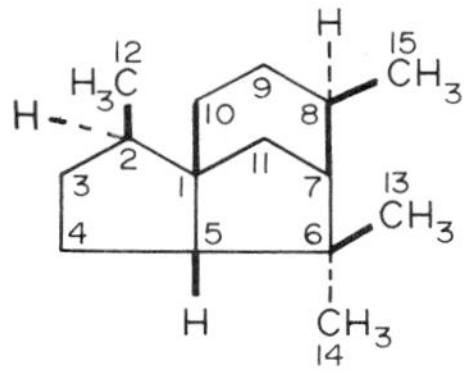

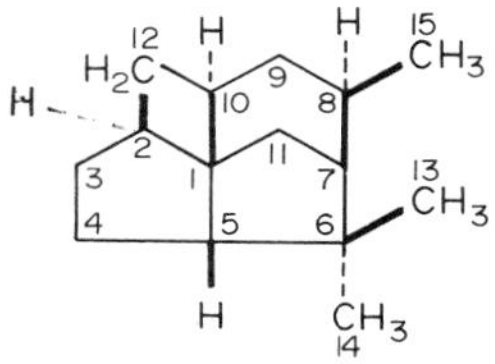

Rule F-4   Modification of the Parent Structure
~~~~~~~

F-4.1   When an additional ring is formed by means of a direct
link between any two atoms of a parent structure, the name of the
parent structure is prefixed by 'cyclo'; this prefix is preceded
by the locants of the positions joined by the new bond and, when
appropriate, the Greek letter (α, β, or ξ) denoting the
configurations at the ends of the new bond and not the hydrogen
atoms, unless that designation is already implicit in the name.

Examples:

Ambrosane

6α,9α–Cycloambrosane

Cedrane   8

10β,12–Cyclocedrane   9

F-4.2   Elimination of one methylene group from a side chain of a
parent structure (including a methyl group) is indicated by the
prefix 'nor', which in all cases is preceded by the locant of the
carbon atom that disappears. When alternatives are possible, the
locant attached to 'nor' is the highest permissible. Elimination
of two or more methylene groups is indicated by the prefix
'dinor', 'trinor', again with citation of all locants.

Examples:

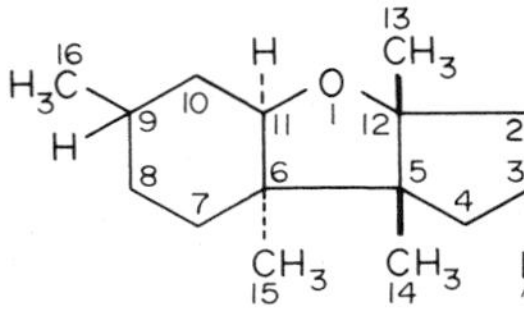

10

Apotrichothecane

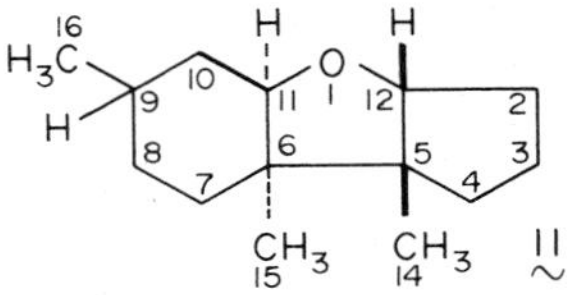

11

13-Norapotrichothecane

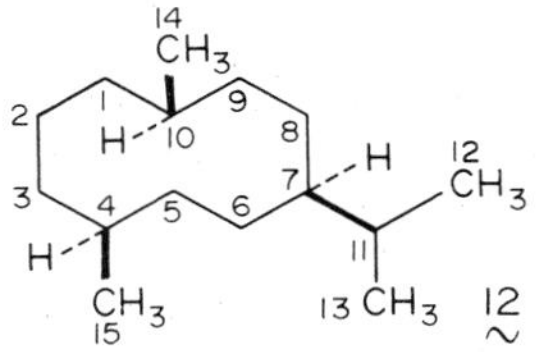

12

Germacrane

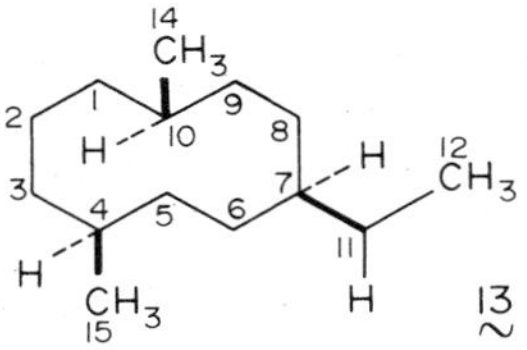

13

13-Norgermacrane

Note: The special usage for monoterpenes (see Rules A-72.1 and A-74.2), whereby the prefix 'nor' (without multiplying prefix) denotes replacement by hydrogen of all methyl groups attached to a ring system, is *abolished* (see Rule C-42.1) and Rules A-72 to A-75 are superseded.

Examples:

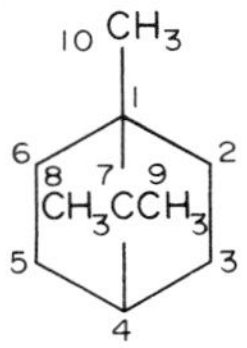

14

Bornane

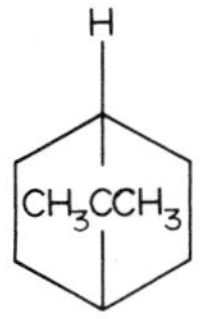

15

10-Norbornane

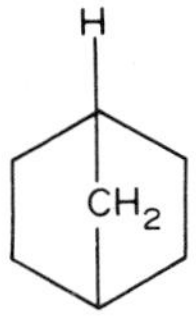

16

8,9,10-Trinorbornane
(previously norbornane)

F-4.3   A sequence of ring atoms, linked together as an unbranched chain and connecting two bridgeheads or junctions, but
not itself containing any bridgeheads or junction, is termed an
atomic ring-sector.

A bond linking together directly two bridgeheads or junctions is termed a *bond ring-sector*.

Example:

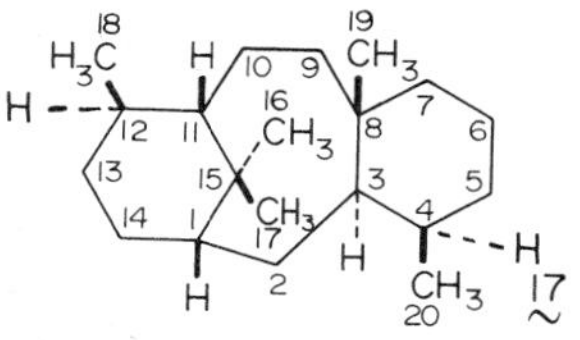

Taxane

In the taxane (17) skeleton the following sequences constitute atomic ring-sectors. The numerals in parentheses are the
locants of the bridgehead atoms; these atoms do not form part of
the sectors.

(1) 2 (3)

(3) 4, 5, 6, 7, (8)

(8) 9, 10, (11)

(11) 12, 13, 14, (1)
(1) 15 (11)

The sequence (3) (8) is a bond ring-sector.

F-4.4   Ring contraction by loss of an unsubstituted methylene
group is indicated by the prefix 'nor'. For loss of two methylene groups, 'dinor' is used.

The methylene group(s) lost is considered to be the highest-
numbered unsubstituted atom of the ring-sector involved and the
locant of that lost methylene group precedes the prefix 'nor' in
the name of the nor-compound. The remainder of the original
numbering is retained.

Examples:

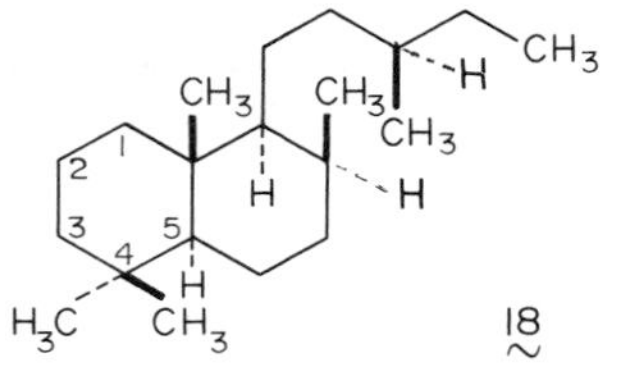

Labdane   18

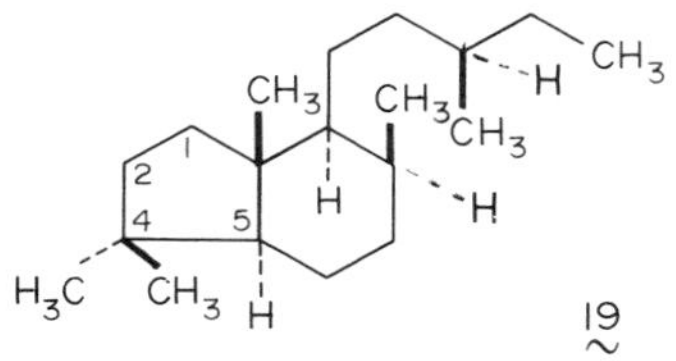

3-Norlabdane   19

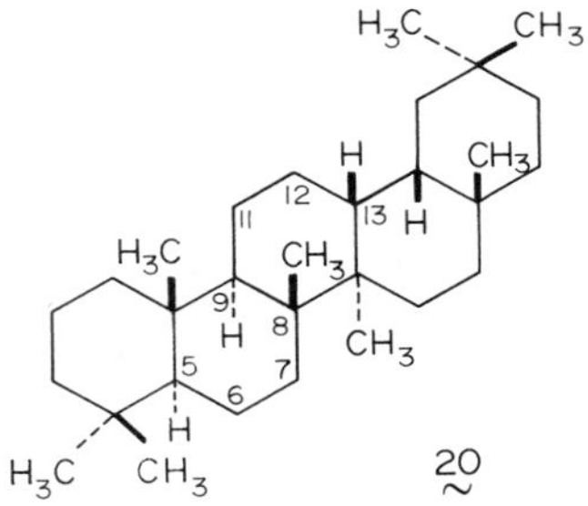

Oleanane   20

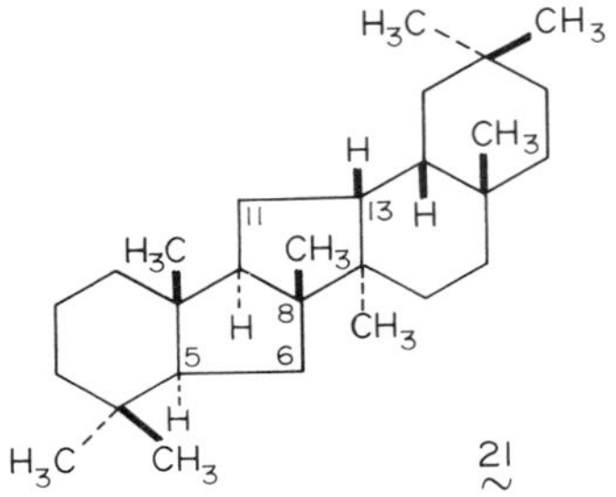

7,12-Dinoroleanane   21

F-4.5   Ring expansion by inclusion of a methylene group into a ring-sector is indicated by the prefix 'homo'. For inclusion of two methylene groups, 'dihomo' is used.

These prefixes are preceded by (complex) locants, indicating the position of the inserted groups, which are chosen as follows:

  (a)   insertion into an *atomic* ring-sector, by adding the letter 'a', 'b', etc. to the locant of the highest numbered atom of the ring-sector *that is not a bridgehead or junction*;

  (b)   insertion into a *bond* ring-sector by adding the letter 'a', 'b', etc. to the pair of locants indicating the two bridgeheads or junctions defining the sector, that of the higher-numbered bridgehead or junction being placed in parentheses.

The original numbering is retained for all other atoms.

Examples:

Parent Structure

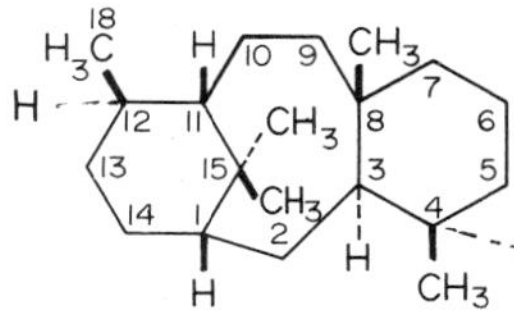

Taxane

Insertion into atomic ring-sectors

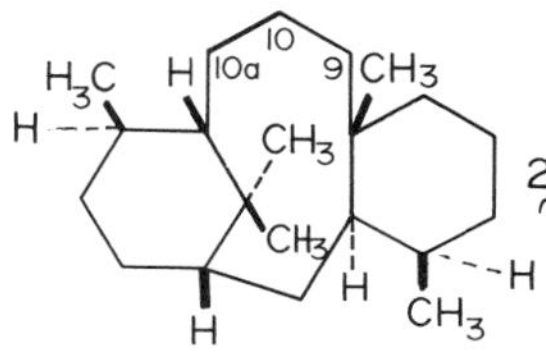

23   10a – Homotaxane

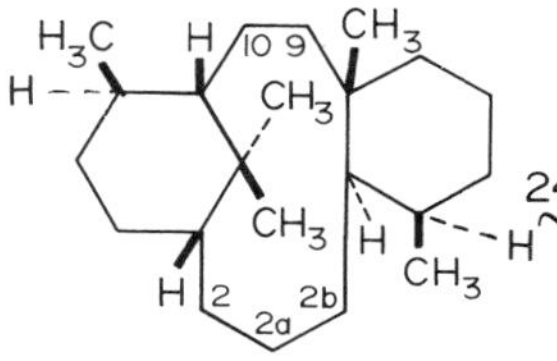

24   2a, 2b – Dihomotaxane

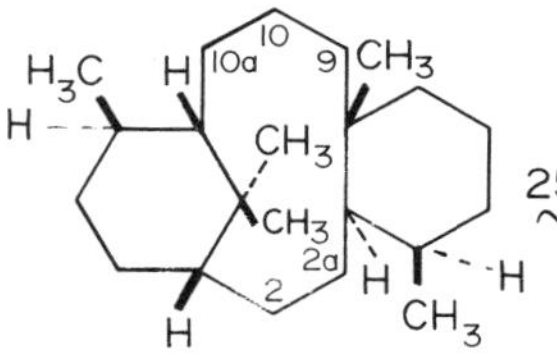

25   2a, 10a – Dihomotaxane

Insertion into bond ring-sectors

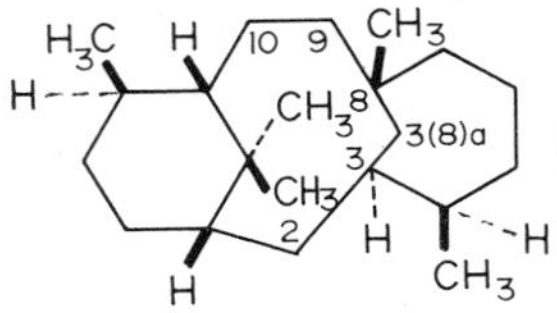

26 3(8)a-Homotaxane

F-4.6   The use of the prefixed 'nor' and 'homo' is limited to
modification of one or two rings only, in any parent structure
having four or more rings, or to one such modification if there
are three or fewer rings in the parent structure.

F-4.7   Fission of a ring, with addition of a hydrogen atom at
each terminal group thus created, is indicated by the prefix
'seco' with the appropriate locants; the original numbering is
retained.

Examples:

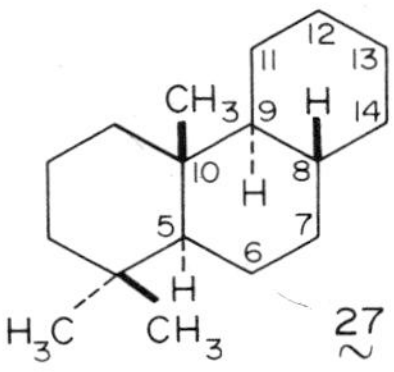

Podocarpane   27

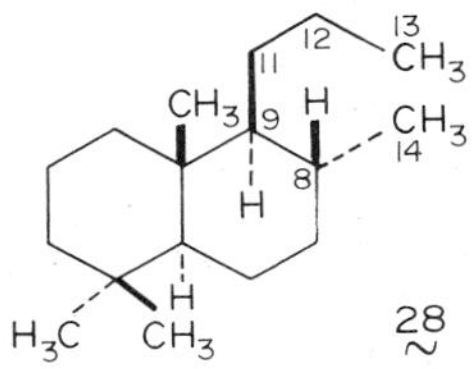

13,14 - Secopodocarpane   28

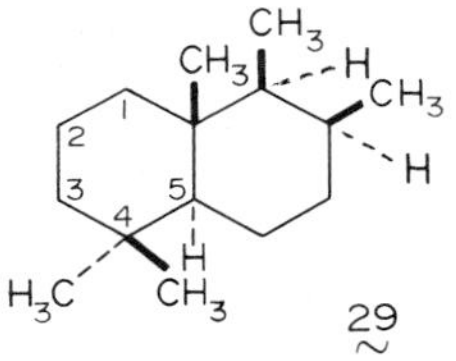

Drimane   29

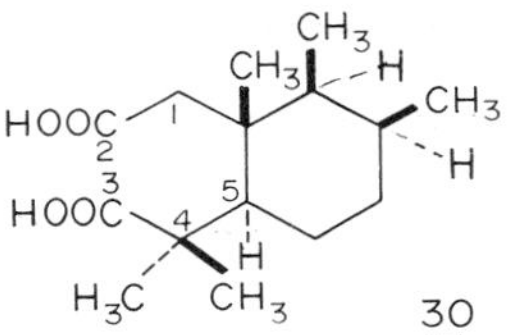

2,3-Secodrimane-2,3-
dioic acid   30

5α−Androstane

17−Nor−15,16−seco−5α−androstane−
15,16−dioic acid

F-4.8 In naming modified structures for which both 'homo-nor'
and 'cyclo-seco' names can be constructed, the 'homo-nor' names
are preferred.

F-4.9 A compound that does not possess a standard parent
structure but may be considered formally to arise from a standard parent structure by bond migration may be given the name
laid down in these Rules for the standard parent structure, to
which is attached a prefix of the form x(y→z)abeo. This prefix
is compiled as follows: A numeral denoting the stationary (un-
changed) end of the migrating bond (x) is followed by parentheses enclosing (i) the locant denoting the original position
(y) from which the other end of this bond has migrated, (ii) an
arrow, and (iii) the locant (z) denoting the new position to
which the bond has moved. The closing parenthesis is followed
by abeo- (Latin, I go away) (italicized) to indicate bond migration. The original numbering is retained for the new compound
and is used for the locants x, y, and z.

Such of the customary letters (R, S, α, β, etc.) as are
necessary are added to specify the resulting stereochemical configurations.

Note: The abeo nomenclature described in this Rule is permissive,
not compulsory. It is most suitable for use in discussions on
reaction mechanism and biogenesis.

Examples:

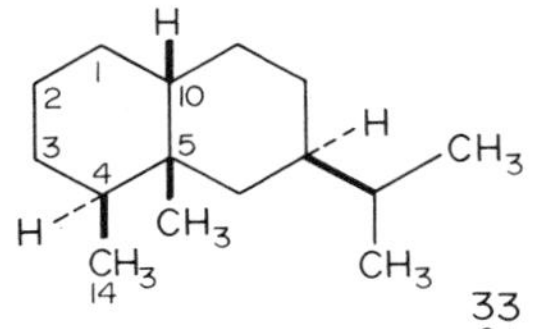

Eremophilane

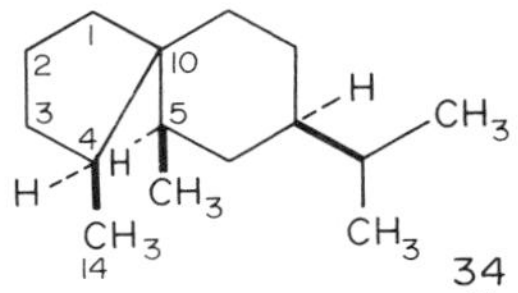

4(5 → 10) *abeo*- Eremophilane

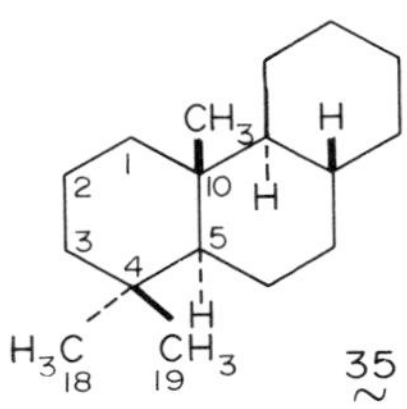

Podocarpane

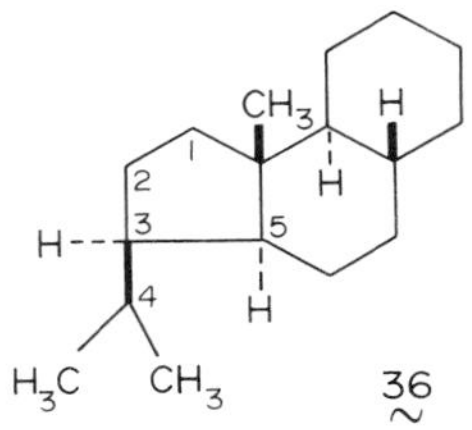

(3αH)-5(4 → 3) *abeo*-Podocarpane

Notes: The use of the prefix *neo* to indicate the bond migration
that converts a *gem*-dimethyl grouping, directly attached to a
ring carbon atom, into an isopropyl group, and of the prefix
friedo for migration of a methyl group from one position to another, has been suggested for terpenes. The continuation of
these usages and their extension to other classes of natural
product are not recommended, the preferred system being the
abeo-nomenclature.

Example 36. The method of citation of the stereochemistry at
C-3 differs from that used in the Steroid Rules 2S-5 *
but is preferred because it separates the indication of the *abeo*
operation from that of the final stereochemistry. If either of
the methyl groups attached to C-4 is substituted, it will be
accorded the lower locant and if stereochemistry at C-4 needs to
be specified, the *R,S* convention is used.

---

* Nomenclature of Steroids (Rules 1971) – Published in *Pure
Appl. Chem.* 31, 283 – 322 (1972).

F-4.10   The removal of a terminal ring, with addition of a
hydrogen atom at each junction atom with the adjacent ring, is
indicated by the prefix 'des', followed by the italic capital
letter designating that ring (Rule F-2.5); substituents and
stereochemistry implied in the trivial name remain unless otherwise stated.

Examples:

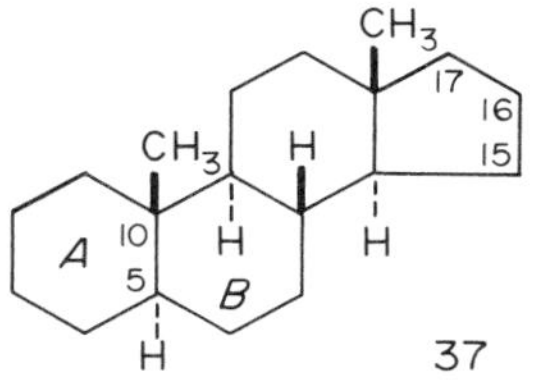

37

5α–Androstane

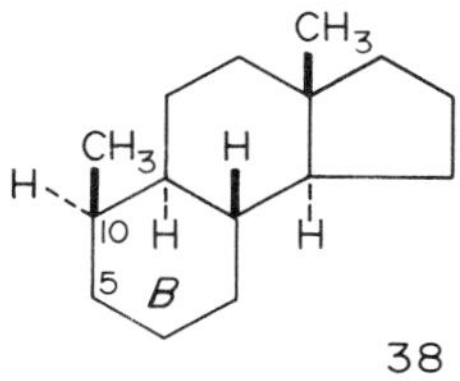

38

Des–*A*–androstane

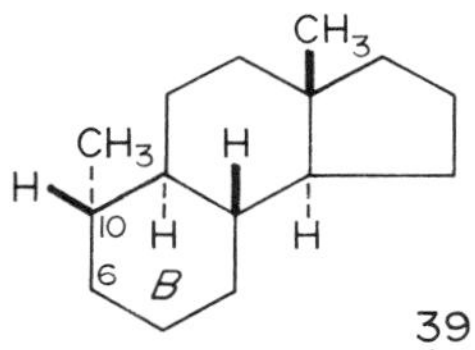

39

(10βH)–Des–*A*–androstane

Note: The prefix 'des' is used here, not 'de', because of the ease
of confusion of the latter, in speech, with 'D'.

F-4.11   If heteroatoms occur in a ring system of a natural
product whose hydrocarbon parent structure has a defined semisystematic name, replacement (oxa-aza) nomenclature may be used,
the numbering of the hydrocarbon parent structure being retained.
   This procedure may be used even when the parent structure
is already a heterocycle, but names so formed do not over-ride
the general principles of Sections A – C.

Examples:

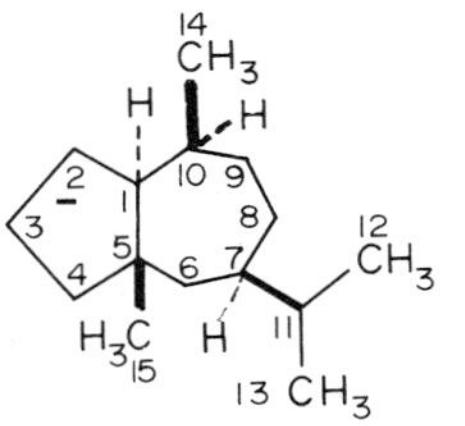

40

Ambrosane

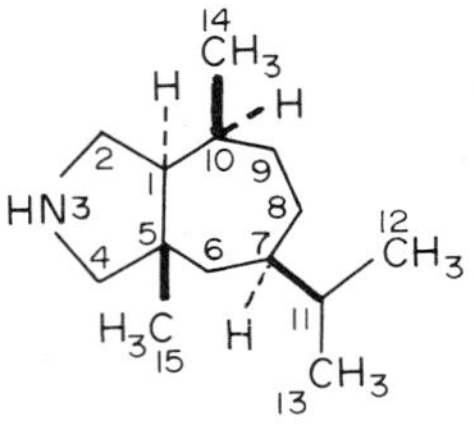

41

3-Aza-ambrosane

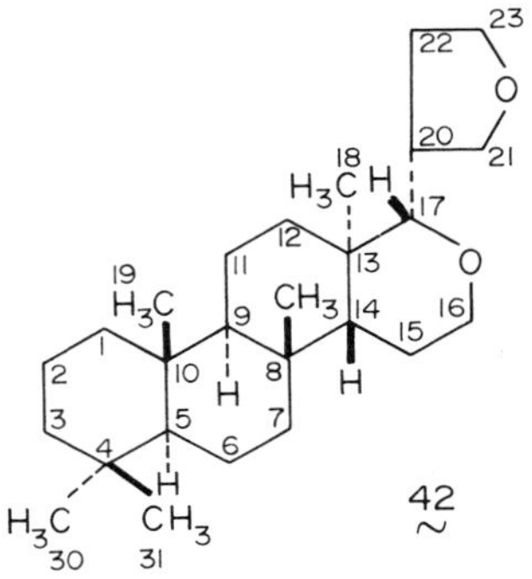

42

Gedunan

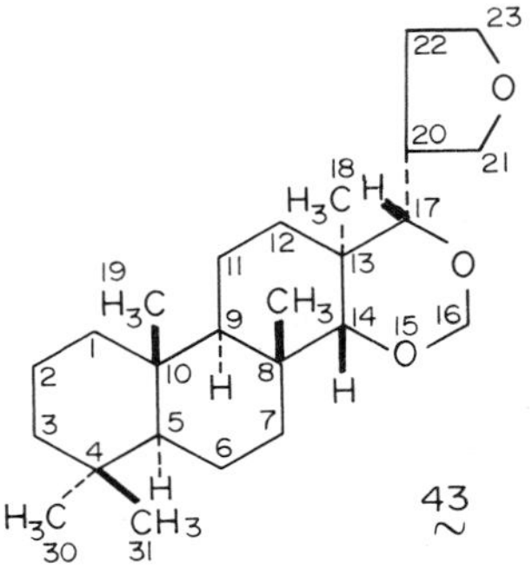

43

15-Oxagedunan

F-4.12   If the parent structure is heterocyclic, its analogues
in which one or more of the heteroatoms are replaced by carbon
may be named by use of the prefix 'carba'.
       The original numbering is retained. If in the parent
structure the heteroatom is un-numbered the carbon atom replacing
it will be numbered by affixing the letter 'a' to the locant of
the lower-numbered of the immediately adjacent atoms.

Examples:

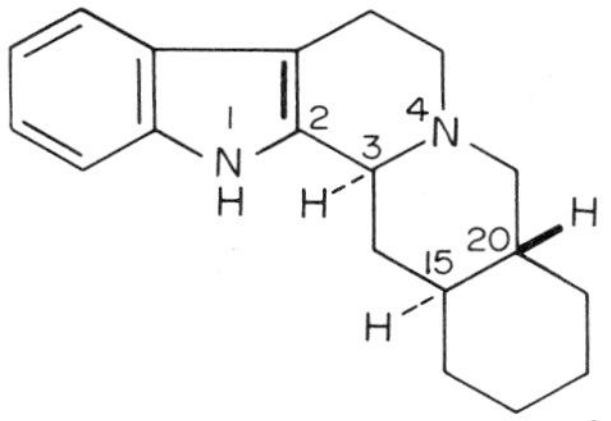

Yohimban
44

4β−4−Carbayohimban
45

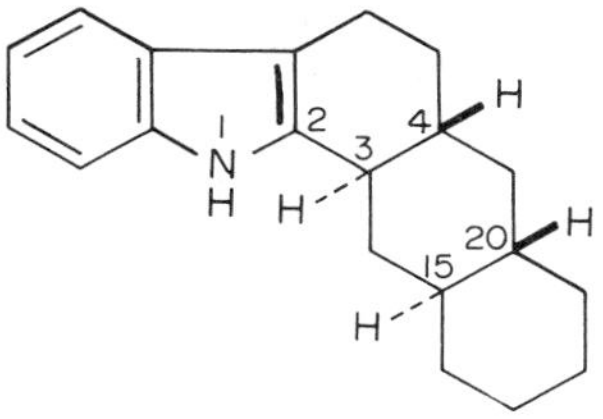

Morphinan
46

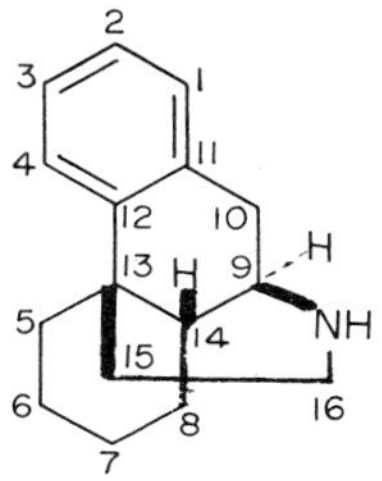

9a − Carbamorphinan
47

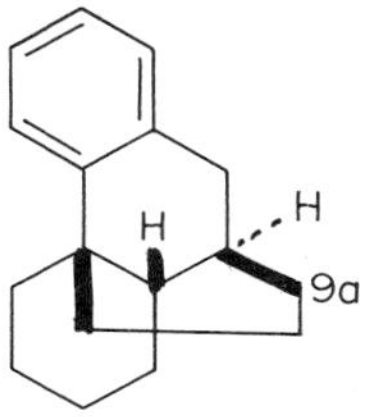

F-4.13   The modifying prefixes of Rules F-4.1 to F-4.10 and the
o̰x̰ā̰-a̰z̰a̰ prefixes of Rules F-4.11 and F-4.12 are non-detachable
(see Rule C-16.11), that is, they are not cited amongst prefixes
denoting substituents, but immediately before the name of the
parent structure.
      The order of citation is as follows:
      (a)   modifying prefixes in alphabetical order,
      (b)   oxa-aza prefixes in the order given in Table I of
            Rule B-1.1, that is, oxa-thia-aza,
      (c)   carba (Rule F-4.12),
      (d)   the name of the parent structure.

507

Rule F- 5   Substituents

Non-hydrocarbon groups and those hydrocarbon groups that are not included in the parent structure used as the basis of the name of the natural product are indicated as prefixes and suffixes strictly according to the rules of Section C.

Examples:

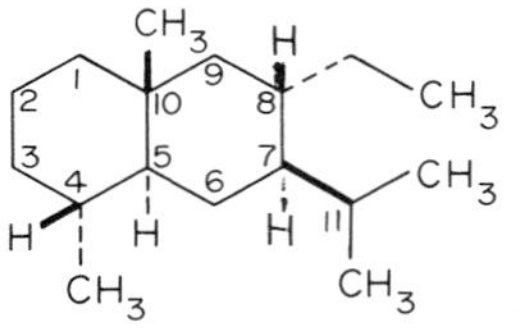

48   8a — Ethyleudesmane

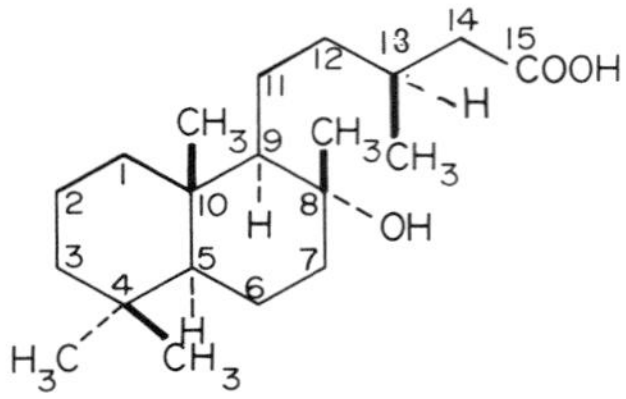

49   8 — Hydroxy — 15 —
labdanoic acid
(trivial name : labdanolic
acid)

Rule F-6   Stereochemistry

F-6.1   The name of the parent structure used as the basis for the derivation of the semisystematic name of a natural product implies, without further specification, the absolute configuration at all chiral elements depicted in these Rules and separate publications for the parent structure.

For steroids, the configuration at all centres *other than* C-5 is generally implied. Configuration at C-5 must be stated explicitly because both 5α- and 5β-configurations commonly occur. For pentacyclic triterpenes, in which the 5α(H) configuration is (almost) universal, this configuration is *implied* in the trivial names.

Examples:

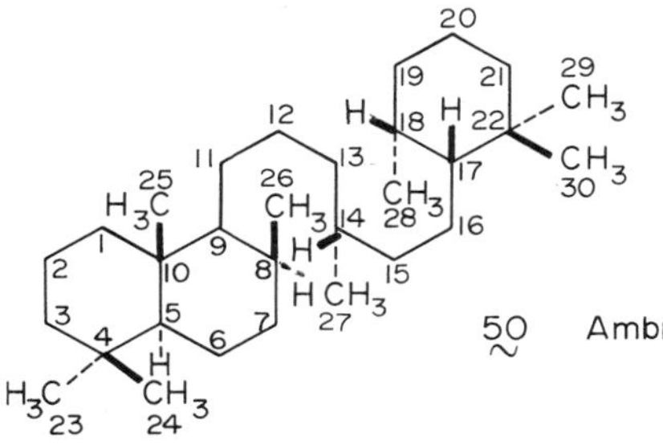

50 Ambrane

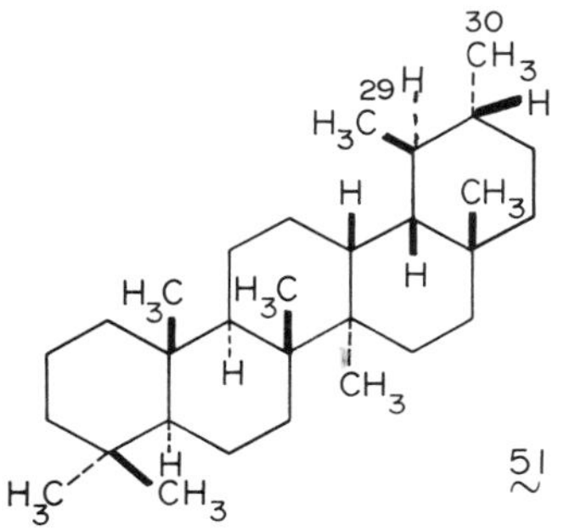

51 Ursane

F-6.2   The configuration at any chiral centre or feature in the
natural product not present in the parent structure will be designated by the αβ procedure if it can be so related to that of
the stereochemical configuration of the parent structure. [The
αβ procedure is described in IUPAC-IUB Recommendations for the
Nomenclature of Steroids   *   , Rule 2S-1.4; see also
Rule E-4.11].

When the rings of a quasi-planar parent structure are denoted
as projections onto the plane of the paper as in the depictions
in these Rules, an atom or group attached to the ring is termed α
if it lies *below* the plane of the paper or β if it lies *above* the
plane of the paper.   Whenever this system is used, the

_____

* See footnote on page 504.

orientations depicted in these Rules for the parent structure must be retained.

Where the αβ procedure cannot be applied the *R* and *S* symbolism of the Sequence Rule is used (see Rule E-4.9).

Example:

Urs – 12–en–3 β–ol
(α – amyrin)

F-6.3   If there is configurational inversion at a minority of the asymmetric centres the configurations of which are implied in a parent name, the configuration of the hydrogen atoms or substituents at the centres are stated by means of a prefix or prefixes α or β (or *R* or *S*), each with its appropriate locant placed before the stem name laid down in these Rules.

Abietane   (13αH)–Abietane

F-6.4   When there is configurational inversion at all the asymmetric centres whose configurations are implied in a name, the italicized prefix *ent-* (a contracted form of *enantio-*) is placed in front of the complete name of the compound. This prefix denotes inversion at all asymmetric centres (including those due to the presence of named substituents) whether or not the configurations of these are cited separately or are implied in the name.

Example:

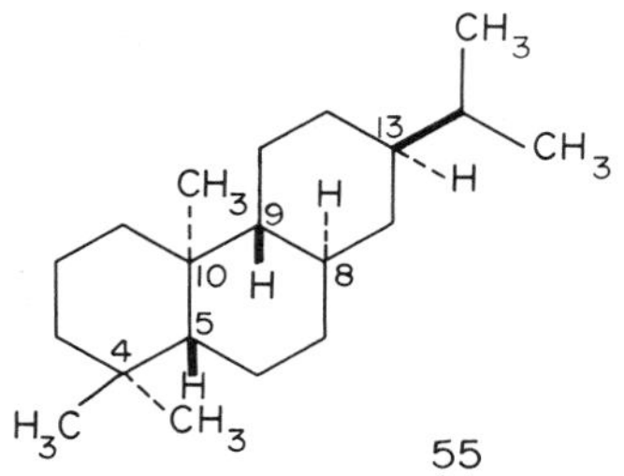

ent – Abietane

F-6.5   The enantiomer of a compound designated as in Rule F-6.3 is given the same name preceded by *ent-*.

F-6.6   Racemates are named by use of the italicized prefix *rac-* (an abbreviation for *racemo-*), placed before the complete name of the compound, the enantiomer chosen for naming being in accord with the above rules.

F-6.7   When the relative, but not the absolute, configurational relationships are known, starred symbols $R*$, $S*$ are used according to Rule E-4.10.

(Approved Recommendations 1978)

INTRODUCTION

These rules provide a general system of nomenclature for organic compounds whose isotopic nuclide (ref. 1)[*]composition deviates from that occurring in nature. [†] Comparative examples of the application of these rules are given in the Appendix, p. 538.

There is one other general system in use for describing isotopically modified compounds. It is based on an extension of the principles proposed by Boughton (ref. 3) for designating compounds containing hydrogen isotopes and is currently in use mainly in the Chemical Abstracts Service index nomenclature system. For a description of its current use, see ref. 4.

The system codified in these present rules provides for recognition of various types of isotopic modification and thus was chosen over the system based on the Boughton principles.

H-1.   SYMBOLS, DEFINITIONS, AND FORMULAE

Rule H-1.1.   Symbols

1.11—Nuclide symbols. The symbol for denoting a nuclide in the formula or name of an isotopically modified compound consists of the atomic symbol for the element and an Arabic numeral in the left superscript position which indicates the mass number of the nuclide (ref. 5a).

1.12—Atomic symbols. The atomic symbols used in the nuclide symbol are those given in the IUPAC Inorganic Nomenclature Rules (ref. 5b). In the nuclide symbol, the atomic symbol is printed in Roman type, italicized atomic symbols being reserved for letter locants, as is customary in organic chemical nomenclature (cf. Rule C-814.4, p.255).

> Note: For the hydrogen isotopes protium, deuterium, and tritium, the nuclide symbols ^1H, ^2H, and ^3H, respectively, are used. The symbols D and T for ^2H and ^3H, respectively, may be used, but not when other modifying nuclides are also present because this may cause difficulties in the alphabetic ordering of the nuclide symbols in the isotopic descriptor. Although the symbols d and t have been and are still used in place of ^2H and ^3H in names formed according to the Boughton system (see Introduction), in no other cases are lower-case letters used as atomic symbols. Therefore, the use of d and t in chemical nomenclature outside of the Boughton system is not recommended.

---

[*] References for Section H are on page 537.

[†] For a discussion of the meaning of 'natural composition', see ref. 2.
   In any context where the accuracy requires it, the natural nuclidic composition used shall be stated.

Rule H-1.2. Definitions and formulae of various types of isotopic modification

1.21—An isotopically *unmodified* compound has a macroscopic composition such that its constituent nuclides are present in the proportions occurring in nature. Its formula and name are written in the customary manner.

Examples:

1. CH_4 Methane

2. CH_3-CH_2-OH Ethanol

1.22—An isotopically *modified* compound has a macroscopic composition such that the isotopic ratio of nuclides for at least one element deviates measurably from that occurring in nature. It is either an isotopically *substituted* compound or an isotopically *labeled* compound.

1.23—An isotopically *substituted* compound has a composition such that essentially all the molecules of the compound have *only* the indicated nuclide(s) at each designated position. For all other positions, the absence of nuclide indication means that the nuclide composition is the natural one.

The formula of an isotopically *substituted* compound is written as usual except that appropriate nuclide symbols are used. When different isotopes of the same element are present at the same position, common usage is to write their symbols in order of increasing mass number.

Examples (for names see Rule H-2.11):

1. $^{14}CH_4$ (^{14}C)Methane

2. $^{12}CHCl_3$ (^{12}C)Chloroform

3. CH_3-CH^2H-OH $(1-^2H_1)$Ethanol

    not

CH_3-C^2HH-OH

1.24—An isotopically *labeled* compound is a mixture of an isotopically unmodified compound with one or more analogous isotopically substituted compound(s).

Note: Although an isotopically labeled compound is really a mixture as far as chemical identity is concerned (in the same way as is an unmodified compound), such mixtures are called "isotopically labeled compounds" for nomenclature purposes.

1.25—An isotopically labeled compound is designated as *specifically labeled* when a *unique* isotopically substituted compound is formally added to the analogous isotopically unmodified compound. In such a case, both position(s) and number of each labeling nuclide are defined.

The structural formula of a specifically labeled compound is written in the usual way, but with the appropriate nuclide symbol(s) and multiplying subscript, if any, enclosed in *square brackets*. Other principles used in writing the formula are described in Rule H-1.23.

Examples:

	Isotopically substituted compound	when added to	Isotopically unmodified compound	gives rise to	Specifically labeled compound
1.	$^{14}CH_4$		CH_4		$[^{14}C]H_4$
2.	$CH_2{}^2H_2$		CH_4		$CH_2[^2H_2]$
3.	$CH_3-CH_2-^{18}OH$		CH_3-CH_2-OH		$CH_3-CH_2-[^{18}O]H$
4.	$CH^2H_2-CH_2-O^2H$		CH_3-CH_2-OH		$CH[^2H_2]-CH_2-O[^2H]$

5.

$$^2H\!\!-\!\!\underset{\underset{CH_3}{|}}{C}\!\!-\!\!H \qquad\qquad H\!\!-\!\!\underset{\underset{CH_3}{|}}{C}\!\!-\!\!H \qquad\qquad [^2H]\!\!-\!\!\underset{\underset{CH_3}{|}}{C}\!\!-\!\!H$$

(each with OH above the central carbon)

Note: Although the formula for a specifically labeled compound does not represent the composition of the bulk material, which usually consists overwhelmingly of the isotopically unmodified compound, it does indicate the presence of the compound of chief interest, the isotopically substituted compound.

A specifically labeled compound is (a) *singly labeled* when the isotopically substituted compound has only one isotopically modified atom, e.g., $CH_3-CH[^2H]-OH$; (b) *multiply labeled* when the isotopically substituted compound has more than one modified atom of the same element at the same position or at different positions, e.g., $CH_3-C[^2H_2]-OH$ and $CH_2[^2H]-CH[^2H]-OH$; or (c) *mixed labeled* when the isotopically substituted compound has more than one kind of modified atom, e.g., $CH_3-CH_2-[^{18}O][^2H]$.

1.26—An isotopically labeled compound is designated as *selectively labeled* when a *mixture* of isotopically substituted compounds is formally added to the analogous isotopically unmodified compound in such a way that the position(s) but not necessarily the number of each labeling nuclide is defined. A selectively labeled compound may be considered as a mixture of specifically labeled compounds.

A selectively labeled compound may be (a) *multiply labeled* when in the unmodified compound there is more than one atom of the same element at the position where the isotopic modification occurs, e.g., H in CH_4, or there are several atoms of the same element at different positions where the isotopic modification occurs, e.g., C in C_4H_8O; or (b) *mixed labeled* when there is more than one labeling nuclide in the compound, e.g., C and O in CH_3-CH_2-OH.

Note: When there is only one atom of an element that can be modified in a compound, only specific labeling can result (see Rule H-1.25).

A selectively labeled compound cannot be described by a unique structural formula; therefore it is represented by inserting the nuclide symbols preceded by any necessary locant(s) (letters and/or numbers) but without multiplying subscripts, enclosed in square brackets directly before the usual formula or, if necessary, before parts of the formula that have an independent numbering. Identical locants are not repeated.

When different nuclides are present, the nuclide symbols are written in alphabetic order according to their symbols, or when the atomic symbols are identical, in order of increasing mass number (see Rules H-2.81 and H-2.82).

Examples:

	Mixture of isotopically substituted compounds	when added to	Isotopically unmodified compound	gives rise to	Selectively labeled compound
1.	$CH_3{}^2H$, $CH_2{}^2H_2$ $CH{}^2H_3$, $C{}^2H_4$ or any two or more of the above		CH_4		$[{}^2H]CH_4$
2.	$CH_3-CH_2-CH{}^2H-CO_2H$ $CH_3-CH_2-C{}^2H_2-CO_2H$		$CH_3-CH_2-CH_2-CO_2H$		$[2-{}^2H]CH_3-CH_2-CH_2-CO_2H$
3.	$CH_3-{}^{14}CH_2-{}^{14}CH_2-CO_2H$ $CH_3-{}^{14}CH_2-CH_2-CO_2H$ $CH_3-CH_2-{}^{14}CH_2-CO_2H$ or any two of the above		$CH_3-CH_2-CH_2-CO_2H$		$[2,3-{}^{14}C]CH_3-CH_2-CH_2-CO_2H$
4.	$CH_3-{}^{14}CH_2-OH$ $CH_3-CH_2-{}^{18}OH$ $CH_3-{}^{14}CH_2-{}^{18}OH$ or any two of the above		CH_3-CH_2-OH		$[1-{}^{14}C,{}^{18}O]CH_3-CH_2-OH$
5.	${}^{14}CH_3-CH_2-CO_2-CH_3$ $CH_3-CH_2-CO_2-{}^{14}CH_3$ ${}^{14}CH_3-CH_2-CO_2-{}^{14}CH_3$ or any two of the above		$CH_3-CH_2-CO_2-CH_3$		$[3-{}^{14}C]CH_3-CH_2-CO_2-[{}^{14}C]$

Note: The method of writing formulae as given by the above rule may also be of use if a compound is represented by its molecular formula rather than its structural formula, e.g., $[^2H]C_2H_6O$.

1.27—In a selectively labeled compound formally arising from mixing several known isotopically substituted compounds with the analogous isotopically unmodified compound, the number or the possible number of labeling nuclide(s) for each position may be indicated by subscripts to the atomic symbol(s). Two or more subscripts referring to the same nuclide symbol are separated by a semicolon. For a multiply labeled or mixed labeled compound (see Rule H-1.26), the subscripts are written successively in the same order as the various isotopically substituted compounds are considered. The subscript zero is used to indicate that one of the isotopically substituted compounds is not modified at the indicated position.

Examples:

Mixture of isotopically substituted compounds	when added to	Isotopically unmodified compound	gives rise to	Selectively labeled compound
1. $CH_2{}^2H-CH_2-OH$ $CH^2H_2-CH_2-OH$		CH_3-CH_2-OH		$[2-^2H_{1;2}]CH_3-CH_2-OH$
2. $CH^2H_2-CH_2-OH$ $CH^2H_2-CH_2-^{18}OH$		CH_3-CH_2-OH		$[2-^2H_{2;2},{}^{18}O_{0;1}]CH_3-CH_2-OH$ *
3. $CH_3-CH_2-^{18}OH$ $CH^2H_2-CH_2-OH$		CH_3-CH_2-OH		$[2-^2H_{0;2},{}^{18}O_{1;0}]CH_3-CH_2-OH$
4. CH_3-CH^2H-OH $CH^2H_2-CH_2-OH$		CH_3-CH_2-OH		$[1-^2H_{1;0},2-^2H_{0;2}]CH_3-CH_2-OH$

1.28—An isotopically labeled compound is designated as *nonselectively labeled* when the position(s) and the number of the labeling nuclide(s) are both undefined.

In such cases the labeling is indicated by inserting the nuclide symbol, enclosed in square brackets, directly before the usual line formula with no locants or subscripts.

Example:

$[^{14}C]CH_3-CH_2-CH_2-CO_2H$

* Repetition of locants is not necessary as it may lead to ambiguity. Therefore, they have been omitted from this edition and a subscript added.

1.29—An isotopically labeled compound may be designated as *isotopically deficient* when the isotopic content of one or more elements has been depleted, i.e., when one or more nuclide(s) is(are) present in less than the natural ratio. Such an isotopically modified compound is denoted in the formula by adding the italicized syllable *def* immediately preceding, without a hyphen, the appropriate nuclide symbol.

Example:

$$[def^{13}C]CHCl_3$$

Note : According to one's viewpoint, one may also use $[^{12}C]CHCl_3$.

H-2. NAMES FOR ISOTOPICALLY MODIFIED COMPOUNDS

Rule H-2.1. Isotopically substituted compounds (cf. Rule H-1.23)

2.11—The name of an isotopically substituted compound is formed by inserting in *parentheses* (curves) the nuclide symbol(s), preceded by any necessary locant(s) (letters and/or numerals), before the name *or preferably* before the denomination of that part of the compound that is isotopically substituted. Immediately after the parentheses there is neither space nor hyphen, except that when the name, or a part of the name, includes a preceding locant, a hyphen is inserted. *

When polysubstitution is possible, the number of atoms substituted is always specified as a right subscript to the atomic symbol(s), even in case of monosubstitution.

Examples:

1. $^{14}CH_4$ (^{14}C)Methane

2. $CH_3{}^2H$ $(^2H_1)$Methane

3. $C^2H_2Cl_2$ Dichloro$(^2H_2)$methane

4. 1-[Amino(^{14}C)methyl]cyclopentanol

5. 1-(Aminomethyl)cyclopentan(^{18}O)ol

    or

    1-(Aminomethyl)(^{18}O)cyclopentanol

---

* In general, in organic nomenclature locants for suffixes, unsaturation, free valences, etc., cited ahead of the parent are considered as part of the name. In this report the practice of citing the isotopic descriptor ahead of such locants is followed; in biochemical usage the isotopic descriptor is often cited after such locants.

6.

N^2H_2

2-Cyclohexen-1-($^2H_2$)amine

or

(N,N-$^2H_2$)-2-Cyclohexen-1-amine

7.

^{131}I NH-CO-CH$_3$

N-[7-(^{131}I)Iodofluoren-2-yl]a-
cetamide

8. $^{14}CH_2$-CO$_2$ C$_2$H$_5$
     $^{14}CH_2$-CO$_2$ Na

Sodium ethyl (2,3-$^{14}C_2$)succinate

Rule H-2.2.   Specifically labeled compounds (cf. Rule H-1.25)

2.21—The name of a specifically labeled compound is formed by inserting in *square brackets* the nuclide symbol(s), preceded by any necessary locant(s) (letters and/or numerals), before the name *or preferably* before the denomination of that part of the compound that is isotopically modified.   Immediately after the brackets there is neither space nor hyphen, except that when the name, or a part of the name, requires a preceding locant, a hyphen is inserted. *

When polylabeling is possible, the number of atoms that have been labeled is always specified as a subscript to the atomic symbol(s), even in the case of monolabeling.   This is necessary in order to distinguish between a specifically and a selectively or nonselectively labeled compound.

The name of a specifically labeled compound differs from the name of the corresponding isotopically substituted compound (see Rule H-2.11) only in the use of *square brackets* surrounding the nuclide descriptor rather than *parentheses* (curves).

Examples:

1.   [^{14}C]H$_4$ [^{14}C]Methane

2.   CH$_3$[^2H]   [$^2H_1$]Methane

3.   CH$_3$
     CH$_3$-CH$_2$-CH-CH=C[$^2H_2$]   3-Methyl[1,1-$^2H_2$]-1-pentene †

---

* See footnote,   p. 518
† Note that here the locant is part of the parent hydrocarbon name; see also
  footnote,   p. 518

4.

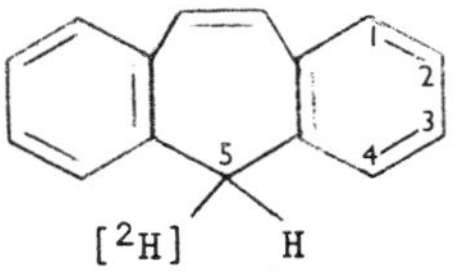

[2H]   H

$[5-^2H_1]-5H$-Dibenzo$[a,d]$cycloheptene *

5.

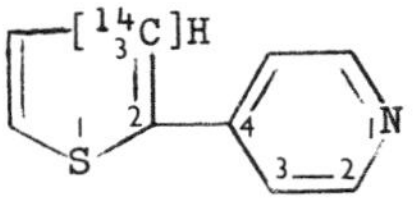

4-($[3-^{14}C]$-2-Thienyl)pyridine *

6. $C[^2H_2]Cl_2$

Dichloro$[^2H_2]$methane

7. $CH_3-CH_2-O[^2H]$

Ethan$[^2H]$ol

or

$[O-^2H]$Ethanol

8.

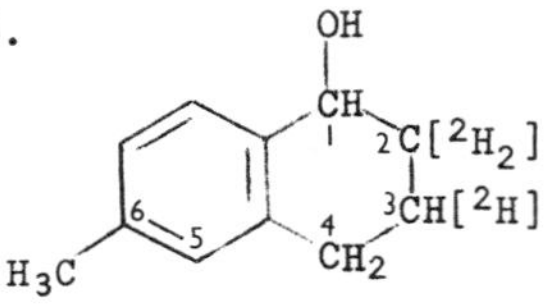

6-Methyl$[2,2,3-^2H_3]$-1,2,3,4-tetra-
hydro-1-naphthol †

9.

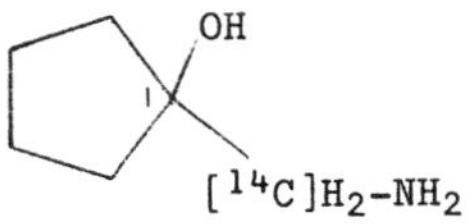

$[^{14}C]H_2-NH_2$

1-(Amino$[^{14}C]$methyl)cyclopentanol

10.

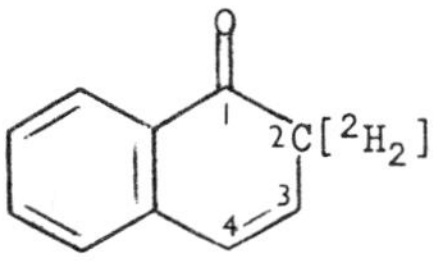

$_2C[^2H_2]$

$[2,2-^2H_2]-1(2H)$-Naphthalenone

---

* Note that here the locant is part of the parent hydrocarbon or radical name;
see also footnote,   p. 518 .

† In cases such as this, treatment of hydro prefixes as nondetachable is preferred; see Rule C-16.11, p. 108.

Names

11.

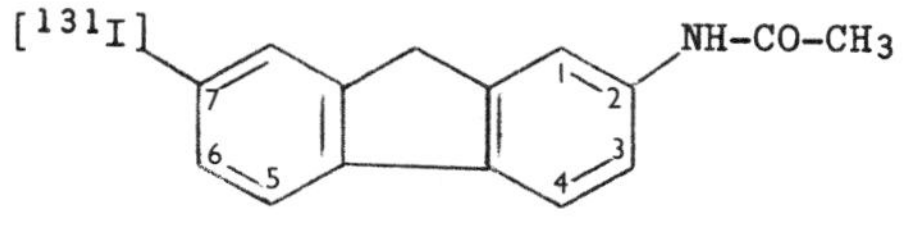

[131I]
NH-CO-CH3
7
6
5
1
2
3
4

N-(7-[^{131}I]Iodofluoren-2-yl)acetamide

12.

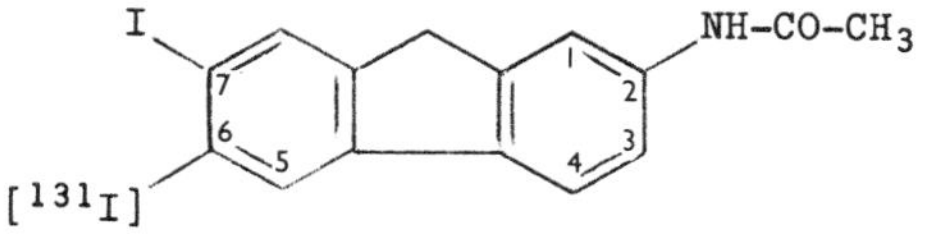

I
NH-CO-CH3
7
6
5
2
3
4
[131I]

N-(6,7-[6-^{131}I]Diiodofluoren-2-yl)acetamide

2.22—**In a name consisting of two or more words, the isotopic designator may be placed before** the appropriate word or part of the word that includes the labeled nuclide(s), unless unambiguous locants are available or are unnecessary.*

Examples:

1. $CH_2[^2H]$-CO_2H — [2-$^2H_1$]**Acetic acid**

2. CH_3-$CO_2[^2H]$ — [*O*-^{2}H]Acetic acid *or* Acetic [^{2}H]acid

3. CH_3-CH_2-CH_2-CH_2-[^{14}C]O_2[^{2}H] — [1-^{14}C]Pentan[^{2}H]oic acid *or* [1-^{14}C, *O*-^{2}H]Pentanoic acid

4. 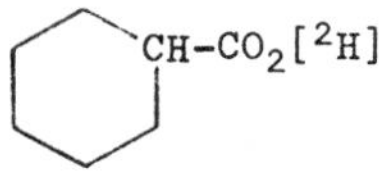

CH-CO2[2H]

Cyclohexane[^{2}H]carboxylic acid *or* [*O*-^{2}H]**Cyclohexanecarboxylic acid**

5. 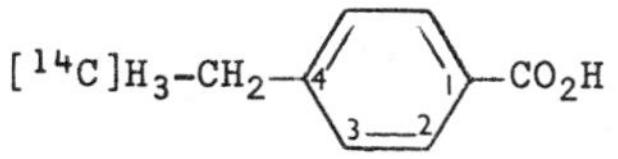

[14C]H3-CH2-
4
1
3
2
-CO2H

4-([2-^{14}C]**Ethyl)benzoic acid**

6. H-[^{14}C]O_2Na — **Sodium** [^{14}C]**formate**

7. CH_3-CH_2-CO_2[^{14}C]H_2-CH_3 — [1-^{14}C]**Ethyl propionate**

* The same rule applies to isotopically substituted compounds (see Rule H-2.11).

8. $CH_3-[^{14}C]H_2-CO_2-CH_2-CH_3$ Ethyl $[2-^{14}C]$propionate

9. $[C_6H_5N_2]^+[^{35}Cl]^-$ Benzenediazonium $[^{35}Cl]$chloride

2.23—In a trivial or semisystematic name consisting of one word, the isotopic designator may be placed before the full trivial name or inserted into the trivial name.

Examples:

1. $(CH_3)_2CH-CH_2-[^{14}C]H(NH_2)-CO_2H$

   $[2-^{14}C]$Leucine

2. $CH_3-[^{35}S]-CH_2-CH_2-CH(NH_2)-CO_2H$

   $[^{35}S]$Methionine

3. $CH_3-CO-NH[^2H]$

   $[N-^2H_1]$Acetamide

   or

   Acet$[^2H_1]$amide

Note: The alternative system based on the Boughton principles (see Introduction) denotes isotopic modification by citing the appropriate symbol and mass number (with subscripts and locants if necessary) *following* the portion of the name to which the symbol refers.

Example:

$$\overset{\displaystyle ^{35}Cl}{\underset{\displaystyle CH_2{}^2H-CH-CH-CH_2-CH_3}{|}} \quad \overset{\displaystyle C^2H_3}{|}$$ 2-(Chloro-^{35}Cl)-3-(methyl-d_3)-pentane-$1-d$

According to the rules recommended here, the name for this compound is : 2-(^{35}Cl)chloro-3-$[(^2H_3)$methyl$](1-^2H_1)$pentane.

Rule H-2.3. Selectively labeled compounds (cf. Rules H-1.26 and H-1.27)

2.31—The name of a *selectively* labeled compound is formed in the same way as the name of a *specifically* labeled compound (see Rule H-2.21), except that the multiplying subscripts following the atomic symbols are generally omitted except as described by Rule H-2.32. Identical locants corresponding to the same element are not repeated.

The name of a selectively labeled compound differs from the name of the corresponding isotopically substituted compound in the use of *square brackets* surrounding the nuclide descriptor rather than parentheses (curves) and in the omission of repeated identical locants and multiplying subscripts.

Examples:

Mixture of isotopically substituted compounds	when added to	is named
1. $\begin{cases} CH_3{}^2H, \ CH_2{}^2H_2 \\ CH^2H_3, \ C^2H_4 \end{cases}$	CH_4	$[^2H]$Methane *not* $[^2H_4]$Methane
2. $\begin{cases} CH_3-CH^2H-OH \\ CH_3-C^2H_2-OH \end{cases}$	CH_3-CH_2-OH	$[1-^2H]$Ethanol *not* $[1,1-^2H_2]$Ethanol

3. $\left\{ \begin{array}{c} [^3H]HC\overset{\displaystyle\bigcirc}{\quad}CH-OH \\[6pt] [^3H_2]C\overset{\displaystyle\bigcirc}{\quad}CH-OH \end{array} \right\}$ $\overset{\displaystyle\bigcirc}{\quad}CH-OH$ $[4-^3H]$Cyclohexanol *not* $[4,4-^3H_2]$Cyclohexanol

4. $\begin{cases} {}^{14}CH_3-CH_2-CO_2-CH_2-CH_3 \\ CH_3-CH_2CO_2-{}^{14}CH_2-CH_3 \end{cases}$ $CH_3-CH_2-CO_2-CH_2-CH_3$ $[1-^{14}C]$Ethyl $[3-^{14}C]$propionate

2.32—In a selectively labeled compound formally arising from mixing several known isotopically substituted compounds with the analogous isotopically unmodified compound , the number or the possible number of labeling nuclide(s) for each position may be indicated by subscripts to the atomic symbol(s) as described in Rule H-1.27.

Examples:

Mixture of isotopically substituted compounds	when added to	is named
1. $\begin{cases} CH_2{}^2H-CH_2-OH \\ CH^2H_2-CH_2-OH \end{cases}$	CH_3-CH_2-OH	$[2-^2H_{1;2}]$Ethanol
2. $\begin{cases} CH_3-CH_2-{}^{18}OH \\ CH^2H_2-CH_2-OH \end{cases}$	CH_3-CH_2-OH	$[2-^2H_{0;2},{}^{18}O_{1;0}]$Ethanol

Rule H-2.4. Nonselectively labeled compounds (cf. Rule H-1.28)

2.41—The name of a nonselectively labeled compound is formed in the same way as the name of a selectively labeled compound (see Rule H-2.31) but contains neither locants nor subscripts in the nuclide descriptor.

Examples:

Chloro[^{3}H]benzene

[^{14}C]Glycerol

Rule H-2.5. Isotopically deficient compounds (cf. Rule H-1.29)

2.51—The name of an isotopically deficient compound may be formed by addi the italiziced syllable *def* immediately preceding, without a hyphen, the appropriate nuclide symbol, both enclosed in square brackets and cited before the name or that part of the name that is isotopically modified.

Example:

[*def*^{13}C]Chloroform

Rule H-2.6. General and uniform labeling

2.61—In the name of a selectively labeled compound in which *all* positions of the designated element are labeled, but not necessarily in the *same isotopic ratio*, the symbol "G" may be used in place of locants to indicate a "general" labeling.

Examples:

Isotopically substituted compounds	when added to		may be designated as
1. mixture of substituted compounds (selective labeling)			
$CH_3-CH_2-CH_2-^{14}CO_2H$			
$CH_3-CH_2-^{14}CH_2-CO_2H$			
$CH_3-^{14}CH_2-CH_2-CO_2H$	$CH_3-CH_2-CH_2-CO_2H$		[G-^{14}C]Butyric acid
$^{14}CH_3-CH_2-CH_2-CO_2H$			
$CH_3-^{14}CH_2-^{14}CH_2-CO_2H$			
etc...			

2. D-Glucose in which all six positions are labeled with ^{14}C, but not necessarily uniformly, may be designated as D-[G-^{14}C]glucose.

2.62—In the name of a selectively labeled compound in which *all* positions of the designated element are labeled in the *same isotopic ratio*, the symbol "U" may be used in place of locants to denote "uniform" labeling.

Examples:

	Isotopically substituted compounds	when added to	may be designated as

1. mixture of substituted compounds (selective labeling)

$$\left\{\begin{array}{l} CH_3-CH_2-CH_2-^{14}CO_2H \\ CH_3-CH_2-^{14}CH_2-CO_2H \\ CH_3-^{14}CH_2-CH_2-CO_2H \\ ^{14}CH_3-CH_2-CH_2-CO_2H \\ \text{in equal amounts} \end{array}\right.$$ $CH_3-CH_2-CH_2-CO_2H$ [U-^{14}C]Butyric acid

2. D-Glucose in which ^{14}C is equally distributed among the six positions may be designated as D-[U-^{14}C]glucose.

Note: In the case of radioactive nuclides, "same isotopic ratio" means "same specific radioactivity".

2.63—In the name of a selectively labeled compound, the symbol "U" (see Rule H-2.62) followed by appropriate locants, may similarly be used to indicate labeling in the *same isotopic ratio at the specified positions*.

Example:

D-Glucose in which ^{14}C is equally distributed among positions 1, 3, and 5 may be designated as D-[U-=1,3,5-^{14}C]glucose.

Rule H-2.7.   Exceptional changes in the names of some unsymmetrically modified compounds

2.71—The name of an isotopically modified compound, substituted or labeled, may differ from the name of the unmodified analog when its structure includes identical units that are not identically modified in equivalent positions. Where there is ambiguity, such different groups must be expressed separately.

Example:

Unmodified compound: Modified compound:

$$CH_3-CH_2 \quad CH_2-CH_3$$
$$CH_3-CH_2-CH_2-CH-CH-CH_2OH$$
$$\qquad\qquad\qquad\quad 3 \quad 2 \quad 1$$

$$CH_3-CH_2 \quad CH_2-CH[^2H_2]$$
$$CH_3-CH_2-CH_2-CH-CH-CH_2OH$$
$$\qquad\qquad\qquad\quad 3 \quad 2 \quad 1$$

2,3-Diethyl-1-hexanol   2-([2,2-$^2H_2$]Ethyl)-3-ethyl-1-hexanol

Rule H-2.8. Order of nuclide symbols

2.81—When isotopes of different elements are present as nuclides in an isotopically modified compound, their symbols are arranged in alphabetical order if they are inserted at the same place in the name.

Examples:

1. Mixed substituted compound

$CH_3^{18}O^2H$ Methan($^2H,^{18}O$)ol

                                        or

                                (O-$^2H,^{18}O$)Methanol

2. Mixed specifically labeled compounds

   a. $CH_3[^{18}O][^2H]$ Methan[$^2H,^{18}O$]ol

                                        or

                                [O-$^2H,^{18}O$]Methanol

   b. $H_2[^{15}N]$-$[^{14}C]O$-$NH[^2H]$ [$^{14}C,N'$-$^2H_1,N$-^{15}N]Urea *

3. Mixed selectively labeled compound

   D-Glucose labeled uniformly
     with ^{14}C and generally
     with ^2H D-[U-^{14}C,G-^2H]Glucose

2.82—When several isotopes of the same element are present as nuclides in an isotopically modified compound, their symbols are arranged in the order of increasing atomic mass number if they are inserted at the same place in the name.

_____

* For numbering priority between N and N', see Rule H-3.21.

Examples:

1. Mixed substituted compound

   $CH_2{}^2H-CH^3H-OH$ $(2-{}^2H_1,1-{}^3H_1)$Ethanol

2. Mixed specifically labeled compounds

   a. $CH_2[{}^2H]-OH, CH_2[{}^3H]-OH$ $[{}^2H_1,{}^3H_1]$Methanol

   b. $CH_3-O[{}^2H], CH_3-O[{}^3H]$ Methan$[{}^2H,{}^3H]$ol

                                             or

                               $[O-{}^2H,O-{}^3H]$Methanol

   c. $CH_2[{}^2H]-CH_2-CH[{}^3H]-CO_2H$ $[4-{}^2H_1,2-{}^3H_1]$Butyric acid

Rule H-2.9. Stereoisomeric isotopically modified compounds

2.91—Two types of stereoisomeric isotopically modified compounds are possible: (a) those in which the stereoisomerism results from isotopic modification, and (b) those whose analogous unmodified compounds are stereoisomers.

The nomenclature of stereoisomers of isotopically modified compounds follows the general methods of stereochemical nomenclature as described in Section E.

Stereochemical affixes are cited at the specified place in the name according to the stereochemical rules. When they must be inserted into the name at the same place as isotopic descriptors, the stereochemical affixes are cited first.

Examples in which stereoisomerism results from isotopic modification:

1.   OH
          |
   ${}^2H$ ➖ C ➖ H $(S)-(1-{}^2H_1)$Ethanol
          |
          CH_3

2.   OH
          |
   $[{}^2H]$ ➖ C ➖ H $(S)-[1-{}^2H_1]$Ethanol
          |
          CH_3

3.

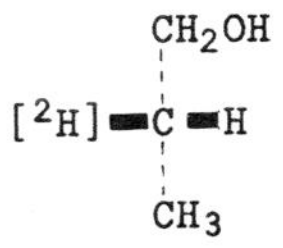

CH$_2$OH

[^{2}H]—C—H

CH$_3$

(S)-[2-^{2}H$_1$]-1-Propanol

4.

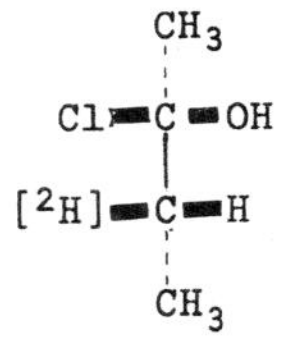

CH$_3$

Cl—C—OH

[^{2}H]—C—H

CH$_3$

(2R,3S)-2-Chloro[3-^{2}H$_1$]-2-
butanol

5.

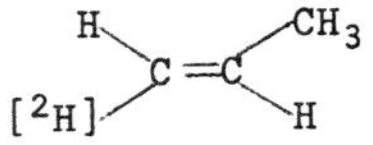

H   CH$_3$
  \   /
   C═C
  /   \
[^{2}H]   H

(E)-[1-^{2}H$_1$]Propene

Examples of isotopically modified stereoisomers:

6.

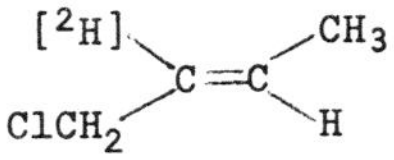

[^{2}H]   CH$_3$
    \   /
     C═C
    /   \
ClCH$_2$ H

(E)-1-Chloro[2-^{2}H]-2-
butene

7.

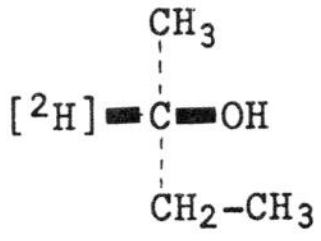

CH$_3$

[^{2}H]—C—OH

CH$_2$–CH$_3$

(S)-[2-^{2}H]-2-Butanol

8.

CH$_3$

HO—C—H

CH$_3$—C[^{2}H]

CH$_3$

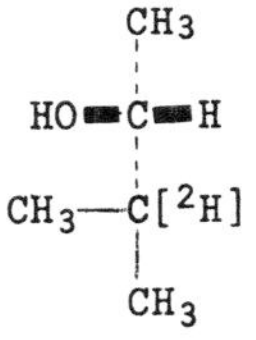

(R)-3-Methyl[3-^{2}H]-2-butanol

9. 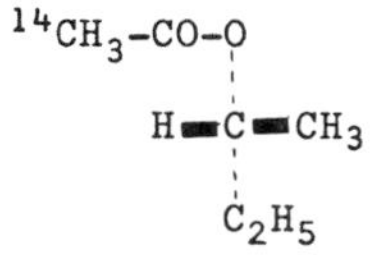 $^{14}CH_3-CO-O$

   H—C—CH_3

   C_2H_5 (R)-sec-Butyl $(2-^{14}C)$acetate

10. 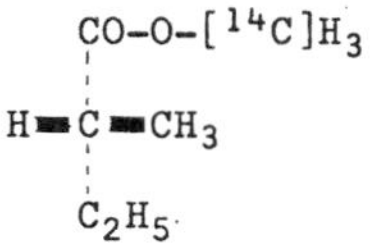   CO-O-$[^{14}C]H_3$

   H—C—CH_3

   C_2H_5 $[^{14}C]$Methyl (R)-2-methylbutyrate

                                 or

                                 (R)-$([^{14}C]$Methyl 2-methylbutyrate)

11. 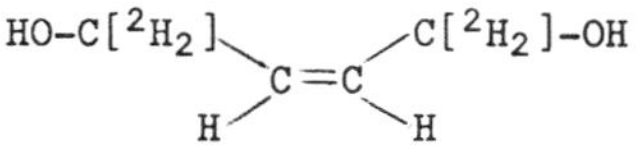   HO-C$[^2H_2]$ $C[^2H_2]$-OH

                    C=C

          H   H (Z)-$[1,1,4,4-^2H_4]$-2-Butene-
                                 1,4-diol

2.92   **Stereochemical affixes** (for example D and L) added according to the rules of special classes, such as carbohydrates, amino acids, steroids, etc., usually refer to the parent substance (or unmodified compound) according to the particular class of compounds. However, isotopic descriptors follow the stereochemical descriptors in these classes, according to biochemical usage.*

Examples:

1.   L-$[3,4-^{13}C,^{35}S]$Methionine

2.   L-$[3-^{14}C,2,3-^2H_2,^{15}N]$Serine

3. 5α-$[17-^2H]$Pregnane

4. $(24R)$-5α-$[24-^2H_1]$Cholestane

5.   2-$([^{18}F]$Fluoro)-2-deoxy-D-glucose

_____

* See footnote, p.518

H-3. NUMBERING OF ISOTOPICALLY MODIFIED COMPOUNDS

Rule H-3.1. Numbering in relation to the unmodified compound

3.11—Numbering of an isotopically modified compound is *not* changed from that of an isotopically unmodified compound. Among the structural features of a compound to be considered successively for numbering as given by the Rule C-15.11, p. 105 and Section E , the presence of nuclides is considered last with the exception of chirality arising from isotopic modification (see Rule H-3.22, example 7).

Examples:

1. $CF_3-CH_2[^2H]$ 1,1,1-Trifluoro$[^2H_1]$ethane

2.

1-Chloro-3-fluoro$[2-^2H]$benzene

3.

2-Methoxy$[3,4,5,6-^3H_4]$phenol

Note: One should be aware that, when isotopically modified compounds are named by the system based on an extension of the Boughton principles (see Introduction, p. 513), lowest locants are assigned to isotopic positions included in the parent structure, which includes unsaturation and principal groups, if any, before other considerations. This sometimes results in the assignment of higher locants for substituents expressed by prefixes, which may give a numbering that differs from the one assigned according to Rule H-3.1.

Example:

$F_3C-CH_2{}^2H$ 2,2,2-Trifluoroethane-*1-d*

Note that here the presence of deuterium (*d*) causes the carbon atom to which it is attached to be assigned the locant 1. For more details and examples see ref. 4.

Rule H-3.2. Priority between isotopically modified and unmodified atoms or
            groups

3.21—When there is a choice between equivalent posssibilities for the
principal chain or senior cyclic system in an isotopically unmodified compound,
the principal chain or senior cyclic system of the analogous isotopically modified compound is chosen so as to include the maximum number of modified atoms
or groups. If a choice still remains, precedence is given to the principal
chain or senior ring that contains a nuclide of higher atomic number. In case
of different nuclides of the same element, precedence is given to a nuclide of
higher mass number.

Examples:

1.

$$CH_3$$
$$|$$
$$\underset{1}{Cl-CH_2}-\underset{2}{CH_2}-\underset{3}{CH_2}-\underset{4}{CH}-\underset{5}{CH_2[^2H]}$$

1-Chloro-4-methyl[5-$^2H_1$]pentane

not

1-Chloro-4-([$^2H_1$]methyl)pentane

2.

$$CH_2[^2H]$$
$$|$$
$$\underset{1}{Cl-CH_2}-\underset{2}{CH_2}-\underset{3}{CH_2}-\underset{4}{CH}-\underset{5}{[^{14}C]H_3}$$

1-Chloro-4-([$^2H_1$]methyl)$[5-^{14}C]$pentane

not

1-Chloro-4-([^{14}C]methyl)$[5-^2H_1]$pentane

3.

$$CH_2[^{79}Br]$$
$$|$$
$$\underset{1}{CH_2[^{81}Br]}-\underset{2}{CH}-\underset{3}{CH_2}-\underset{4}{CH_3}$$

1-([^{81}Br]Bromo)-2-([^{79}Br]bromome-
thyl)butane
not
1-([^{79}Br]Bromo)-2-([^{81}Br]bromome-
thyl)butane

3.22—When there is a choice between equivalent numberings in an isotopically unmodified compound, the starting point and direction of numbering of the
analogous isotopically modified compound are chosen so as to give lowest locants
to the *modified* atoms or groups considered together in one series in ascending
numerical order (see Rule C-15.11 p. 105). If a choice still remains, precedence
for the lowest locants is given to a nuclide of higher atomic number. In the case
of different nuclides of the same element, precedence is given to the nuclide
of higher mass number.

Examples:

1. $\underset{1}{CH_3}-\underset{2}{[^{14}C]H_2}-CH_2-CH_3$

[2-^{14}C]Butane

not

[3-^{14}C]Butane

2. $CH_3-C[^2H_2]-[^{14}C]H_2-CH_3$
   (positions 1 2 3 4)

   $[3-^{14}C,2,2-^2H_2]$Butane
   not
   $[2-^{14}C,3,3-^2H_2]$Butane

3. $CH_3-[^{14}C]H_2-CH[^2H]-CH_3$
   (positions 1 2 3 4)

   $[2-^{14}C,3-^2H_1]$Butane
   not
   $[3-^{14}C,2-^2H_1]$Butane

4.

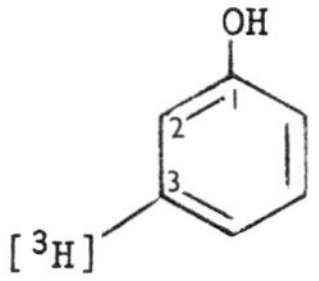

$[3-^3H]$Phenol

5.

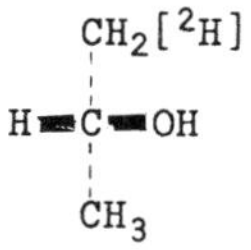

$(R)-[1-^2H_1]$-2-Propanol

6.

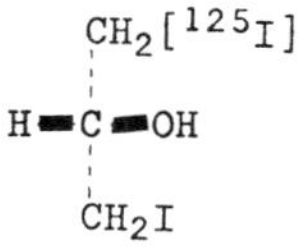

$(S)-1,3-[1-^{125}I]$Diiodo-2-pro-panol

7.

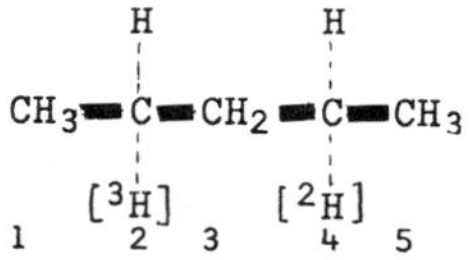

$(2S,4R)-[4-^2H_1,2-^3H_1]$Pentane
not
$(2R,4S)-[2-^2H_1,4-^3H_1]$Pentane

Note: In the last example above, the numbering follows Rule H-3.22 rather than the stereochemical preferences as described in Rules of Section E , which give preference to R groups over S groups for lowest locants.

H-4.   LOCANTS FOR NUCLIDES IN ISOTOPICALLY MODIFIED COMPOUNDS

Rule H-4.1.   Omission or introduction of locants

4.11—When there is no ambiguity, locants may be omitted from the isotopic designator in the name of an isotopically modified compound.

Examples:

1. $C[^2H_3]-CN$ $[^2H_3]$Acetonitrile

2. $CH_3-NH[^2H]$ Methyl$[^2H_1]$amine

3. $CH_3-CH_2-O[^2H]$ Ethan$[^2H]$ol

4. $[^2H]O-CH_2-CH_2-O[^2H]$ 1,2-Ethane$[^2H_2]$diol

5. $CH_2[^2H]-O-C[^2H_2]-S-CH_2-OOH$ $[([^2H_1]$Methoxy$[^2H_2]$methyl)thio]methyl hydroperoxide

4.12—When ambiguity may occur, specific positions of nuclides should be indicated in the isotopic designator by appropriate locants, letters and/or numerals, preceding the nuclide symbol(s).

Examples (see also Rules H-2.11, H-2.21, H-2.31):

1. $C[^2H_3]-NH_2$ $[1,1,1-^2H_3]$Methylamine

2. $CH_3-CH[^2H]-OH$ $[1-^2H_1]$Ethanol

3. $HO-CH[^2H]-CH[^2H]-OH$ $[1,2-^2H_2]$-1,2-Ethanediol

4.
$$C[^2H_3]-\overset{\overset{\text{O}}{\|}}{C}-C[^2H_2]-CH_2-CH_3$$
$[1,1,1,3,3-^2H_5]$-2-Pentanone

5.
$$CH_3-\overset{\overset{\text{SH}}{|}}{C}H-CH[^2H]-CH_3$$
$[3-^2H_1]$-2-Butanethiol

6.
$[^2H]$—(ring)—CO-O-O-CO—(ring)—$[^2H]$

Di($[4-^2H]$benzoyl) peroxide

7.

$$C^2HO$$
$$H\!\!-\!\!C\!\!-\!\!O^2H$$
$$CH_2OH$$

(R)-(O^2,1-$^2H_2$)Glyceraldehyde

or

(R)-(O-^2H)Glycer(^2H)aldehyde

8.

$$CHO$$
$$H\!\!-\!\!C\!\!-\!\!O[^2H]$$
$$CH_2O[^2H]$$

(R)-[O^2,O^3-$^2H_2$]Glyceraldehyde

9. $[^2H]S-CH_2-CH(N[^2H_2])-CO_2[^2H]$

<u>DL</u>-[N,N,O,S-$^2H_4$]Cysteine

**Rule H-4.2. Location of nuclides on positions of compounds that are normally
not assigned locants (letters or numerals)**

4.21—When a nuclide occupies a position that is not numbered, an italicized prefix or Greek letter may be used to denote its position.

Examples:

1. $[^2H_3][^{14}C]-S-CH_2-CH_2-CH(NH_2)-COOH$

<u>DL</u>-[*methyl*-(^{14}C,$^2H_3$)]Methionine

2.

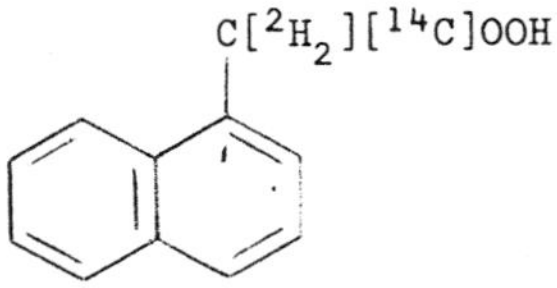

$$C[^2H_2][^{14}C]OOH$$

1-Naphthalene[*carboxy*-^{14}C,α,α-$^2H_2$]a-
cetic acid

3.

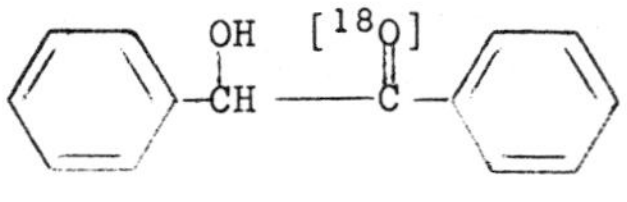

$$OH \quad [^{18}O]$$
$$-CH\!\!-\!\!-\!\!C-$$

[*carbonyl*-^{18}O]Benzoin

4.

$$NH \qquad NH_2$$
$$H_2[^{15}N]-[^{14}C]-NH-(CH_2)_3-C-COOH$$
$$H$$

<u>L</u>-[*guanidino*-^{14}C,N^3-^{15}N]Arginine

4.22—When a nuclide occupies a position that is not numbered or when its position cannot be easily defined according to Rules H-4.12 and H-4.21, the nuclide symbol may be included in the entire symbol of the group through which it is linked to the main part of the structure.

Examples:

1. $CH_3-CH_2-S-[^{34}S]-S-(CH_2)_2-CO_2H$

3-(Ethyl[S-^{34}S-S]trithio)pro-
  pionic acid

or

3-(Ethyl[2-^{34}S]trisulfanyl)pro-
  pionic acid *

2. $CH_3-CH_2-[^{34}S]-S-S-(CH_2)_2-CO_2H$

3-(Ethyl[^{34}S-S-S]trithio)pro
  pionic acid

or

3-(Ethyl[3-^{34}S]trisulfanyl)pro-
  pionic acid *

3.

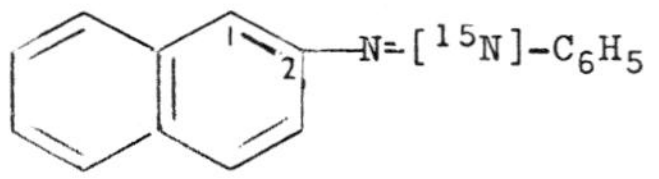

Naphthalene-2-[N=^{15}N]azo benzene

4.

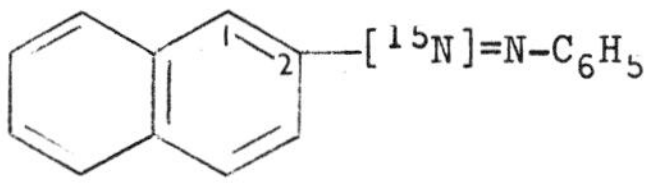

Naphthalene-2-[^{15}N=N]azo benzene

4.23—Italicized nuclide symbols and/or capital italic letters may be used as locants to distinguish between different nuclides of the same element.

Examples:

1.

$CH_3-CH_2-\overset{O}{\overset{\|}{C}}-[^{18}O]C_2H_5$

^{18}O-Ethyl [$^{18}O_1$]propionate

---

* See Rule C-515.2, p. 216.

Examples (cont.):

2. $[^{18}O]$
    $CH_3-CH_2-\overset{\|}{C}-OC_2H_5$ *O*-Ethyl $[^{18}O_1]$propionate

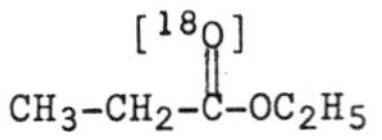

3. $\overset{O}{\|}$
    $CH_3-O-\overset{\|}{C}-[^{18}O]-CH_2-CH_3$ ^{18}O-Ethyl *O*-methyl $[^{18}O_1]$car-
    bonate

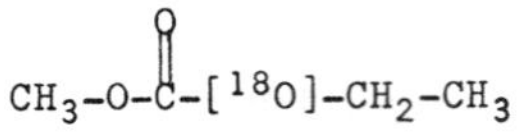

4. $[^{18}O]$
    $CH_3-CH_2-O-\overset{\|}{C}-[^{18}O]-CH_3$ *O*-Ethyl ^{18}O-methyl $[^{18}O_2]$car-
    bonate

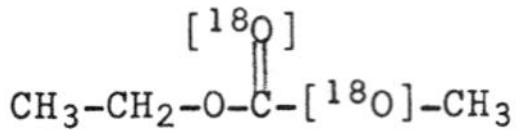

5.

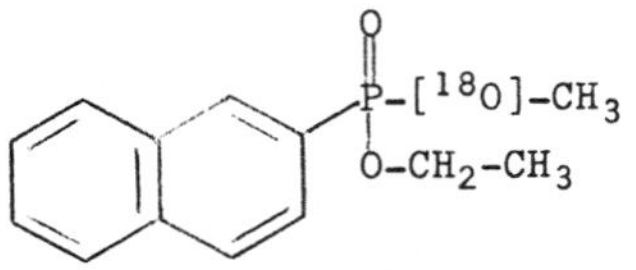

$P-[^{18}O]-CH_3$
$O-CH_2-CH_3$

O-Ethyl ^{18}O-methyl
2-naphthyl$[^{18}O_1]$phosphonate

6.

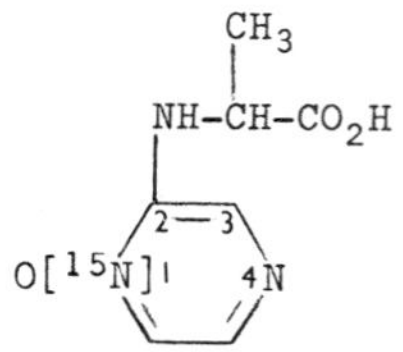

CH_3
$NH-CH-CO_2H$

$O[^{15}N]$ N

N-($[1-^{15}N]$-2-Pyrazinyl)alanine
^{15}N-oxide

References

1. I.U.P.A.C. Manual of Symbols and Terminology for Physicochemical Quantities and Units, 1973 Edition, Butterworths, London, 1975, Rules 7.1 and 7.2, p. 24.

2. I.U.P.A.C. Commission on Atomic Weights, *Pure and Applied Chemistry*, 37, 591-603 (1974).

3. (a) W. A. Boughton, Naming Hydrogen Isotopes, *Science*, 79, 159-60 (1934); (b) E. J. Crane, Nomenclature of the Hydrogen Isotopes and Their Compounds, *Science*, 80, 86-9 (1934); (c) American Chemical Society, Report of Committee on Nomenclature, Spelling, and Pronunciation, Nomenclature of the Hydrogen Isotopes and Their Compounds, *Ind. Eng. Chem.* (News Ed.), 13, 200-1 (1935).

4. "The Naming and Indexing of Chemical Substances for Chemical Abstracts During the Ninth Collective Period (1972-1976)", ¶220, p. 111I, a reprint of Section IV of the Introduction to the Chemical Abstracts Volume 76 Index Guide.

5. I.U.P.A.C. Nomenclature of Inorganic Chemistry, 2nd Edition (1970), Butterworths, London, 1971: (a) Rule 1.31, p. 11; (b) Rule 1.1, p. 10.

APPENDIX

Comparative Examples of Formulae and Names for Isotopically Modified Compounds

Type of Compound	Formula	Name
Unmodified	CH_3-CH_2-OH	Ethanol
Isotopically substituted	$C^2H_3-CH_2-O^2H$	$(2,2,2-^2H_3)$Ethan(^2H)ol *or* $(O,2,2,2-^2H_4)$Ethanol
Specifically labeled	$C[^2H_3]-CH_2-O[^2H]$	$[2,2,2-^2H_3]$Ethan$[^2H]$ol *or* $[O,2,2,2-^2H_4]$Ethanol
Selectively labeled	a. $[O,2-^2H]CH_3-CH_2-OH$ b. $[2-^2H_{2;2}, {}^{18}O_{0;1}]CH_3-CH_2-OH$	a. $[O,2-^2H]$Ethanol b. $[2-^2H_{2;2}, {}^{18}O_{0;1}]$Ethanol
Nonselectively labeled	$[^2H]CH_3-CH_2-OH$	$[^2H]$Ethanol
Isotopically deficient	$[def^{13}C]CH_3-CH_2-OH$	$[def^{13}C]$Ethanol

INDEX

See also the list of radical names on pages 305-322.
An initial or a terminal hyphen indicates that the syllables occur only
at the end or at the beginning, respectively, of the names of a compound.